Mechanical Behavior and Damage Fracture Mechanism of Deep Rocks

Sheng-Qi Yang

Mechanical Behavior and Damage Fracture Mechanism of Deep Rocks

Springer

Sheng-Qi Yang
State Key Laboratory for Geomechanics and Deep
Underground Engineering
School of Mechanics and Civil Engineering
China University of Mining and Technology
Xuzhou, China

ISBN 978-981-16-7738-0 ISBN 978-981-16-7739-7 (eBook)
https://doi.org/10.1007/978-981-16-7739-7

This Springer imprint is published by the registered company Springer Nature Singapore Pte Ltd.
The registered company address is: 152 Beach Road, #21-01/04 Gateway East, Singapore 189721, Singapore

Foreword I

Gathering resources (metals, rare-earth minerals, oil, gas, and more) in a manner that is both affordable and sustainable is arguably the greatest challenge for our modern way of life. Declining ore grades (the concentration of minerals in rock) mean we consume an ever-increasing amount of energy during extraction, emit more carbon dioxide, and leave huge amounts of waste rock. In order to find sustainable mineral and energy extraction technologies and to be consistent with the UN's Sustainability Development Goals, we need to have a deep understanding of rock and mineral ore bodies which are deep below the surface and subjected to complex geological stress conditions.

For deep rock underground engineering applications, the geological environment is very challenging, with high geostress, high temperature, and high seepage pressure, which present severe contests for the safe and efficient exploitation of deep energy sources. The problems of the mechanical parameters of deep rocks in complex geological environments make it more and more difficult to ensure the stability of deep rock underground engineering. However, with the use of advanced rock mechanics testing systems and X-ray micro-CT observations, the author of this book has carried out many valuable experimental investigations to fill the knowledge gap.

The recipient of a 2014 Endeavor Research Fellowship from the Australian government, Prof. Sheng-Qi Yang, commenced these studies in my 3GDeep Earth Energy group at Monash University as a visiting professor. As his host supervisor, I have been very impressed with Prof. Yang's contributions to rock mechanics. An exemplary worker, he has pursued his project diligently and has progressed extremely well with his research results. Based on a close consideration of this book, I am pleased to note its logical organization and clear presentation, which make for easy and enjoyable reading.

The book is an integration by the authors of research results over recent years. The project obtained support from the National Natural Science Foundation of China and the Natural Science Foundation of Jiangsu Province for Distinguished Young Scholars. Many of the experimental results have been published in major international journals, such as *International Journal of Rock Mechanics and Mining Sciences*, *Rock Mechanics and Rock Engineering*, *Engineering Geology*, *Experimental Mechanics*, *International Journal of Solids and Structures*, *Geophysical Journal International*, and *Geothermics*. The international peer-review process for these English-language journals is an additional guarantee of quality.

In comparison with other works on the topic, this book clearly succeeds in at least four aims: (1) to investigate the influence of the loading path on the triaxial strength and deformation behavior of sandstone and marble; (2) to analyze the effect of borehole size on the triaxial strength and deformation behavior of hollow sandstone; (3) to explore the influence of high temperature on the triaxial deformation and permeability behavior of sandstone and granite; (4) to reveal the damage fracture mechanism of deep rocks using spatial AE techniques and X-ray micro-CT observations. This book will be an essential reference for postgraduate students, researchers, and technicians in the fields of geotechnical engineering and engineering geology.

All things considered, I believe that the publication of this book offers a sound experimental basis for developments in deep rock mechanics. It enhances our understanding of the damage failure mechanisms in deep rock underground engineering related to energy and resources. I am pleased to commend this book to all interested readers.

June 2021

Prof. P. G. Ranjith
Professor of Sustainable Development of Energy and Resources
Fellow of Australian Academy of Technology and Engineering
Director of Deep Earth Energy Research Laboratory, Monash University
Melbourne, Australia

Foreword II

This monograph presents a detailed compendium of the role of progressive failure and related damage on the evolving load bearing capacity and fluid transmission characteristics of brittle rocks—defining both their response during subsurface excavation and subsequent utilization and their functionality as reservoirs for supplying and for the sequestering of energy fluids. It explores the use of advanced and integrated characterization techniques, those of acoustic emission and X-ray CT and of advanced methods of damage mechanics to deconvolve the response to complex stress paths and far-from-equilibrium environmental conditions of temperature and stress. As such, it offers an important and valuable resource for students, practitioners and researchers involved in the complex response of rocks to brittle failure and the design of structures on and in them. The author is to be commended on both the breadth and depth of this study and the comprehensiveness of its presentation.

November 2021

Prof. Derek Elsworth
Professor of College of Earth and Mineral Sciences
The Pennsylvania State University
Fellow of American Academy of Engineering
USA

Preface

Mechanical behavior and damage fracture mechanism of rocks are affected by a lot of factors such as rock type, loading path, temperature, and water pressure. For deep rock underground engineering, the geological environment will become more complicated, having obvious high geostress, high temperature, and high seepage pressure, which put forward severe challenges for the safe and efficient exploitation of deep energies. However, at present, the problems of mechanical parameters of deep rocks in complex geological environments become more and more difficult on how to predict the unstable failure of deep rock underground engineering. Therefore, based on the background of deep underground engineering, it is necessary to conduct the investigations on the strength, deformation, permeability, and damage failure behavior of rocks related to deep energy applications, which are very significant for assuring the stability and safety of deep underground rock engineering.

This book includes seventeen chapters. Chapter 1 summarizes the effect of stress state, loading path and high temperature on the failure mechanical behavior of rocks, and the damage failure mechanism of rocks by AE and X-ray CT observations. Chapters 2, 3, and 4 deal with the strength, deformation, and damage failure behavior of red sandstone under triaxial cyclic loading, simple and complex loading path. Chapter 5 makes a systematical analysis on the mechanical, acoustic, and fracture behaviors of yellow sandstone under triaxial monotonic and cyclic loading. Chapter 6 explores the triaxial strength and deformation failure behavior of coarse marble under six different loading paths. Chapter 7 investigates the deformation failure characteristics of crystalline marble under triaxial cyclic loading. Chapter 8 analyzes experimentally the strength, deformability, and X-ray micro-CT observations of deeply buried marble under different confining pressures. Chapter 9 carries out a numerical investigation on the deformation and damage failure behavior of mudstone under single-stage and multi-stage triaxial compression on the basis of the experimental results. Chapters 10, 11, and 12 analyze systematically the deformation, peak strength, crack damage behavior, and fracture evolution mechanism of hollow sandstone by using X-ray micro-CT observations and 3D numerical simulations. Chapter 13 studies the thermal damage and uniaxial mechanical behavior of granite after exposure to different high-temperature treatments. Chapter 14 investigates the triaxial mechanical and permeability behavior of sandstone after exposure to different high-temperature treatments. Chapter 15 analyzes the effect of high temperature on the permeability evolution and failure response of granite under triaxial compression. Chapter 16 compares experiment and grain-based modeling results on damage failure behavior of granite under different confining pressures. Chapter 17 explores the effect of diameter on strength and failure behavior of coal specimens under conventional triaxial compression.

This book was supported by the National Natural Science Foundation of China (Grant Nos. 41272344, 42077231), Fundamental Research Funds for the Central Universities (Grant Nos. 2020ZDPYMS34, 2021ZDPYJQ002) and the Natural Science Foundation of Jiangsu Province for Distinguished Young Scholars (Grant No. BK20150005), which are greatly appreciated. At the same time, Prof. P. G. Ranjith from Monash University (Australia) made a foreword for this book, which was also greatly appreciated. Since 2014, I visit Monash University (Clayton campus) in Melbourne for many times as a visiting professor, my host supervisor Prof. P. G. Ranjith often discussed with me and brought me a lot of valuable insights, which greatly improved this research. Here, I would like to express my sincere thanks to Prof. P. G. Ranjith. Besides, Prof. Derek Elsworth from the Pennsylvania State University also made a foreword for this book. He also put forward a lot of good comments on thermal damage and permeability behavior of rock under triaxial compression, which was also greatly appreciated.

Finally, I would also like to express my sincere thanks for Prof. Yang Ju from China University of Mining and Technology (Beijing), Prof. Tao Xu from Northeastern University, which have greatly improved this book. My Ph.D. students Yan-Hua Huang, Wen-Ling Tian, and Peng Xu assisted me to finish some experimental and numerical research works in Chapters 5, 7, 9, 11, 14, and 15 of this book, which were also appreciated. I am extremely grateful to my wife Xiao-Qin Chen working in China University of Mining and Technology for the many years of personal support and understanding. I would like to thank my daughter Ai-Chen Yang for bringing me a lot of joyfulness and happiness.

December 2021

Prof. Dr. Sheng-Qi Yang
Professor of Geotechnical Engineering
State Key Laboratory for Geomechanics
and Deep Underground Engineering
School of Mechanics and Civil Engineering,
China University of Mining and Technology
Xuzhou, Jiangsu, China

Funding Information

National Natural Science Foundation of China (41272344, 42077231)
Fundamental Research Funds for the Central Universities (2020ZDPYMS34, 2021ZDPYJQ002)
Natural Science Foundation of Jiangsu Province for Distinguished Young Scholars (BK20150005)

Contents

About the Author

Dr. Sheng-Qi Yang was born in December 1978. In 2003, he started his Ph.D. research, in Hohai University, Nanjing, P. R. China, and got Geotechnical Engineering of Doctor's degree in April 2006. In 2007–2008, he continued his postdoctoral work in Ecole Polytechnique de Paris, France. In 2014–2015, he obtained Endeavor Research Fellowship in Australia and commenced his research in the department of civil engineering, Monash University, as a visiting professor. From January 2012, he has been promoted a full Professor and a Ph.D. supervisor in China University of Mining and Technology. In November, 2016, he was elected the vice dean of School of Mechanics and Civil Engineering, China University of Mining and Technology. Since 2016, he conducted a lot of international cooperation research in Nanyang Technological University (Singapore), The University of Nottingham (UK), Monash University (Australia), and Imperial College London (UK) as a visiting professor. He was also one of the 2019 and 2020 Chinese most cited researchers in the field of civil and structural engineering.

In 2013, he was awarded the Program for New Century Excellent Talents in University from Ministry of Education. In 2015, he obtained the support from Natural Science Foundation of Jiangsu Province for Distinguished Young Scholars, and in 2018, he was elected for the Six Talents Peak Project in Jiangsu Province. He also obtained the Youth Science and Technology award from Chinese Society for Rock Mechanics and Engineering and Sunyueqi Foundation Council. In 2011, he also obtained the second prize of Science and Technology Progress from Ministry of Education. In 2015 and 2018, he obtained the first prize of Natural Science from Ministry of Education for two times. In 2018, he obtained the second prize of Science and Technology Progress from Jiangsu Province. In the past five years, he takes charge of more than ten key scientific projects including four National Natural Science Foundation of China (NSFC). His research interest focuses mainly on deep rock mechanics and underground engineering.

To date, Prof. Sheng-Qi Yang has published one Chinese monograph with the title *Study on the Mechanical Behavior of Fissured Rock and Its Time-Dependent Effect Analysis* (Beijing: Science Press, 2011) and one English monograph with the title *Strength Failure and Crack Evolution Behavior of Rock Materials Containing Pre-existing Fissures* (Springer Press, 2015). He has also published about 200 papers as the first and corresponding author, in which more than 110 English papers (indexed by SCI) were published in *International Journal of Rock Mechanics and Mining Sciences, Rock Mechanics and Rock Engineering, Tunneling and Underground Space Technology, Journal of Structural Geology, Engineering Geology, Computers and Geotechnics, Geothermics, International Journal of Fracture, Engineering Fracture Mechanics, Canadian Geotechnical Journal*, etc. He has applied for more than 30 national invented patents, in which 23 patents have been awarded. He is the editorial board member of several journals *Rock Mechanics and Rock Engineering, Engineering Geology, Geomatics, Natural Hazards and Risk, Geomechanics and Geophysics for Geo-Energy and Geo-Resources, Energies, Journal of Basic Science and Engineering*, and *Engineering Science Journal*, etc. He has also made more than twenty lectures in some important academic conferences, e.g., 2014 International Symposium on Soft Rock, etc.

Chapter 1
Introduction

All kinds of rock underground engineering related to energies (such as fossil energy, nuclear energy, shale gas, geothermal energy, etc.) are usually located in a complex triaxial stress state, which is affected by a lot of factors such as rock type, anisotropy, loading path, temperature, water pressure, etc. With the depletion of shallow energies, underground engineering will inevitably move to deep exploration. In this way, the geological environment of deep rock mass engineering will become more complicated, and more and more engineering disasters and accidents are caused by high geostress, high temperature, and high seepage pressure, which put forward severe challenges for the safe and efficient exploitation of deep energies. However, at present, the problems of mechanical parameters of deep rocks in complex geological environments become more and more difficult on how to predict the unstable failure of deep rock underground engineering. Therefore, based on the background of deep underground engineering, it is necessary to conduct the investigations on the strength, deformation, permeability and damage failure behavior of rocks related to deep energy applications, which are very significant for assuring the stability and safety of deep underground rock engineering.

1.1 Effect of Stress State on the Failure Mechanical Behavior of Rocks

Rock under high geostress are commonly encountered in engineering practice (Feng et al. 2017), especially in the coal mine (Tan et al. 2017). Sandstone is a sedimentary rock that is widely used in rock engineering. The knowledge of the failure mechanical behavior of sandstone is important in rock engineering related to deep energy. For this reason, in the last few decades, many experimental studies have investigated the mechanical behavior of sandstone, including its strength, deformation behavior, failure characteristics and time-dependent behavior (Wong et al. 1997; Baud et al. 2000; Wu et al. 2000; Klein et al. 2001; Jeng et al. 2002; Bésuelle et al. 2003; Fortin

S.-Q. Yang, *Mechanical Behavior and Damage Fracture Mechanism of Deep Rocks*,
https://doi.org/10.1007/978-981-16-7739-7_1

et al. 2006; Tembe et al. 2008; Tsai et al. 2008; Zorlu et al. 2008; Heap et al. 2009a, 2009b; Tavallali and Vervoort 2010; Huang et al. 2010; Yang and Jiang 2010, 2011; Yang et al. 2011).

Uniaxial compression and uniaxial tension tests (or Brazilian tests) are conventional methods that are often used to investigate the strength, deformation behavior and failure characteristics of sandstone because its nature of their formation can be inherently anisotropic (Benson et al. 2005). Al-Harthi (1998) investigated the effect of anisotropy on the uniaxial compressive strength (UCS) of Ranyah sandstone (Saudi Arabia), which found that the strength of Ranyah sandstone varied tremendously with orientation (a maximum UCS of 225 MPa and a minimum of 125 MPa). A W-shaped strength anisotropy curve was obtained as a result of plotting the UCS values against the angle between the loading direction and the bedding. Tavallali and Vervoort (2010) conducted indirect uniaxial tension tests on a layered sandstone. They showed that the variation in tensile strength and the failure mode are both a function of the inclination angle between the layer plane and the loading direction. Dynamic cyclic stressing can lead to the degradation of rock physical properties. Heap et al. (2010) performed increasing-amplitude, cyclic-stressing experiments on both Bentheim and Darley Dale sandstone. They found that there is a gradual deterioration in static elastic moduli after each increasing-amplitude cycle. Bagde and Petroš (2005) also performed dynamic uniaxial compressive cyclic-stressing experiments on sandstone. Their experimental results showed that the fatigue strength of their sandstone decreased with loading frequency. Nespereira et al. (2010) carried out petrographic analyses and UCS tests on Salamanca sandstone. They showed that the correlation factor between the mechanical behavior and the petrographic parameters revealed that the cement content is not the only factor governing the UCS, and the composition of the cement may play an important role in the UCS of clay-bearing sandstones. Due to the difficulty of direct tensile testing, Lin et al. (2009) conducted several three-point bending tests on Berea sandstone to study fracture initiation and the associated damage zone by using 2D acoustic emission (AE) locations and electronic speckle pattern interferometry (ESPI). Their experiments indicated that specimen failure was initiated by a large fracture though a localized damage zone.

However, deep underground engineering rock masses, for example, are usually located within triaxial stress state conditions; therefore, it is more relevant to investigate the mechanical characteristics of sandstone under triaxial compression. Wong et al. (1997) studied the transition from brittle faulting and cataclastic flow in six different sandstones. At low confining pressures, shear-induced dilation and brittle faulting was observed. However, at higher confining pressures, shear-enhanced compaction and cataclastic flow was observed. Gatelier et al. (2002) presented an extensive experimental study on the mechanical behavior of anisotropic porous sandstone, and analyzed the influence of the structural anisotropy on the progressive development of pre-peak damage. Baud et al. (2000) investigated water-weakening in the cataclastic flow regime on Berea, Boise, Darley Dale, and Gosford sandstones. They found that the presence of water reduced the strength of the sandstones and altered the failure mode at elevated pressures. Heap et al. (2009a) studied time-dependent brittle deformation in Darley Dale sandstone and found that the rate of creep deformation during constant stress experiments was heavily dependent on the applied differential stress. In a later study, Heap et al. (2009b) showed that an increase in temperature from 20 to 75 °C increased the rate of creep deformation by up to three orders of magnitude in Darley Dale, Crab Orchard, and Bentheim sandstone. Bésuelle et al. (2000) conducted triaxial compression experiments on Vosges sandstone with two slenderness ratios, and observed a strong positive dilatancy at lower confining pressures, which decreased and became negative at higher confining pressures. Yang et al. (2012) performed an experimental study on red sandstone under conventional triaxial compression and reducing confining pressure, and analyzed the influence of confining pressure and loading path on the strength, deformability, failure behavior and acoustic emission locations. Yang and Jing (2013) evaluated the strength and deformation behavior of red sandstone under simple and complex loading paths. By adopting a spatial AE monitoring system, Yang et al. (2014b) analyzed the internal damage evolution characteristics of red sandstone during multi-stage triaxial deformation.

As it is well-known to us, excavation in such rock carries great risk as unloading would lead to serious rock failure (He et al. 2010; Huang and Li 2014; Xiao et al. 2018), and thus an excavation damage zone (EDZ) will be formed (Eberhardt 2011; Tang et al. 2018). Therefore, knowledge of rock deformation and damage behaviour due to excavation disturbance of rock underground engineering is especially significant for the design of deep tunnel support (Wang et al. 2018; Feng et al. 2018).

Figure 1.1a shows a typical mudstone tunnel at a depth of 960 m in the Yuncheng coal mine, Shandong province. During and after the excavation of a deep tunnel, the surrounding rock mass often undergoes a complicated stress change, such as the decrease of radial stress and the increase of circumferential stress. The vertical stress of the coal in front of the work face will usually increase, whereas the horizontal stress usually decreases during the process of underground coal mining, The stress

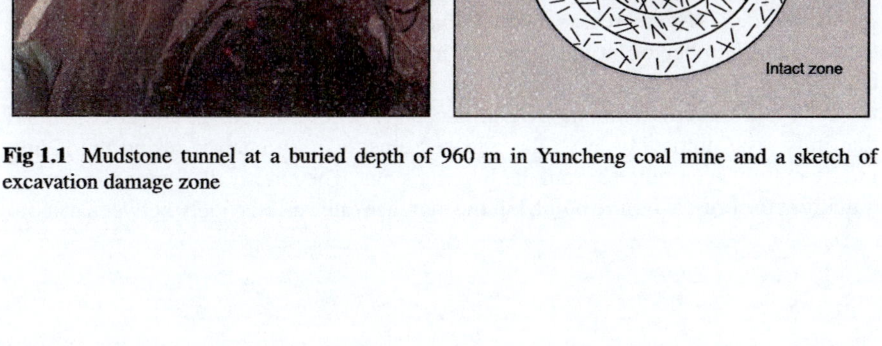

Fig 1.1 Mudstone tunnel at a buried depth of 960 m in Yuncheng coal mine and a sketch of excavation damage zone

change in the tunnel and work face corresponds to the unloading confining pressure and loading axial stress under laboratory conditions (Huang et al. 2001; Xue et al. 2017), which results in different extents of damage to the surrounding rock mass (Fig. 1.1b). It should be noted that Fig. 1.1b is only a sketch of an EDZ. The EDZ is affected not only by stress redistribution (Eberhardt 2011) but also by the excavation method: explosives, road header, etc. When the same excavation method is used, different stress redistributions will result in a different EDZ (Li et al. 2014). As the damage extent increases from low to high, the rock mass will have more fractures, and the numbers of cracks in the surrounding rock mass will significantly increase. Furthermore, compared with the intact zone, the damaged rock often undergoes a post-peak response. At increased depths, the post-peak rock will show reduced strength. The anchorage object of deep underground engineering is actually a rock mass that has been damaged to various extents, as shown in Fig. 1.1b. However, due to the strength reduction of the post-peak rock mass, deep tunnelling often induces large deformation failures. Therefore, it is very important to study the strength reduction and mechanical damage behaviour of rock specimens with various fractured extents to assure the stability and the safety of deep tunnels.

Several studies have pointed that the EDZ was constructed by engineering unloading in the excavations of a tunnel, demonstrating that the rock mechanical behaviour and failure characteristics under unloading conditions were significantly different from those under loading condition (Xie et al. 2004; Huang and Li 2014). The influence of unloading confining pressure on the mechanical properties, failure process and failure mode of rock has been investigated through experiment (Huang et al. 2001; Xie and He 2004; Adushkin et al. 2016). He et al. (2010) investigated the acoustic emission (AE) characteristics under true-triaxial single-face dynamic unloading tests, and found that as stress increases, the AE signals tend to shift more towards a signature of high amplitude. Zhao et al. (2014) performed a series of tests under different unloading rates, and indicated that the degree of violence during failure and associated AE energy release are dependent on the unloading rate. Xu et al. (2015) proposed a method for determining Poisson's ratio of rock by changing the axial stress and unloading the lateral stress. Du et al. (2016) indicated that slabbing occurred in granite and sandstone, whereas the cement mortar underwent shear failure under the true-triaxial unloading test. Li et al. (2015, 2018) carried out true triaxial tests on cubic granite, sandstone and cement mortar specimens under unloading the minor principal stress. They show that the strength and stiffness of rock, sample height-to-width ratios and intermediate principal stress have an impact on the failure modes. Zhou et al. (2018) performed the complex magnetic resonance tests on the unloaded specimens to investigate the meso-damage evolution of the unloading rock. With increases in the unloading ratio, the small pores increased, and the size and number of large pores also increased. Rock deformation and fracture are evolutionary processes describing the exchange of energy and mass between the rock and the environment (Xie et al. 2004). Therefore, Li et al. (2017) found that in triaxial unloading tests, the released elastic strain energy during unloading failure mainly accumulates from the loading stage. The energy dissipation increases quickly at the critical failure point, but the increase rate varied widely between triaxial loading and unloading conditions. The experimental results also indicated that strain energy conversion was influenced by the initial confining pressure and unloading rate (Huang and Li 2014). Xue et al. (2017) revealed that the increase in the unloading rate reduced the elastic and dissipation energy, and accelerated the rock damage rate.

1.2 Effect of Loading Path on the Failure Mechanical Behavior of Rocks

Since Jaeger (1967) pointed out "the possibility of rock failure dependent to the loading path is worth of an argumentative problem" in 1967, the effect of loading path on mechanical behavior of rock has been investigated in past several decades to understand and explore the fracture mechanism of various rock engineering under different loading paths (Swanson and Brown 1971; Crouch 1972; Xu and Geng 1986; Wang et al. 2008; Gen 1982; Yao et al. 1980; Shen et al. 2003; Xiong and Zhou 2006; Lee et al. 1999; Yang et al. 2000; Ferfera et al. 1997; Cai 2008). Up to now, the main experimental results on the influence of loading path on mechanical behavior of rock have included the following three kinds.

- Influence of loading path on the strength and deformation behavior of rock.
- Influence of loading path on the acoustic emission of rock.
- Influence of loading path on failure behavior of rock.

The above first kind of experiment result is widely obtained in previous studies, mainly focusing on the influence of loading path on the strength, volume changes, BDT (brittle-ductile transition) behavior of rock.

On the effect of loading path on strength of rock, there have been two kinds of contradictory opinions. Some typical results were presented as follows. Swanson and Brown (1971) investigated the strength of granite under conventional triaxial compression (Path I), confining pressure reduction (Path III) and proportional loading, which showed that strength of granite were independent of the above loading path, as shown in Fig. 1.2. Crouch (1972) obtained similar conclusion by studying the effect of loading path on the peak strength of norite. However, Xu and Geng (1986) studied the various loading paths causing strength, deformation and failure in hard and soft rocks, which found that the effect of two loading paths (Path I and III) on the peak strength was related to lithologic character. The strength of hard rock under Path III was a little lower than that under Path I, but the strength of soft rock under Path III was higher than that under Path I. Wang et al. (2008) discussed the effect of Path I and III on the strength parameters of marble, which showed that the peak strength under Path III were all lower than that under Path I and the strength parameters of marble under Path I and III were obviously various. The cohesion of marble under Path III was distinctly lower than that under Path I, but the internal friction angle of marble under Path III had no obvious difference with that under Path I.

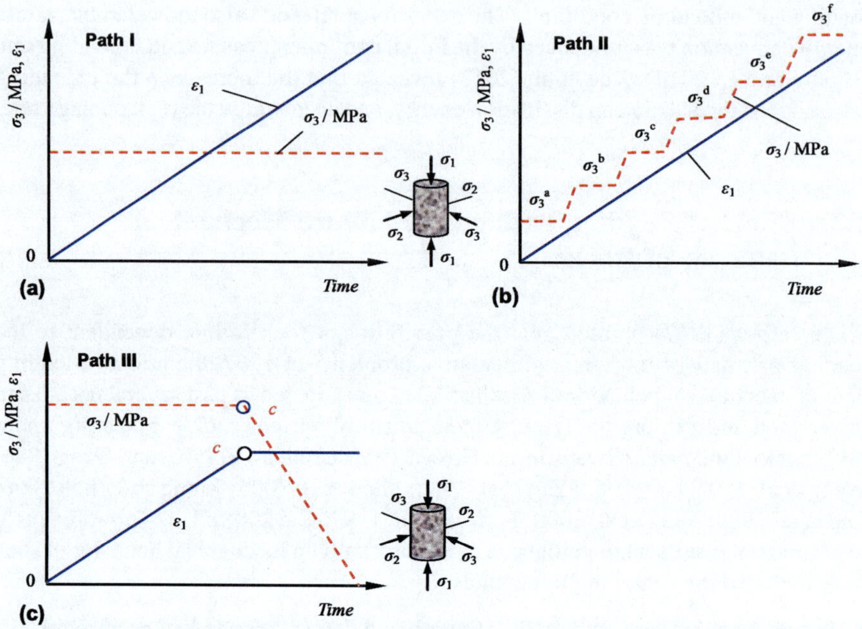

Fig. 1.2 There different loading paths in the past study

On the effect of loading path on deformation behavior of rock, some typical experimental results can be summarized as follows. Gen (1982) investigated the volume changes of gabbro and granite under various loading paths, which found that the dilatancy effects of the same rock were quite different in various loading paths. The super-dense and ultra-dilatant states were all observed during testing. Yao et al. (1980) carried out the experimental study on BDT of gabbro and marble under two loading paths (Path I and III). Under Path I and III, the gabbro exhibited brittle behavior, but under Path III, the gabbro was more brittle than that under Path I. With the increase of confining pressure, the marble exhibited BDT behavior. But under Path III, some brittle fracture of marble could appear even though under higher confining pressure. Shen et al. (2003) performed triaxial test of red sandstone under different loading paths, which analyzed the effect of different loading path on the deformation of rock samples. He concluded that the deformation of rock under different loading paths had nonlinear property, which was unlike the test results under Path I. Xiong and Zhou (2006) carried out in-situ true triaxial tests on rock mass under complex loading paths, i.e. (1) σ_1 decreased, σ_2 and σ_3 maintain constant in the same course; (2) σ_1 decreased, σ_2 maintain constant and σ_3 decreased in the same course; (3) σ_1 decreased, σ_2 maintain constant and σ_3 increased in the same course. The results showed that when a principal stress decreased and another one increased in the same process, the deformation was nonlinear and anisotropic. The elastic modulus kept approximately invariable in the loading direction and speed-up decreasing in the unloading direction in accordance with the decreasing of the effective principal stresses.

Besides above experimental results on deformation behaviors under various loading paths, Lee et al. (1999) preformed a series of hollow cylindrical triaxial tests to investigate the potential effect of loading path on the mechanical behavior of the Pei-Tou sandstone, which showed that Kim and Lade's model was shown to be more accurate than the Mohr–Coulomb theory in predicting the failure surface according to the limited test data. Yang et al. (2000) investigated the effect of loading path on creep and relaxation behavior of salt rock, which showed that the steady state creep strain rate was strongly dependent on confining pressure in a small range (0 ~ 3 MPa), but weakly dependent on loading history. Ferfera et al. (1997) carried out the experimental study on monophasic permeability changes under various loading paths to evidence the influence of the effective mean stress and the influence of the deviatoric stress.

Acoustic emission (AE) is a manifestation of releasing the strain energy in the form of elastic wave, which can be used to predict and forecast the failure process of rock engineering and the precursor of earthquake occurrence (Lavrov 2001; Chen 1979; Su et al. 2009; Xiang et al. 2008; Li et al. 2007). Chen (1979) investigated the influence of two loading paths on the acoustic emission of gabbro and granite under triaxial compression. Under Path I, the average acoustic emission rate gradually increased; while under Path III, an abrupt increase of acoustic emission occurred much later. The total number of acoustic emission events in the fracture process under Path III was much less than that under Path I. The different feature of acoustic emission in various loading modes was due to the existence of the unloading process. Su et al. (2009) analyzed the influence of three loading paths (i.e. uniaxial compression, Path I and III) on the acoustic emission of coal sample, which indicated that the AE characteristics of coal sample were different in deformation and failure process under different loading paths. Compared with Path I, the maximum values of AE counts and energy were much larger, and the results indicated that the failure of coal sample under Path III were more violent. Xiang et al. (2008) carried out acoustic emission experimental study of failure process of hard rock under excavating and supporting stress paths. The experimental results indicated that the whole process of AE activity could be divided into three active stages (crevasse crack closing stage, unloading damage stage and destruction stage) and two relatively quiescence stages (elastic stage and fracture stable development stage), which were similar with rock mass behaviors and support effect in project field so as to provide test basis and guide instruction for engineering practice. Li et al. (2007) conducted the numerical simulation for the AE behavior of rock by using the RFPA^{2D} (Rock Failure Process Analysis) software to understand the characteristics of Kaiser Effect along different stress paths.

There are less experimental results on the failure behavior of rock under different loading paths, which can be presented as follows. Yao (1981) studied the micro-cracks behavior of gabbro under different stress loading paths, i.e. (1) Path I; (2) Path III; (3) Increased the minimum principal stress and kept the rock sample intact (Path A). He observed that there existed obvious differences in micro-fractures between the three paths. Under Path III, most of the micro-cracks in plagioclase were shortened whose lengths were less than the minor axis of the plagioclase grains. These cracks were concentrated in rather narrow regions. Under Path A, the number of inclined micro-cracks increased, and the micro-cracks in the plagioclase grains became much longer and much more in number than that under Path I and III. Zhang et al. (2007) adopted rock triaxial test apparatus to model different stress unloading paths according to excavation methods. Five sandstone samples were used to carry out the tests under Path I and III (including confining pressure reduction before peak strength and after peak strength), which showed that the failure under Path III took on brittle failure; moreover the extent of failure of rock under Path III (confining pressure reduction before peak strength) was larger than that under Path III (confining pressure reduction after peak strength). However due to limit experimental results, the above conclusion has still needed to be reinforced.

1.3 Effect of High Temperature on the Failure Mechanical Behavior of Rocks

In nuclear waste disposal, underground coal gasification, heat mining in enhanced geothermal energy and volcano flank stability, a deep understanding of the influence of temperature on the brittle response of rock is a major issue of interest. Compared with conventional fossil fuels such as coal and oil, geothermal heat offers great advantages in cost, reliability and environmental friendliness (Martín-Gamboa et al. 2015; Axelsson et al. 2010), given the release of acid rain and greenhouse gases created by the use of fossil fuel (Lund et al. 2005), and geothermal heat is now recommended and identified as a renewable and alternative energy source (Kumari et al. 2017). As a common deeply buried rock, deep sandstone reservoirs with adequate temperatures are developed to serve as geothermal reservoirs (Gong et al. 2011). For exploitation of geothermal energy, four pre-requisites are needed, namely, a heat source, heat-carrying fluid, permeable or fractured rock that acts as reservoir and cap rock, and an impermeable insulation cover (Chandrasekharam et al. 2010; Shao et al. 2014).

The mechanical and permeability behaviors of reservoir rock and cap rock have a great influence on deep geothermal energy extraction (Angelo et al. 2008). An increase in the permeability of reservoir rock favors heat-carrying fluid extraction, but an increase in the permeability of cap rock leads to loss of heat-carrying fluid, which is not beneficial for full exploitation of a geothermal resource. For this reason, analysis of the mechanical properties of geothermal reservoir and cap rock under high pressure and temperature conditions (geothermal conditions) can be helpful in basic predictions. Laboratory experiments conducted at high temperature can better represent practical situations than those conducted with high-temperature treatment, and if subjected to the experimental technical conditions, few triaxial compression tests of rock at high temperature (above 500 °C) can be conducted, much less analysis of micro-structure and permeability. As a result, experiments performed on rock after

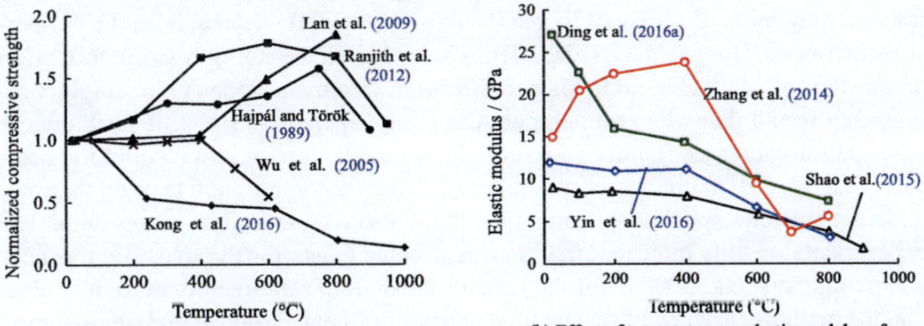

(a) Effect of temperature on normalized compressive strength of rock

(b) Effect of temperature on elastic modulus of rock

Fig. 1.3 Uniaxial compressive strength and elastic modulus of rock after heat treatment

thermal treatment are conducted as an alternative to reduce the difficulty and costs of laboratory experiments.

In recent decades, the influence of temperature on physical (Mambou et al. 2014), thermal (Tang et al. 2015) and mechanical properties (Li et al. 2016); microstructures (Török et al. 2005); failure mechanisms (Shao et al. 2015); and thermo-mechanical models (Yao et al. 2016) have been investigated. In addition, extensive laboratory research has been performed on different types of rock materials exposed to high temperatures including marble (Peng et al. 2016a, 2016b), sandstone (Kong et al. 2016; Zhu et al. 2016), claystone (Tian et al. 2014), shale (Morteza et al. 2014), granite (Liu et al. 2014; Yin et al. 2016; Yang et al. 2017), salt rock (Liang et al. 2006), and mudstone (Luo et al. 2011), among others.

Figure 1.3 shows the normalized peak strength and elastic modulus versus high temperature from references (Shao et al. 2015; Kong et al. 2016; Yin et al. 2016; Ranjith et al. 2012; Zhang et al. 2014; Lan et al. 2009; Hajpál and Török 1989; Ding et al. 2016a, 2016b). Figure 1.3 also indicates that rock properties can be highly dependent on temperature, and the characteristics of temperature dependency vary with the rock types. Even for the same type of rock material, the critical temperature is not a constant, and might result from initial rock features such as micro-cracks and microstructure. Moreover, the effect of the testing temperature on the mechanical behavior of rock under different test methods (such as uniaxial compression, the Brazilian test and the three-point bedding test) has also been widely studied. Taking sandstone material as an example, the experimental results from sandstone can be briefly described as follows. The results of uniaxial compression tests on heat treated sandstone demonstrated that the compressive strength and elastic modulus for sandstone are increased at 500 °C, before decreasing up to the maximum temperature of 950 °C (Ranjith et al. 2012). Sirdesai et al. (2016) performed a series of Brazilian test on heated sandstone specimens and found that the physical properties and tensile strength are closely related to the treatment temperature. Mahanta et al. (2016) examined Indian sandstones after thermal treatment at different temperatures under three-point bedding tests. The experimental results showed that the fracture toughness of rock first increases and subsequently decreases as the temperature increases. Ding et al. (2016a, 2016b) investigated the strength and deformation parameters of sandstone after high temperature treatment under unloading conditions and found that when the temperature is higher than 400 °C, the peak ductile deformation increased rapidly with increasing temperature or initial confining pressure. Li et al. (2016) conducted a number of dynamic compression tests on thermal-treated sandstone specimens using split Hopkinson pressure bar test system. The experimental results indicated that a critical value exists for the dynamic mechanical parameters, e.g. 500 °C for the tested sandstone. From the literature, it is clear that the evolutionary law of mechanical behavior of heated rock at high temperature shows different responses under different geological and stress conditions even for the same types of rocks (Meng et al. 2006).

Analytical and numerical modeling of thermo-mechanical processes in sandstones have shown that variations in rock physical properties (including bulk density, porosity and compressional wave velocity) should not be expected between 100–200 °C, but become more significant above 400 or 500 °C (Tian et al. 2012). Indeed, thermal cracking experiments have shown that the greatest changes are likely to be achieved once the temperature exceeds 573 °C, i.e., the α/β transition of quartz (Glover et al. 1995). Li et al. (2011) investigated the change in the longitudinal wave velocity and mechanical behavior of sandstone after different high temperature treatments using triaxial unloading experiments. Heap et al. (2012) reported their experimental results on the influence of high temperatures on the strength of three tuffs that are commonly used for building in the Neapolitan region of Italy. Their results showed that a reduction in strength was only observed for one tuff; the other two were unaffected by high temperatures. The cause of this strength discrepancy was found to be a product of the initial mineralogical composition, or more specifically, the presence of thermally unstable zeolites within the initial rock matrix. Ranjith et al. (2012) carried out uniaxial compressive strength (UCS) testing of Hawkesbury sandstone at various temperatures between 25 and 950 °C. Their results demonstrated a mechanical dependence on temperature whereby the UCS and elastic modulus for the sandstone increase with increasing temperature for temperatures less than 500 °C and decrease with increasing temperature for temperatures greater than 500 °C. Yang et al. (2014a) performed a systematic experiment analyzing the strength, deformation behavior, crack initiation and crack coalescence stress of red sandstone containing either a single fissure or two parallel fissures after exposure to different high temperatures (25~900 °C). The results demonstrated that, in all cases, the strength and stiffness of red sandstone was increased at 300 °C, before decreasing up to the maximum temperature of 900 °C. However, the peak strain at failure always showed an increase when the temperature was increased.

Worldwide, granite material is an excellent medium for nuclear waste disposition rock projects due to its lower permeability and better integrity. Therefore, many researchers have focused on the mechanical behavior of granite under all kinds of stress conditions and external environments. Tullis and Yund (1977) reported their experimental results for granite at temperatures between 25 and 1000 °C. Their results showed that the UCS and elastic modulus of granite decreased with increasing temperature. Xu et al. (2008, 2009) conducted a study on the mechanical effects of temperature on granite that involved UCS testing at temperatures between 25 and 1200 °C. Their results showed that the mechanical properties of their granite varied only slightly with increasing temperature for experiments performed at temperatures between 25 and 800 °C, but that both UCS and elastic modulus decreased considerably with increasing temperature for experiments at temperatures greater than 800 °C. Zhao et al. (2013) experimentally studied the deformation, peak and post-peak strength characteristics of granite from the Beishan area (a preferable region for high-level radioactive waste disposal in China) under uniaxial and triaxial compression. Using a three-dimensional (3-D) acoustic emission (AE) monitoring system, Chen et al. (2014) experimentally studied the cracking process of Beishan granite under compressive stress and its effect on the hydro-mechanical properties. Shao et al. (2014) analyzed the effect of cooling rate on the mechanical behavior of heated Strathbogie granites with different grain sizes. The results showed that the UCS of granite decreases with increasing temperature beyond 400 °C under both slow and rapid cooling. The elastic modulus initially showed an increase with increasing temperature until 400 °C and then showed a decrease with further temperature increase until 800 °C for granites with different grain sizes and under both slow and rapid cooling. Using AE and scanning electron microscopy (SEM) systems, Shao et al. (2015) investigated the effect of temperature on the mechanical behavior of Strathbogie granite, and their results showed that the stress–strain curves of this granite showed plastic and post-peak behavior for temperatures above 800 °C and the brittle-plastic transition was observed to occur between 600 and 800 °C under uniaxial compression.

The physical characteristics and mechanical behaviour of granite after thermal treatment have been investigated for many years; however, high temperature can lead to the expansion of mineral grains, dehydration and chemical reaction of grains in the granite specimen, and transformation of the permeability evolution and failure behaviour. Here, the physical characteristics of granite mainly refer to the colour, volume, mass, density, thermal conductivity coefficient, wave velocity, mineral composition, and distribution of microcracks. Thermal treatment can induce the initiation of microcracks in granite and other works, while the confining pressure can restrain the propagation and coalescence of microcracks (Wang et al. 1989). X-ray micro-CT (Computed Tomography) may be adopted to observe the distribution of microcracks in granite. The results show that granite achieves a maximum anisotropy and heterogeneity at $T = 500$ °C (Fan et al. 2018). The propagation of microcracks with increasing temperature can be divided into four stages: 50 °C $\leq T \leq$ 250 °C, microcracks propagate stably; 250 °C $\leq T \leq$ 350 °C, crack formation accelerates; 300 °C $\leq T \leq$ 350 °C, cracks enter a transition stage; and 350 °C $\leq T \leq$ 700 °C, cracks enter a final development stage (Jiang et al. 2018). It has been reported that the heating–cooling process induces significant microcracking in granite, which are manifest as an increase in the porosity and gas permeability (Zhang et al. 2018). Using optical microscopy and X-ray micro-CT, Yang et al. (2017) interpreted the mechanisms of high-temperature treatment impacting the mechanical properties of granite.

Real-time high-temperature tests and tests following high-temperature treatment are the two main methods that are used to investigate the influence of the high temperature on the failure and mechanical behaviours of granite. Real-time high temperature tests are more approximate in representing response at the in-situ stress state. Uniaxial and triaxial compression tests are two test methods that can be used to investigate the failure and mechanical behaviours of granite under real-time high temperature conditions. The highest temperature of ~ 1200 °C can be reached under uniaxial compression conditions (Shao et al. 2015). The uniaxial compressive strength (UCS), deformation, fracture, and acoustic emission (AE) characteristics of granite under real-time high temperature conditions have been previously investigated (Xu et al. 2018). Triaxial compression tests under real-time high-temperature conditions are based on the use of hydraulic fluid to provide the confining pressure and have difficulty in reaching temperatures > 600 °C (Zhao et al. 2017). Triaxial compression tests can be used to investigate the strength and shear parameters of granite under real-time high-temperature conditions. Combined with AE, damage analysis can be performed to define the strength response (Chen et al. 2018). Using the mechanical parameters obtained from triaxial compressive tests, the fracture characteristics anticipated around HLW repositories may then be simulated (Guo et al. 2018).

Tests following high temperature treatment have been used for many years to investigate the influence of temperature on the mechanical behaviour of granite. Wang et al. (2017) compared the mechanical behaviour of granite under real-time high-temperature conditions both before and after thermal treatment. Their experimental results showed that real-time high-temperature conditions have a larger effect on the mechanical behaviour of granite than when viewing behaviour only after thermal treatment. However, the variation in the mechanical behaviour depending on the temperature of real-time high-temperature and post-thermal treatment conditions is similar. Therefore, tests of the response, post-thermal treatment are commonly used to investigate the effect of the temperature on the mechanical behaviour of granite. The degradation in the mechanical parameters as a result of thermal treatment is typically related to the mineral composition and microstructure of the granite (Chen et al. 2017b). Combining these observations of response with X-ray CT imaging shows that temperature has a significant effect on failure mode (Yang et al. 2017). Variations in the compressive strength, elastic modulus, tensile strength, and fracture toughness of granite have been analysed as functions of the temperature with four stages of behaviour identified. These stages are in the ranges of room temperature–200 °C, 200 °C–400 °C, 400 °C–600 °C, and > 600 °C, corresponding to undamaged, micro-damaged, secondarily damaged, and strongly damaged stages, respectively (Hu et al. 2018). High temperature also has a significant effect on peak strength, deformational characteristics, and failure modes of granite with pre-existing fissures (Huang et al. 2017). These properties are important for the evaluation of stability in rock engineering (Yang et al. 2019).

1.4 Damage Failure Mechanism of Rocks by AE and X-ray CT Observations

Acoustic emission (AE) technique is an efficient monitoring method, which can be used to predict and forecast the failure process of rock engineering and the precursor of earthquake occurrence (Lockner 1993; Rudajev et al. 1996; Spies and Eisenblatter 2001; Shiotant 2006; Ishida et al. 2010; Yang 2012). However, in the previous studies, a greater number of experiments concentrated mainly on one-dimensional (1 D) and two dimensional (2 D) AE behavior (e.g. AE counts, AE energy, etc.) of rock material (Alkan et al. 2007; Jing et al. 2008; Moradian et al. 2010, 2012; Yang and Jing 2011). Zang et al. (1996) investigated the AE behavior of dry and wet sandstone from Germany during uniaxial compression, which obtained the relation between the AE counts, AE energy and the microscopic damage of sandstone. But, engineering rock mass is actually a space structure, therefore it is more significant to explore three-dimensional (3-D) AE behavior (e.g. spatial AE distribution, etc.) of rock material in the process of the deformation (Lei et al. 2004; Thompson et al. 2006; Li et al. 2010; Xie et al. 2011; Yang et al. 2012; Schubnel et al. 2007). Schubnel et al. (2007) obtained a spatial AE distribution of Fontainebleau sandstone specimen (diameter = 40 mm, and height = 88 mm) under conventional triaxial compression. Li et al. (2010) acquired the changes in spatial correction length of AE events during fracture process of granite and marble specimens under uniaxial compression with the single-link cluster method, which showed that 3-D location of AE events directly reflected the process of initiation, propagation and evolution of micro-cracks and the stress re-distribution in the specimens. Xie et al. (2011) performed experiments on bedded rock salt specimens under uniaxial compression and indirect tension by a rock mechanics test system and the AE test system. Based on the basic theory of fractal, they proposed a column covering method to analyze the fractal property of the spatial AE distribution during the rock damage and failure process. Thompson et al. (2006) presented new observations of fracture nucleation from three triaxial compression experiments on intact specimen of Westerly granite using AE monitoring technique, which demonstrated the fracture process for quasistatic fracture, a slowly developing unstable fracture and an unstable fracture that developed near instantaneously. Lei et al. (2004) obtained the detailed time–space distribution of AE events during the catastrophic fracture of rock specimens containing a pre-existing joint or potential fracture plane under triaxial compression using a high-speed 32-channel waveform recording system. Benson et al. (2007) investigated the imaging slow failure in triaxially deformed basalt using 3-D AE location and X-ray computed tomography, which found excellent agreement between AE locations and post-test images from X-ray microtomography scanning that delineated deformation zone architecture. Baud et al. (2004) investigated the phenomenon of compaction localization and mechanical deformation of porous sandstone with real-time damage evolution using AE activity. Townend et al. (2008) studied the formation of compaction bands in Diemelstadt sandstone using 3D AE locations recorded by a Hyperion Giga-RAM recorder. Their results demonstrated that compaction bands nucleate at the specimen edge and propagate across the specimen at a rate of 0.08 mm/s. Fortin et al. (2009) investigated shear localization, compaction localization, and cataclastic compaction in Bleurswiller sandstone using 3D AE locations. They found that compaction localization and cataclastic compaction generate similar AE signatures to shear localization.

As is well-known, X-ray CT is the best non-invasive non-destructive method to characterize complex pore structures, and has been applied to investigate the damage and fracture behavior within the rock material in many past studies (Ge et al. 2001; Hirono et al. 2003; Feng et al. 2004; Sufian and Russell 2013; Zhao et al. 2014; Meier et al. 2014; Yang et al. 2015).

Using a conventional medical X-ray CT scanning system, Kawakata et al. (1999) analyzed 3-D observations of faulting in Westerly granite under uniaxial and triaxial conditions. Zhou et al. (2008) obtained CT images and CT values for the process of unloading, micro-cracking and dilation up to the failure of limestone specimens under different unloading levels. Using an X-ray CT system, Sufian and Russell (2013) investigated the microstructural pore changes and energy dissipation in Gosford sandstone during pre-failure loading, but the stress condition was limited to uniaxial compression. Based on X-ray CT observations of Gosford sandstone under uniaxial compression, Zhao et al. (2014) analyzed the mechanisms causing the strain-rate dependency of the uniaxial tensile strength of Gosford sandstone using the distinct lattice spring model (DLSM). Using an X-ray micro CT scanning system, Yang et al. (2015) analyzed in detail the internal damage characteristics of sandstone material after uniaxial and triaxial compression failure. Combined with two-stage uniaxial compression tests of gas shales, Cala et al. (2016) used an X-ray CT system to determine a possible connection between mechanical parameters, the geometry of failure mechanism and structural as well as textural features.

1.5 Main Contents in This Book

In this book, the author used several rock mechanics servo-controlled testing systems, AE techniques and X-ray micro CT observations, etc. to make a systematic investigation on the mechanical behavior and damage fracture mechanism of deep rocks. The main highlights in this book can be summarized four aspects:: (1) to investigate the influence of the loading path on the triaxial strength and deformation behavior of sandstone and marble; (2) to analyze the effect of borehole size on the triaxial strength and deformation behavior of hollow sandstone; (3) to explore the influence of high temperature on the triaxial deformation and permeability behavior of sandstone and granite; (4) to reveal the damage fracture mechanism of deep rocks using spatial AE techniques and X-ray micro CT observations. The detailed contents in this book can be described as following according to each chapter.

In Chapter 1, the effect of stress state, loading path and high temperature on the failure mechanical behavior of rocks are summarized. At the same time, the damage failure mechanism of rocks by using AE and X-ray CT observations were also summarized.

In Chap. 2, the strength, deformation, and damage failure behavior of red sandstone under two loading paths were investigated. This research investigates the influence of confining pressure on the deformation parameters of sandstone (in particular, volumetric contraction and dilatancy parameters) to construct a relationship between confining pressure and strength parameters (including peak and residual strength, crack initiation threshold, crack damage threshold, and the stress at zero volumetric strain) of sandstone. The evolution of damage in our sandstone specimens using the spatial and temporal distribution of AE counts was also analyzed.

In Chap. 3, the strength, deformation, and damage failure behavior of red sandstone under triaxial cyclic loading were investigated. Based on the experimental results, triaxial monotonic and cyclic mechanical parameters of sandstone specimens under different confining pressures are first compared. Then, the influence of cycle number on the crack damage threshold of sandstone is analyzed. Next, using an X-ray micro CT scanning system, the internal damage characteristics of sandstone material after triaxial monotonic and cyclic failure are explored. Finally, the influence of cycle number on the elastic modulus, Poisson's ratio and strain damage of sandstone under different confining pressures is discussed in detail.

In Chap. 4, the strength, deformation, and damage failure behavior of red sandstone under simple and complex loading paths were investigated. The emphasis of this chapter is focused on the investigation of the effect of confining pressure on strength and deformation parameters of red sandstone material under simple loading path, to analyze the strength parameters of red sandstone under two complex loading paths, to compare the strength and deformation difference of red sandstone between simple and complex loading paths, and to put forward a kind of new revised method to predict peak strength of red sandstone under simple loading path using obtained triaxial strength under complex loading path with only one specimen.

Chapter 5 is to make a systematical analysis on the mechanical, acoustic and fracture behaviors of yellow sandstone under triaxial monotonic and cyclic loading. The influence of confining pressure on the mechanical parameters of the sandstone is first analyzed, and the influence of the cycle number on the deformation and crack damage parameters of the sandstone is also investigated. During the loading process, real-time AE and ultrasonic velocity monitoring techniques are used to explore the acoustic behavior of yellow sandstone under 3D stress, and the effect of the cycle number on the AE behaviors is explored. Finally, by analyzing cross-sections and 3-D reconstructed images of the CT scans, the fracture mechanism of sandstone is revealed.

In Chap. 6, the triaxial strength and deformation failure behavior of coarse marble under six different loading paths are explored. This chapter is focused on to investigate the influence of loading path on the strength parameters (cohesion and internal friction angle) of coarse marble material, to analyze the re-fracture strength and deformation behavior of flawed and damaged rock specimen under different loading paths, and to discuss the failure mode of coarse marble under different loading paths.

In Chap. 7, the deformation failure characteristics of crystalline marble under triaxial cyclic loading are investigated. The results of a series of triaxial monotonic and cyclic compression tests on marble under different confining pressures were reported. Based on the experimental results of the marble under triaxial monotonic loading, the effect of confining pressure on the strength and deformation failure behavior of the marble specimens is first compared. Using the experimental results of the marble under triaxial cyclic loading, the influence of cycle number on the strain damage behavior and elastic modulus of the marble is analyzed. Finally, on the basis of the experimental results of the marble under complex cyclic loading, the strength and deformation parameters of the marble specimens with various damage extents are evaluated.

In Chap. 8, the strength, deformability and X-ray micro CT observations of deeply-buried marble under different confining pressures are experimentally analyzed. The results of a series of uniaxial and conventional triaxial compression tests on deeply-buried marble under different confining pressures ranging from 0 to 50 MPa were reported. Based on the experimental results, we first investigate the influence of the confining pressure on the strength and deformation parameters of marble. Then, using an X-ray micro CT scanning system, the internal damage characteristics of the deformed marble specimens are analyzed in detail.

Chapter 9 is to carry out a numerical investigation on the deformation and damage failure behavior of mudstone under single-stage and multi-stage triaxial compression on basis of the experimental results. This chapter has three contents: (1) to calibrate the micro-parameters in PFC^{2D} by single-stage triaxial compression of mudstone, (2) to analyse the influence of the extent of damage on the cohesion and internal friction angle of the mudstone specimen, and (3) to reveal the damage evolution mechanism of the mudstone specimen under triaxial compression.

In Chap. 10, the deformation, peak strength and crack damage behavior of hollow sandstone were analyzed systematically. The results of a series of conventional triaxial compression tests using hollow sandstone with various hole diameters are reported. Based on the experimental results, the influences of confining pressure and hole diameter on the deformation parameters of hollow sandstone are first discussed. Then, the effects of hole diameter and confining pressure on the peak strength parameters of the hollow sandstone are investigated in detail. Finally, the effects of hole diameter and confining pressure on crack damage behavior of the hollow sandstone are analyzed in detail.

In Chap. 11, the main aims were to conduct a series of X-ray micro-CT observations and 3D numerical simulations on hollow sandstone specimens with various borehole diameters, compare the experimental and numerical results for hollow sandstone specimens under different confining pressures, analyze the effects of borehole diameter and confining pressure on the peak strength and failure behavior of hollow sandstone, and explore the influence of the confining pressure on the fracture evolution mechanism of hollow sandstone with various borehole diameters.

In Chap. 12, the triaxial compression experimental results of compressed hollow-cylinder sandstone under two different confining pressures of 16 and 35 MPa were reported. Based on the experimental results, the internal damage failure characteristics of the compressed hollow-cylinder sandstone are analysed in detail using an X-ray micro CT scanning system. Then, the three-dimensional reconstructions of X-ray CT images of compressed hollow-cylinder sandstone are also performed. Finally, to better explain the internal crack evolution mechanism of hollow-cylinder sandstone during the entire loading, a conceptual model of hollow-cylinder sandstone material under the action of confining pressure is proposed.

In Chap. 13, the thermal damage and uniaxial mechanical behavior of granite after exposure to different high temperature treatments are investigated. The results of a series of uniaxial compression tests on granite subjected to various high temperature treatments ranging from 25 to 800 °C are reported. Based on the experimental results, the influence of high temperature on the crack damage, strength and deformation parameters of granite material is first investigated. Then, the effect of high temperature on the thermal damage characteristics of granite is analyzed. Finally, the internal crack and failure process of deformed granite specimens with different high temperatures are analyzed in detail.

In Chap. 14, the triaxial mechanical and permeability behavior of sandstone after exposure to different high temperature treatments are studied. First, the physical properties of the sandstone specimens were analyzed after high temperature treatment. Second, the influences of thermal treatment temperature on the strength and failure behaviors were analyzed in detail. Furthermore, the permeability evolution behavior of sandstone during the complete stress–strain process after exposure to different high temperature treatments was also investigated. Finally, the microscopic mechanism of high temperature on the physical, mechanical and permeability behaviors of the tested sandstone were revealed.

In Chap. 15, the effect of high temperature on the permeability evolution and failure response of granite under triaxial compression is analyzed. First, the effect of high temperature on the mechanical characteristics and permeability evolution behaviour of granite was analysed in detail. Subsequently, the effect of high temperature on triaxial deformation and AE behaviour of granite was investigated and the peak strength and crack damage threshold was analysed. Finally, mechanisms responsible for these impacts of high temperature influences the triaxial strength parameters of granite are discussed.

In Chap. 16, the influence of confining pressure (from 2 to 50MPa) on the crack damage, peak strength and deformation parameters of granite material is first investigated experimentally. Then, based on clustering algorithm, a new GBM method is put forward to simulate the damage mechanical behaviour of granite under different confining pressures. Finally, using an X-ray micro-CT scanning system and spatial acoustic emission (AE) monitoring, the internal crack and failure process of deformed granite specimens with different confining pressures are analysed in detail.

In Chap. 17, the strength and failure behaviors of coal specimens with different diameters are investigated to bring forward the reference for the choose of the cohesion and internal friction angle of coal specimen in mining engineering, The emphasis of this chapter is focused on to investigate the influence of the diameter on strength and failure behavior of coal material, to put forward a new evaluation criterion of confirming the cohesion and internal friction angle of coal material, to analyze the effect of the diameter on the cohesion and internal friction angle of coal specimen.

References

Al-Harthi AA (1998) Effect of planar structures on the anisotropy of Ranyah sandstone Saudi Arabia. Eng Geol 50(1/2):49–57

Adushkin VV, Kocharyan GG, Ostapchuk AA (2016) Parameters determining the portion of energy radiated during dynamic unloading of a section of rock massif. Dokl Earth Sci 467(1):275–279

Alkan H, Cinar Y, Pusch G (2007) Rock salt dilatancy boundary from combined acoustic emission and triaxial compression tests. Int J Rock Mech Min Sci 44:108–119

Alexander L (2001) Kaiser effect observation in brittle rock cyclically loaded with different loading rates. Mech Mater 33:669–677

Angelo M, Daniele B, Giordano M et al. (2008) The Tianjin geothermal field (north-eastern China): water chemistry and possible reservoir permeability reduction phenomena. Geothermics 37:400–428

Axelsson G (2010) Sustainable geothermal utilization-case histories; definitions; research issues and modelling. Geothermics 39:283–291

Bagde MN, Petroš V (2005) Fatigue properties of intact sandstone specimens subjected to dynamic uniaxial cyclical loading. Int J Rock Mech Min Sci 42(2):237–250

Bésuelle P, Desrues J, Raynaud S (2000) Experimental characterization of the localisation phenomenon inside a Vosges sandstone in a triaxial cell. Int J Rock Mech Min Sci 37(8):1223–1237

Baud P, Zhu W, Wong TF (2000) Failure mode and weakening effect of water on sandstone. J Geophys Res 105(B7):16371–16389

Benson PM, Meredith PG, Platzman ES, White RE (2005) Pore fabric shape anisotropy in porous sandstones and its relation to elastic wave velocity and permeability anisotropy under hydrostatic pressure. Int J Rock Mech Min Sci 42(7/8):890–899

Benson PM, Thompson BD, Meredith PG, Vinciguerra S, Young RP (2007) Imaging slow failure in triaxially deformed Etna basalt using 3D acoustic-emission location and X-ray computed tomography. Geophys Res Lett 34:L03303. https://doi.org/10.1029/2006GL028721

Baud P, Klein E, Wong TF (2004) Compaction localization in porous sandstones: spatial evolution of damage and acoustic emission activity. J Struct Geol 26(4):603–624

Cai M (2008) Influence of stress path on tunnel excavation response-Numerical tool selection and modeling strategy. Tunn Undergr Space Technol 23:618–628

Cala M, Cyran K, Stopkowicz A, Kolano M, Szczygielski M (2016) Preliminary application of X-ray computed tomograph on characterization of polish gas shale mechanical properties. Rock Mech Rock Eng. https://doi.org/10.1007/s00603-016-1045-6

Chandrasekharam D, Chandrasekhar V (2010) Geothermal energy resources, India: country update. In: Proceedings of the world geothermal congress. Bali, Indonesia, pp 25–29

Chen Y (1979) Acoustic emission of rocks under triaxial compression with various stress paths. Internal Int J Rock Mech Mining Sci Geomech Abstracts 16(6):401–405

Chen L, Liu JF, Wang CP, Liu J, Su R, Wang J (2014) Characterization of damage evolution in granite under compressive stress condition and its effect on permeability. Int J Rock Mech Min Sci 71:340–349

Chen YL, Wang SR, Ni J, Azzam R, Fernandez-Steeger TM (2017) An experimental study of the mechanical properties of granite after high temperature exposure based on mineral characteristics. Eng Geol 220:234–242

Chen G, Wang J, Li J, Li T, Zhang H (2018) Influence of temperature on crack initiation and propagation in granite. Int J Geomech 18(8):04018094

Crouch SL (1972) A note on post-failure stress-strain path dependence in Norite. Int J Rock Mech Min Sci 9(2):197–204

Ding QL, Ju F, Mao XB, Ma D, Yu BY, Song SB (2016) Experimental investigation of the mechanical behavior in unloading conditions of sandstone after high-temperature treatment. Rock Mech Rock Eng 49:2641–2653

Ding QL, Ju F, Song SB, Yu BY, Ma D (2016) An experimental study of fractured sandstone permeability after high-temperature treatment under different confining pressures. J Natural Gas Sci Eng 34:55–63

Du K, Tao M, Li XB, Zhou J (2016) Experimental study of slabbing and rockburst induced by true-triaxial unloading and local dynamic disturbance. Rock Mech Rock Eng 49(9):1–17

Eberhardt E (2001) Numerical modelling of three-dimension stress rotation ahead of an advancing tunnel face. Int J Rock Mech Min Sci 38:499–518

Fan LF, Gao JW, Wu ZJ, Yang SQ, Ma GW (2018) An investigation of thermal effects on micro-properties of granite by X-ray CT technique. Appl Therm Eng 140:505–519

Feng XT, Pei SF, Jiang Q, Zhou YY, Li SJ, Yao ZB (2017) Deep fracturing of the hard rock surrounding a large underground cavern subjected to high geostress: in situ observation and mechanism analysis. Rock Mech Rock Eng 50(8):2155–2175

Feng XT, Yao ZB, Li SJ, Wu SY, Yang CX, Guo HS (2018) In situ observation of hard surrounding rock displacement at 2400-m-deep tunnels. Rock Mech Rock Eng 51(3):873–892

Feng XT, Chen SL, Zhou H (2004) Real-time computerized tomography (CT) experiments on sandstone damage evolution during triaxial compression with chemical corrosion. Int J Rock Mech Min Sci 41(2):181–192

Fortin J, Stanchits S, Dresen G, Gueguen Y (2009) Acoustic emissions monitoring during inelastic deformation of porous sandstone: comparison of three modes of deformation. Pure Appl Geophys 166(5–7):823–841

Ferfera FMR, Sardau JP, Bouteca M et al. (1997) Experimental study of monophasic permeability changes under various stress paths. Int J Rock Mech Mining Sci 34(3–4), paper No. 037

Fortin J, Stanchits S, Dresen G, Gueguen Y (2006) Acoustic emission and velocities associated with the formation of compaction bands in sandstone. J Geophys Res. https://doi.org/10.1029/2005JB003854

Gatelier N, Pellet F, Loret B (2002) Mechanical damage of ananisotropic porous rock in cyclic triaxial tests. Int J Rock Mech Mining Sci 39:335–354

Gen NG (1982) Volume changes of rocks in various stress paths. Northwestern Seismol J 4(1):115–124

Glover PWJ, Baud P, Darot M, Meredith PG, Boon SA, Leravalec M (1995) α/β phase transition in quartz monitored using acoustic emissions. Geophys J Int 120(3):775–782

Gong HW, Chen ZG, Kong DF (2011) Primary Analysis of sandstone pore geothermal reserve reinjection based on Lindian geothermal field. Heilongjiang Sci Technol Water Conservancy 05:10–12 (in Chinese)

Guo LL, Zhang YB, Zhang YJ et al. (2018) Experimental investigation of granite properties under different temperatures and pressures and numerical analysis of damage effect in enhanced geothermal system. Renew Energy 126:107–125

Ge XR, Ren JX, Pu YB, Ma W, Zhu YL (2001) Real-in time CT test of the rock meso-damage propagation law. Sci China (series E) 44(3):328–336

Hirono T, Takahashi M, Nakashima S (2003) In situ visualization of fluid flow image within deformed rock by X-ray CT. Eng Geol 70:37–46

Hajpál M, Török Á (1989) Petrophysical and mineralogical studies of burnt sandstones. In: 2nd international PhD symposium in civil engineering, Budapest. pp 1–9

Heap MJ, Baud P, Meredith PG, Bell AF, Main IG (2009) Time-dependent brittle creep in Darley Dale sandstone. J Geophys Res. https://doi.org/10.1029/2008JB006212

Heap MJ, Baud P, Meredith PG (2009) The influence of temperature on brittle creep in sandstones. Geophys Res Lett. https://doi.org/10.1029/2009GL039373

Heap MJ, Faulkner DR, Meredith PG, Vinciguerra S (2010) Elastic moduli evolution and accompanying stress changes with increasing crack damage: implications for stress changes around fault zones and volcanoes during deformation. Geophys J Int 183:225–236

Huang S, Xia KW, Yang F, Feng XT (2010) An experimental study of the rate dependence of tensile strength softening of Longyou sandstone. Rock Mech Rock Eng 43(6):677–683

Huang D, Li Y (2014) Conversion of strain energy in triaxial unloading tests on marble. Int J Rock Mech Min Sci 66(1):160–168

Huang RQ, Wang XN, Chan LS (2001) Triaxial unloading test of rocks and its implication for rock burst. Bull Eng Geol Environ 60(1):37–41

He MC, Miao JL, Feng JL (2010) Rock burst process of limestone and its acoustic emission characteristics under true-triaxial unloading conditions. Int J Rock Mech Min Sci 47(2):286–298

He MC, Zhou L, Li DJ et al. (2008) Experimental research on hydrophilic characteristics of mudstone in deep well. Chin J Rock Mech Eng 27(6):1113–1120 (In Chinese)

Hu J, Sun Q, Pan X (2018) Variation of mechanical properties of granite after high-temperature treatment. Arab J Geosci 11(2):43

Huang YH, Yang SQ, Tian WL, Zhao J, Ma D, Zhang CS (2017) Physical and mechanical behavior of granite containing pre-existing holes after high temperature treatment. Arch Civ Mech Eng 17(4):912–925

Ishida T, Kanagawa T, Kanaori Y (2010) Source distribution of acoustic emissions during an in-situ direct shear test : Implications for an analog model of seismogenic faulting in an inhomogeneous rock mass. Eng Geol 110:66–76

Jaeger JC (1967) Brittle fracture of rocks. In: Proceedings of eighth symposium on rock mechanics, Port City Press. Baltimore, pp 3–57

Jiang G, Zuo J, Li L, Ma T, Wei X (2018) The evolution of cracks in Maluanshan granite subjected to different temperature processing. Rock Mech Rock Eng 51(6):1683–1695

Jeng FS, Weng MC, Huang TH, Lin ML (2002) Deformational characteristics of weak sandstone and impact to tunnel deformation. Tunn Undergr Space Technol 17(3):263–274

Jing HW, Zhang ZY, Xu GA (2008) Study of electromagnetic and acoustic emission in creep experiments of water-containing rock samples. J China Univer Mining Technol 18:42–45

Kawakata H, Cho A, Kiyama T, Yanagidani T, Kusunose K, Shimada M (1999) Three-dimensional observations of faulting process in Westerly granite under uniaxial and triaxial conditions by X-ray CT scan. Tectonophysics 313:293–305

Klein E, Baud P, Reuschle T, Wong TF (2001) Mechanical behaviour and failure mode of Bentheim sandstone under triaxial compression. Phys Chem Earth (a) 26:21–25

Kong B, Wang EY, Li ZH, Wong XR, Liu XF, Li N, Yang YL (2016) Electromagnetic radiation characteristics and mechanical properties of deformed and fractured sandstone after high temperature treatment. Eng Geol 209:82–92

Kumari WGP, Ranjith PG, Perera MSA et al. (2017) Mechanical behaviour of Australian Strathbogie granite under in-situ stress and temperature conditions: an application to geothermal energy extraction. Geothermics 65:44–59

Lan JL (2009) A study of the mechanical properties of sandstone after cyclic thermal action (Master thesis). National Cheng Kung University, Taiwan

Li M, Mao XB, Cao LL, Pu H, Mao RR, Lu AH (2016) Effects of thermal treatment on the dynamic mechanical properties of coal measures sandstone. Rock Mech Rock Eng 49:3525–3539

Liang WG, Xu SG, Zhao YS (2006) Experimental study of temperature effects on physical and mechanical characteristics of salt rock. Rock Mech Rock Eng 39:469–482

Liu S, Xu JY (2014) Mechanical properties of Qinling biotite granite after high temperature treatment. Int J Rock Mech Min Sci 71:188–193

Lund JW, Freeston DH, Boyd TL (2005) Direct application of geothermal energy: 2005 worldwide review. Geothermics 34(6):691–727

Luo JA, Wang LG (2011) High-temperature mechanical properties of mudstone in the process of underground coal gasification. Rock Mech Rock Eng 44:749–754

Lee DH, Juang CH, Chen JW et al. (1999) Stress paths and mechanical behavior of a sandstone in hollow cylinder tests. Int J Rock Mech Min Sci 36:857–870

Li YH, Yuan RF, Zhao XD (2007) Effect of different stress paths on Kaiser effect of rock acoustic emission. J Northeastern Univer (natural Science) 28(4):576–579

Li DY, Sun Z, Xie T, Li XB, Ranjith PG (2017) Energy evolution characteristics of hard rock during triaxial failure with different loading and unloading paths. Eng Geol 228:270–281

Li XB, Du K, Li DY (2015) True triaxial strength and failure modes of cubic rock specimens with unloading the minor principal stress. Rock Mech Rock Eng 48(6):2185–2196

Li XB, Feng F, Li DY, Du K, Ranjith PG, Rostami J (2018) Failure characteristics of granite influenced by sample height-to-width ratios and intermediate principal stress under true-triaxial unloading conditions. Rock Mech Rock Eng 51(5):1321–1345

Li XJ, Yang WM, Wang LG, Butler IB (2014) Displacement forecasting method in brittle crack surrounding rock under excavation unloading incorporating opening deformation. Rock Mech Rock Eng 47(6):2211–2223

Li YH, Liu JP, Zhao XD, Yang YJ (2010) Experimental studies of the change of spatial correction length of acoustic emission events during rock fracture process. Int J Rock Mech Min Sci 47:1254–1262

Lei XL, Masuda K, Nishizawa O et al. (2004) Detailed analysis of acoustic emission activity during catastrophic fracture of faults in rock. J Struct Geol 26:247–258

Lockner D (1993) The role of acoustic emission in the study of rock fracture. Int J Rock Mech Mining Sci Geomech Abstracts 30(7):883–899

Mahanta B, Singh TN, Ranjith PG (2016) Influence of thermal treatment on mode I fracture toughness of certain Indian rocks. Eng Geol 210:103–114

Mambou LLN, Ndop J, Ndjaka JMB (2014) Theoretical investigations of mechanical properties of sandstone rock specimen at high temperatures. J Min Sci 50(1):69–80

Martín-Gamboa M, Iribarren D, Dufour J (2015) On the environmental suitability of high-and low-enthalpy geothermal systems. Geothermics 53:27–37

Meier T, Rybacki E, Backers T, Dresen G (2014) Influence of bedding angle on borehole stability: a laboratory investigation of transverse isotropic oil shale. Rock Mech Rock Eng. https://doi.org/10.1007/s00603-014-0654-1

Meng ZP, Li MS, Lu PQ, Tian JQ, Lei Y (2006) Temperature and pressure under deep conditions and their influences on mechanical properties of sandstone. Chin J Rock Mech Eng 25(6):1177–1181

Morteza M, Richard GW (2016) Strength and post-peak response of Colorado shale at high pressure and temperature. Int J Rock Mech Min Sci 84:34–46

Moradian ZA, Ballivy G, Rivard P, Gravel C, Rousseau B (2010) Evaluating damage during shear test of rock joints using acoustic emissions. Int J Rock Mech Min Sci 47:590–598

Moradian ZA, Ballivy G, Rivard P (2012) Correlating acoustic emission sources with damaged zones during direct shear test of rock joints. Can Geotech J 49(6):710–718

Nespereira J, Blanco JA, Yenes M, Pereira D (2010) Irregular silica cementation in sandstones and its implication on the usability as building stone. Eng Geol 115(3/4):167–174

Peng J, Rong G, Cai M, Yao M, Zhou C (2016) Comparison of mechanical properties of undamaged and thermal - damaged coarse marbles under triaxial compression. Int J Rock Mech Min Sci 83:135–139

Peng J, Rong G, Cai M, Yao M, Zhou C (2016) Physical and mechanical behaviors of a thermal-damaged coarse marble under uniaxial compression. Eng Geol 200:88–93

Ranjith PG, Viete DR, Chen BJ, Perera MSA (2012) Transformation plasticity and the effect of temperature on the mechanical behaviour of Hawkesbury sandstone at atmospheric pressure. Eng Geol 151:120–127

Rudajev V, Vilhelm J, Kozak J (1996) Statistical precursors of instability of loaded rock samples based on acoustic emission. Int J Rock Mech Mining Sci Geomech Abstract 33(7):743–748

Swanson RS, Brown WS (1971) An observation of loading path independence of fracture in rock. Int J Rock Mech Min Sci 8(3):277–281

Shen MY, Shi ZM, Zhang L (2003) Deformation properties of samples under different loading paths. Chin J Rock Mech Eng 22(8):1234–1238

Su CD, Gao BB, Nan H, Li XJ (2009) Experimental study on acoustic emission characteristics during deformation and failure processes of coal samples under different stress paths. Chin J Rock Mech Eng 28(4):757–766

Shao SS, Wasantha PLP, Ranjith PG, Chen BK (2014) Effect of cooling rate on the mechanical behavior of heated Strathbogie granite with different grain sizes. Int J Rock Mech Min Sci 70:381–387

Shao SS, Ranjith PG, Wasantha PLP, Chen BK (2015) Experimental and numerical studies on the mechanical behaviour of Australian Strathbogie granite at high temperatures: an application to geothermal energy. Geothermics 54:96–108

Sirdesai NN, Singh TN, Ranjith PG, Singh R (2017) Effect of varied durations of thermal treatment on the tensile strength of red sandstone. Rock Mech Rock Eng 50:205–213

Spies T, Eisenblatter J (2001) Acoustic emission investigation of microcrack generation at geological boundaries. Eng Geol 61:181–188

Shiotani T (2006) Evaluation of long-term stability for rock slope by means of acoustic emission technique. NDT&E Int 39:217–228

Schubnel A, Jones G, Thompson B, Fortin J, Gueguen Y, Young RP (2007) Pore pressure induced rupture and aftershocks on intact and fractured sandstone in the laboratory. Geophys Res Lett 34. https://doi.org/10.1029/2007GL031076

Sufian A, Russell AR (2013) Microstructural pore changes and energy dissipation in Gosford sandstone during pre-failure loading using X-ray CT. Int J Rock Mech Min Sci 57:119–131

Tang FR, Wang LG, Lu YL, Yang XQ (2015) Thermophysical properties of coal measure strata under high temperature. Environ Earth Sci 73(10):6009–6018

Tian H, Ziegler M, Kempka T (2014) Physical and mechanical behavior of claystone exposed to temperatures up to 1000°C. Int J Rock Mech Min Sci 70:144–153

Török Á, Hajpál M (2005) Effect of temperature changes on the mineralogy and physical properties of sandstones: a laboratory study. Restor Build Monum 11(4):1–8

Tang B, Cheng H, Tang Y, Yao Z, Li M, Zheng T (2018) Excavation damaged zone depths prediction for TBM-excavated roadways in deep collieries. Environ Earth Sci 77(5):165

Tembe S, Baud P, Wong TF (2008) Stress conditions for the propagation of discrete compaction bands in porous sandstone. J Geophys Res. https://doi.org/10.1029/2007JB005439

Tsai LS, Hsieh YM, Weng MC, Huang TH, Jeng FS (2008) Time-dependent deformation behaviors of weak sandstones. Int J Rock Mech Min Sci 45(2):144–154

Tavallali A, Vervoort A (2010) Effect of layer orientation on the failure of layered sandstone under Brazilian test conditions. Int J Rock Mech Min Sci 47(2):313–322

Tan X, Chen W, Liu H, Chan AHC, Tian H, Meng X (2017) A combined supporting system based on foamed concrete and u-shaped steel for underground coal mine roadways undergoing large deformations. Tunn Undergr Sp Tech 68:196–210

Tullis J, Yund RA (1977) Experimental deformation of dry Westerly granite. J Geophys Res 82:5705–5718

Townend E, Thompson BD, Benson PM, Meredith PG, Baud P, Young RP (2008) Imaging compaction band propagation in Diemelstadt sandstone using acoustic emission locations. Geophys Res Lett. https://doi.org/10.1029/2008GL034723

Thompson BD, Young RP, Lockner DA (2006) Fracture in Westerly granite under AE feedback and constant strain rate loading: Nnucleation, Quasi-static propagation, and the transition to unstable fracture propagation. Pure Appl Geophys 163:995–1019

Xiao XH, Xiao PW, Dai F, Li HB, Zhang X, Zhou JW (2018) Large deformation characteristics and reinforcement measures for a rock pillar in the houziyan underground powerhouse. Rock Mech Rock Eng 51(2):1–18

Xiang TB, Feng XT, Chen BR, Jing Q (2008) True triaxial and acoustic emission experimental study of failure process of hard rock under excavating and supporting stress paths. Rock and Soil Mech 29(Supp.):500–506

Xie HP, Peng RD, Ju Y (2004) Energy dissipation of rock deformation and fracture. Chin J Rock Mech Eng 23(21):3565–3570 (In chinese)

Xie HP, Liu JF, Ju Y, Li J, Xie LZ (2011) Fractal property of spatial distribution of acoustic emissions during the failure process of bedded rock salt. Int J Rock Mech Min Sci 48:1344–1351

Xie HQ, He CH (2004) Study of the unloading characteristics of a rock mass using the triaxial test and damage mechanics. Int J Rock Mech Min Sci 41(3):366

Xiong SH, Zhou HM (2006) Deformation properties of rock mass of TGP permanent shiplock slopes under complex stress paths. Chin J Rock Mech Eng 25(Supp. 2):3636–3641

Xu XT, Huang RQ, Li H, Huang QX (2015) Determination of poisson's ratio of rock material by changing axial stress and unloading lateral stress test. Rock Mech Rock Eng 48(2):853–857

Xu XL, Gao F, Shen XM, Xie HP (2008) Mechanical characteristics and microcosmic mechanisms of granite under temperature loads. J China Univ Min Technol 18:413–417 (In Chinese)

Xu XL, Kang ZX, Ji M, Ge WX, Chen J (2009) Research of microcosmic mechanism of brittle-plastic transition for granite under high temperature. Procedia Earth Planet. Sci. 1:432–437

Xu DJ, Geng NG (1986) The various stress paths causing deformation and failure in rocks. Rock and Soil Mech 7(2):17–25 (in Chinese)

Xue Y, Ranjith PG, Gao F, Zhang DC, Cheng HM, Chong ZH (2017) Mechanical behaviour and permeability evolution of gas-containing coal from unloading confining pressure tests. J Nat Gas Sci Eng 40:336–346

Xu XL, Zhang ZZ (2018) Acoustic emission and damage characteristics of granite subjected to high temperature. Adv Mater Sci Eng. https://doi.org/10.1155/2018/8149870

Wang Z, He A, Shi G, Mei G (2017) Temperature effect on AE energy characteristics and damage mechanical behaviors of granite. Int J Geomech 18(3):04017163

Wong TF, David C, Zhu W (1997) The transition from brittle faulting to cataclastic flow in porous sandstones: Mechanical deformation. J Geophys Res 102(B2):3009–3025

Wu Z, Qian BD, Chen LJ, Luo YJ (2005) Experimental study on mechanical character of sandstone of the upper plank of coal bed under high temperature. Chin J Rock Mech Eng 24:1863–1867 (in Chinese)

Wu XY, Baud P, Wong TF (2000) Micromechanics of compressive failure and spatial evolution of anisotropic damage in Darley Dale sandstone. Int J Rock Mech Min Sci 37(1/2):143–160

Wang Q, Pan R, Li SC, Wang HT, Jiang B (2018) The control effect of surrounding rock with different combinations of the bolt anchoring lengths and pre-tightening forces in underground engineering. Environ Earth Sci 77(13):501

Wang HF, Bonner BP, Carlson SR, Kowallis BJ, Heard HC (1989) Thermal stress cracking in granite. J Geophys Res Solid Earth 94(B2):1745–1758

Wang B, Zhu JB, Wu AQ et al. (2008) Experimental study on mechanical properties of Jinping marble under loading and unloading stress paths. Chin J Rock Mech Eng 27(10):2138–2145 (in Chinese)

Yang SQ, Jiang YZ (2010) Triaxial mechanical creep behavior of sandstone. Min Sci Technol 20(3):339–349

Yang SQ, Jing HW (2011) Strength failure and crack coalescence behavior of brittle sandstone samples containing a single fissure under uniaxial compression. Int J Fract 168(2):227–250

Yang SQ, Jing HW, Li YS, Han LJ (2011) Experimental investigation on mechanical behavior of coarse marble under six different loading paths. Exp Mech 51(3):315–334

Yang SQ (2012) Strength and deformation behavior of red sandstone under multi-stage triaxial compression. Can Geotech J 49(6):694–709

Yang SQ, Jing HW, Wang SY (2012) Experimental investigation on the strength, deformability, failure behavior and acoustic emission locations of red sandstone under triaxial compression. Rock Mech Rock Eng 45(4):583–606

Yang SQ, Jing HW (2013) Evaluation on strength and deformation behavior of red sandstone under simple and complex loading paths. Eng Geol 164:1–17

Yang SQ, Jing HW, Huang YH, Ranjith PG, Jiao YY (2014a) Fracture mechanical behavior of red sandstone containing a single fissure and two parallel fissures after exposure to different high temperature treatments. J Struct Geol 69(Part A):245–264

Yang SQ, Ni HM, Wen S (2014b) Spatial acoustic emission evolution of red sandstone during multi-stage triaxial deformation. J Central South Univer 21(8):3316–3326

Yang SQ, Ranjith PG, Huang YH, Yin PF, Jing HW, Gui YL, Yu QL (2015) Experimental investigation on mechanical damage characteristics of sandstone under triaxial cyclic loading. Geophys J Int 201:662–682

Yang SQ, Ranjith PG, Jing HW, Tian WL, Ju Y (2017) An experimental investigation on thermal damage and failure mechanical behavior of granite after exposure to different high temperature treatment. Geothermics 65:180–197

Yang SQ, Huang YH, Tian WL, Yin PF, Jing HW (2019) Effect of high temperature on deformation failure behavior of granite specimen containing a single fissure under uniaxial compression. Rock Mech Rock Eng 52:2087–2107

Yao XX, Geng NG, Chen Y (1980) The effect of stress path on brittleness and ductility of rocks. Acta Geophysica Sinica 23(3):312–319

Yao XX (1981) Observation of micro-cracks within gabbro in experiments along various stress paths. Acta Geophysica Sinica 3(1):49–54

Yao MD, Rong G, Zhou CB et al. (2016) Effects of thermal damage and confining pressure on the mechanical properties of coarse marble. Rock Mech Rock Eng 49:2043–2054

Yin TB, Shu RH, Li XB, Wong P, Liu XL (2016) Comparison of mechanical properties in high temperature and thermal treatment granite. Trans Nonferrous Metals Soc China 26:1926–1937

Yang CH, Bai SW, Wu YM (2000) Stress level and loading path effect on time-dependent properties of salt rock. Chin J Rock Mech Eng 19(3):270–275

Zang A, Wagner CF, Dresen G. Acoustic emission, micro-structure, and damage model of dry and wet sandstone stressed to failure [J] Journal of Geophysical Research, 1996, 101(17): 507- 517, 521.

Zhang HB, Song XG, Huang MS, Shang QS (2007) Research on failure features of rocks under different stress unloading path. J Shandong Univer (engineering Science) 37(6):83–86

Zhao XG, Wang J, Cai M, Cheng C, Ma LK, Su R (2014a) Influence of unloading rate on the strainburst characteristics of beishan granite under true-triaxial unloading conditions. Rock Mech Rock Eng 47(2):467–483

Zorlu K, Gokceoglu C, Ocakoglu F, Nefelioglu HA, Acikalin S (2008) Prediction of uniaxial compression strength of sandstones using petrography-based models. Eng Geol 96(3/4):141–158

Zhou KP, Liu TY, Hu ZX (2018) Exploration of damage evolution in marble due to lateral unloading using nuclear magnetic resonance. Eng Geol 244:75–85

Zhao GF, Russell AR, Zhao XB, Khalili N (2014b) Strain rate dependency of uniaxial tensile strength in Gosford sandstone by the distinct lattice spring model with X-ray micro CT. Int J Solids Struct 51:1587–1600

Zhou XP, Zhang YX, Ha QL (2008) Real-time computerized tomography (CT) experiments on limestone damage evolution during unloading. Theor Appl Fract Mech 50:49–56

Zhang LY, Mao XB, Liu RX, Guo X, Ma D (2014) The mechanical properties of mudstone at high temperatures: an experimental study. Rock Mech Rock Eng 47:1479–1484

Zhang WQ, Sun Q, Hao SQ (2016) Experimental study on the variation of physical and mechanical of rock after high temperature treatment. Appl Therm Eng 98:1297–1304

Zhu TT, Jing HW, Su HJ, Yin Q, Du MR, Han GS (2016) Physical and mechanical properties of sandstone containing a single fissure after exposure to high temperatures. Int J Min Sci Technol 26:319–325

Zhang F, Zhao J, Hu D, Skoczylas F, Shao J (2018) Laboratory investigation on physical and mechanical properties of granite after heating and water-cooling treatment. Rock Mech Rock Eng 51(3):677–694

Zhao XG, Cai M, Wang J, Ma LK (2013) Damage stress and acoustic emission characteristics of the Beishan granite. Int J Rock Mech Min Sci 64:258–269

Zhao Y, Feng Z, Zhao Y, Wan Z (2017) Experimental investigation on thermal cracking, permeability under HTHP and application for geothermal mining of HDR. Energy 132:305–314

Chapter 2
Strength, Deformation, Failure Behavior and Acoustic Emission Locations of Red Sandstone Under Triaxial Compression

In the past, most triaxial experiments on sandstone have been performed using the axial displacement rate as the controlling feedback signal. However, since specimen failure in the brittle regime is often very rapid under this mode of control, it is often difficult to obtain reliable post-peak deformation data. In this chapter, we will attempt to improve our knowledge of post-peak behavior in sandstone using the circumferential displacement rate as the controlling feedback mechanism. Further, in previous experimental research on sandstone, the effect of the loading path on the strength, failure behavior, crack initiation threshold, crack damage threshold and 3D damage evolution process under triaxial compression are less discussed.

Therefore in this chapter, we have performed triaxial experiments on sandstone under two loading paths (i.e. conventional triaxial compression experiments and "reducing confining pressure" experiments (to be described in detail in the next section) using the circumferential displacement rate as the controlling feedback mechanism (Yang et al. 2012). Our research investigates the influence of confining pressure on the deformation parameters of sandstone (in particular, volumetric contraction and dilatancy parameters) to construct a relationship between confining pressure and strength parameters (including peak and residual strength, crack initiation threshold, crack damage threshold, and the stress at zero volumetric strain) of sandstone. We also analyze the evolution of damage in our sandstone specimens using the spatial and temporal distribution of AE counts.

2.1 Experimental Material and Testing Procedures

2.1.1 Experimental Material

The material used throughout this study was red sandstone (collected from the Shandong province of China). Red sandstone (Fig. 2.1) is a fine- to medium-grained, feldspathic sandstone with a connected porosity of 8.8% and a bulk density of 2360 kg/m^3.

S.-Q. Yang, *Mechanical Behavior and Damage Fracture Mechanism of Deep Rocks*,
https://doi.org/10.1007/978-981-16-7739-7_2

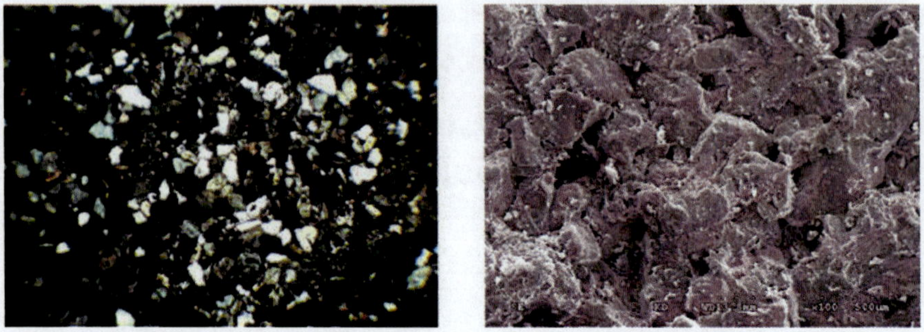

(a) Optical microscopy (×40) (b) SEM (×100)

Fig. 2.1 Optical microscopy and SEM photomicrographs of red sandstone

The modal composition is 45% feldspar (with a grain radius of 0.15–0.25 mm), 13.5% quartz (with a grain radius of 0.2–0.3 mm), 31.5% debris (with a grain radius of 0.25–0.4 mm), and 10% agglutinate (which consisted of gypsum, barite and iron integument).

2.1.2 Testing Equipment

All experiments were performed in an MTS815 Flex Test GT triaxial rock mechanics experimental system (Fig. 2.2). The maximum loading capacity of the servo-controlled system is 4700 kN, and the maximum confining pressure is 140 MPa. During experimentation, we measured the axial force (using a load cell with a loading capacity of 2600 kN) and the axial deformation with an axial LVDT (Linear Variable

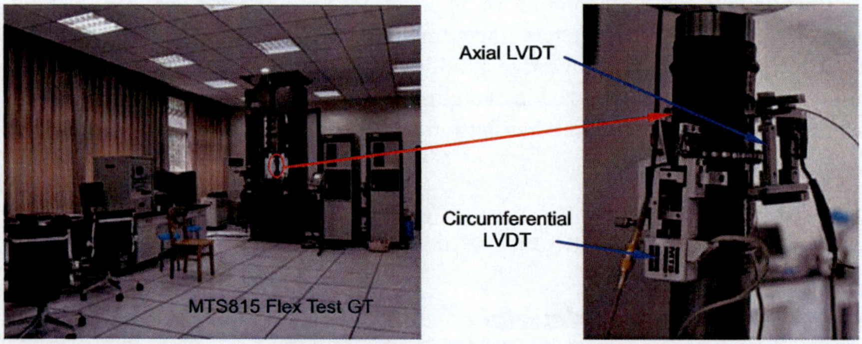

Fig. 2.2 MTS815 Flex Test GT triaxial rock mechanics experimental system and strain measurement

Differential Transducer) with a range of 5 mm (Fig. 2.2). The circumferential deformation was measured using an LVDT attached to a chain wrapped tightly around the specimen (Fig. 2.2), for which the maximum displacement capacity was 8 mm. To eliminate the effect of end friction effects (see Mogi 1966; Hawkes and Mellor 1970), the axial and circumferential LVDT were located in the central part of the specimen (Fig. 2.2).

2.1.3 Testing Procedure

In this study, we performed a series of triaxial compression experiments using a circumferential displacement rate of 0.06 mm/min. To reiterate, we used the circumferential displacement as the controlling feedback signal (see also Amann et al. 2011; Youn and Tonon 2010), rather than the axial displacement feedback commonly used in rock deformation experiments (see Paterson and Wong 2005). For the purpose of this study, we performed two types of triaxial test: (1) conventional triaxial experiments, and (2) "reducing confining pressure" experiments.

The conventional triaxial experiments (Fig. 2.3a) were carried out under different confining pressures of 5, 20, 35, 50 and 65 MPa and consisted of the following two steps. Firstly, the confining pressure was increased to the desired value at a constant rate of 0.1 MPa/s (to ensure that the specimen was under uniform hydrostatic stresses (*oa*)). Finally, the deviatoric stress (*ab* → *bc*) was applied to the surface of the specimen at a constant circumferential displacement rate of 0.06 mm/min until failure.

Our "reducing confining pressure" experiments (Fig. 2.3b) were conducted using the following three steps. First, the hydrostatic pressure was applied to the specimen at a rate of 0.1 MPa/s until the desired value was reached (*oa*). The deviatoric stress was applied up to a pre-determined level (either 70 or 90% of the triaxial compressive strength (TCS) determined from our conventional triaxial tests) at a constant

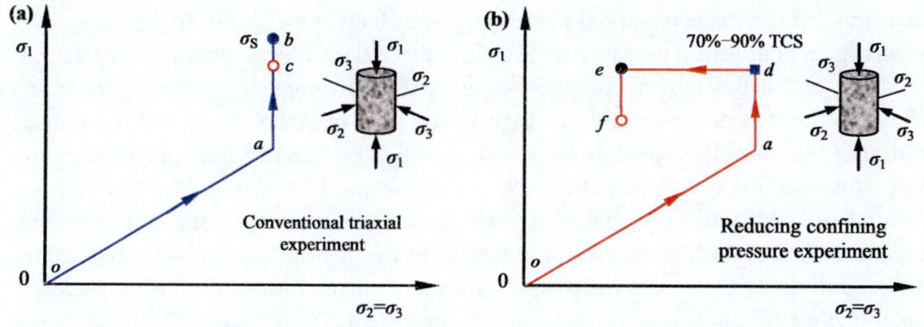

Fig. 2.3 The two different loading paths used in this study, presented in the principal stress space (σ_3, σ_1). Conventional triaxial experiment: *oa* → *ab* → *bc*; "reducing confining pressure experiment": *oa* → *ad* → *de* → *ef*

circumferential displacement control rate of 0.06 mm/min (*ad*). Finally, the major principal stress (σ_1) was kept constant, but the confining pressure was reduced at a rate of 0.1 MPa/s until the specimen failure occurred ($de \rightarrow ef$).

All triaxial experiments were carried out on cylindrical specimens 50 mm in diameter and 100 mm in length in accordance with the ISRM standard (Fairhurst and Hudson 1999). The specimens were all cored from the same large rectangular block and in the same orientation. All the experiments were performed on dry specimens at room temperature.

2.1.4 AE Measuring Procedure

As it is well-known, the AE technique is predominantly related to the release of elastic energy within rock materials. AE hits and/or energy are often monitored to detect the onset (termed C', see Wong et al. 1997) and the evolution of microcracking and to analyze the spatial and temporal progression of internal cracks (e.g., Lockner 1991). In this study, the output of AE was recorded by a PCI-2 AE 3-D measuring system made by PAC (Physical Acoustic Corporation) using eight AE sensors with a frequency of 6 kHz. The amplitude threshold of the sensors was set to 33 dB.

2.2 Triaxial Experimental Results of Red Sandstone

2.2.1 Deformation Behavior Under Conventional Triaxial Compression

The triaxial stress–strain curves for red sandstone under a variety of confining pressures are presented in Figs. 2.4 and 2.5. The synopsis plots of Fig. 2.5 demonstrate that an increase in confining pressure served to increase the strength of red sandstone (in a similar manner as observed previously, see Heap et al. 2009). In all cases, even at the highest confining pressure of 65 MPa, the failure mode remained brittle (i.e., the ability of the rock to resist load decreases with permanent strain, see Rutter 1986). However, there is an observable change in the post-peak behavior with increasing confining pressure. Post-peak behavior switched from class I (the slope of the post-peak stress–strain curve is negative, see Wawersik and Fairhust 1970) at a confining pressure of 5 MPa to class II (where the slope is positive) at confining pressures above 5 MPa. Further, there was a gradual increase in strain-softening behavior and the magnitude of the stress drop was reduced with increasing confining pressure. The influence of confining pressure on the deformation parameters of red sandstone is described in more detail in Sect. 2.2.3.

The stress–strain curve of red sandstone at 5 MPa (Fig. 2.4a) shows initially non-linear (concave) deformation, usually attributed to the closure of pre-existing

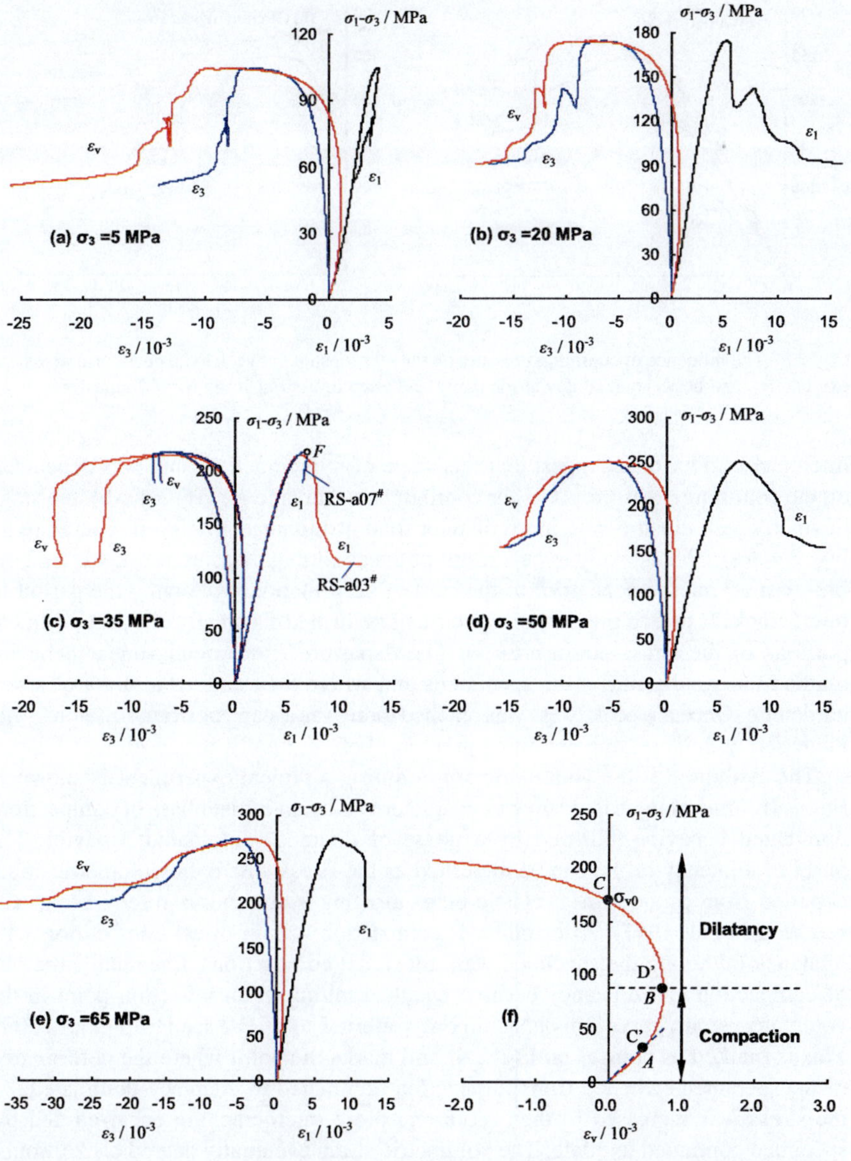

Fig. 2.4 Stress–strain curves for red sandstone from conventional triaxial compression experiments. ε_1 and ε_3 represents the axial and circumferential strain, respectively. ε_v refers to the volumetric strain, which is calculated from the sum of the axial strain and twice the circumferential strain (i.e. $\varepsilon_v = \varepsilon_1 + 2\varepsilon_3$)

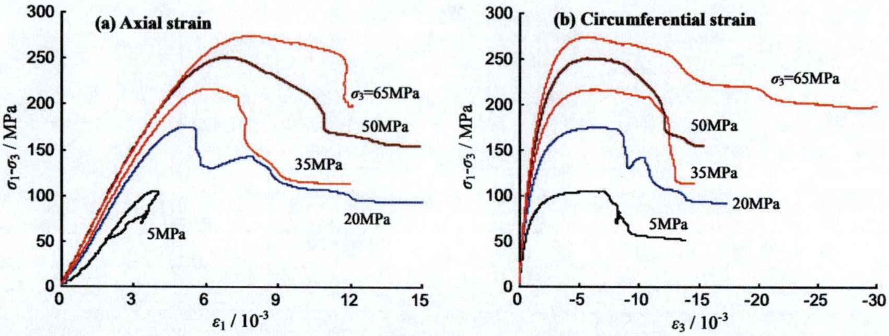

Fig. 2.5 The influence of confining pressure on the relationship between axial deviatoric stress and axial strain, and between axial deviatoric stress and circumferential strain for red sandstone

microcracks. Our data suggest that this stage of microcrack closure was dependent on the confining pressure. At lower confining pressures (e.g., Fig. 2.4 a-b), the stage of microcrack closure was more distinct than at higher confining pressures (e.g., Fig. 2.4 d-e). This is likely to be a result of the confining pressure acting to close the pre-existing microcracks prior to the start of deformation. Following the period of microcrack closure (if present) the specimens deformed elastically (the pseudo-linear portions of the stress–strain curves). The departure from pseudo-linear behavior marked the yield point of the specimens and where they entered a phase of strain hardening. Once the peak stress was reached there was a stage of strain softening (that increased with increasing confining pressure) before, finally, the specimens failed.

The evolution of the volumetric strain during a typical experiment is shown in Fig. 2.4f. The volumetric strain was characterized by an initial phase of compaction-dominated behavior followed by a phase of dilatancy-dominated behavior. The onset of dilatancy or the can be identified as the stress where the volumetric strain departed from pseudo-linear elastic behavior (this is commonly referred to as C', see Wong et al. 1997). After point *A*, compaction was in direct competition with dilatancy (although the specimen was still in net compaction). Eventually, the rate of compaction and dilatancy became equal, resulting in an inflection point in the volumetric strain curve (this is commonly referred to as D', see Heap et al. 2009). This is marked as point *B* on Fig. 2.4f and marks the point where the deformation of the specimen switched from compaction-dominated to dilatancy-dominated. As the stress was increased further, more and more microcracking occurred and the specimen continued to dilate. The volumetric strain eventually passed back through zero strain (σ_{v0}) before it continued to increase until specimen failure. Table 2.1 lists the axial and circumferential strain of red sandstone at C', D' and the stress at zero volumetric strain under conventional triaxial compression.

To investigate the reproducibility of our experimental setup and our experimental material we performed a repeat experiment at a confining pressure of 35 MPa. Figure 2.4c shows the axial stress–strain curves for two red sandstone specimens RS-a03# and RS-a07# (after the peak stress, specimen RS-a07# was unloaded at

Table 2.1 Axial and circumferential strain of red sandstone at C', D' and the stress at zero volumetric strain from conventional triaxial compression experiments. $\varepsilon_{1c'}$ and $\varepsilon_{3c'}$ represent the axial and circumferential strain at $\sigma_{c'}$, respectively. $\varepsilon_{1d'}$ and $\varepsilon_{3d'}$ represent the axial and circumferential strain at $\sigma_{d'}$, respectively. ε_{1v0} and ε_{3v0} represent the axial and circumferential strain at σ_{v0}, respectively

σ_3 / MPa	$\varepsilon_{1c'}$ / 10^{-3}	$\varepsilon_{3c'}$ / 10^{-3}	$\varepsilon_{1d'}$ / 10^{-3}	$\varepsilon_{3d'}$ / 10^{-3}	ε_{1v0} / 10^{-3}	ε_{3v0} / 10^{-3}
5	0.51	−0.03	1.65	−0.34	2.79	−1.39
20	0.53	−0.06	1.62	−0.43	3.20	−1.60
35 (RS-a03#)	0.60	−0.09	1.73	−0.49	3.80	−1.90
35 (RS-a07#)	0.58	−0.06	1.98	−0.60	4.04	−2.02
50	0.62	−0.12	2.08	−0.71	4.30	−2.15
65	0.68	−0.12	2.74	−0.94	5.49	−2.74

point *F*). The stress–strain curves and the peak stress were nearly identical, adding to the veracity of our dataset.

Figure 2.6 shows influence of confining pressure (Fig. 2.6a) and specimen variability (Fig. 2.6b) on the relationship between the circumferential strain and the axial strain of red sandstone prior to the peak stress. Firstly, it can be seen that the circumferential strain increased nonlinearly with the axial strain. At lower stress levels, the circumferential strain increased slowly with the axial strain, but at higher stress levels, the rate of circumferential strain was higher than that of axial strain. The confining pressure and specimen variability had no obvious influence on the relationship between the circumferential strain and the axial strain at the lower stress levels. However, at higher stress levels, the rate of circumferential strain with axial strain was smaller at higher confining pressures (Fig. 2.6a). This suggests that higher confining pressures suppressed the circumferential deformation of the specimens. We observed only a small deviation in the circumferential-axial strain curves for two

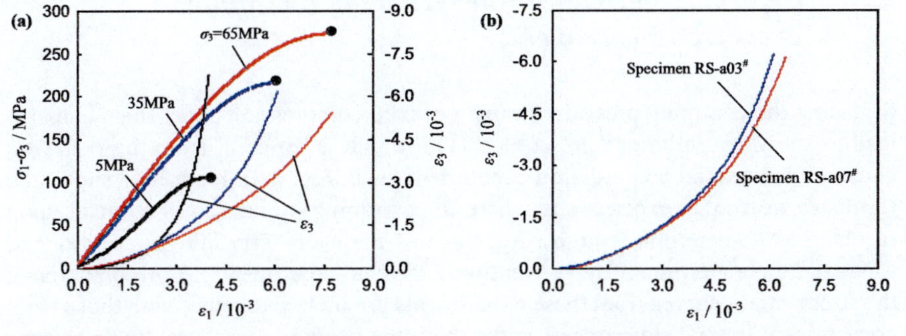

Fig. 2.6 Relationship between the circumferential strain and the axial strain of red sandstone prior to the peak stress: **a** the effect of confining pressure and **b** the effect of specimen variability. The filled, black circles represent the peak stress of the specimen

Table 2.2 Young's modulus, Poisson's ratio and peak strain of red sandstone from conventional triaxial compression experiments

σ_3 / MPa	E_S / GPa	E_{50} / GPa	υ	ε_{1c} / 10^{-3}	ε_{3c} / 10^{-3}	ε_{vc} / 10^{-3}
5	35.9	27.4	0.28	3.99	−6.77	−9.54
20	44.6	42.0	0.34	5.18	−6.49	−7.77
35 (RS-a03#)	47.3	46.3	0.36	6.10	−6.22	−6.33
35 (RS-a07#)	46.9	45.4	0.34	6.43	−6.16	−5.89
50	51.0	51.6	0.37	6.77	−5.67	−4.57
65	50.1	50.7	0.34	7.75	−5.48	−3.20

specimens deformed under the same pressure conditions (Fig. 2.6b), again adding to the veracity of our dataset.

Figure 2.6 also indicates that Poisson's ratio was not a constant and that it changed with the axial deviatoric stress. Thus, it was more difficult to determine Poisson's ratio because of obvious nonlinearity between the circumferential strain and the axial strain, even though it was observed during elastic deformation. Therefore, in this study, we have calculated the Poisson's ratio using the following method. We have used the average of the absolute value of the ratio of circumferential strain and axial strain in the range of axial differential stress from 30 to 70% TCS, as shown in Table 2.2. Table 2.2 lists Young's modulus, Poisson's ratio and the peak strain of red sandstone from our conventional triaxial compression experiments. E_S and E_{50} represent the elastic modulus and the deformation modulus of the sandstone (Yang and Jiang 2010), respectively. υ represents Poisson's ratio of rock. ε_{1c}, ε_{3c} and ε_{vc} are respectively defined as the axial, circumferential and volumetric strain values at the peak stress.

2.2.2 *Deformation Behavior in "reducing Confining Pressure" Experiments*

Reducing the confining pressure during a triaxial compression experiment can also lead to specimen failure (Yang et al. 2011), which is different from the path of a conventional triaxial compression experiment (as in Fig. 2.3). Figure 2.7 shows the results from triaxial experiments where the confining pressure was reduced upon reaching a pre-determined high percentage of the short-term failure stress of red sandstone (as determined by conventional triaxial experiments). For comparison, the stress–strain curves from these experiments are plotted together with those from conventional triaxial experiments under the same confining pressure. It can be seen that, between the point where the confining pressure reduction was initiated and the failure of the specimen, the axial deviatoric stress still increased, even when σ_1 was kept constant (Fig. 2.8). After failure, the axial supporting capacity dropped rapidly to the value of the residual strength.

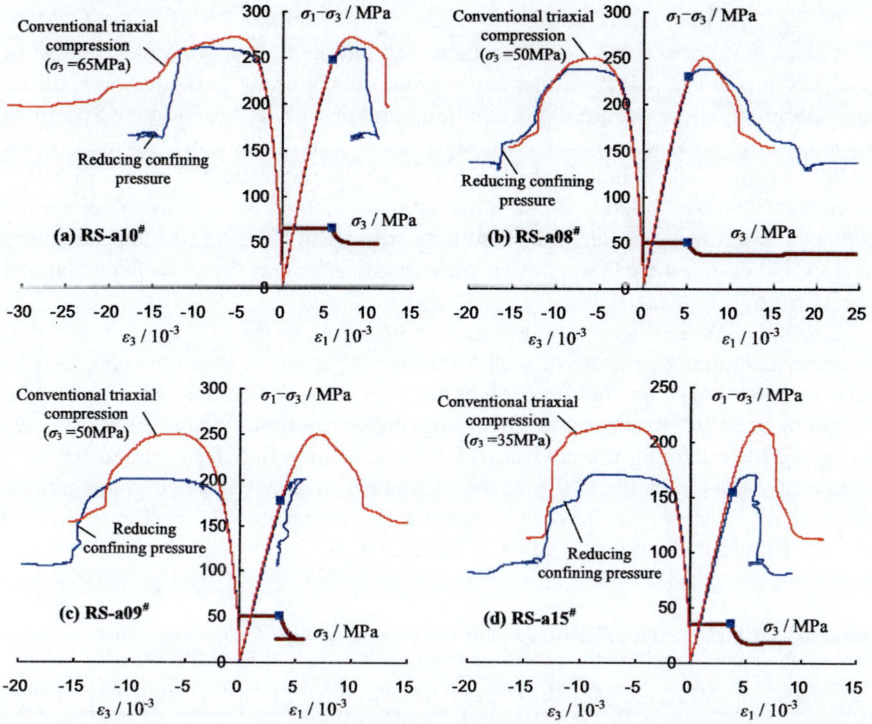

Fig. 2.7 Results from conventional triaxial experiments (red lines) and "reducing confining pressure" experiments (blue lines). The filled, blue squares represent the position where the confining pressure was reduced

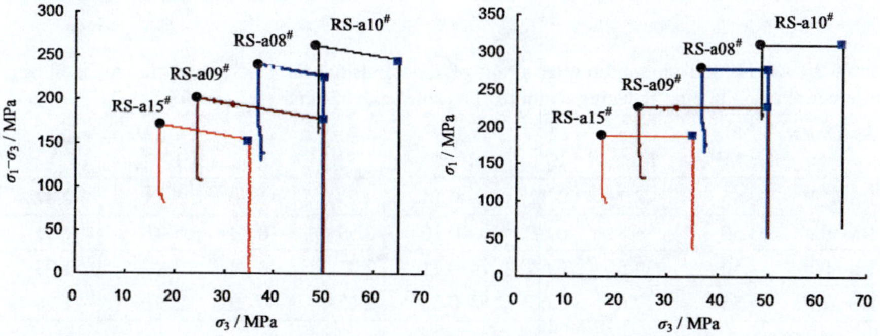

Fig. 2.8 Relationship between axial differential stress, axial stress and confining pressure for red sandstone in our "reducing confining pressure" experiments (see text for details). The filled, blue squares represent the position where the confining pressure was reduced, and the filled, black circles indicate the failure point of the specimen

In Fig. 2.7a, the initial confining pressure σ_{3i} of specimen RS-a10# was 65 MPa. The axial deviatoric stress was then increased and kept at approximately 90% of the TCS at $\sigma_3 = 65$ MPa. At this stress level, the confining pressure was reduced and specimen failure occurred at a confining pressure of 48.7 MPa (σ_{3f}). Specimen RS-a08# (Fig. 2.7b) had an initial confining pressure of 50 MPa and was also loaded to 90% of the TCS (although, this time 90% of the peak stress at $\sigma_3 = 50$ MPa). Specimen RS-a08# failed at a confining pressure of 36.8 MPa, lower than for the previous experiment with an initial confining pressure of 65 MPa. The analysis shows that for lower σ_{3i}, the σ_{3f} was lower, for specimens held at the same percentage of the TCS.

Specimen RS-a09# ($\sigma_{3i} = 50$ MPa and held at 70% of the TCS at $\sigma_3 = 50$ MPa) failed at a confining pressure of 24.5 MPa (Fig. 2.7c), lower than the value for RS-a08# ($\sigma_{3i} = 50$ MPa and held at 90% of the TCS at $\sigma_3 = 50$ MPa). So, if the initial confining pressure was the same, increasing the percentage of the failure stress acted to reduce the confining pressure at which the specimen failed. Specimen RS-a15# ($\sigma_{3i} = 35$ MPa and held at 70% of the TCS at $\sigma_3 = 35$ MPa) failed at a confining pressure of 17 MPa (Fig. 2.7d). The deformation parameters, as in Tables 2.1 and 2.2, are listed for these experiments in Tables 2.3 and 2.4.

Table 2.3 Young's modulus, Poisson's ratio and peak strain of red sandstone in our "reducing confining pressure" experiments (see text for details)

Specimen	σ_{3i} / MPa	σ_{3f} / MPa	E_S / GPa	E_{50} / GPa	υ	ε_{1c} / 10^{-3}	ε_{3c} / 10^{-3}	ε_{vc} / 10^{-3}
RS-a10#	65	48.7	50.1	49.7	0.32	8.15	−8.79	−9.42
RS-a08#	50	36.8	49.3	48.7	0.33	7.18	-7.75	−8.32
RS-a09#	50	24.5	49.5	49.4	0.34	5.61	−9.04	−12.5
RS-a15#	35	17.0	43.8	42.6	0.29	5.49	−7.69	−9.89

Table 2.4 Axial and circumferential strain of red sandstone at C', D' and the stress at zero volumetric strain in our "reducing confining pressure" experiments (see text for details)

Specimen	σ_{3i} / MPa	σ_{3f} / MPa	$\varepsilon_{1c'}$ / 10^{-3}	$\varepsilon_{3c'}$ / 10^{-3}	$\varepsilon_{1d'}$ / 10^{-3}	$\varepsilon_{3d'}$ / 10^{-3}	ε_{1v0} / 10^{-3}	ε_{3v0} / 10^{-3}
RS-a10#	65	48.7	0.72	−0.14	2.99	−1.00	5.73	−2.87
RS-a08#	50	36.8	0.63	−0.10	2.59	−0.84	5.01	−2.51
RS-a09#	50	24.5	0.63	−0.11	2.31	−0.74	3.85	−1.93
RS-a15#	35	17.0	0.63	−0.08	2.42	−0.68	3.91	−1.96

2.2.3 *Influence of the Confining Pressure on the Deformation Parameters*

Using the deformation parameters listed in Tables 2.1, 2.2, 2.3 and 2.4, the relationship between the deformation parameters of red sandstone and the confining pressure can now be investigated. Figure 2.9 illustrates the influence of the confining pressure on Young's modulus and Poisson's ratio. It can be seen that the Young's modulus of red sandstone increased nonlinearly with increasing confining pressure. At $\sigma_3 =$ 5 MPa, the elastic modulus and the deformation modulus were 35.9 and 27.4 GPa, respectively. Whereas when σ_3 was increased to 50 MPa, the elastic modulus and the deformation modulus were increased to 51.0 and 51.6 GPa, respectively. At pressures above 50 MPa, the Young's modulus remained constant. Furthermore, the difference between the elastic modulus and the deformation modulus tended to decrease as σ_3 was increased, which was the result of the more pronounced stage of microcrack closure at lower confining pressures. At $\sigma_3 = 35$ MPa, the Young's modulus showed a slight variation, the result of natural specimen variability between specimens cored from the same block. However, at higher confining pressures of 50 and 65 MPa, the effect of specimen variability on the Young's modulus was significantly reduced. The Poisson's ratio of red sandstone was unaffected by confining pressure (Fig. 2.9b). In the range of tested σ_3, Poisson's ratio varied from 0.28 to 0.37, and the average of Poisson's ratio was approximately 0.33. Thus, the dispersion extent (the ratio of the D-value between the maximum value and the minimum value to the average value) was approximately 27.3%.

The influence of σ_3 on the corresponding strains (listed in Tables 2.1, 2.2, 2.3 and 2.4) is presented in Fig. 2.10. We can conclude that the peak axial strain of red sandstone increased linearly, but the peak circumferential strain had a slight decreasing trend as a function of increasing σ_3 (Fig. 2.10a). Furthermore, the sensitivity of the peak axial strain on σ_3 was higher than that of the peak circumferential strain. However, in "reducing confining pressure" experiments, the peak strains had

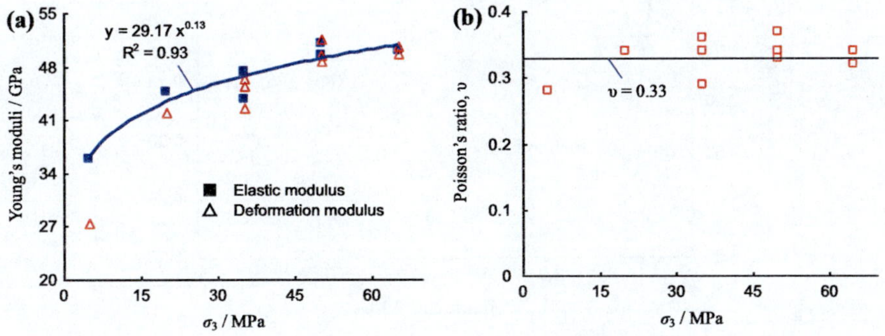

Fig. 2.9 Influence of the confining pressure on the Young's modulus and Poisson's ratio of red sandstone

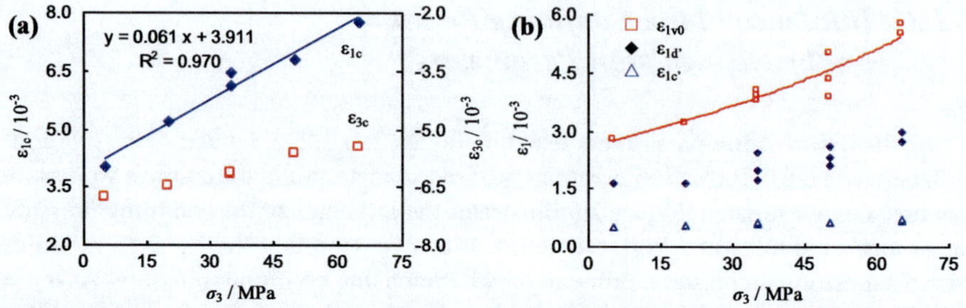

Fig. 2.10 Influence of the confining pressure on the corresponding strains (listed in Tables 3.1–3.4) in red sandstone

no clear linear relationship, especially for the peak circumferential strain. Also, the peak strain in our "reducing confining pressure" experiments was higher than that for our conventional triaxial compression experiments, for the same confining pressure.

Based on the data listed in Tables 2.1 and 2.4, the effect of σ_3 on the axial and circumferential strains at C', D' and the stress at zero volumetric strain will now be discussed (Fig. 2.10b). From Fig. 2.10b, it can be seen that $\varepsilon_{1c'}$, $\varepsilon_{1d'}$ and ε_{1v0} all increased nonlinearly with increasing σ_3, different from the effect of σ_3 on the peak axial strain. In detail, the sensitivity of σ_3 on ε_{1v0} was higher than that for $\varepsilon_{1d'}$, while the sensitivity of σ_3 on $\varepsilon_{1d'}$ was higher than that for $\varepsilon_{1c'}$.

Figure 2.11 illustrates the relationship between the axial strain and the circumferential strain at C', D', and the stress of zero volumetric strain. From Fig. 2.11, we can conclude that the $\varepsilon_{3c'}$, $\varepsilon_{3d'}$ and ε_{3v0} of red sandstone increased linearly with increasing $\varepsilon_{1c'}$, $\varepsilon_{1d'}$ and ε_{1v0}, respectively.

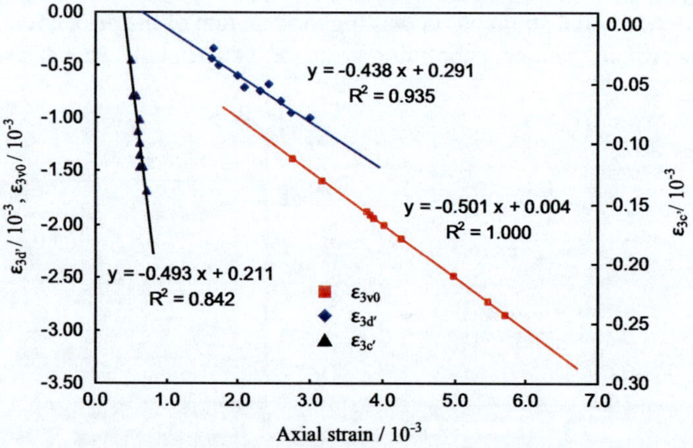

Fig. 2.11 Relationship between the axial strain and the circumferential strain of red sandstone at C', D', and the stress of zero volumetric strain

2.3 Triaxial Strength and Failure Behavior of Red Sandstone

The linear Mohr–Coulomb criterion can be expressed in terms of the maximum axial supporting capability σ_S and the minimum principal stress σ_3 (i.e., confining pressure) (Mohr 1914; Yang et al. 2011):

$$\sigma_S = \sigma_0 + q\sigma_3 \tag{2.1}$$

where σ_0 is usually regarded as the UCS of rock material. The σ_0 and q are related to the cohesion C and the internal friction angle Φ of rock material, which can be expressed in the following forms, respectively:

$$\sigma_0 = 2C \cos \phi/(1 - \sin \phi) \tag{2.2}$$

$$q = (1 + \sin \phi)/(1 - \sin \phi) \tag{2.3}$$

The Hoek–Brown criterion is an empirical strength criterion that has been widely applied in rock mechanics and engineering. The basic equation describing the peak triaxial compressive strength of a wide range of isotropic rock materials can be written as (Hoek and Brown 1980, 1997):

$$\sigma_S = \sigma_3 + (m\sigma_c\sigma_3 + s\sigma_c^2)^{0.5} \tag{2.4}$$

where σ_c is the UCS of the intact rock material and m and s are all material constants for a specific rock. When the parameter m is larger, the rock is stronger. The parameter s reflects the fractured extent, ranging from 0 to 1. When the parameter s is closer to 1, the rock is more intact.

2.3.1 *Peak Strength and Residual Strength Behavior*

Tables 2.5 and 2.6 list the peak stress (σ_p) and the residual strength (σ_{cr}) values for red sandstone from our conventional triaxial compression experiments (obtained from the stress–strain curves of Figs. 2.4 and 2.5). Over the entire confining pressure range, σ_S increased from 110.1 to 338.6 MPa and σ_{sr} from 57 to 262.6 MPa. These findings are different from those for Vosges sandstone (Bésuelle et al. 2000), i.e. the peak strength decreased with increasing confining pressure within the range 50 to 60 MPa.

By adopting the linear Mohr–Coulomb criterion (Eq. (2.1)) and the nonlinear Hoek–Brown criterion (Eq. (2.4)), the influence of σ_3 on the peak strength of red sandstone under triaxial compression can be presented as in Fig. 2.12. From Fig. 2.12,

Table 2.5 Peak strength and residual strength values for red sandstone in conventional triaxial compression experiments

σ_3 / MPa	σ_{cr} / MPa	σ_p / MPa	σ_{sr} / MPa	σ_S / MPa	Failure mode
5	52.0	105.1	57.0	110.1	Mixed tension and shear fracture
20	91.3	174.7	111.3	194.7	Mixed tension and shear fracture
35 (RS-a03[#])	113.2	216.1	148.2	251.1	Single shear fracture
35 (RS-a07[#])	—	218.0	—	253.0	—
50	154.3	251.2	204.3	301.2	Shear fracture with local slippage lines
65	197.6	273.6	262.6	338.6	Shear fracture with double slippage planes

Note $\sigma_S = \sigma_p + \sigma_3$; $\sigma_{sr} = \sigma_{cr} + \sigma_3$

Table 2.6 Peak strength and residual strength values for red sandstone in our "reducing confining pressure" experiments (see text for details)

Specimen	σ_{3i} / MPa	σ_{3f} / MPa	σ_{cr} / MPa	σ_p / MPa	σ_{sr} / MPa	σ_S / MPa	Failure mode
RS-a10[#]	65	48.7	164	261.2	212.7	309.9	Shear fracture with local tension crack
RS-a08[#]	50	36.8	133	239.2	169.8	276.0	Complicated fracture
RS-a09[#]	50	24.5	106	201.4	130.5	225.9	Complicated fracture
RS-a15[#]	35	17.0	81	170.7	98.0	187.7	Complicated fracture

we can see out that the peak strength of red sandstone under triaxial compression possessed distinctly non-linear behavior with increasing σ_3, which indicates that the nonlinear Hoek–Brown criterion better reflects the peak strength properties better than the linear Mohr–Coulomb criterion. With an increase of confining pressure, the failure mode of red sandstone (Fig. 2.12) changed from mixed tension and shear fracture (single shear fracture) to shear fracture with double slippage planes, which further testified to the fact that the non-linear Hoek–Brown criterion is more suitable to be used on our data. Moreover, the linear Mohr–Coulomb criterion predicts the peak strength to be higher than its measured value at low and high confining pressures, and lower at intermediate confining pressures (Fig. 2.12).

It can also be seen that the peak strength of red sandstone was not directly dependent on the loading path (Fig. 2.12), meaning that our "reducing confining pressure" experiments can also be used to obtain the strength behavior of the rock. Note that, at a confining pressure of 35 MPa, the difference in the peak strength was only 1.9 MPa for two different curves. Table 2.7 presents more details on the peak strength parameters of red sandstone using the non-linear Hoek–Brown criterion. In Table 2.7, the parameters σ_c and m are back-calculated by assuming $s = 1.000$. The parameters σ_c

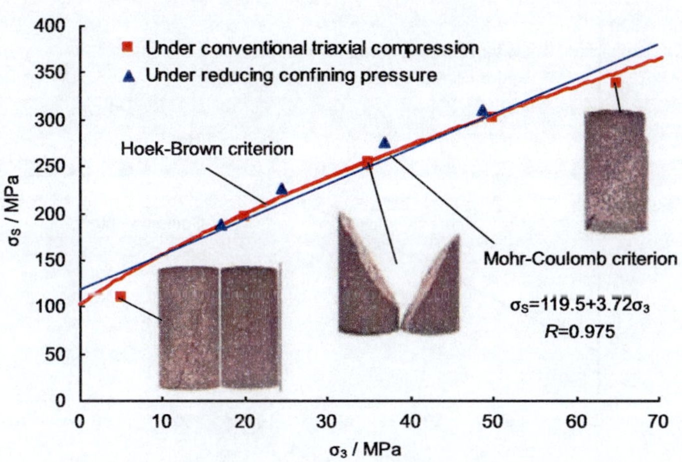

Fig. 2.12 Peak strength analysis of red sandstone under triaxial compression. The symbols represent the experimental results and the lines represent the theoretical values in accordance with Mohr–Coulomb or Hoek–Brown criterion. (*R* is the correlation coefficient of determination of the regression)

Table 2.7 Peak strength parameters of red sandstone using the non-linear Hoek–Brown criterion

Loading path	UCS / MPa	σ_c / MPa	*m*	*s*	*a*	*R*
Conventional triaxial compression	70	90.8	11.76	1.000	0.5	0.995
"Reducing confining pressure"	70	97.4	12.71	1.000	0.5	0.994
United regression for all of the specimens	70	101.7	10.66	1.000	0.5	0.980

and *m* are a little higher for the "reducing confining pressure" experiments than for the conventional triaxial compression experiments. However, the uniaxial compressive strength σ_c (back-calculated using the non-linear Hoek–Brown criterion) was distinctly higher than the measured value of about 70 MPa obtained from a uniaxial compressive experiment.

The residual strength parameters of red sandstone under triaxial compression using the linear Mohr–Coulomb criterion are presented in Table 2.8 and Fig. 2.13. Good linear regression coefficients of $R = 0.995$ and 0.999 were determined for the conventional triaxial experiments and the "reducing confining pressure" experiments, respectively. Even though there was an obvious non-linear relationship between

Table 2.8 Residual strength parameters of red sandstone in using the linear Mohr–Coulomb criterion

Loading path	σ_0 / MPa	q	*C* / MPa	Φ / (°)	*R*
Conventional triaxial compression	39.03	3.36	10.65	32.8	0.995
"Reducing confining pressure"	39.94	3.55	10.60	34.1	0.999

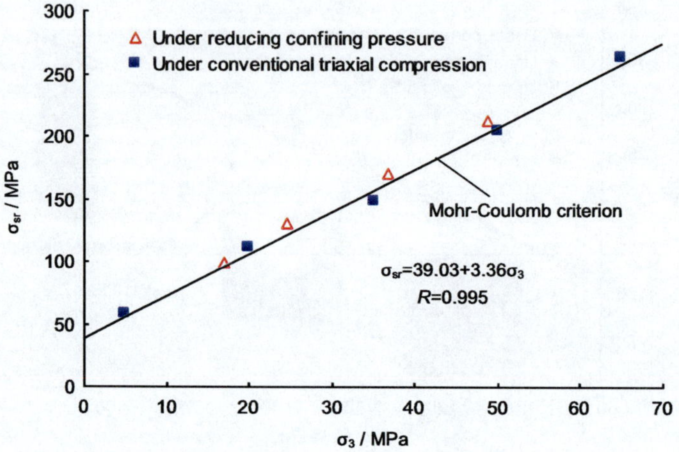

Fig. 2.13 Residual strength analysis of red sandstone under triaxial compression. All of the symbols represent the experimental results and the lines represent the theoretical values in accordance with Mohr–Coulomb criterion. (*R* is the correlation coefficient of determination of the linear regression.)

the peak strength and the confining pressure, the residual strength clearly had a linear relation with σ_3 (Fig. 2.13). Moreover, this relationship was not dependent on the loading path, which results mainly from supporting the axial capacity of the sandstone by friction slippage at the residual strength stage. Thus, the residual cohesion values for red sandstone, obtained for two different loading paths, were all approximately 10.6 MPa. Also, the difference in the residual internal friction angle between our conventional triaxial compression tests and our "reducing confining pressure" tests was very small (only 1.3°), and can therefore be ignored in *in-situ* engineering practices.

Table 2.9 shows the peak strength parameters of red sandstone obtained using the linear Mohr–Coulomb criterion. Comparing these values with those of Table 2.8, it is clear that the peak cohesion of red sandstone was 28.36 MPa for our conventional triaxial compression tests and 32.63 MPa for our "reducing confining pressure" tests (which are both considerably higher than residual cohesion of ~ 10.6 MPa). However, the peak internal friction angle (which has an average value of 35.7°) obtained from the two different loading paths was very close to the residual internal friction angle (which has an average value of 33.5).

Table 2.9 Peak strength parameters of red sandstone using the linear Mohr–Coulomb criterion

Loading path	σ_0 / MPa	q	*C* / MPa	*Φ* / (°)	*R*
Conventional triaxial compression	110.0	3.76	28.36	35.4	0.985
"Reducing confining pressure"	127.9	3.84	32.63	35.9	0.995
United regression for all the specimens	119.5	3.72	30.98	35.2	0.975

2.3.2 Evaluation on Three Characteristics Stresses

C', D' and the stress at zero volumetric strain for red sandstone under triaxial compression are listed in Table 2.10. It can be seen that, over the whole range of tested pressures (5 to 65 MPa), $\sigma_{c'}$ varied from 10.2 to 37.3 MPa, $\sigma_{d'}$ from 43.2 to 147.5 MPa, and σ_{v0} from 83.5 to 246.5 MPa. Figure 2.14 presents the influence of σ_3 on C', D' and the stress at zero volumetric strain, together with the relationship between σ_S and σ_3. It can be seen that σ_{si}, σ_{sd} and σ_{s0} all increased linearly (with good linear regression coefficients of $R = 0.995$–0.998) with the confining pressure. Notably, however, the σ_{s0} of two specimens (i.e., RS-a09# and RS-a15#) were both lower than other specimens at the same confining pressure. This is a result of reducing the confining pressure before the point of zero volumetric strain.

The sensitivity of the stress at zero volumetric strain on σ_3 was clearly higher than that for D', while the sensitivity of D' on σ_3 was clearly higher than that for C'. Table 2.11 lists the strength parameters of red sandstone obtained using the stress at zero volumetric strain and D' and the linear Mohr–Coulomb criterion. The σ_0 had values of 77.22 and 30.74 MPa and q was 3.64 and 2.70 for the regression analysis using the stress at zero volumetric strain and D', respectively. Using Eqs. (2.2) and (2.3), the values of C and Φ were calculated using the regression analysis for the stress at zero volumetric strain to be 20.24 MPa and 34.7°, respectively. These values were higher than the values of C and Φ (9.35 MPa and 27.4°) calculated using the regression analysis for D'.

Figure 2.15 shows the relationship between the peak strength and C', D', and the stress at zero volumetric strain (from the conventional triaxial compression tests). From Fig. 2.15, it can be seen that with the increase of peak strength, C', D' and the stress at zero volumetric strain for red sandstone all increased nonlinearly. A logarithmic equation is proposed to describe the nonlinear relationship between the

Table 2.10 C', D' and the stress at zero volumetric strain for red sandstone under triaxial compression

σ_3 / MPa	$\sigma_{c'}$ / MPa	$\sigma_{d'}$ / MPa	σ_{v0} / MPa	σ_{si} / MPa	σ_{sd} / MPa	σ_{s0} / MPa
5	10.2	43.2	83.5	15.2	48.2	88.5
20	20.3	66.4	133.6	40.3	86.4	153.6
35 (RS-a03#)	26.6	79.5	169.5	61.6	114.5	204.5
35 (RS-a07#)	24.5	89.5	174.4	59.5	124.5	209.4
50	34.0	108.4	202.8	84.0	158.4	252.8
65	36.6	138.7	242.2	101.6	203.7	307.2
65 (RS-a10#)	37.3	147.5	246.5	102.3	212.5	311.5
50 (RS-a08#)	29.9	125.8	222.0	79.9	175.8	272.0
50 (RS-a09#)	31.6	114.5	182.1	81.6	164.5	232.1
35 (RS-a15#)	25.3	102.9	159.2	60.3	137.9	194.2

Note $\sigma_{s0} = \sigma_{v0} + \sigma_3$; $\sigma_{sd} = \sigma_{d'} + \sigma_3$; $\sigma_{si} = \sigma_{c'} + \sigma_3$

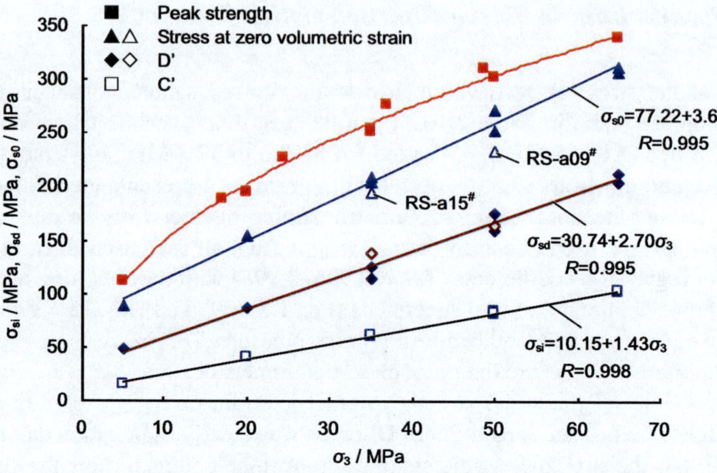

Fig. 2.14 The influence of the confining pressure on C', D' and the stress at zero volumetric strain under triaxial compression. The symbols represent the experimental results and the curves represent the theoretical values found using the Mohr–Coulomb criterion

Table 2.11 Strength parameters of red sandstone obtained using C', D' and the stress at zero volumetric strain by the linear Mohr–Coulomb criterion

Different regression analysis	σ_0 / MPa	q	C / MPa	Φ / (°)	R
Using the stress at zero volumetric strain	77.22	3.64	20.24	34.7	0.995
Using D'	30.74	2.70	9.35	27.4	0.995
Using C'	10.15	1.43	4.24	10.2	0.998

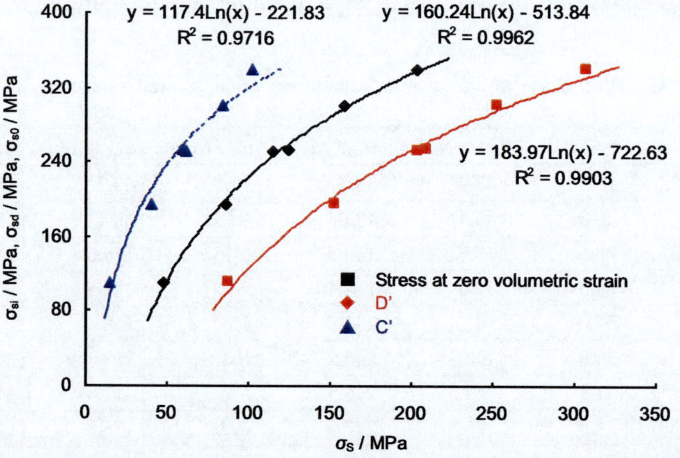

Fig. 2.15 Relation between the peak strength and C', D' and the stress at zero volumetric strain for the conventional triaxial compression tests. The symbols represent the experimental results

peak strength and the stress at zero volumetric strain, C' and D' (all of which have good nonlinear regression coefficients). By using Fig. 2.15, we can approximately predict the position of C', D' and the stress at zero volumetric strain using the peak strength of rock, even without measuring the circumferential strain curve.

2.3.3 Failure Behavior Under Triaxial Compression

Figures 2.16, 2.17, 2.18, 2.19 and 2.20 show the failure modes of red sandstone specimens from our conventional triaxial compression experiments. The confining pressure had exerted a significant influence on the failure mode.

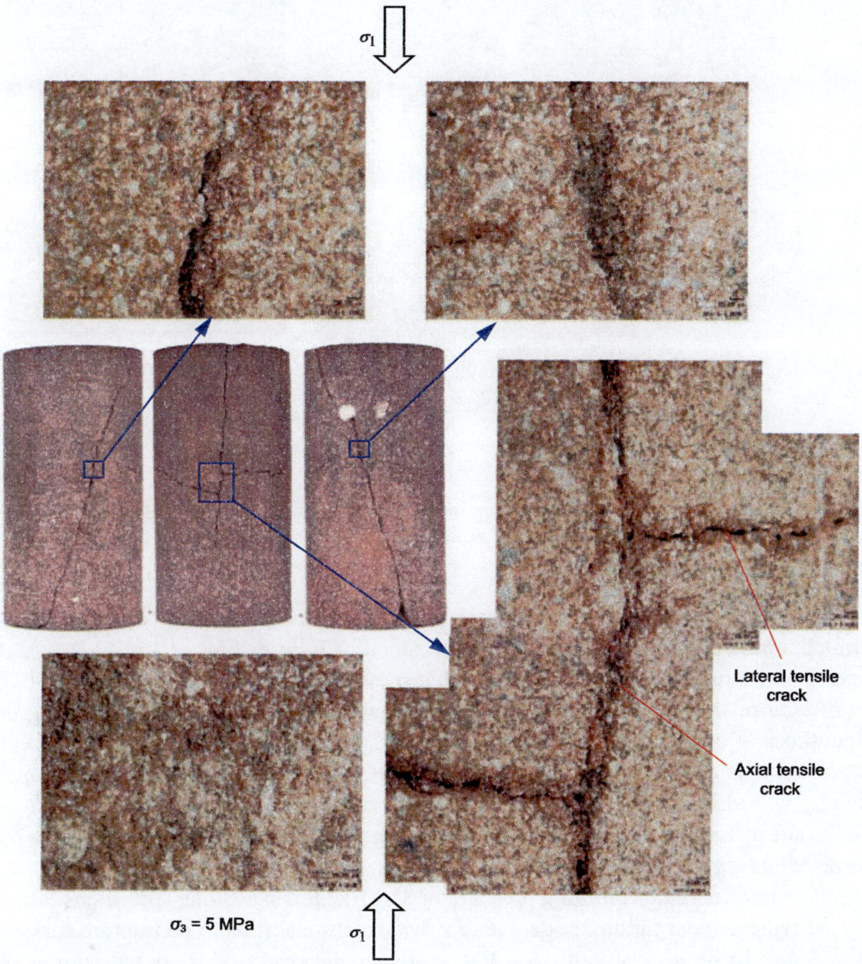

Fig. 2.16 Mixed tension and shear fracture mode of red sandstone from a conventional triaxial compression experiment at a confining pressure of 5 MPa

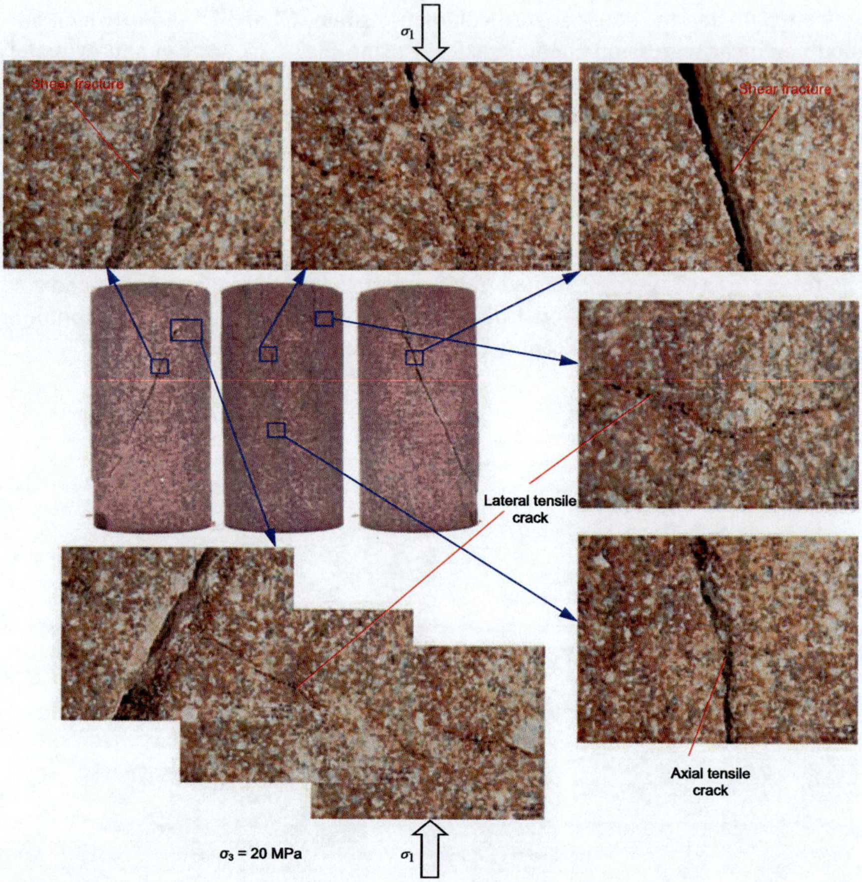

Fig. 2.17 Mixed tension and shear fracture mode of red sandstone from a conventional triaxial compression experiment at a confining pressure of 20 MPa

At confining pressures between 5 and 20 MPa, our sandstone specimens all demonstrated a mixed tension and shear fracture mode of failure (Figs. 2.16–2.17). Axial and lateral tensile cracks were observed at $\sigma_3 = 5$ and 20 MPa (Figs. 2.16–2.17). Furthermore, the widths of the tensile cracks in specimen at $\sigma_3 = 5$ MPa were bigger than those at $\sigma_3 = 20$ MPa. Our post-mortem specimens also indicated that it was easier for tension failure to occur at lower confining pressures. In Fig. 2.16, frictional sliding was also observed along the direction of the shear fracture plane. The angle of the shear fracture plane at $\sigma_3 = 5$ MPa was approximately 74°, higher than for $\sigma_3 = 20$ MPa (approximately 68°).

At an intermediate confining pressure of 35 MPa, red sandstone specimens failed by the typical shear failure mode with a single fracture surface. The fracture surface was found to be very smooth and flat, which is the result of frictional sliding of

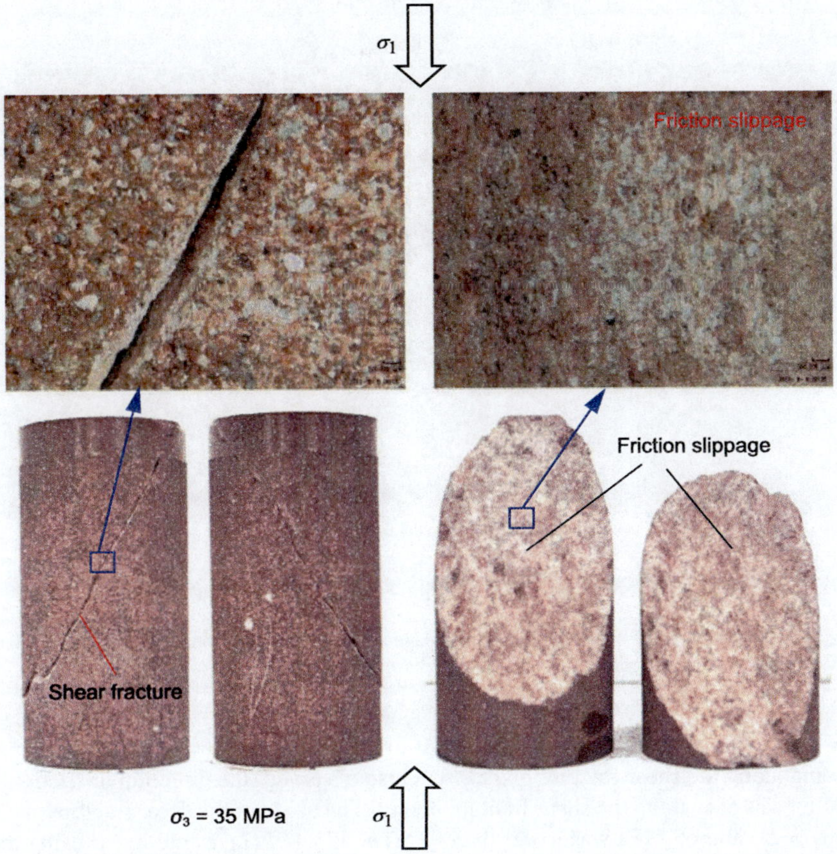

Fig. 2.18 Single shear fracture mode of red sandstone from a conventional triaxial compression experiment at a confining pressure of 35 MPa

the macroscopic fracture surfaces along the direction of the shear fracture plane (Fig. 2.18). The angle of the shear fracture plane at $\sigma_3 = 35$ MPa was approximately 62°, approximately equal to that predicted using the linear Mohr–Coulomb criterion (i.e., 63° (45° + Φ/2)).

However, at higher confining pressures of 50 and 65 MPa, the failure modes of red sandstone were different. At $\sigma_3 = 50$ MPa, the sandstone exhibited a shear failure mode, with several closed shear bands (Fig. 2.19) and an angle of shear fracture approximately 57°. The shear frictional sliding was distinct, and the direction of sliding was along the shear fracture plane. At $\sigma_3 = 65$ MPa, the sandstone exhibited a shear failure mode with double slippage planes (Fig. 2.20), and the angle of the shear fracture plane was approximately 55°. It should be noted that the shear fracture surfaces under higher confining pressures of 50 and 65 MPa were not as smooth as for $\sigma_3 = 35$ MPa. A closed shear band and an open shear fracture were both observed

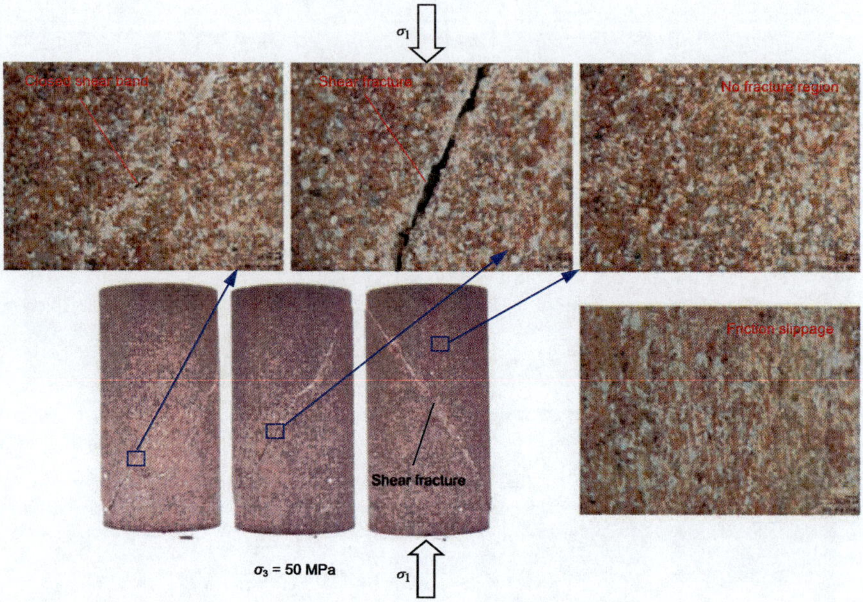

Fig. 2.19 Shear fracture mode with local slippage lines of red sandstone from a conventional triaxial compression experiment at a confining pressure of 50 MPa

for $\sigma_3 = 65$ MPa. A considerable amount of fault gouge, a result of the frictional sliding along the shear fracture planes, was also observed (the direction of frictional sliding was also along the shear fracture plane). The angle of the shear fracture plane at $\sigma_3 = 50$ and 65 MPa was lower than 63° (i.e. $45° + \Phi/2$), as predicted using the linear Mohr–Coulomb criterion.

For our "reducing confining pressure" experiments, the failure mode of red sandstone (Fig. 2.21) was significantly different from that obtained in our conventional triaxial compression experiments, it was much more complex. Further, the width of the fracture surfaces appeared to be greatly increased.

At $\sigma_{3i} = 65$ MPa and $\sigma_{3f} = 48.7$ MPa, the sandstone exhibited a shear fracture failure mode with local tensile cracks. The angle of the shear fracture plane was approximately 57°. At $\sigma_{3i} = 50$ MPa and $\sigma_{3f} = 36.8$ MPa, a more complicated failure mode was observed (Fig. 2.21b). The main failure mode of specimen RS-a08# was by a shear fracture (e.g., 1–3 on Fig. 2.21b). Shear fracture plane 1 was the reverse of shear fracture plane 2, while shear fracture plane 2 was linked with shear fracture plane 3. Furthermore, in specimen RS-a08#, some axial and lateral tensile cracks were also observed, perhaps due to decreasing the confinement on the specimen.

We also saw a complicated failure mode at $\sigma_{3i} = 50$ MPa and $\sigma_{3f} = 24.5$ MPa (Fig. 2.21c). From Fig. 2.21c, the main failure mode of specimen RS-a09# was also by means of a shear fracture. Shear fracture plane 1 was the reverse of shear fracture

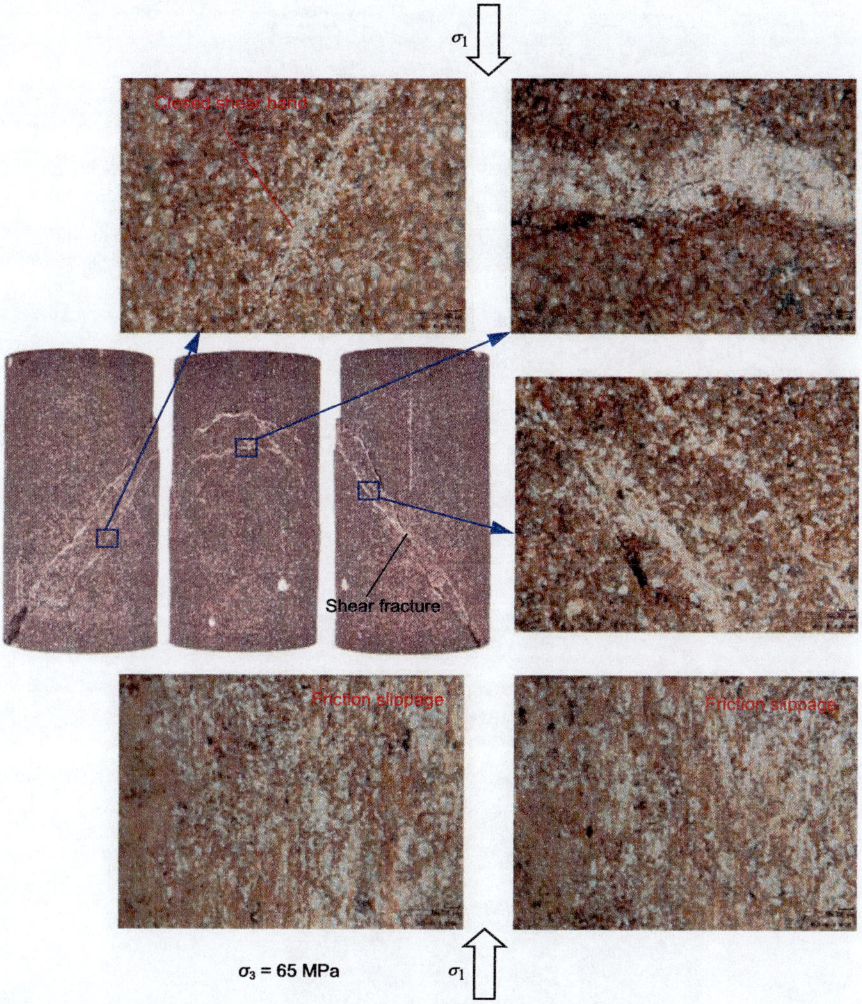

Fig. 2.20 Shear fracture mode with double slippage planes of red sandstone from a conventional triaxial compression experiment at a confining pressure of 65 MPa

plane 3, while shear fracture plane 2 was linked with shear fracture plane 1. Tensile cracks were again observed (see A and B in Fig. 2.22). The initial confining pressure of two specimens RS-a08# and RS-a09# was the same, but the final confining pressure upon failure for specimen RS-a09# was lower. Therefore, in specimen RS-a09#, more small tensile cracks were required to initiate in the process of final failure.

At $\sigma_{3i} = 35$ MPa and $\sigma_{3f} = 17$ MPa, the sandstone primarily took on a shear failure mode with a single shear fracture plane. However, at the point of localization, there were some axial and lateral tensile cracks (see C and D in Fig. 2.22). Again, these

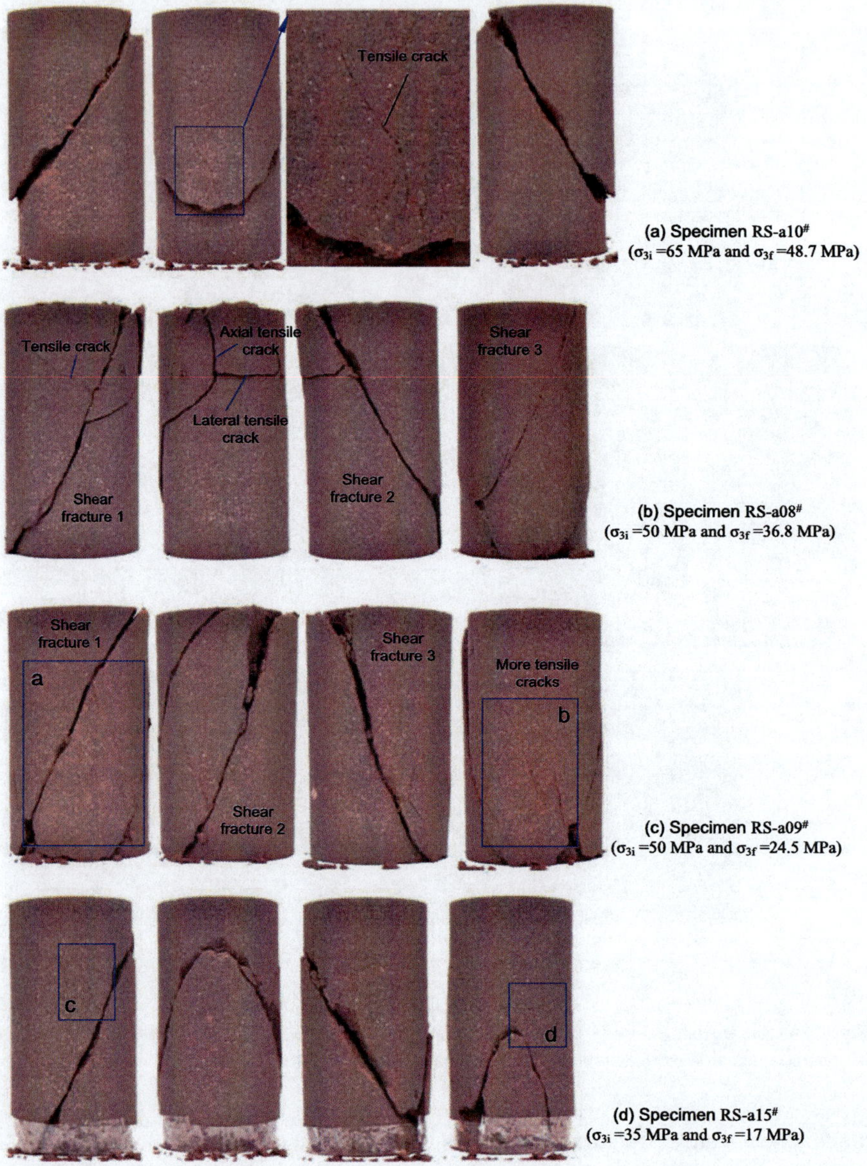

Fig. 2.21 Failure mode of red sandstone deformed in "reducing confining pressure" experiments, with different initial conditions

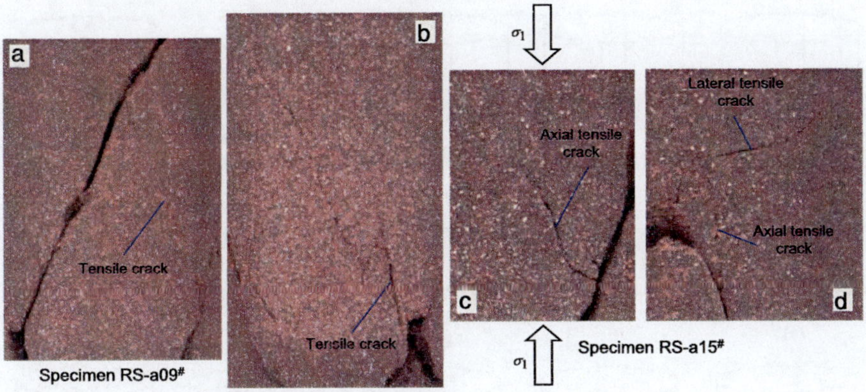

Fig. 2.22 Local magnification of tensile cracks in red sandstone from "reducing confining pressure" experiments. The denoted letters A-D correspond to those in Fig. 3.21

are likely to be the result of the specimen failing by reducing the confining pressure, they were not observed in the specimens from the conventional triaxial compression experiments.

2.4 Spatial AE Locations Behavior of Red Sandstone

Figure 2.23 shows the relationship between axial deviatoric stress, AE counts and the strain of red sandstone at a confining pressure of 35 MPa. Fig. 2.24 shows the

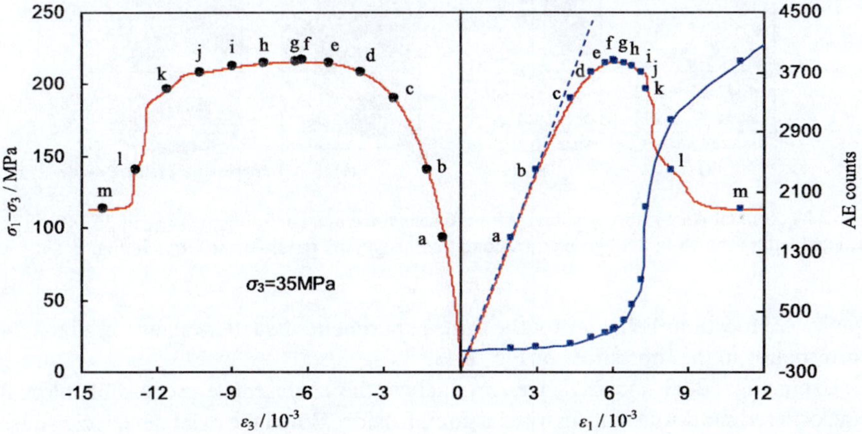

Fig. 2.23. Relationship between axial differential stress, AE counts and strain for red sandstone at a confining pressure of 35 MPa

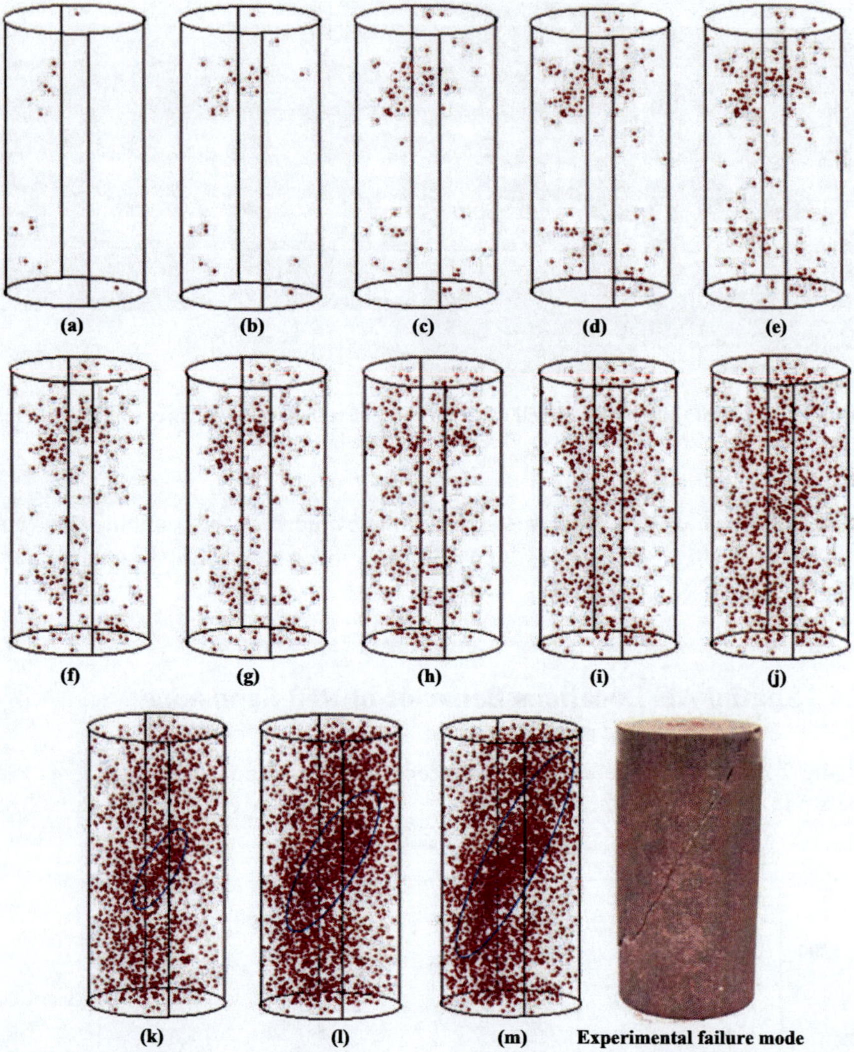

Fig. 2.24 Spatial AE locations behavior for red sandstone at a confining pressure of 35 MPa. The denoted letters shown in the figure correspond to those on the stress–strain curve in Fig. 2.23

spatial AE locations behavior for the same experiment; the letters shown in Fig. 2.24 corresponds to the time slices of Fig. 2.23.

Using Figs. 2.23 and 2.24, we can analyze the coalescence process of internal cracks in red sandstone during triaxial compression. When the axial deviatoric stress was loaded to points a and b, the AE counts in the specimen were very rare, and they resulted from small micro-failure occurrences during elastic deformation. With increasing circumferential deformation, the axial deviatoric stress–strain curve began to depart from the domain of elastic deformation. When the axial deviatoric stress reached point c (σ_1-σ_3 = 189.4 MPa = 87.6% σ_p, ε_1 = 4.41 × 10^{-3}), the AE counts in the specimen reached 80, which meant that the specimen contained a small amount of internal damage. When the specimen was loaded to point d (σ_1-σ_3 = 208.2 MPa = 96.3% σ_p, ε_1 = 5.21 × 10^{-3}), the AE counts within the specimen reached 149, indicating that more microcracking took place within the specimen. Before point d, the AE counts were located in the top and bottom regions of the specimen, and there were almost no AE counts in the central region of the specimen.

When the axial deviatoric stress was loaded to point e ($\sigma_1-\sigma_3$ = 214.7 MPa = 99.4% σ_p, ε_1 = 5.75 × 10^{-3}), the AE counts reached 207 in the specimen. There were also some AE counts in the central region of the specimen. Once the specimen had reached point f ($\sigma_1-\sigma_3$ = 216.1 MPa = 100% σ_p, ε_1 = 6.10 × 10^{-3}), the AE counts in the specimen totaled 267. The AE counts were now accumulating throughout the entire specimen. However, the specimen was still macroscopically coherent, even though it had some internal damage failure. After the peak strength, from point f to j, the axial deviatoric stress on the specimen slowly dropped from 216.1 to 207.9 MPa as the circumferential strain increased from 6.22 × 10^{-3} to 10.27 × 10^{-3} (Fig. 2.23), but the axial strain of the specimen only increased to 7.17 × 10^{-3} from 6.10 × 10^{-3}. During the reduction of axial deviatoric stress from point f to j, even though it involved a small stress drop, the circumferential deformation of the specimen was increasing faster than the axial deformation (indicating that there was a large amount of dilatancy). The AE counts increased to 926 at point j and more AE events were observed throughout the specimen (Fig. 2.24j).

Between point j and point k, the axial deviatoric stress significantly decreased from 207.9 to 196.4 MPa and the axial strain slightly increased from 7.17 × 10^{-3} to 7.39 × 10^{-3}. However, the accumulated AE counts in the specimen increased rapidly to 1883. At this moment, the specimen experienced a serious damage failure. The axial stress dropped rapidly to 140.2 from 196.4 MPa upon reaching point l, and the AE counts increased to 3056. The main failure mode with the single shear fracture plane could be seen from the elliptical region in Fig. 2.24l. The axial deviatoric stress was then loaded to point m ($\sigma_1-\sigma_3$ = 113.2 MPa, ε_1 = 11.15 × 10^{-3}), and the specimen was located at the stage of residual strength. At this time, the AE counts from the specimen reached 3843 and the macroscopic shear fracture plane was initiated, which could be observed from Fig. 2.24m. The failure mode of the specimen formed by AE location analysis showed a good agreement with the experimental failure mode.

2.5 Conclusions

In this chapter, we have presented an experimental investigation of the strength, deformation behavior, failure characteristics and spatial AE locations of red sandstone under triaxial compression. In our study, we used the circumferential displacement as the controlling feedback signal (as opposed to the more usual axial displacement feedback) to obtain better triaxial compression experimental results for red sandstone. Based on our data, the following conclusions can be drawn.

(1) Our results show that the post-peak axial deformation of red sandstone was changed as the confining pressure was increased from 5 to 65 MPa. The Young's modulus increased nonlinearly with confining pressure, but Poisson's ratio remained unchanged. In the range of tested σ_3 (from 5 to 65 MPa), the Poisson's ratio ranged from 0.28 to 0.37, and the average Poisson's ratio was approximately 0.33.
(2) According to the observed compactive and dilatancy behavior, the influence of the confining pressure on the corresponding deformation parameters of red sandstone can be discussed further. The peak axial strain increased linearly with the confining pressure, but the peak circumferential strain only showed a modest increase. However, during our "reducing confining pressure" experiments, the peak strain showed no clear linear relation, especially for the peak circumferential strain. In addition, the peak strain during our "reducing confining pressure" experiments was higher than that for our conventional triaxial compression experiments for the same confining pressure. The axial and circumferential strain at C', D', and the stress at zero volumetric strain all increased nonlinearly with the confining pressure. However, the $\varepsilon_{3c'}$, $\varepsilon_{3d'}$ and ε_{3v0} increased linearly with increasing $\varepsilon_{1c'}$, $\varepsilon_{1d'}$ and ε_{1v0}, respectively.
(3) The strength and failure behavior of red sandstone under triaxial compression can also be analyzed. The nonlinear Hoek–Brown criterion better reflected the peak strength properties than the linear Mohr–Coulomb criterion. However, even though there was an obvious non-linear relationship between the peak strength and the confining pressure, the residual strength of red sandstone exhibited a clear linear relationship with the confining pressure (best described using the linear Mohr–Coulomb criterion). The peak and residual strengths were not directly related to two different loading paths. C', D' and the stress at zero volumetric strain all increased linearly with confining pressure. The strength parameters obtained using C' and D' were obviously lower than those obtained by the peak strength.
(4) In our conventional triaxial compression experiments, with increasing confining pressure, the failure mode of red sandstone changed from mixed tension and shear fracture (single shear fracture) to shear fracture with double slippage planes. However, the failure mode in our "reducing confining pressure" experiments was found to be more complicated, which results mainly from the unstable failure characteristics of the rock during the reduction in confining pressure.
(5) Finally, based on the obtained spatial AE locations at a confining pressure of 35 MPa, a detailed analysis of the evolutionary process of internal cracks in red sandstone was established for the entire loading process. These results may be significant for predicting and forecasting failure processes in the field of deep underground rock mass engineering.

References

Amann F, Button EA, Evans KF, Gischig VS, Blumel M (2011) Experimental study of the brittle behavior of clay shale in rapid unconfined compression. Rock Mech Rock Eng 44(4):415–430

Bésuelle P, Desrues J, Raynaud S (2000) Experimental characterization of the localisation phenomenon inside a Vosges sandstone in a triaxial cell. Int J Rock Mech Min Sci 37(8):1223–1237

Fairhurst CE, Hudson JA (1999) Draft ISRM suggested method for the complete stress-strain curve for the intact rock in uniaxial compression. Int J Rock Mech Min Sci 36(3):279–289

Hawkes I, Mellor M (1970) Uniaxial testing in rock mechanics laboratories. Eng Geol 4:177–285

Heap MJ, Baud P, Meredith PG, Bell AF, Main IG (2009) Time-dependent brittle creep in Darley Dale sandstone. J Geophys Res. https://doi.org/10.1029/2008JB006212

Hoek E, Brown ET (1980) Underground excavation in rock. Institution of Mining and Metallurgy, London

Hoek E, Brown ET (1997) Practical estimates of rock mass strength. Int J Rock Mech Min Sci 34(8):1165–1186

Lockner D, Byerlee J, Kuksenko V, Ponomarev A, Sidorin A (1991) Quasi-static fault growth and shear fracture energy in granite. Nature 350:39–42

Mogi K (1966) Some precise measurements of fracture strength of rocks under uniform compressive stress. Rock Mech Eng Geol 4:41–55

Mohr O (1914) Abhandlungen aus dem Gebiete der Technischen Mechanik, 2nd edn. Ernst, Berlin

Paterson MS, Wong TF (2005) Experimental rock deformation—the brittle field. Springer, New York

Rutter E (1986) On the nomenclature of mode of failure transitions in rocks. Tectonophysics 122:381–387

Wawersik WR, Fairhurst C (1970) A study of brittle rock fracture in laboratory compression experiments. Int J Rock Mech Min Sci 7:561–575

Wong TF, David C, Zhu W (1997) The transition from brittle faulting to cataclastic flow in porous sandstones: mechanical deformation. J Geophys Res 102(B2):3009–3025

Yang SQ, Jiang YZ (2010) Triaxial mechanical creep behavior of sandstone. Min Sci Technol 20(3):339–349

Yang SQ, Jing HW, Li YS, Han LJ (2011) Experimental investigation on mechanical behavior of coarse marble under six different loading paths. Exp Mech 51(3):315–334

Yang SQ, Jing HW, Wang SY (2012) Experimental investigation on the strength, deformability, failure behavior and acoustic emission locations of red sandstone under triaxial compression. Rock Mech Rock Eng 45(4):583–606

Youn H, Tonon F (2010) Multi-stage triaxial test on brittle rock. Int J Rock Mech Min Sci 47(4):678–684

Chapter 3
Mechanical Damage Characteristics of Red Sandstone Under Triaxial Cyclic Loading

In all kinds of rock engineering, due to the influence of strong blasting and excavation disturbance, the rock mass often experiences cyclic loading, which leads to the continuous damage failure of tunnel (Zhu et al. 2010; Xu et al. 2012). Previous experimental studies have demonstrated that the mechanical behavior of rocks under cyclic loading is significantly different from those under monotonic loading (Rao and Ramana 1992; Bagde and Petros 2005; Heap et al. 2009a, b; Ma et al. 2013; Wang et al. 2013). Therefore, it is very important to investigate the mechanical damage characteristics of rock subjected to cyclic loading to better understand the unstable failure mechanism of rock engineering.

Sandstone, a type of sedimentary rock, is ubiquitous in underground engineering. In the past, the mechanical behavior of sandstone under cyclic loading has been experimentally studied (Bieniawski 1969). Gatelier et al. (2002) carried out an extensive laboratory tests on the mechanical behavior of an anisotropic porous sandstone under uniaxial and triaxial cyclic loading, which revealed two main mechanisms: compaction and microcracking. In the unaxial tests, both mechanisms were shown to be strongly influenced by the inclination of loading with respect to the isotropy planes. However, with increased confining pressures, the influence of anisotropy was significantly reduced. On the basis of experimental results for intact sandstone under static and dynamic uniaxial cyclic loading, Bagde and Petros (2005) found that Young's modulus, UCS and stress energy increased with the loading frequency. Xu et al. (2009) carried out an experimental study on the damage process of sandstone by adopting the spatial AE technique. The experimental results showed that different damage types were located at different cyclic phases during the entire loading process. From initial damage in the static state phase to the early core region of cracks, the cracks expanded stably from the early core region at the beginning of the cycle and rapidly at the end of the cycle prior to the macroscopic failure of the rock. By conducting a series of laboratory tests, Liu and He (2012) and Liu et al. (2011) assessed the effects of frequency on the dynamic properties of sandstone specimens subjected to cycling loading under different confining pressures. Song et al. (2013)

S.-Q. Yang, *Mechanical Behavior and Damage Fracture Mechanism of Deep Rocks*,
https://doi.org/10.1007/978-981-16-7739-7_3

studied the damage evolution process and the crack development of sandstone using uniaxial cyclic testing and digital image correlation.

However, few X-ray micro CT observations have been adopted to reveal the damage mechanism of sandstone under triaxial cyclic loading. Therefore, in this chapter, we report the results of a series of triaxial cyclic compression tests on sandstone under different confining pressures (Yang et al. 2015). Based on the experimental results, triaxial monotonic and cyclic mechanical parameters of sandstone specimens under different confining pressures are first compared. Then, the influence of cycle number on the crack damage threshold of sandstone is analyzed. Next, using an X-ray micro CT scanning system, the internal damage characteristics of sandstone material after triaxial monotonic and cyclic failure are explored. Finally, the influence of cycle number on the elastic modulus, Poisson's ratio and strain damage of sandstone under different confining pressures is discussed in detail.

3.1 Tested Rock Material and Testing Method

3.1.1 Red Sandstone Material and Testing Equipment

The material used throughout this study was sandstone (a marine sedimentary rock) collected from the Shandong province in China. The sandstone is formed from cemented grains that may either be fragments of a pre-existing rock or monominerallic crystals. The cements binding these grains together are typically calcite, clays, and silica. According to the results of X-ray diffraction (XRD), the minerals in the sandstone material are feldspar, quartz, smectite, calcite and volcanic rock fragments (see Fig. 3.1). A detailed composition of this rock is described as follows: the content (weight %) of feldspar is 50%, quartz 20%, smectite 15%, calcite 10%

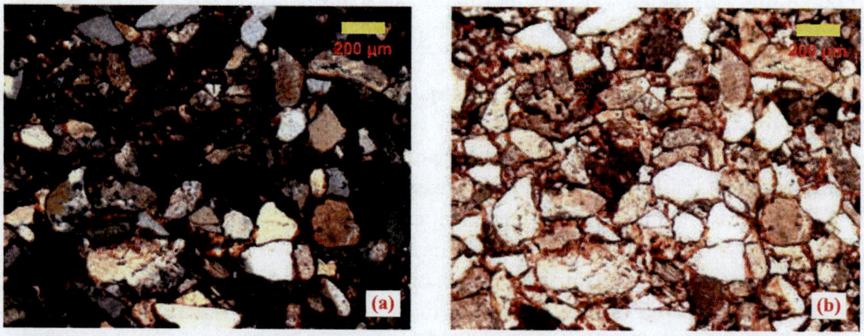

Fig. 3.1 Thin section images of sandstone used in this research. **a** Crossed polarization photograph of microstructure. **b** Plane polarization photograph of microstructure

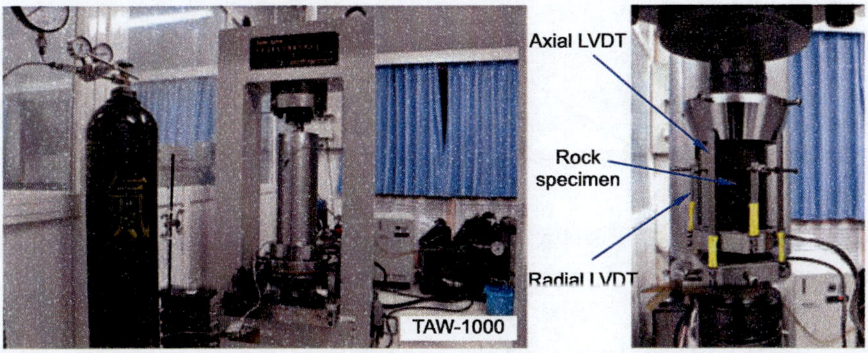

Fig. 3.2 TAW-1000 triaxial rock mechanics experimental system and strain measurement

and volcanic rock fragments 5%. The sandstone in this research is a fine- to medium-grained, feldspathic rock material with a connected porosity of 5.3% and a bulk density of 2450 kg/m^3. All specimens were cored from the same block of material to an actual diameter of 49.3 mm and approximate 100 mm in length. All the experiments were performed on dry specimens at room temperature. The UCS of the sandstone was 83.14 MPa, and the elastic modulus (E_S) and Poisson's ratio (υ) of the sandstone under uniaxial compression were 13.95 GPa and 0.180, respectively. However, the peak axial strain (ε_{1c}) and peak radial strain (ε_{3c}) of the sandstone under uniaxial compression were 8.462×10^{-3} and -4.813×10^{-3}, respectively.

In the present research, a TAW-1000 servo-controlled rock mechanics experimental system (Fig. 3.2) was used to complete all experiments. The maximum loading capacity of the servo-controlled system is 1000 kN, and the maximum confining pressure is 60 MPa. During the experimentation, the axial force and the axial deformation were measured using an axial linear variable differential transducer (LVDT) with a range of 8 mm. The circumferential deformation was measured using an LVDT attached to a chain wrapped tightly around the specimen, for which the maximum displacement capacity was 4 mm. To eliminate the influence of end friction effects (Mogi 1966; Hawkes and Mellor 1970), the axial and circumferential LVDTs were located in the central part of the specimen.

3.1.2 *Two Types of Triaxial Tests*

To investigate the evolution of the mechanical damage of the sandstone material, we performed two types of triaxial tests: (1) monotonic, and (2) cyclic.

The triaxial monotonic experiments (Fig. 3.3) (i.e. the conventional triaxial experiment) were carried out under different confining pressures (σ_3) of 5, 10, 20 and 30 MPa and consisted of the following two steps. Firstly, the confining pressure was increased to the desired value at a constant rate of 0.5 MPa/s to ensure that the

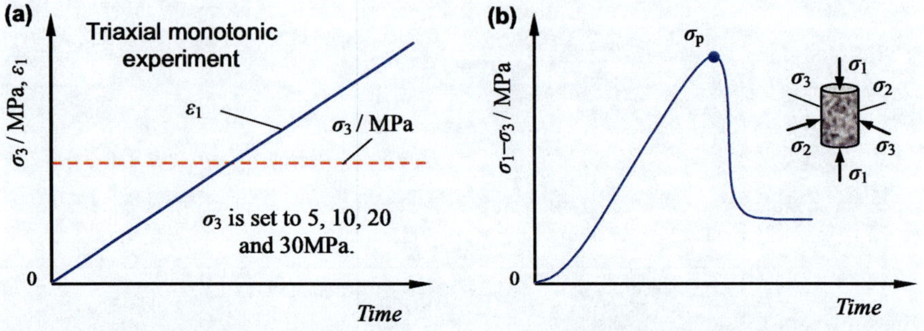

Fig. 3.3 Monotonic loading path applied in the present research. **a** Variation of confining pressure and axial strain with time. **b** Variation of axial deviatoric stress with time

specimen was under uniform hydrostatic stresses. The deviatoric stress (σ_1–σ_3) was then applied to the surface of the specimen at a constant axial displacement rate of 0.08 mm/min until failure. In Fig. 3.3, ε_1 means the axial strain.

The triaxial cyclic experiments (Fig. 3.4) were conducted using the following three steps. First, the hydrostatic pressure was applied to the specimen at a rate of 0.5 MPa/s until the desired value was reached. Second, the specimens were loaded to the first displacement value (corresponding to the first deviatoric stress level) at a controlled axial displacement rate of 0.08 mm/min, and then the first deviatoric stress level was unloaded to zero at a controlled axial force of 0.5 KN/s. In each subsequent cycle the second deviatoric stress level was increased and specimens again unloaded to zero. Finally, stress-cycling was continued in this way until specimens eventually failed. It should be noted that in our study, after the peak strength, two or three cyclic loading were still continued in order to analyze the mechanical behaviors of fractured rock material, which is a different procedure to that used in previous studies (Heap, et al. 2009b; Trippetta et al. 2013). Figure 3.5 shows a typical experimental result

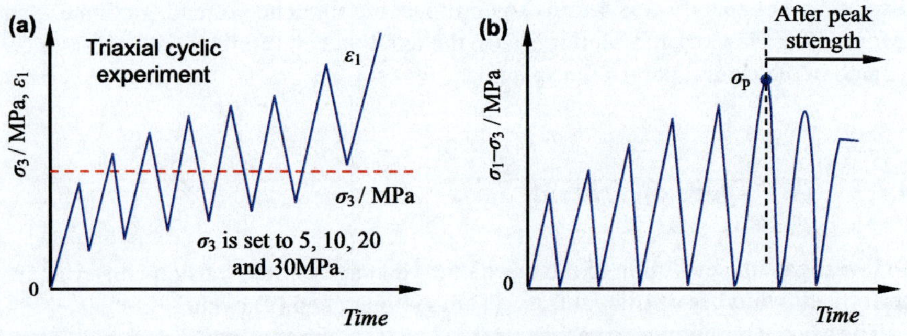

Fig. 3.4 Cyclic loading path applied in the present research. **a** Variation of confining pressure and axial strain with time. **b** Variation of axial deviatoric stress with time

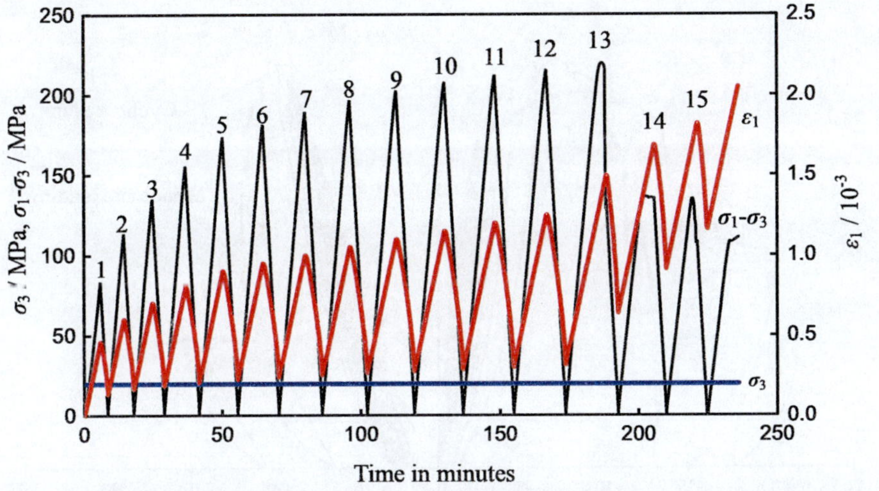

Fig. 3.5 A typical experimental result of cyclic loading path from increasing amplitude, cyclic stressing triaxial experiment on sandstone specimen that failed during the 15th loading cycle. The numbers in this figure represent the cycle number. The red line represents the axial strain, the blue line represents the confining pressure and the black line represents the axial deviatoric stress

of the triaxial cyclic loading path for a sandstone specimen that failed on the 15th loading cycle.

3.2 Analysis of Triaxial Cyclic Experimental Results of Red Sandstone

3.2.1 Comparison Between Monotonic and Cyclic Experimental Results

Figure 3.6 shows typical triaxial cyclic experimental results of sandstone at $\sigma_3 = 5$ and 20 MPa. In Fig. 3.6, the triaxial monotonic experimental curves of sandstone are also given for comparison purposes. It should be noted that even after the peak stress, the specimens still continued to perform cyclic loading until the residual strength, which was usually difficult to carry out due to the sudden brittle failure of rock material after the peak strength.

Based on the triaxial experimental results of sandstone under monotonic and cyclic loading (Fig. 3.6), Fig. 3.7 presents the comparison of peak strength between monotonic and cyclic loading. From Fig. 3.6, we can see an obvious difference between monotonic and cyclic triaxial strength, and the difference changes with different confining pressures. Table 3.1 lists the comparison between monotonic and cyclic

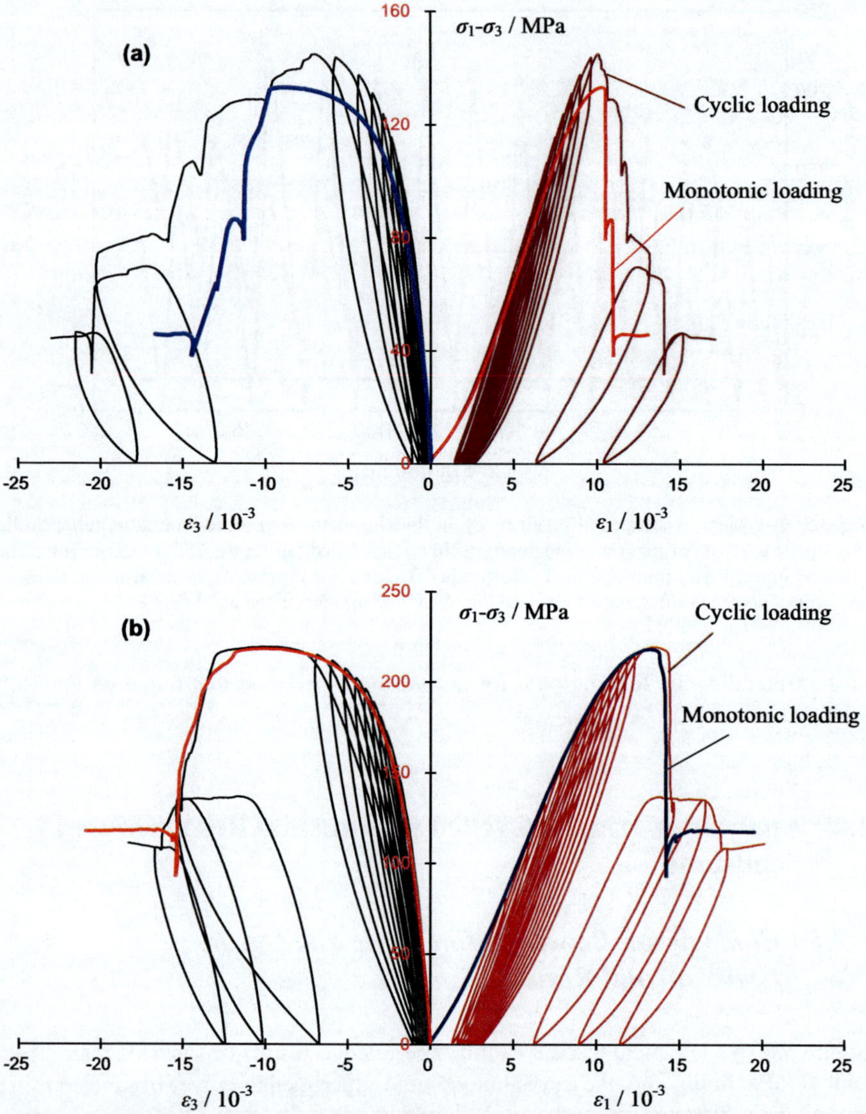

Fig. 3.6 Comparison of monotonic and cyclic stress–strain curves of sandstone. **a** $\sigma_3 = 5$ MPa. **b** $\sigma_3 = 20$ MPa

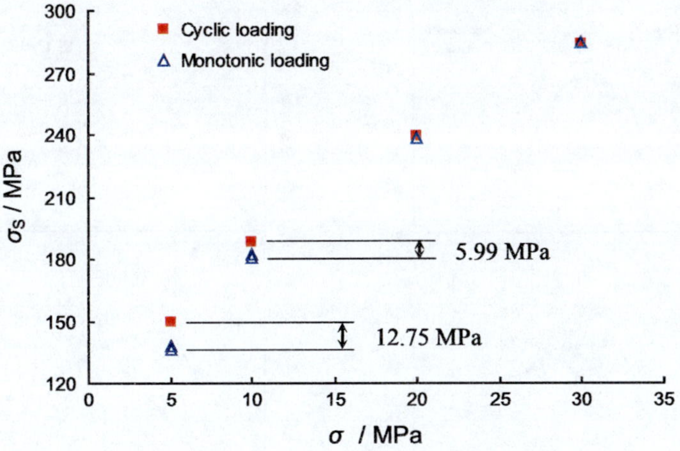

Fig. 3.7 Comparison of monotonic and cyclic triaxial strength of sandstone specimen under different confining pressures. The denoted values mean the difference between monotonic and cyclic triaxial strength

Table 3.1 Comparison of monotonic and cyclic triaxial strength (unit: MPa) of sandstone

Loading path	$\sigma_3 = 5$ MPa	$\sigma_3 = 10$ MPa	$\sigma_3 = 20$ MPa	$\sigma_3 = 30$ MPa
Monotonic loading	137.24	181.97	238.30	284.25
Cyclic loading	149.99	187.96	239.55	283.31

triaxial strength of sandstone. It should be noted that in Table 3.1, the monotonic triaxial strength of sandstone is the average value of the tested specimens. From Table 3.1, it can be seen at $\sigma_3 = 5$ and 10 MPa, the cyclic triaxial strength of sandstone is higher than the monotonic triaxial strength. For $\sigma_3 = 5$ MPa, the difference between the cyclic and monotonic triaxial strength of sandstone is 12.75 MPa. As the confining pressure increases from 5 to 10 MPa, the difference between the cyclic and monotonic triaxial strength of sandstone decreases from 12.75 MPa to 5.99 MPa. However, at $\sigma_3 = 20$ and 30 MPa, the cyclic triaxial strength of sandstone is approximately equal to the monotonic triaxial strength, which indicates that at higher confining pressure, cyclic loading does not affect the triaxial strength of sandstone compared with monotonic loading.

In order to analyze the difference of elastic modulus obtained by monotonic and cyclic loading testing, the deviatoric stress-axial strain of sandstone under cyclic unloading is compared with the monotonic loading curve of sandstone, and the results are shown in Fig. 3.8. Under cyclic loading condition, the elastic modulus of sandstone can be confirmed by the average slope at the stage of elastic deformation, and the dotted line shown in Fig. 3.8. Table 3.2 lists the comparison between the monotonic and cyclic triaxial elastic modulus of sandstone. From Table 3.2, it can be seen that at $\sigma_3 = 5$ and 10 MPa, the cyclic triaxial elastic modulus of sandstone is

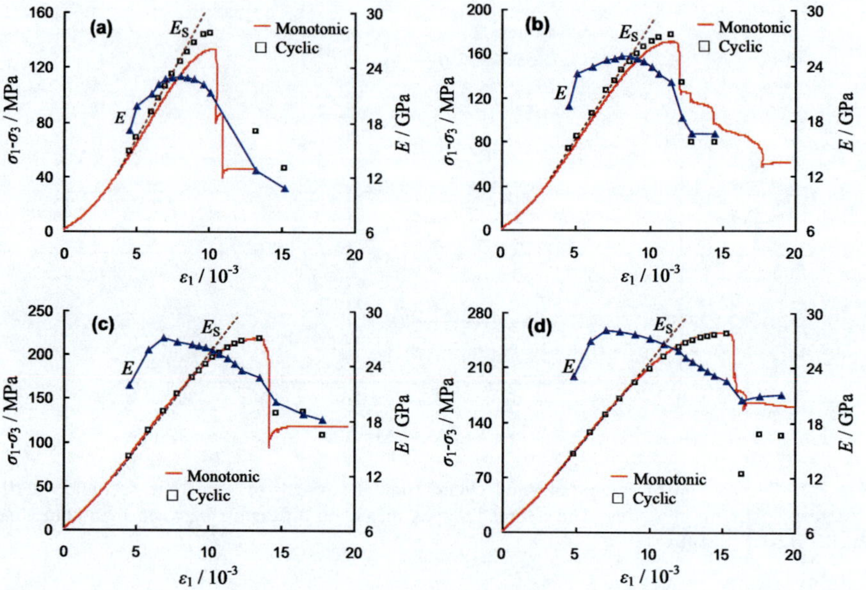

Fig. 3.8 Comparison of monotonic and cyclic triaxial elastic modulus of sandstone specimen under different confining pressures. E represents Young's modulus and E_S represents the elastic modulus. **a** $\sigma_3 = 5$ MPa. **b** $\sigma_3 = 10$ MPa. **c** $\sigma_3 = 20$ MPa. **d** $\sigma_3 = 30$ MPa

Table 3.2 Comparison of monotonic and cyclic triaxial elastic modulus (unit: GPa) of sandstone

Loading path	$\sigma_3 = 5$ MPa	$\sigma_3 = 10$ MPa	$\sigma_3 = 20$ MPa	$\sigma_3 = 30$ MPa
Monotonic loading	17.89	20.32	21.87	22.64
Cyclic loading	19.40	20.92	21.63	22.71

higher than the monotonic triaxial elastic modulus. For $\sigma_3 = 5$ MPa, the difference between the cyclic and monotonic triaxial elastic modulus of sandstone is 1.51 GPa As the confining pressure increases from 5 to 10 MPa, the difference between the cyclic and monotonic triaxial elastic modulus of sandstone decreases from 1.51 to 0.60 GPa. However, at $\sigma_3 = 20$ and 30 MPa, the cyclic triaxial elastic modulus of sandstone is approximately equal to the monotonic triaxial elastic modulus, which indicates that at higher confining pressure, cyclic loading does not influence the triaxial elastic modulus of sandstone.

The influence of confining pressure on the difference between monotonic and cyclic triaxial strength and elastic modulus of sandstone results from the following reason. Under the lower confining pressures, with the increase of axial deformation, the rock specimen allows local yield to occur (Yang et al. 2011). Therefore, after local yield, extruded and fractured material in the rock can fill in the internal voids of specimens and strengthens the supporting structure in the rock material, as shown in Fig. 3.8a, b. Therefore, the cyclic triaxial strength and elastic modulus of sandstone are higher than the monotonic triaxial strength and elastic modulus at lower confining pressures. However, with the increase of confining pressure, even though local yield still occurs with increasing deformation, extruded and fractured materials in the rock find it more and more difficult to fill in the internal voids due to the obvious decrease of porosity as the confining pressure increases. After the confining pressure is increased to 20 MPa, the porosity of sandstone will decrease less easily with increasing confining pressure and the supporting structure in the rock material cannot be strengthened any more (Fig. 3.8c, d), which results in approximate equality between the cyclic and monotonic triaxial strength and elastic modulus.

3.2.2 Evolution of Crack Damage Threshold

The crack damage threshold of rock specimens (Wong et al. 1997; Fairhurst and Hudson 1999; Heap et al. 2009a) can be confirmed by the corresponding axial deviatoric stress where the volumetric deformation of the specimen switches from compaction-dominated to dilatancy-dominated (Yang et al. 2012; Yang and Jing 2013). Figure 3.9 shows the typical evolution of the crack damage threshold of a sandstone specimen under triaxial cyclic loading.

In accordance with the relation between deviatoric stress-volumetric strains as shown in Fig. 3.9, the values of the crack damage threshold (σ_{cd}) of sandstone specimen under cyclic loading were confirmed, and the results are listed in Table 3.3. The σ_{cd} of sandstone specimen under monotonic loading is also listed in Table 3.3 to enable a comparison. Based on the data listed in Table 3.3, the relation between crack damage threshold of sandstone specimen and cycle number is shown in Fig. 3.10.

Figure 3.10 indicates that when cycle number is less than 3, the sandstone specimen does not switch from compaction-dominated to dilatancy-dominated, i.e. the σ_{cd} of the specimen cannot be obtained. At the same confining pressure, the first σ_{cd} of the sandstone specimen under cyclic loading is approximately equal to that under monotonic loading, which indicates that the discreteness of the tested sandstone specimens is very small in the present research. With the increase of cycle number, the crack damage threshold of sandstone specimens first increases. For example at $\sigma_3 =$ 20 MPa, the σ_{cd} value of sandstone increases from 131.14 to 154.58 MPa, because the closed cracks after unloading are more difficult to re-open with the increase of cycle number in the early cycles. Then, the σ_{cd} value decreases from 154.58 to 60.05 MPa, which can be explained as follows. On one hand, the closed cracks after unloading are easier to re-open with the increase of cycle number; on the other hand, some new cracks in the specimen are initiated, and propagate and coalesce with the increase of cycle number. Finally, when the specimen enters the stage of residual strength, the σ_{cd} value remains unaffected by cycle number.

Figure 3.11 depicts the influence of confining pressure on the evolution of the crack damage threshold of sandstone. In general, the σ_{cd} value of sandstone specimen under higher confining pressure is greater than that under lower confining pressure.

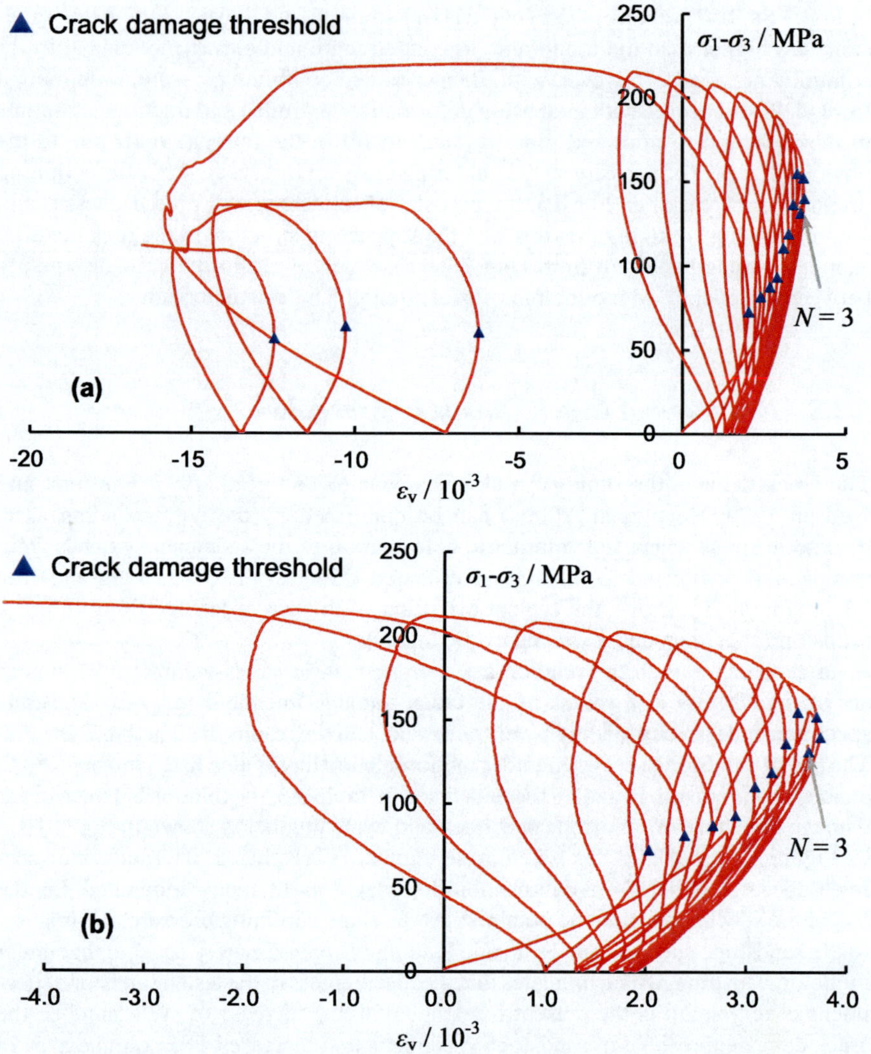

Fig. 3.9 Typical evolution of crack damage threshold of sandstone specimen under triaxial cyclic loading ($\sigma_3 = 20$ MPa). **a** Relation between axial deviatoric stress and volumetric strain. **b** Local magnification

Moreover, the confining pressure does not affect the evolution characteristics of the crack damage threshold. Figure 3.12 illustrates the relation between cycle number and the corresponding volumetric strain ($\varepsilon_{cd,\,v}$) reaching the crack damage threshold of sandstone specimen under different confining pressures. Figure 3.12 shows that in the early cycles, the confining pressure has no great influence on the relation between $\varepsilon_{cd,\,v}$ and cycle number; whereas in later cycles, the confining pressure has

Table 3.3 Values of crack damage threshold (σ_{cd}) of sandstone specimen under monotonic and cyclic loading (Unit: MPa)

Cycle number	$\sigma_3 = 5$ MPa	$\sigma_3 = 10$ MPa	$\sigma_3 = 20$ MPa	$\sigma_3 = 30$ MPa
Monotonic loading	74.88	102.50	138.35	167.11
3	78.98	98.47	131.14	–
4	87.17	105.29	139.79	163.78
5	86.62	114.34	151.93	174.81
6	85.64	98.76	154.58	179.40
7	75.79	89.23	136.15	171.48
8	64.17	81.25	118.77	140.64
9	55.01	64.84	110.44	126.10
10	42.53	61.54	92.95	117.55
11	36.36	54.23	86.99	111.76
12	27.95	49.51	81.23	106.38
13	22.03	38.35	72.42	99.29
14	–	38.52	60.05	94.05
15	–	45.44	63.42	68.12
16	–	–	56.28	72.02
17	–	–	–	66.24

Note The crack damage threshold of sandstone specimen under monotonic loading is the average value

a significant effect on the relation between $\varepsilon_{cd, v}$ and cycle number, especially after the peak strength. At the same cycle number, if the confining pressure is higher, the volumetric strain reaching the crack damage threshold will be lower due to the limitation of high confining pressure.

3.3 Mechanism of Mechanical Damage of Red Sandstone

3.3.1 X-ray Micro CT Equipment and Scanning Procedure

X-ray CTscanning of the sandstone specimen was carried out using a Nanotom 160 high-resolution micro-CT at a spatial resolution of 30 μm. The X-ray beam penetrating the specimen is measured by an array of detectors. The X-ray is produced by electrons striking a Mo-W alloy target in an X-ray tube. The electron current was 80 μA, the accelerating voltage was 140 kV, and the scanning time was 4 s. The degree of X-ray attenuation depends on the density and the atomic number of the materials in the specimens. Material with higher density and higher atomic number generally causes higher attenuation of the X-rays. X-ray projection data from various

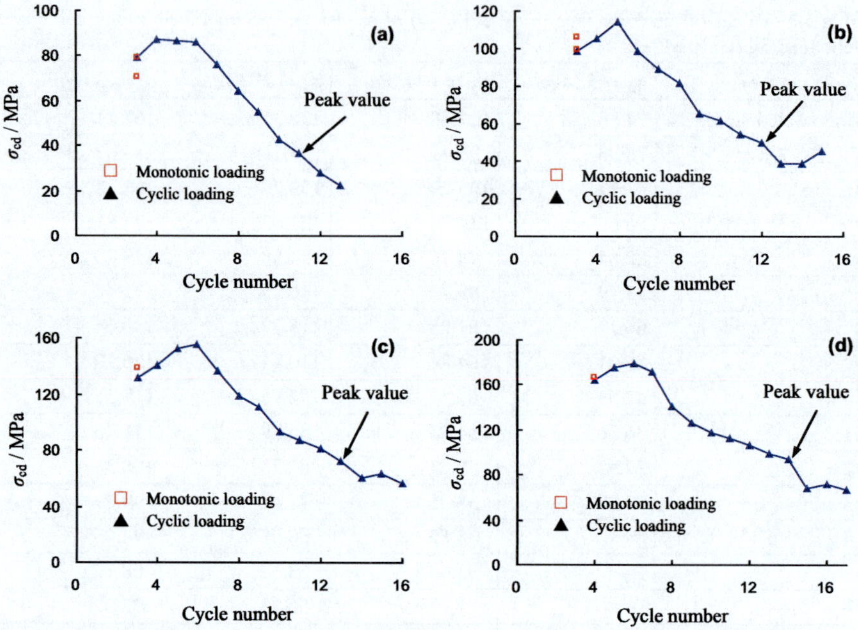

Fig. 3.10 Relationship between crack damage threshold of sandstone specimen and cycle number. **a** $\sigma_3 = 5$ MPa. **b** $\sigma_3 = 10$ MPa. **c** $\sigma_3 = 20$ MPa. **d** $\sigma_3 = 30$ MPa

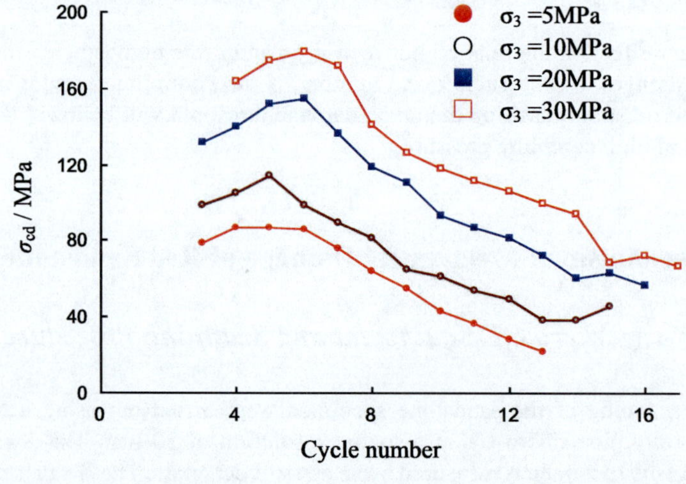

Fig. 3.11 Influence of confining pressure on the evolution of crack damage threshold of sandstone specimen

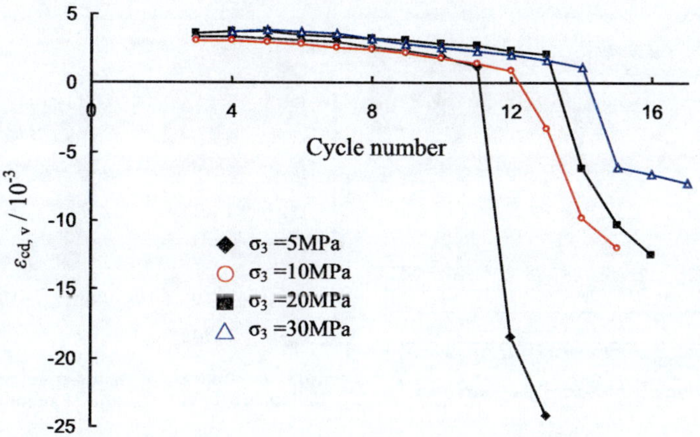

Fig. 3.12 Relation between cycle number and the corresponding volumetric strain reaching crack damage threshold of sandstone specimen under different confining pressures

directions are obtained by the 360° rotation of the X-ray source (Fig. 3.13a). We collected a two-dimensional image at intervals of 0.18°, and thus 2000 slice images could be obtained for one specimen. A two-dimensional image representing the linear distribution of X-ray attenuation was reconstructed using Fourier transformation of the projection data. A 3-D data set of the specimen was obtained by stacking consecutive two-dimensional images.

It should be noted that only a cylinder 50 mm diameter and 68 mm length can be regarded as the scanning region, as shown in Fig. 3.13b, c. For the front vertical cross-section shown in Fig. 3.13d, the slice images were captured at 0.5 mm intervals and the scanning sequence was from front to back. For the horizontal cross-section shown in Fig. 3.12e, the slice images were captured at 0.03 mm intervals, and each image was 0.03 mm thick. Therefore, a total of 2266 slice images were obtained in the X-ray micro CT testing of each sandstone specimen. These slice images can be used to explore the internal damage of sandstone material under monotonic and cyclic loading.

3.3.2 Internal Damage of Sandstone Under Triaxial Monotonic and Cyclic Loading

Figures 3.14 and 3.15 depict the comparison of X-ray micro CT scanning surface images and actual surface crack photographs of sandstone specimen after triaxial monotonic and cyclic loading (σ_3 = 5 MPa), respectively. In Figs. 3.14 and .3.15, black regions with lower CT numbers mean cracks, and other regions with higher CT

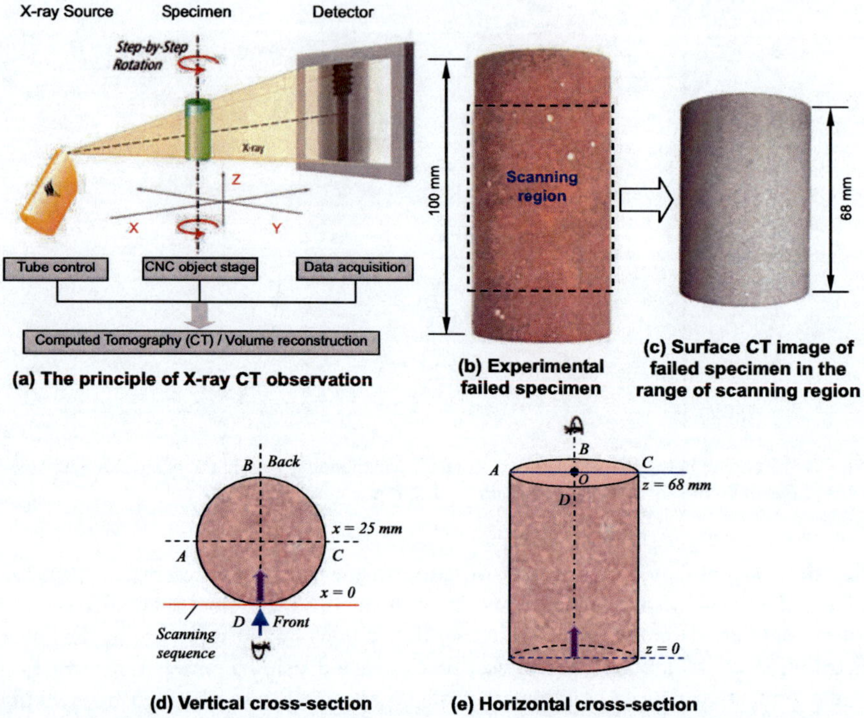

Fig. 3.13 The principle of X-ray micro CT observation, illustration of scanning region, and illustration of scanning direction on vertical and horizontal cross-section. **a** The principle of X-ray CT observation; **b** experimental failed specimen; **c** surface CT image of failed specimen in the range of scanning region; **d** vertical cross-section; **e** Horizontal cross-section

numbers indicate no surface failure. From Figs. 3.14 and .3.15, it can be seen that X-ray CT scanning surface images approximate the actual surface crack images, which demonstrates that X-ray micro CT scanning can be well used to explore internal damage in sandstone material.

Figures 3.16 and 3.17 respectively show horizontal cross-sections with different heights and front vertical cross-sections with different depths of sandstone specimen after triaxial monotonic failure ($\sigma_3 = 5$ MPa). From Fig. 3.16, we can analyze the effect of height (z) on the extent of internal damage to sandstone after triaxial monotonic failure. At $z = 5 \sim 15$ mm, the horizontal cross-section is located in the bottom region of the specimen, only one main crack across the diameter direction is observed and the system of cracks is very simple, which can also be demonstrated in the bottom region shown in Fig. 3.14. At $z = 15 \sim 45$ mm, the horizontal cross-section is located in the middle region of the specimen. Two bigger cracks are observed, resulting from the crack propagation and coalescence inside the specimen with the increase of deformation. The left crack in the specimen does not almost take place moving, whereas the right crack in the specimen moves gradually towards the right boundary

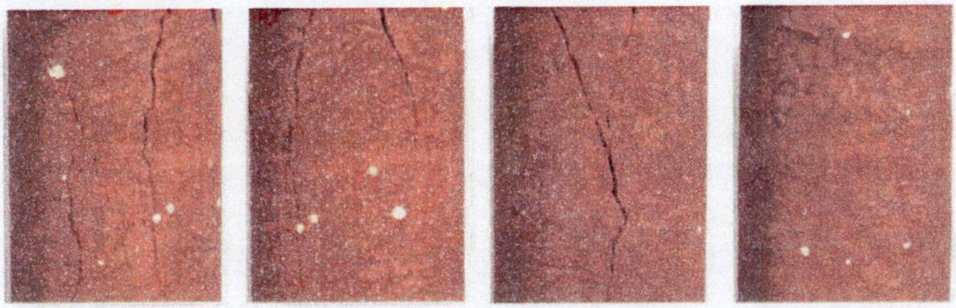

(a) Actual surface crack photographs

(b) X-ray CT scanning surface images

Fig. 3.14 Comparison of X-ray CT scanning surface images and actual surface crack photographs of sandstone specimen after triaxial monotonic loading ($\sigma_3 = 5$ MPa). **a** Actual surface crack photographs. **b** X-ray CT scanning surface images

at an angle of about 30° with the horizontal direction. However, at $z = 45 \sim 60$ mm, when the horizontal cross-section is located in the top region of the specimen, even though no new cracks are observed, the width of the two cracks begins to increase compared with those at $z = 15 \sim 45$ mm. Figure 3.17 depicts the effect of depth (x) on the extent of internal damage to the sandstone material after triaxial monotonic loading. From Fig. 3.17, it is clear that two large cracks with "Y" type are observed and the system of cracks is very simple.

For the two large cracks shown in Figs. 3.14, 3.16 and 3.17, it is very difficult to distinguish which crack is shear failure or axial splitting tensile failure, only in accordance with the angle of the 2-D crack with the direction of major principal stress. For example, in Fig. 3.17, from CT images at $x = 5$, 25 and 45 mm, we can draw contradictory conclusions on shear or tensile fracture. Therefore, we reconstructed 3-D images of the sandstone specimens before and after triaxial monotonic failure ($\sigma_3 = 5$ MPa). In the 3-D images, fracture regions with lower CT numbers are black, and other regions with relatively higher CT numbers are transparent, as shown in Fig. 3.18. From Fig. 3.18, we are able to confirm that 3-D crack 1 is axial splitting tensile fracture, and 3-D crack 2 is shear fracture.

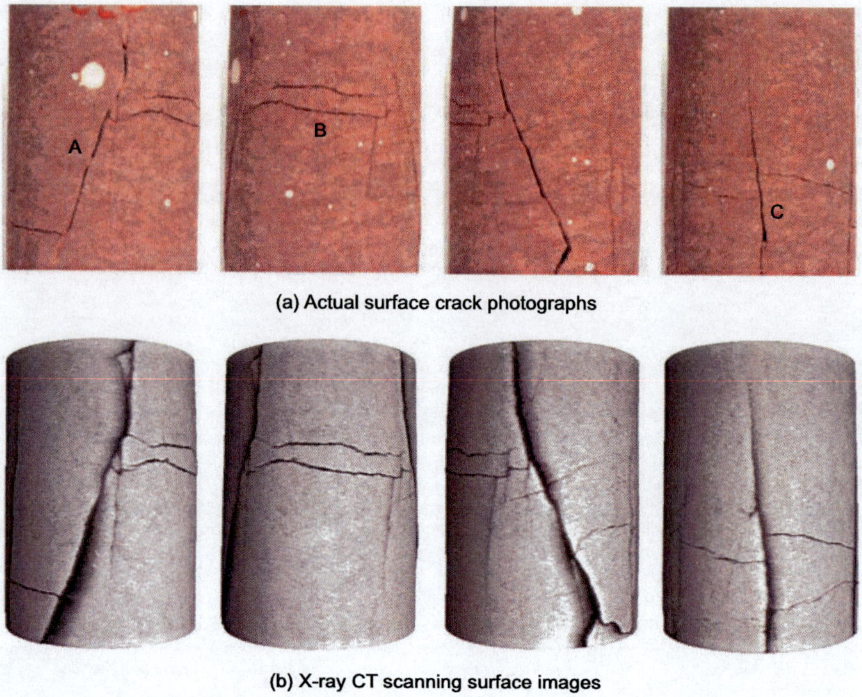

Fig. 3.15 Comparison of X-ray CT scanning surface images and actual surface crack photographs of sandstone specimens after triaxial cyclic loading ($\sigma_3 = 5$ MPa). **a** Actual surface crack photographs. **b** X-ray CT scanning surface images

Figures 3.19 and 3.20 respectively, show horizontal cross-sections with different heights and front vertical cross-sections with different depths for sandstone specimens after triaxial cyclic loading ($\sigma_3 = 5$ MPa). In general, in accordance with Figs. 3.19 and 3.20, it can be seen that under triaxial cyclic loading, the system of cracks is more complicated than that under triaxial monotonic loading for the same confining pressure (Figs. 3.16 and 3.17), which is also demonstrated from the comparison of surface cracks shown in Figs. 3.14 and.3.15, respectively. Under triaxial cyclic loading, the specimen takes on a typical single shear fracture, and at the same time, axial and lateral tensile cracks are also observed, whereas under triaxial monotonic loading, no lateral cracks are observed. These shear fracture, axial and lateral tensile cracks under triaxial cyclic loading can be all further inferred from microscopic observations on the fracture surface of the failed specimen shown in Fig. 3.21.

In order to make a quantitative evaluation of the fracture extent of internal damage of sandstone specimen after triaxial monotonic and cyclic compression failure, each horizontal CT image shown in Figs. 3.16 and 3.19 was processed using Matlab software (Gui et al. 2012) and changed to binarized images (Fig. 3.22). In Fig. 3.22,

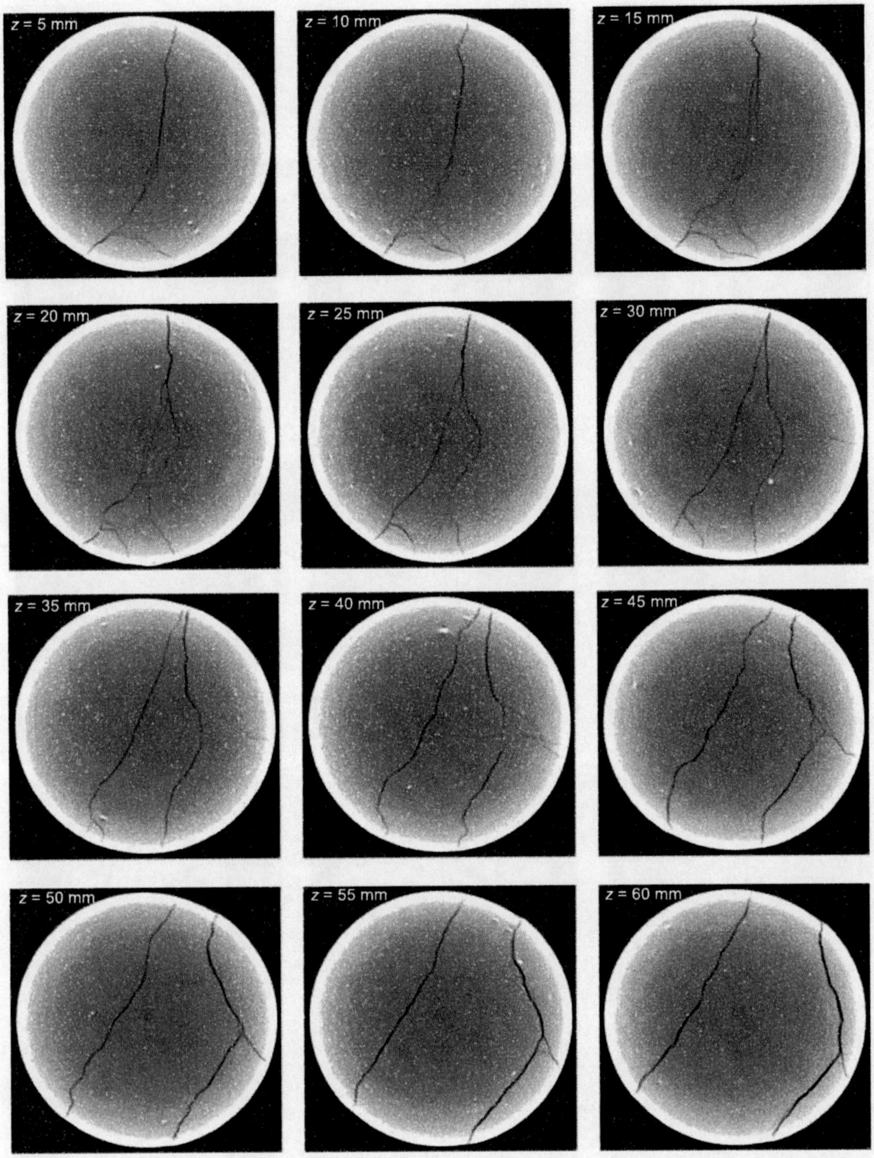

Fig. 3.16 Horizontal cross-sections with different heights of sandstone specimens after triaxial monotonic loading ($\sigma_3 = 5$ MPa)

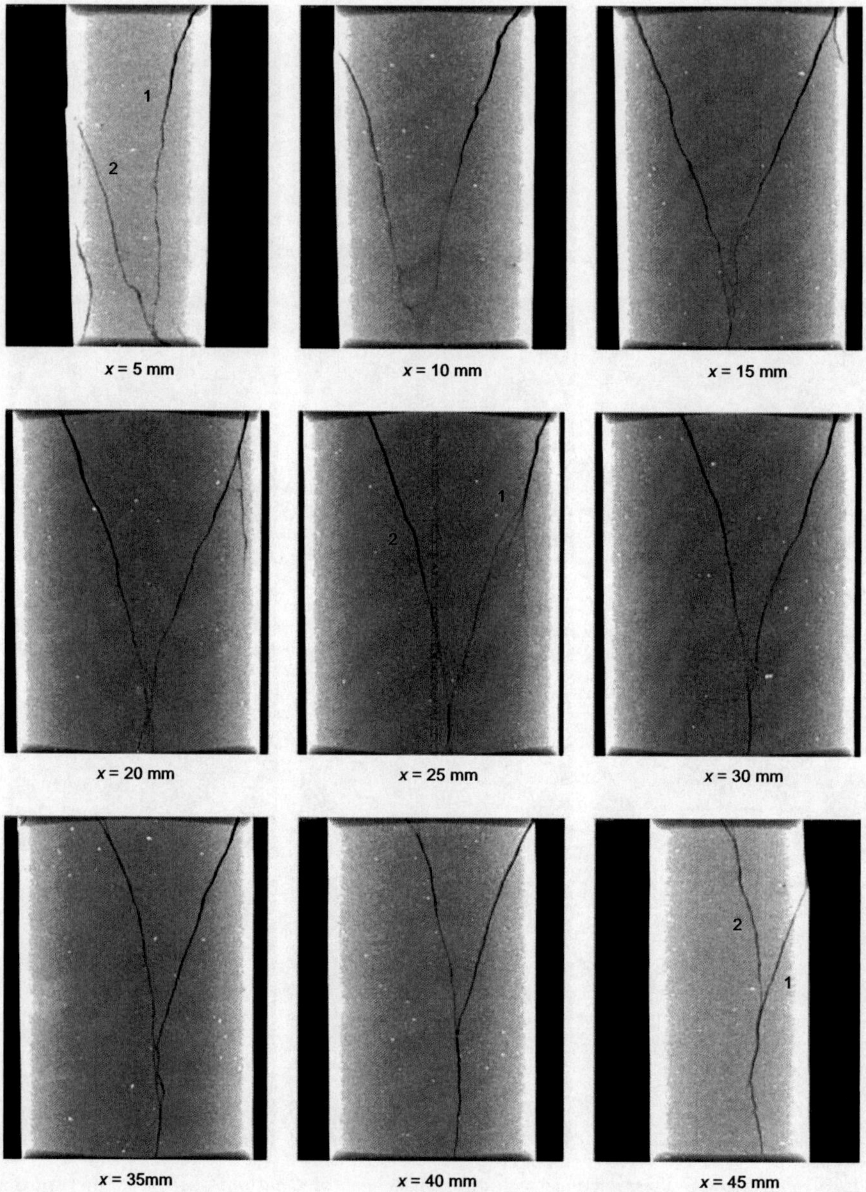

Fig. 3.17 Front vertical cross-sections with different depths of sandstone specimens after triaxial monotonic loading ($\sigma_3 = 5$ MPa)

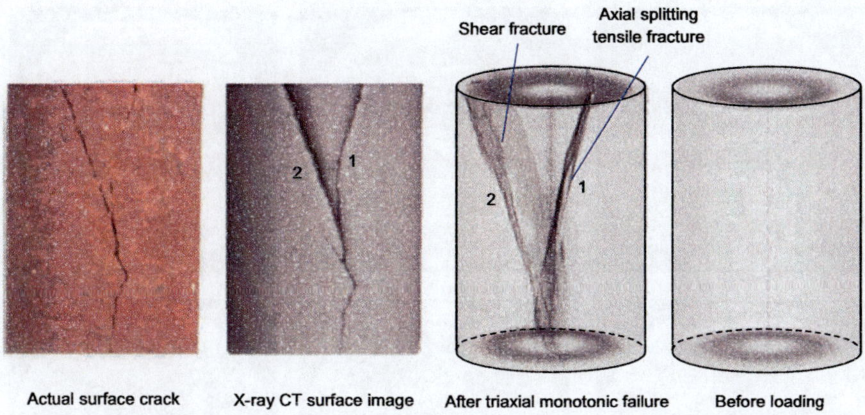

Fig. 3.18 Three-dimensional images of sandstone specimens before and after triaxial monotonic failure ($\sigma_3 = 5$ MPa). The numbers in the figure correspond to those shown in Fig. 3.17

the white regions represent the cracks, whereas the black regions represent intact rocks without any failure. Generally, the crack system in Fig. 3.22 approximates to the experimental results in Figs. 3.16 and 3.19. On the basis of binarized images of all horizontal cross-sections of sandstone specimens after triaxial monotonic and cyclic compression failure, we obtained the crack area and aperture extent (i.e. the percentage ratio of crack area to entire area) for each horizontal cross-section. As a result, we were able to investigate quantitatively the effect of height (z) on the extent of internal damage of sandstone material after triaxial monotonic and cyclic compression failure.

Figure 3.23 shows the evolution of crack area and aperture extent along different heights (z) of sandstone specimen after triaxial monotonic and cyclic compression failure. From Fig. 3.23a, it can be seen that after triaxial monotonic compression failure, the crack area of sandstone specimens increases from 12.50 mm^2 at $z = 5$ mm to 27.80 mm^2 at $z = 65$ mm. In contrast, after triaxial cyclic compression failure, the crack area of sandstone specimens varies in a smaller range from 34.51 mm^2 at $z =$ 5 mm to 39.01 mm^2 at $z = 40$ mm, and then decreases rapidly from 39.01 mm^2 at $z =$ 40 mm to 7.21 mm^2 at $z = 65$ mm. From Fig. 3.23b, we can see that the evolution of aperture extent approximates to that of the crack area. With the increase of height, the aperture extent of sandstone specimens after triaxial monotonic compression failure increases from 0.47% at $z = 5$ mm to 1.05% at $z = 65$ mm. However, after triaxial cyclic compression failure, the aperture extent of sandstone specimens varies in a smaller range from 1.29% at $z = 5$ mm to 1.46% at $z = 40$ mm, and then decreases rapidly from 1.46% at $z = 40$ mm to 0.27% at $z = 65$ mm. Furthermore, at $z = 5 \sim$ 40 mm, the crack area and aperture extent of sandstone after cyclic failure are higher than those after monotonic failure. At $z = 40 \sim 50$ mm, the crack area and aperture extent of sandstone after cyclic failure approximate to those after monotonic failure.

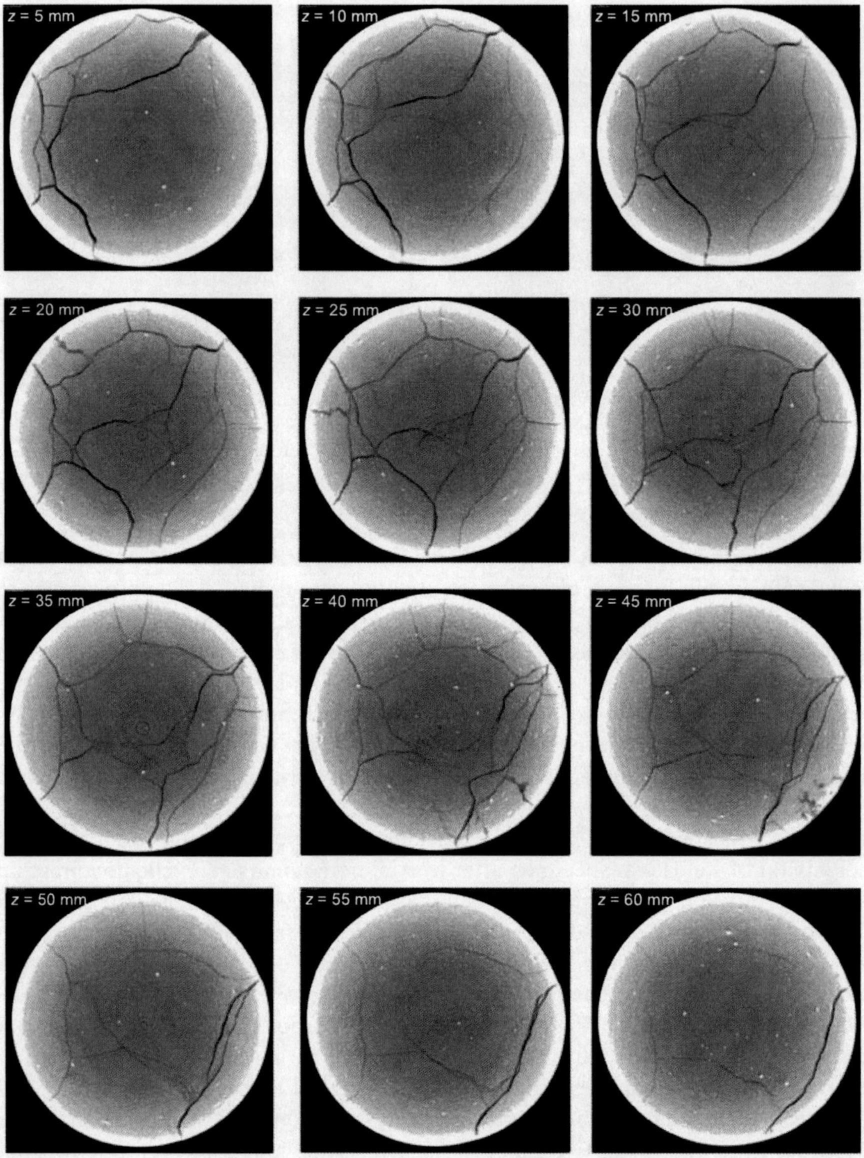

Fig. 3.19 Horizontal cross-sections with different heights of sandstone specimens after triaxial cyclic loading ($\sigma_3 = 5$ MPa)

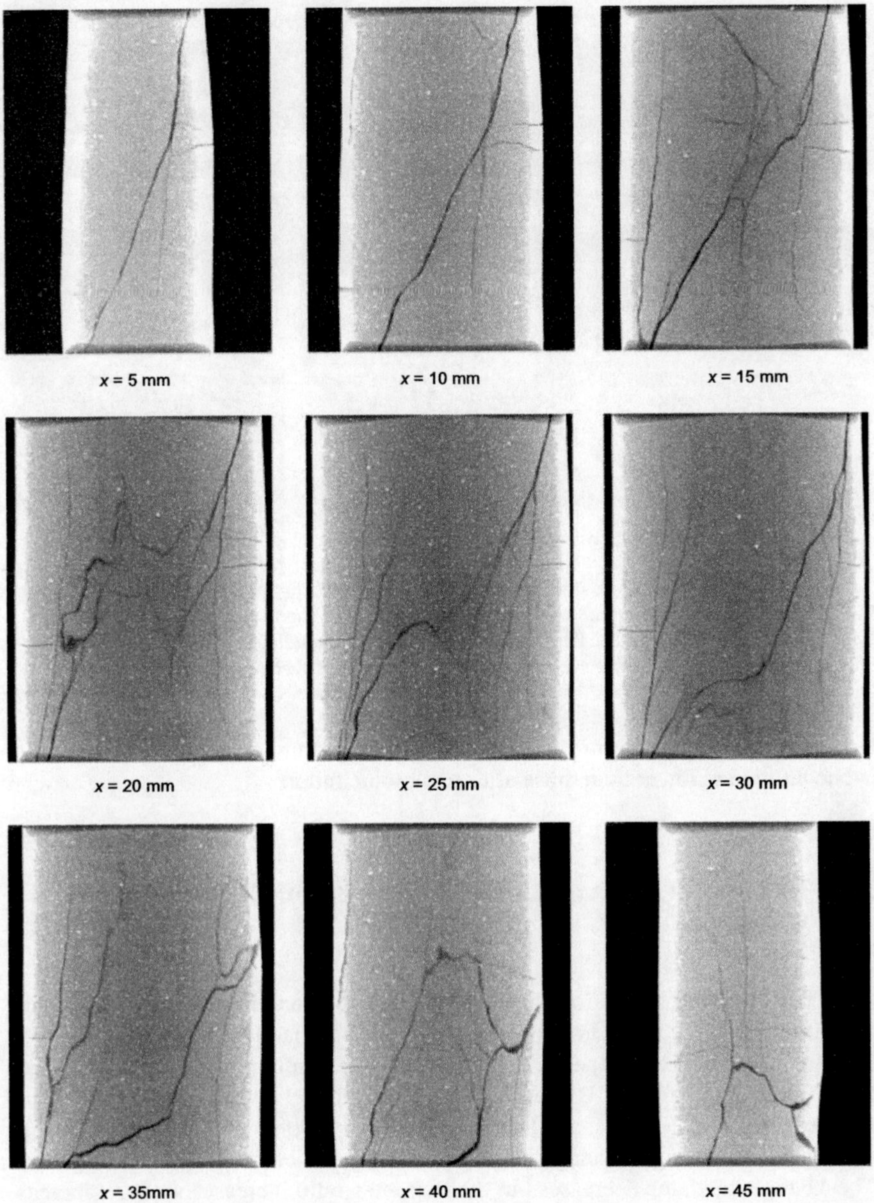

Fig. 3.20 Front vertical cross-sections with different depths of sandstone specimens after triaxial cyclic loading ($\sigma_3 = 5$ MPa)

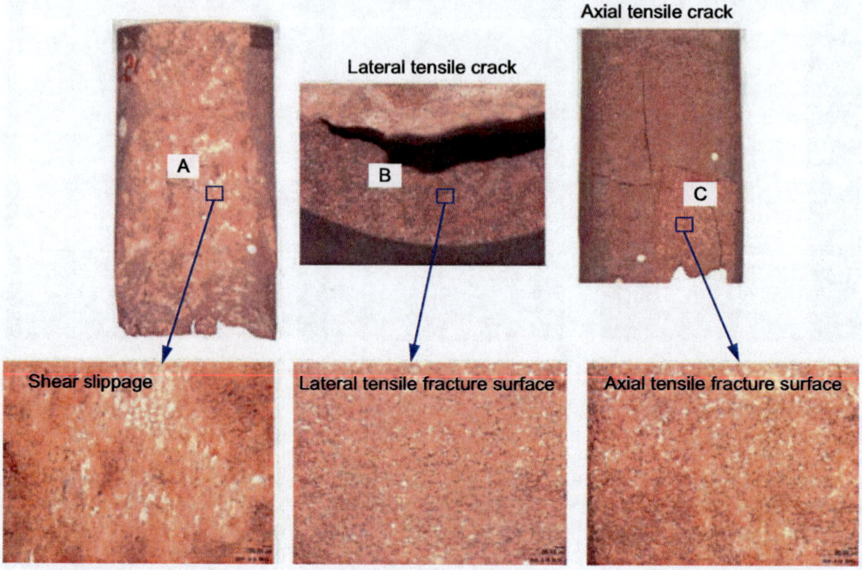

Fig. 3.21 Microscopic observations on the fracture surface of sandstone specimens under triaxial cyclic loading ($\sigma_3 = 5$ MPa). The letters in this figure are corresponding to those shown in Fig. 3.14a. Shear slippage indicates the shear crack, and tensile fracture surface means the tensile crack

However, at $z = 50 \sim 65$ mm, the crack area and aperture extent of sandstone after cyclic failure are lower than those after monotonic failure.

3.4 Discussion on Evolution of Deformation Damage of Red Sandstone

In the past, many investigations have been conducted on the effect of cyclic loading on the elastic modulus and Poisson's ratio of rock material. However, and it has been found that the effect is dependent not only on the confining pressure but also on the softness and hardness. By carrying out uniaxial cyclic tests for hard basalt and granite, Heap et al. (2009b), and Heap and Faulkner (2008) demonstrated that cyclic stressing changes significantly the deformation parameters of basalt and granite. The Young's modulus decreases but the Poisson's ratio increases with the increase of cycle number. Moreover the variance of basalt in both E and υ did not depend on the action of high temperature treatment (Heap et al. 2009b). Fuenkajorn and Phueakphum (2010) performed a series of uniaxial cyclic tests on salt. They found that the elastic modulus of salt decreased slightly during the first few cycles, and tended to remain constant until failure. Trippetta et al. (2013) reported laboratory measurements of Young's modulus and Poisson's ratio of soft gypsum and hard

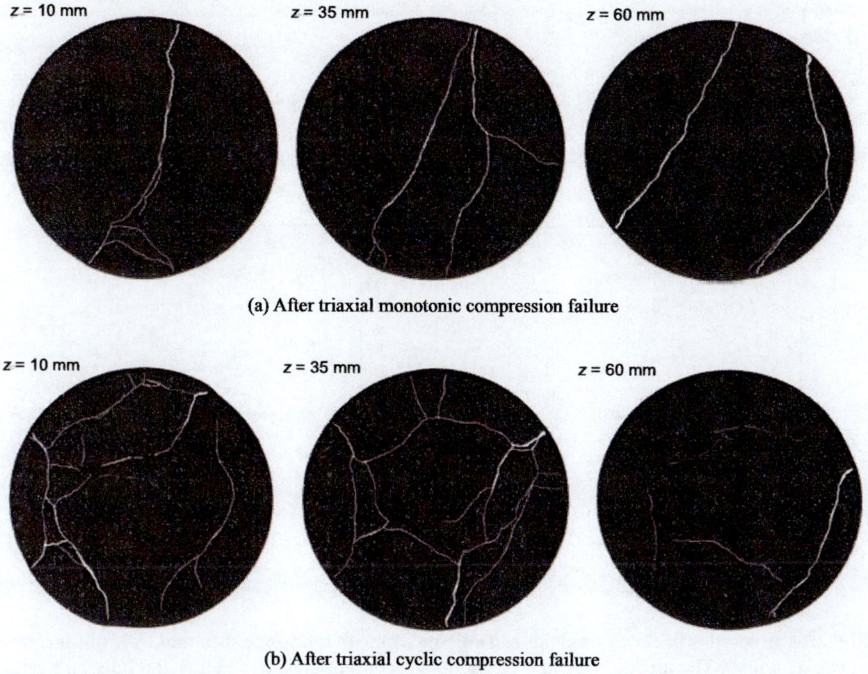

Fig. 3.22 Typical binarized pictures of horizontal cross-sections of sandstone specimens after triaxial monotonic and cyclic compression failure ($\sigma_3 = 5$ MPa)

Fig. 3.23 Evolution of crack area and aperture extent along different heights (z) of sandstone specimens after triaxial monotonic and cyclic compression failure ($\sigma_3 = 5$ MPa)

dolostones under uniaxial cyclic loading. A complex variation in E and υ was found as follows. In the early cycles, large increases in E and υ were observed, followed by essentially constant values, before E decreased and υ increased approaching failure. The above analysis shows that under uniaxial compression, the elastic modulus and

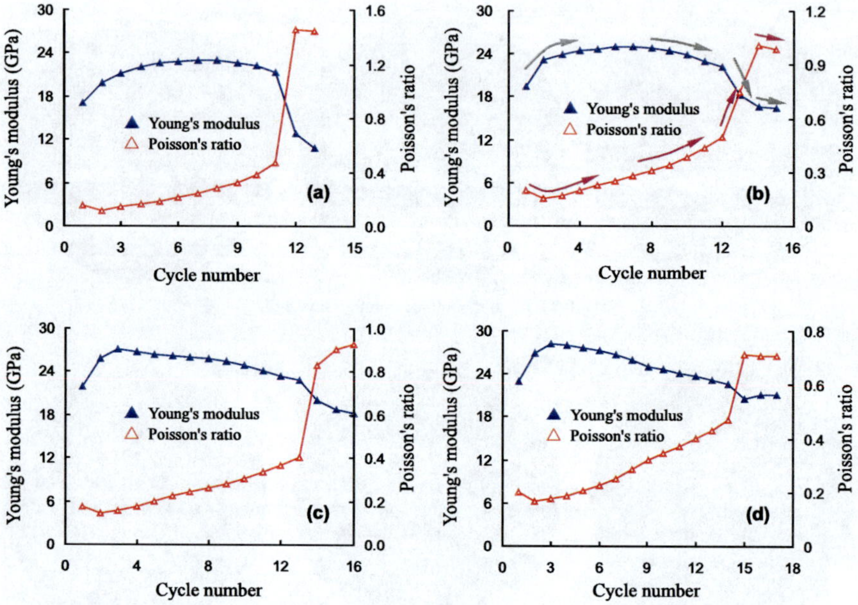

Fig. 3.24 Evolution of elastic modulus and Poisson's ratio of sandstone specimens with the increase of cycle number. The arrows show the evolution direction of elastic modulus and Poisson's ratio. **a** $\sigma_3 = 5$ MPa; **b** $\sigma_3 = 10$ MPa; **c** $\sigma_3 = 20$ MPa; **d** $\sigma_3 = 30$ MPa

Poisson's ratio of rocks with different softness and hardness have various evolution characteristics.

It should be noted that our research on sandstone specimen under triaxial cyclic loading differs from that reported in previous studies (Heap et al. 2009b; Heap and Faulkner 2008; Trippetta et al. 2013). Figure 3.24 presents the evolution of elastic modulus and Poisson's ratio with increasing cycle number for sandstone specimens (derived from Fig. 3.6). The relationship between elastic modulus and axial strain of sandstone specimen for each cycle is also presented in Fig. 3.8 to demonstrate the influence of stress level on the evolution of elastic modulus. In accordance with Figs. 3.8 and 3.24, the evolution can be characterized as having four stages (Fig. 3.25): (a) Stage I: material strengthening; (b) Stage II: material degradation; (c) Stage III: material failure; (d) Stage IV: structure slippage.

During Stage I, the elastic modulus (E) increases obviously, whereas the Poisson's ratio (υ) first decreases and then increases slightly. The increase in E indicates that the specimens become stiffer and more difficult to deform. In previous studies of rocks under uniaxial cyclic loading (Heap, et al. 2009b; Trippetta et al. 2013), the increase in elastic modulus usually contributes to the closure of pre-existing pores and fissures in the rock material. However, it should be noted that in our research, the unloading stress level of the sandstone specimen for the first cycle is located at the stage of elastic deformation in the monotonic loading curve (Fig. 3.8). Therefore,

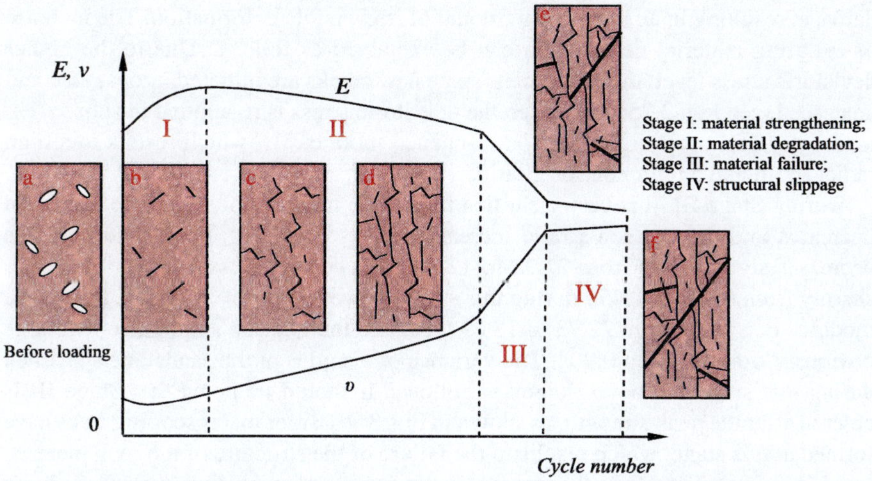

Fig. 3.25 A sketch on four stages characterizing the evolution of elastic modulus and Poisson's ratio of sandstone specimens with the increase of cycle number under confining pressure

the closure of pre-existing pores and fissures cannot indicate clearly the evolution characteristics of the E and υ of sandstone material. New insights on the evolution can be presented as follows. On one hand, with the increase of cycling stress, some mesoscopic elements with lower strength and stiffness reach first their maximum supporting capacity, yield softening, and produce plastic deformation during the yielding of specimens (Yang et al. 2008). Some extruded and fractured elements can then fill in the internal voids of sandstone specimens. On the other hand, some pre-existing pores and fissures can also cause the closure (see Fig. 3.25a, b). The above two factors strengthen the supporting structure of sandstone material, which results in the increase of E and the decrease of ν. Afterwards, with further increase of cycling stress, the rate of increase of elastic modulus becomes slower, and the Poisson's ratio begins to increase, which indicates that each increment of axial strain results in a larger increment of radial strain.

During Stage II, we observe a significant increase in υ, but a decrease in E, which is different from previously reported results of cyclic stressing experiments on softer rocks (Trippetta et al. 2013). However, our results are in good agreement with those on hard rocks including sandstone, basalt and granite (Heap and Faulkner, 2008; Heap et al., 2009b, 2010). At $\sigma_3 = 5$ MPa, the elastic modulus decreased slightly from 22.92 to 21.25 GPa, but the Poisson's ratio increased significantly from 0.237 to 0.462 during increasing amplitude cycling stressing to failure. However, at $\sigma_3 = 20$ MPa, the elastic modulus decreased greatly from 27.01 to 22.73 GPa, and the Poisson's ratio also increased obviously from 0.154 to 0.401. The decrease in E can be attributed to a progressive degradation in the stiffness of the rock due to the increase of crack damage with increasing cycle number. At higher stress level, each cycle applied to the specimen will produce an additional increment of crack damage, resulting in an increasing amount of irreversible deformation. The increase of υ during material degradation can be explained as follows. Due to the higher deviatoric stress level, in early cycles, some new cracks are initiated, propagated and coalesced (see Fig. 3.25c, d). When the deviatoric stress is re-applied in subsequent loading cycles, these cracks can be re-opened more easily, which results in a high radial strain and a high Poisson's ratio.

During Stage III, it is very clear that the elastic modulus of sandstone specimen decreases and the Poisson's ratio increases. At $\sigma_3 = 5$ MPa, the elastic modulus decreases significantly from 21.25 to 12.71 GPa, but the Poisson's ratio increases sharply from 0.462 to 1.455 during this stage. However, at $\sigma_3 = 20$ MPa, the elastic modulus decreases from 22.73 to 18.69 GPa, and the Poisson's ratio also increases obviously from 0.401 to 0.904. The variance of E and υ of the sandstone specimen during this stage can be explained as follows. It should be noted that Stage III is entered after the peak strength, as shown in Fig. 3.9. Larger macroscopic cracks have formed at this stage, which results in the failure of the structure of the rock material (see Fig. 3.25e). Therefore, the increase in the number of macroscopic cracks reduces the axial stiffness and simultaneously allows higher radial strain for the same level of applied deviatoric stress.

During Stage IV, the elastic modulus decreases slightly from 12.71 to 10.81 GPa, but the Poisson's ratio varies only slightly from 1.455 to 1.443 at $\sigma_3 = 5$ MPa. In contrast, at $\sigma_3 = 20$ MPa, the elastic modulus changes from 18.69 to 18.09 GPa, and the Poisson's ratio varies from 0.904 to 0.927. This indicates that neither the elastic modulus nor the Poisson's ratio of sandstone are dependent on cycle number during this stage, for the following reason. Due to the constriction of confining pressure, the specimen has entered the stage of residual strength, and mainly supports the axial capacity by the friction slippage in the macroscopic failure plane (see Fig. 3.25f), which results in no obvious variance in both the E and υ of sandstone specimens.

It should be noted that for the same cycle number, the elastic modulus of sandstone specimen before peak strength increases nonlinearly with the increase of confining pressure (Yang et al. 2012; Yang and Jing 2013), whereas at higher confining pressures of 20 and 30 MPa, the elastic modulus is unaffected by confining pressure. In contrast, after the peak strength, with the increase of cycle number, the elastic modulus of sandstone specimen decreases more slowly for higher confining pressures, which indicates that the confining pressure can also increase the axial stiffness of fractured rock material. However, in early cycles, the confining pressure has almost no effect on the relation between the Poisson's ratio of sandstone specimen and cycle number; but in later cycles, the Poisson's ratio of sandstone specimens is significantly dependent on the confining pressure. The Poisson's ratio of sandstone specimen under higher confining pressures is lower than that under lower confining pressures, which results from lower radial strains due to the constriction of higher confining pressures.

In order to quantitatively evaluate the damage extent of sandstone specimen under triaxial cyclic loading, the axial and radial damage variable are defined in accordance with the irreversible deformation (plastic deformation) for each cycle, which are expressed in Eqs. (3.1) and (3.2), respectively.

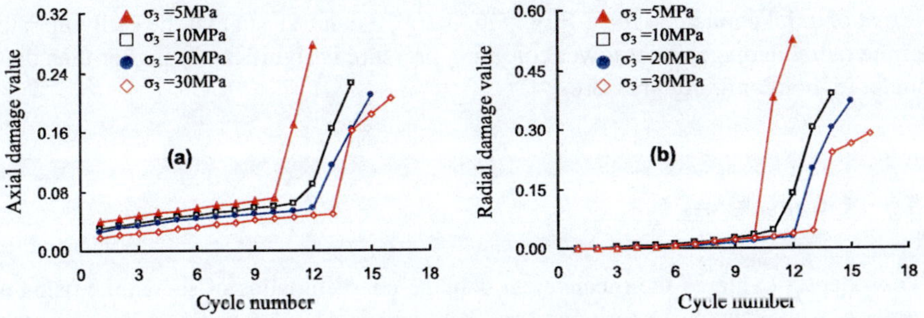

Fig. 3.26 Influence of confining pressure on axial and radial damage evolution of sandstone. **a** Relation between axial damage value and cycle number; **b** Relation between radial damage value and cycle number

$$D_{axial} = \frac{\left(\varepsilon_1^p\right)_N}{\sum_{N=1}^{m}\left(\varepsilon_1^p\right)_N} \tag{3.1}$$

$$D_{radial} = \frac{\left(\varepsilon_3^p\right)_N}{\sum_{N=1}^{m}\left(\varepsilon_3^p\right)_N} \tag{3.2}$$

where, D_{axial} and D_{radial} represent the axial and radial damage variable of rock respectively. ε_1^p and ε_3^p, means the irreversible deformation of rock and N indicates cycle number. The letter m represents the maximum cycle number.

Using Eqs. (3.1) and (3.2), the axial and radial damage value of sandstone specimen for each loading–unloading cycle can be confirmed in accordance with the deviatoric stress–strain curves shown in Fig. 3.6. Figure 3.26 depicts the comparison between the axial and radial damage evolution of sandstone specimen under confining pressure. From Fig. 3.26, it can be seen that at the same confining pressure, the axial damage value is higher than the radial damage value before the peak strength; whereas the radial damage value is higher than the axial damage value after the peak strength. Furthermore, in the early cycles, the evolving rate of the radial damage is lower than that of the axial damage, but in the later cycles, the evolving rate of the radial damage is significantly higher than that of the axial damage, which indicates that the evolution of the mechanical parameters of sandstone specimen results mainly from the radial deformation damage.

It should be noted that the confining pressure has also a great effect on the axial and radial damage evolution of sandstone specimen. With the increase of cycle number, the axial and radial damage value of sandstone specimens all first increase slightly before the peak strength and then increase dramatically after the peak strength. From Fig. 3.26a, we can see that the sandstone specimen with higher confining pressure has smaller damage value, which indicates that the confining pressure can reduce the extent of axial damage to rock. From Fig. 3.26b, it can be seen that the evolving rate of the radial damage under lower confining pressure is significantly higher than that under higher confining pressure.

3.5 Conclusions

This chapter explores the mechanical damage characteristics of sandstone using a series of triaxial cyclic experiments and X-ray micro CT observations. Based on the experimental results, the following conclusions can be drawn:

(1) At lower confining pressures, the triaxial strength and elastic modulus of sandstone specimen under cyclic loading are all higher than those under monotonic loading. However, after confining pressure is increased to 20 MPa, the triaxial strength and elastic modulus of sandstone under cyclic loading are approximately equal to those under monotonic loading. The evolution of Young's modulus and Poisson's ratio of sandstone specimen with cycle number can be characterized as having four stages: (a) Stage I (material strengthening): the elastic modulus increases obviously, whereas the Poisson's ratio firstly decreases and then increases slightly; (b) Stage II (material degradation): significant increase in Poisson's ratio and decrease in the elastic modulus are observed; (c) Stage III (material failure): the elastic modulus of sandstone specimens decreases and the Poisson's ratio increases; (d) Stage IV (structure slippage): there is no obvious variance in both the elastic modulus and Poisson's ratio of sandstone specimens.

(2) With the increase of cycle number, the crack damage threshold of sandstone specimen firstly increases, and then significantly decreases, and finally remains constant. Furthermore, the above variance does not depend on the confining pressure. At the same cycle number, the sandstone specimen with higher confining pressure has a larger crack damage threshold, which indicates that the sandstone is more difficult to switch from compaction-dominated to dilatancy-dominated under higher confining pressure. The axial damage value of sandstone is higher than the radial damage value before the peak strength; whereas the radial damage value is higher than the axial damage value after the peak strength. Furthermore, in the early cycles, the evolving rate of the radial damage is lower than that of axial damage, but in the later cycles, the evolving rate of the radial damage is significantly higher than that of axial damage, which indicates that the evolution of mechanical parameters of sandstone specimens results mainly from radial deformation damage.

(3) After triaxial monotonic and cyclic compression failure ($\sigma_3 = 5$ MPa), the sandstone specimens were analyzed using an X-ray micro CT scanning system. Firstly, the surface X-ray CT scanning surface images of sandstone specimens were in good agreement with actual surface crack photographs, which demonstrates that X-ray micro CT scanning can be used to explore internal damage to

rock material. Based on the horizontal and vertical cross-sections of sandstone specimens, the internal damage in sandstone specimen after triaxial monotonic and cyclic compression failure was evaluated quantitatively by analyzing the crack area and aperture extent for each horizontal cross-section. Generally, the system of crack planes under triaxial cyclic loading is much more complicated than that under triaxial monotonic loading. More axial and lateral tensile cracks in the specimens under cyclic loading were observed than under monotonic loading. The system of cracks is also in good agreement with the evolution of crack area and aperture extent with the height.

References

Bagde MN, Petros V (2005) The effect of machine behavior and mechanical properties of intact sandstone under static and dynamic uniaxial cyclic loading. Rock Mech Rock Eng 38(1):59–67

Bieniawski ZT (1971) Deformation behavior of fractured rock under multiaxial compression. In: Te'eni M (ed) Structure, solid mechanics and engineering design. In: Proceedings of southampton 1969 civil engineering materials conference. Wiley, Lodon, pp 589–598

Fairhurst CE, Hudston JA (1999) Draft ISRM suggested method for the complete stress-strain curve for the intact rock in uniaxial compression. Int J Rock Mech Min Sci 36(3):279–289

Fuenkajorn K, Phueakphum D (2010) Effects of cyclic loading on mechanical properties of Maha Sarakham salt. Eng Geol 112:43–52

Gatelier N, Pellet F, Loret B (2002) Mechanical damage of an anisotropic porous rock in cyclic triaxial tests. Int J Rock Mech Min Sci 39:335–354

Hawkes I, Mellor M (1970) Uniaxial testing in rock mechanics laboratories. Eng Geol 4:177–285

Heap MJ, Faulkner DR (2008) Quantifying the evolution of static elastic properties as crystalline rock approaches failure. Int J Rock Mech Min Sci 45:564–573

Heap MJ, Baud P, Meredith PG, Bell AF, Main IG (2009) Time-dependent brittle creep in Darley Dale sandstone. J Geophys Res. https://doi.org/10.1029/2008JB006212

Heap MJ, Vinciguerra S, Meredith PG (2009) The evolution of elastic moduli with increasing crack damage during cyclic stressing of a basalt from Mt. Etna Volcano Tectonophys 47:153–160

Liu EL, He SM, Xue XH, Xu J (2011) Dynamic properties of intact rock samples subjected to cyclic loading under confining pressure conditions. Rock Mech Rock Eng 44:619–634

Liu EL, He SM (2012) Effects of cyclic dynamic loading on the mechanical properties of intact rock samples under confining pressure conditions. Eng Geol 125:81–91

Ma LJ, Liu XY, Wang MY, Xu HF, Hua RP, Fan PX, Jiang SR, Wang GA, Yi QK (2013) Experimental investigation of the mechanical properties of rock salt under triaxial cyclic loading. Int J Rock Mech Min Sci 62:34–41

Mogi K (1966) Some precise measurements of fracture strength of rocks under uniform compressive stress. Rock Mech Eng Geol 4:41–55

Rao MVMS, Ramana YV (1992) A study of progressive failure of rock under cyclic loading by ultrasonic and AE monitoring techniques. Rock Mech Rock Eng 25(4):237–251

Song HP, Zhang H, Kang YL, Huang GY, Fu DH, Qu CY (2013) Damage evolution study of sandstone by cyclic uniaxial test and digital image correlation. Tectonophysics 608:1343–1348

Trippetta F, Collettini C, Meredith PG, Vinciguerra S (2013) Evolution of the elastic moduli of seismogenic Triassic Evaporites subjected to cyclic stressing. Tectonophysics 592:67–79

Xu J, Tang XJ, Li SC, Tao YQ, Jiang YD (2009) Space-time evolution rules study on acoustic emission location in rock under cyclic loading. Front Archit Civ Eng China 3(4):422–427

Xu T, Tang CA, Zhao J, Li LC, Heap MJ (2012) Modelling the time-dependent rheological behavior of heterogeneous brittle rocks. Geophys J Int 189(3):1781–1796

Wang ZC, Li SC, Qiao LP, Zhao JG (2013) Fatigue behavior of granite subjected to cyclic loading under triaxial compression condition. Rock Mech Rock Eng 46:1603–1615

Wong TF, David C, Zhu W (1997) The transition from brittle faulting to cataclastic flow in porous sandstones: mechanical deformation. J Geophys Res 102(B2):3009–3025

Yang SQ, Jiang YZ, Xu WY, Chen XQ (2008) Experimental investigation on strength and failure behavior of pre-cracked marble under conventional triaxial compression. Int J Solids Struct 45:4796–4819

Yang SQ, Jing HW, Li YS, Han LJ (2011) Experimental investigation on mechanical behavior of coarse marble under six different loading paths. Exp Mech 51:315–334

Yang SQ, Jing HW, Wang SY (2012) Experimental investigation on the strength, deformability, failure behavior and acoustic emission locations of red sandstone under triaxial compression. Rock Mech Rock Eng 45:583–606

Yang SQ, Jing HW (2013) Evaluation on strength and deformation behavior of red sandstone under simple and complex loading paths. Eng Geol 164:1–17

Yang SQ, Ranjith PG, Huang YH, Yin PF, Jing HW, Gui YL, Yu QL (2015) Experimental investigation on mechanical damage characteristics of sandstone under triaxial cyclic loading. Geophys J Int 201(2):662–682

Zhu WC, Li ZH, Zhu L, Tang CA (2010) Numerical simulation on rockburst of underground opening triggered by dynamic disturbance. Tunn Undergr Space Technol 25(5):587–599

Chapter 4
Strength and Deformation Behavior of Red Sandstone Under Simple and Complex Loading Paths

Real engineering rock mass experienced by loading path is more complex, which makes that mechanical behavior of rock under complex loading paths has been more and more significant. The re-fracture strength and deformation behavior of fractured rock mass are not clear under complex loading paths; therefore it is necessary to explore the strength and deformation failure behavior of rock under complex loading path, which is helpful to controlling the unstable failure of underground rock engineering structures.

Therefore in this chapter, in order to investigate the influence of complex loading path on strength and deformation behavior of red sandstone, a series of triaxial compression experiments were carried out under simple and complex loading paths by the rock mechanics servo-controlled testing system (Yang and Jing 2013). The emphasis of this chapter is focused on the investigation of the effect of confining pressure on strength and deformation parameters of red sandstone material under simple loading path, to analyze the strength parameters of red sandstone under two complex loading paths, to compare the strength and deformation difference of red sandstone between simple and complex loading paths, and to put forward a kind of new revised method to predict peak strength of red sandstone under simple loading path using obtained triaxial strength under complex loading path with only one specimen, which is very significant to deep underground structural engineering science.

4.1 Tested Rock Material and Testing Procedure

4.1.1 Red Sandstone Material and Specimen Preparation

The material used throughout this research was red sandstone, which was collected from Shandong province of China. The mineral components of tested red sandstone material were feldspar, quartz, debris and agglutinate according to optical microscopy and SEM results. The mass content of agglutinate in red sandstone material consisted

S.-Q. Yang, *Mechanical Behavior and Damage Fracture Mechanism of Deep Rocks*,
https://doi.org/10.1007/978-981-16-7739-7_4

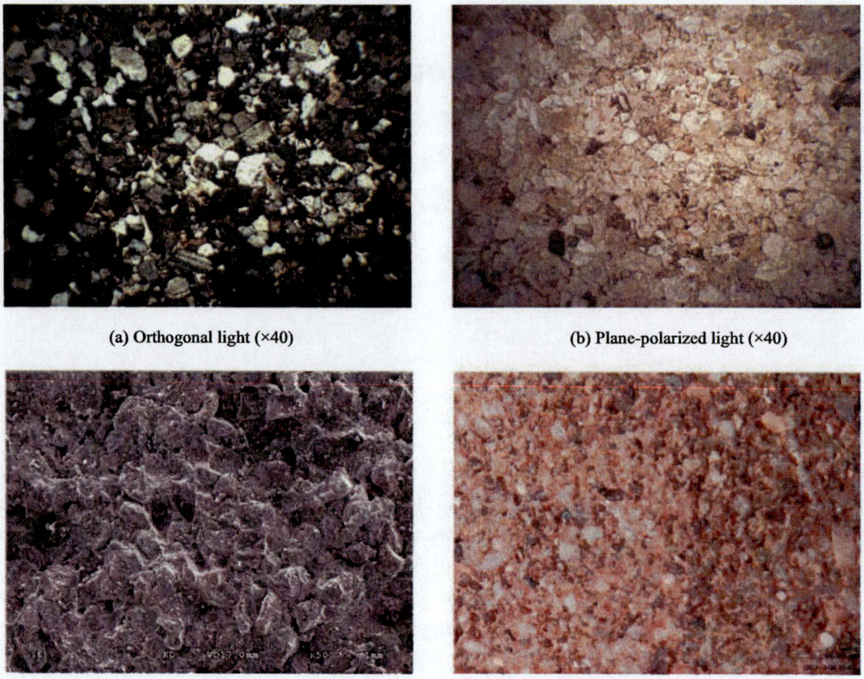

(a) Orthogonal light (×40) (b) Plane-polarized light (×40)

(c) SEM (×50) (d) Microscope result (×50)

Fig. 4.1 Microscopic structure of red sandstone in the present study

of gypsum, barite and iron integument. Red sandstone (Fig. 4.1) is a fine- to medium-grained, feldspathic sandstone with a connected porosity of 8.9%.

Table 4.1 listed tested red sandstone specimens and conditions in this research. All triaxial experiments were carried out on cylindrical specimens with 55 mm in diameter and 110 mm in length. The specimens were all cored from the same large rectangular block and in the same orientation. The experiments were all carried out on dry specimens at room temperature. The average dry density of tested red sandstone was approximately 2375 kg/m^3.

4.1.2 Testing Equipment

All compression experiments for red sandstone specimens were carried out on an MTS815 rock servo- controlled system with the maximum loading capacity of 2000 KN and the maximum displacement capacity of 25 mm. This servo-controlled system can test specimens in load or displacement control while the data from LVDT (Linear Variable Displacement Transducer) are recorded and analyzed in real-time.

Table 4.1 Tested red sandstone specimens and conditions in this research

Specimen	D·/mm	L·/mm	M·/g	ρ/(kg/m^3)	σ_3/MPa	Testing design
RS-J13#	55.4	109.2	632.5	2403	5	Simple loading path
RS-J4#	55.4	107.2	605.7	2344	5	Simple loading path
RS-J1#	55.4	110.0	636.3	2400	5	Simple loading path
RS-J2#	55.4	109.4	622.3	2360	10	Simple loading path
RS-J10#	55.4	105.7	593.7	2330	15	Simple loading path
RS-J21#	55.4	108.7	609.1	2324	20	Simple loading path
RS-J11#	55.4	108.5	612.0	2340	25	Simple loading path
RS-J12#	55.4	110.1	629.2	2371	35	Simple loading path
RS-J19#	55.4	104.4	631.7	2510	35	Simple loading path
RS-J5#	55.4	108.8	623.1	2376	35	Simple loading path
RS-J6#	55.4	109.6	626.1	2370	5 → … → 15	Complex loading path A
RS-J8#	55.4	109.8	630.8	2383	5 → … → 15	Complex loading path A
RS-J3#	55.4	109.5	618.8	2344	5 → … → 35	Complex loading path A
RS-J17#	55.4	108.1	609.2	2338	5 → … → 35	Complex loading path A
RS-J15#	55.4	107.2	614.6	2378	35 → … → 5	Complex loading path B
RS-J16#	55.4	106.1	604.7	2364	35 → … → 5	Complex loading path B

Note *D* diameter, *L* length, *M* mass, ρ density, σ_3 confining pressure

In this research, we measured the axial force with the loading capacity of 600 KN and the axial deformation with the displacement capacity of 5 mm. However, the circumferential deformation was measured by a LVDT attached to a chain wrapped tightly around the specimen, which the maximum displacement capacity can reach to 8 mm. During experiments, loads and deformations of tested red sandstone specimens were recorded simultaneously. Moreover, two rigid steel cylinders (55 mm in diameter and 20 mm in length) were placed between the loading frame and rock specimen, which decreased distinctly the influence of the end friction effects on the testing results of red sandstone specimens with the length to diameter ratio ~2.0.

4.1.3 Designed Loading Paths

One simple loading path and two complex loading paths, i.e. complex loading path A and B, are carried out for red sandstone specimen. Simple loading path (Fig. 4.2)

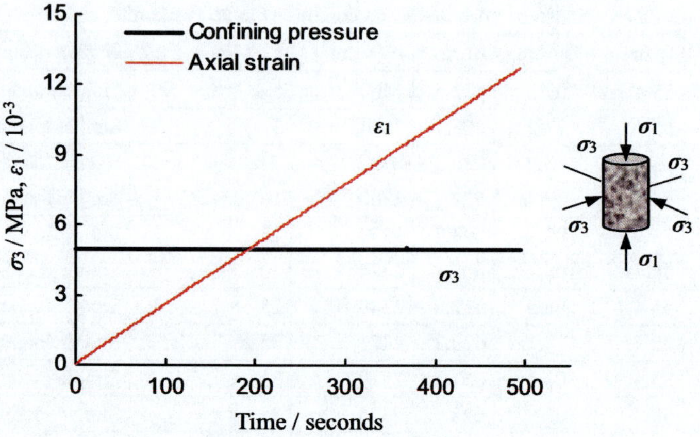

Fig. 4.2 Simple loading path designed in the present study

(i.e. conventional triaxial compression experiment) consisted of the following two loading steps. The confining pressure was firstly applied to a desired value at a constant rate of 0.1 MPa/s, which ensured the specimen is under uniform hydrostatic stresses. And then the deviatoric stress was loaded to the surface of specimen at a constant axial displacement control rate of 0.0028 mm/s (the axial strain rate is approximately 2.55×10^{-5}/s) until the failure took place. Six confining pressures of 5, 10, 15, 20, 25 and 35 MPa were used under simple loading path.

Complex loading path A (Fig. 4.3a) was carried out by the following steps. Firstly, the hydrostatic pressure was applied to the specimen at a rate of 0.1 MPa/s, and then the 1st confining pressure was kept as 5 MPa. The deviatoric stress was loaded at a constant axial strain control rate of 2.55×10^{-5}/s to post-peak deformation value at the 1st confining pressure. Secondly, the axial deviatoric stress was unloaded to zero at an axial strain rate of 0.014 mm/s. And then, the deviatoric stress was

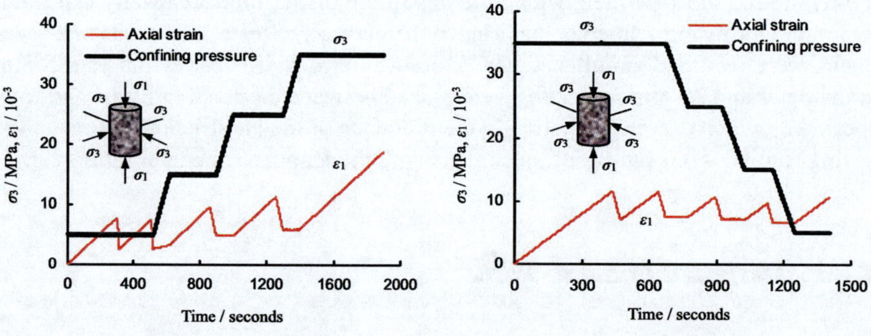

(a) Complex loading path A (b) Complex loading path B

Fig. 4.3 Complex loading paths designed in the present study reloaded to the specimen at a constant axial strain control rate of 2.55×10^{-5}/s until the specimen continued into the stage of post failure and the deviatoric stress dropped by approximately 2% from its peak. Afterwards, the deviatoric stress was unloaded to zero at an axial strain rate of 0.014 mm/s and the confining pressure was hydrostatically increased to the 2nd level, i.e. 15 MPa. At the 2nd confining pressure of 15 MPa, the deviatoric stress was increased to the specimen until the specimen dropped to 98% peak stress after peak strength. After that, we repeated above steps at the 3rd confining pressure ($\sigma_3 = 25$ MPa). At the 4th confining pressure of 35 MPa, the deviatoric stress was loaded to the surface of specimen at a constant axial strain control rate of 2.55×10^{-5}/s until the failure took place.

Compared with complex loading path A, complex loading path B (Fig. 4.3b) is only different from the variance of confining pressure. Under complex loading path A, the red sandstone specimen was tested by increasing the confining pressure from 5 to 35 MPa at an interval of 10 MPa. However, under complex loading path B, the red sandstone specimen was tested by reducing the confining pressure from 35 to 5 MPa at an interval of 10 MPa. Note two complex loading paths are designed to investigate the re-fractured mechanical behavior of pre-cracked red sandstone material with different post-peak stress drops. Moreover, under complex loading path A and B, for each confining pressures, 2% axial stress drop after peak strength can be controlled automatically in MTS815 experimental loading system. In accordance with above steps, we can obtain four peak strengths with only one specimen for each complex loading path.

4.2 Strength and Deformation Behavior of Red Sandstone Under Simple Loading Path

4.2.1 Deformation Behavior of Red Sandstone Under Simple Loading Path

The triaxial deviatoric stress–strain curves for red sandstone under simple loading path are presented in Fig. 4.4. From Fig. 4.4, it can be seen that the post-peak behavior of red sandstone appears brittle even at the highest confining pressure of 35 MPa. In accordance with triaxial experimental results of red sandstone under simple loading path, the influence of confining pressure on strength and deformation behavior of red sandstone can be described in more detail in the following sections.

In accordance with Fig. 4.4, it can be seen that the stress–strain curves at the lower confining pressures show initially a non-linear (concave) deformation, which usually results from the closure of pre-existing microcracks. Moreover, this stage of microcrack closure depends on the confining pressure. With the increase of confining pressure, the stage of microcrack closure becomes less and less distinct. This is likely to be a result o the confining pressure acting to close the pre-existing microcracks prior to the start of deformation (Yang et al. 2012). After the stage of microcrack

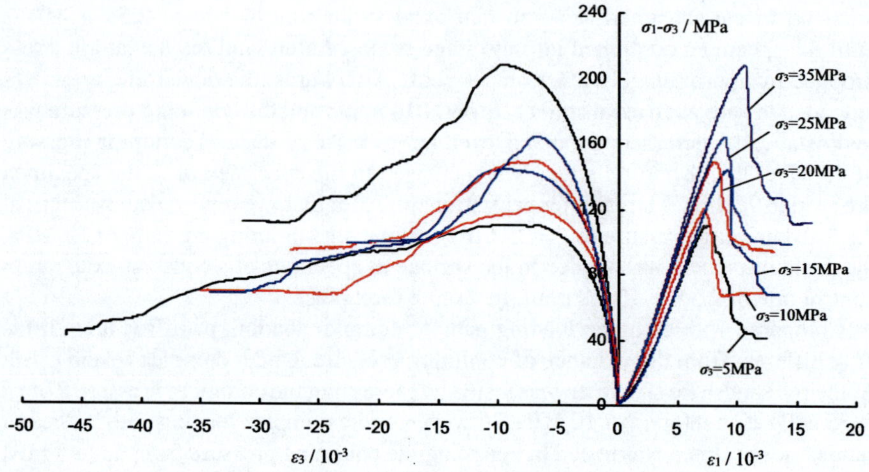

Fig. 4.4 Triaxial experimental results of red sandstone under simple loading path

closure, the stage of elastic deformation dominates the linear portions of the stress–strain curves. With continuous increase of axial deformation, the stress–strain curves begin to depart from the linear behavior, which marks the yielding of the specimens. At this time, the specimens enter a phase of strain hardening. However, when the peak strength is reached, there is a stage of strain softening. In the end, the specimens enter the stage of residual strength.

Figure 4.5 illustrates the effect of specimen variability on triaxial stress–strain curves of red sandstone. From Fig. 4.5, it can be seen that specimen variability has an inappreciable influence on the deformation behavior before the peak strength at $\sigma_3 = 5$ or 35 MPa. Moreover, after the peak strength, red sandstone specimen has similar strain softening characteristics, which is not closely related to specimen

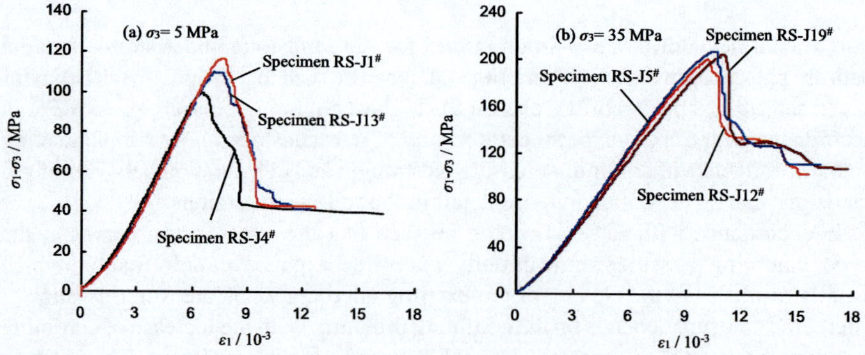

Fig. 4.5 Effect of specimen variability on triaxial stress–strain curves of red sandstone in this research

variability. Heterogeneity in rock material can also affect the peak strength behaviour even though at higher confining pressure (e. g. $\sigma_3 = 35$ MPa). The difference of peak strength of different specimens at lower confining pressures is always larger than that at the higher confining pressures due to different stress states available in the rock material because of grain shape and size heterogeneity. However, in a word, the discreteness of tested red sandstone specimens is very small; therefore we can analyze the relationship between the mechanical parameters of red sandstone and the confining pressure.

In order to analyze the influence of confining pressure on the elastic modulus (E_S) of red sandstone, we have used the average slope of the more-or-less straight line portion of the axial stress–axial strain curve in the range of axial differential stress from 30 to 70% TCS (Triaxial compressive strength) to calculate the E_S, as shown in Table 4.2. Based on the data of the E_S listed in Table 4.2, we obtain the influence of confining pressure on elastic modulus of red sandstone specimen in the range of tested confining pressure from 5 to 35 MPa, as illustrated in Fig. 4.6. Note, in Fig. 4.6, the average value of red sandstone at $\sigma_3 = 5$ and 35 MPa are used. From Fig. 4.6, it can be seen that elastic modulus of red sandstone increases nonlinearly with increasing confining pressure. At $\sigma_3 = 5$ MPa, the elastic modulus is 18.89 GPa. Whereas when σ_3 is increased to 25 MPa, the elastic modulus is increased to 22.13 GPa. However, when σ_3 is increased from 25 to 35 MPa, the elastic modulus of red sandstone keeps basically a constant, which only increases from 22.13 GPa to 22.94 GPa.

Table 4.2 Strength and deformation values of red sandstone specimen under simple loading path

Specimen	σ_3/MPa	E_S/GPa	σ_{cr}/MPa	σ_{sr}/MPa	σ_{cd}/MPa	σ_{sd}/MPa	σ_p/MPa	σ_S/MPa
RS-J13#	5	18.81	41.5	46.5	38.79	43.79	110.1	115.1
RS-J4#	5	18.88	42.8	47.8	41.33	46.33	116.8	121.8
RS-J1#	5	18.97	39.1	44.1	41.64	46.64	99.8	104.8
RS-J2#	10	20.04	69.1	79.1	50.68	60.68	120.5	130.5
RS-J10#	15	20.40	67.6	82.6	59.28	74.28	143.8	158.8
RS-J21#	20	21.64	94.7	114.7	51.99	71.99	149.1	169.1
RS-J11#	25	22.13	98.6	123.6	57.53	82.53	164.1	189.1
RS-J12#	35	23.14	103.3	138.3	69.48	104.48	201.1	236.1
RS-J19#	35	21.86	108.0	143.0	81.62	116.62	205.0	240.0
RS-J5#	35	23.81	111.8	146.8	86.97	121.97	207.9	242.9

Note $\sigma_{sr} = \sigma_{cr} + \sigma_3$; $\sigma_{sd} = \sigma_{cd} + \sigma_3$; $\sigma_S = \sigma_p + \sigma_3$

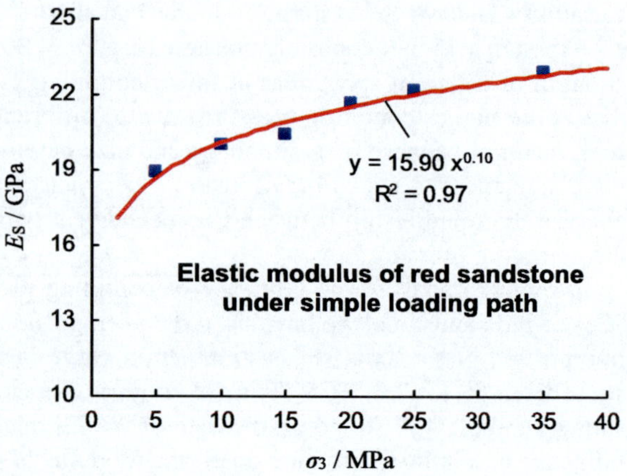

Fig. 4.6 Effect of confining pressure on elastic modulus of red sandstone under simple loading path

4.2.2 *Strength Behavior of Red Sandstone Under Simple Loading Path*

As it is well-known to all, Mohr–Coulomb criterion is a common strength criterion, which have been applied widely to rock mechanics and engineering practice. Hence, in this research, Mohr–Coulomb criterion is selected to analyze the relation between triaxial strength of red sandstone and the confining pressure. The linear Mohr–Coulomb criterion can be expressed with the following equation, i.e. Equation (4.1).

$$\sigma_S = \sigma_0 + q\sigma_3 = \frac{2C\cos\phi + \sigma_3(1+\sin\phi)}{1-\sin\phi} \tag{4.1}$$

where σ_0 is usually regarded as the uniaxial compressive strength (UCS) of rock material. The σ_0 and q are related to the cohesion C and the internal friction angle Φ of rock material.

Table 4.2 lists the strength values of red sandstone in the present study. In Table 4.2, σ_{cr} and σ_p are respectively the residual strength and peak strength, which are confirmed in accordance with the axial deviatoric stress-axial strain curves of red sandstone specimen as shown in Fig. 4.7a. But σ_{cd} is the critical damage threshold of the specimen (Wong et al. 1997; Fairhurst and Hudson 1999; Heap et al. 2009), i.e. the corresponding axial deviatoric stress where the volumetric deformation of the specimen switches from compaction-dominated to dilatancy-dominated, as shown in Fig. 4.7b.

Based on peak and residual strength values of red sandstone under simple loading path listed in Table 4.2, by adopting linear Mohr–Coulomb criterion (Eq. (4.1)), the influence of σ_3 on the peak and residual strength for red sandstone under simple loading path can be presented in Fig. 4.8. From Fig. 4.8, it can be seen that the

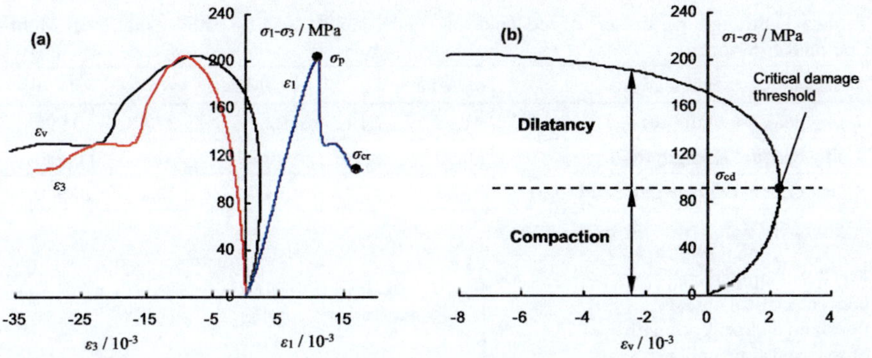

Fig. 4.7 Confirm on peak strength, residual strength, and critical damage threshold of red sandstone. ε_1 and ε_3 represents the axial and circumferential strain, respectively. ε_v refers to the volumetric strain, which is calculated from the sum of the axial strain and twice the circumferential strain (i.e. $\varepsilon_v = \varepsilon_1 + 2\varepsilon_3$)

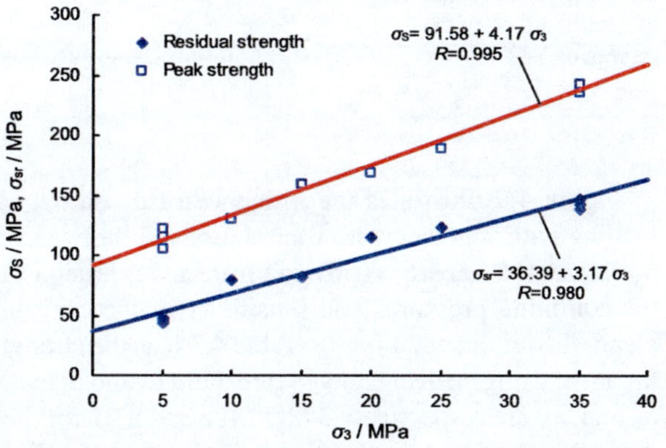

Fig. 4.8 Comparison between peak and residual strength of red sandstone under simple loading path. All the symbols represent the experimental results and the line represents the theoretical value in accordance with Mohr–Coulomb criterion. (*R* is the correlation coefficient of determination of the regression.)

peak and residual strength of red sandstone under simple loading path shows distinct linear behavior in the range of σ_3 from 5 to 35 MPa, which indicates that the peak and residual strength behavior of red sandstone are in a good agreement with the linear Mohr–Coulomb criterion.

In accordance with Fig. 4.8, the values of C and Φ of red sandstone can be calculated by the linear Mohr-Coulomb criterion, as listed in Table 4.3. For the peak strength, the values of C and Φ of red sandstone are respectively 22.42 MPa and 37.8°. But for the residual strength, the values of C and Φ of red sandstone are respectively 10.22 MPa and 31.4°. Therefore, the peak cohesion of red sandstone is approximately 54.4% higher than the residual cohesion. However, the peak internal friction angle of red sandstone is only 16.9% higher than the residual internal friction angle.

Table 4.3 Strength parameters of red sandstone under simple loading path by the linear Mohr–Coulomb criterion

Different regression analysis	σ_0/MPa	q	C/MPa	Φ/(°)	R
Using peak strength (σ_S)	91.58	4.17	22.42	37.8	0.995
Using residual strength (σ_{sr})	36.39	3.17	10.22	31.4	0.980
Using critical damage threshold (σ_{sd})	34.87	2.21	11.73	22.1	0.980

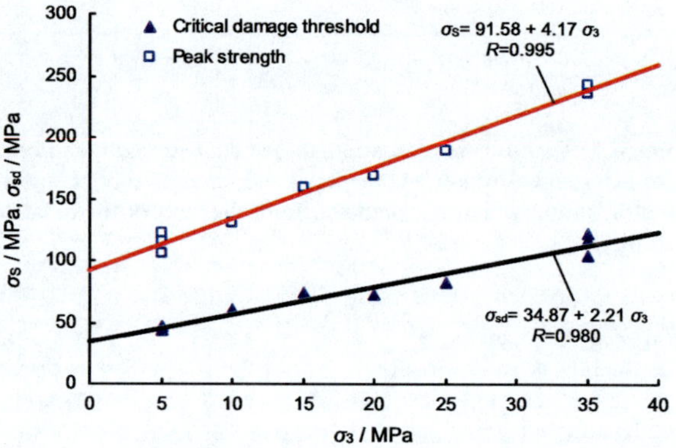

Fig. 4.9 Comparison between critical damage threshold and peak strength of red sandstone under simple loading path. All the symbols represent the experimental results and the line represents the theoretical value in accordance with Mohr–Coulomb criterion. (*R* is the correlation coefficient of determination of the regression.)

Figure 4.9 illustrates the influence of σ_3 on σ_{sd} of red sandstone under simple loading path together with the relationship between σ_S and σ_3. It can be seen that σ_{sd} increased linearly (with good linear regression coefficient of $R = 0.980$) with the confining pressure. The sensitivity of the critical damage threshold on σ_3 was clearly lower than that for σ_S. Table 4.3 lists the strength parameters of red sandstone obtained using critical damage threshold by the linear Mohr–Coulomb criterion. The σ_0 and q were respectively 34.87 MPa and 2.21 for the regression analysis using σ_{sd}. According to Eq. (4.1), the values of C and Φ of red sandstone were calculated using the regression analysis for the critical damage threshold to be 11.73 MPa and 22.1°. These values were lower than the values of C and Φ calculated using the regression analysis for σ_S. However, the value of C using the regression analysis for σ_{sd} was approximately equal to that for σ_{sr}, but the value of Φ using the regression analysis for σ_{sd} was distinctly lower than that for σ_{sr}.

4.2.3 Failure Mode of Red Sandstone Under Simple Loading Path

Figure 4.10 shows the failure mode of red sandstone under simple loading path. From Fig. 4.10, it is very clear that with the increase of confining pressure, the failure mode of red sandstone under simple loading path changes from single tension fracture to

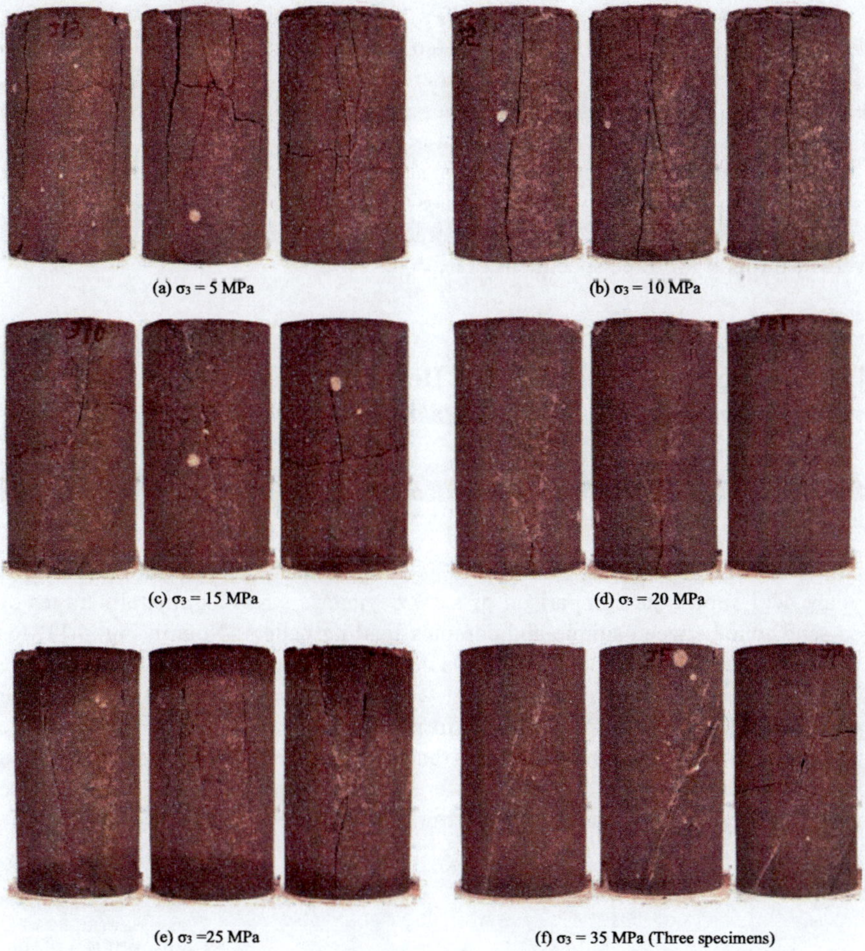

Fig. 4.10 Failure modes of red sandstone under simple loading path

mixed tension and shear fracture, and then to single shear fracture, which has been validated according to macroscopic and microscopic observation for tested confining pressure from 5 to 35 MPa.

At the confining pressure of 5 MPa, our sandstone specimens demonstrated a single tension fracture mode of failure (Fig. 4.10a). More axial and lateral tensile cracks were observed at $\sigma_3 = 5$ MPa, which also indicated that it was easier for tension failure to occur at $\sigma_3 = 5$ MPa. At $\sigma_3 = 10$ MPa, the red sandstone specimen took on a typical axial splitting tension failure with some minor lateral tensile cracks (Fig. 4.10b). At $\sigma_3 = 15$ MPa, the shear fracture modes were observed with a steep angle of 78° (Fig. 4.10c); at the same time, some axial and lateral tensile cracks were also found at this confining pressure.

When σ_3 was increased to 20–25 MPa, the shear failure mode with multiple shear fracture planes were found for tested red sandstone specimens (Fig. 4.10d-e); moreover some axial splitting tension cracks were also observed, but no lateral tension cracks were found. The above description also indicated that it was more difficult for lateral tension cracks to occur at $\sigma_3 = 20–25$ MPa.

However, when σ_3 reached to 35 MPa, single shear fracture dominated the final failure mode of red sandstone and the shear fracture angle was approximately 74° (Fig. 4.10f). Except for the single shear fracture, some local lateral tension cracks were also found for three specimens at $\sigma_3 = 35$ MPa.

4.3 Strength and Deformation Behavior of Red Sandstone Under Complex Loading Path A

4.3.1 Deformation Behavior of Red Sandstone Under Complex Loading Path A

Figure 4.11 shows the comparison of axial deviatoric stress-axial strain curves of red sandstone between simple and complex loading path A. Note in Fig. 4.11, for complex loading path A, only two levels of confining pressures, i.e. 5 and 15 MPa were applied to the specimen. From Fig. 4.11, it could be seen that at the confining pressure of 5 MPa, peak strength and deformation behaviors of two specimens were all approximately equal, which indicated that tested specimens used under simple and complex loading path had almost no difference. Thus, we can analyse the influence of two different loading paths on peak strength parameters.

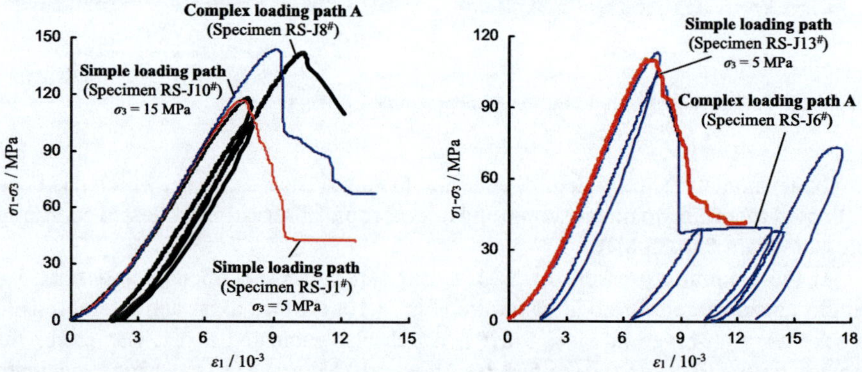

Fig. 4.11 Comparison of axial deviatoric stress-axial strain curves of red sandstone between simple and complex loading path A

Under complex loading path A, specimen RS-J8# was firstly unloaded at $\sigma_3 =$ 5 MPa when the deviatoric stress dropped by approximately 5% from its peak. After the 1st confining pressure, specimen RS-J8# was loaded at a constant axial strain control rate of 2.55×10^{-5}/s until the specimen continued into the stage of post failure. At this time, peak strength of specimen RS-J8# under complex loading path A was 157.3 MPa at $\sigma_3 = 15$ MPa, which was approximately equal to that of specimen RS-J10# (158.8 MPa) under simple loading path. The above descriptions reveal that complex loading path A is feasible and reasonable, which can be used to confirm the peak strength of rock with only one specimen. However, under complex loading path A, specimen RS-J6# was applied three cyclic loading–unloading after peak strength at $\sigma_3 = 5$ MPa. From three cyclic loading–unloading stress–strain curves, we could see that red sandstone had a distinct plastic deformation after unloading. Even though when specimen RS-J6# came into the stage of residual strength, the peak strength of the specimen still could be increased to 88.6 MPa when the confining pressure was increased to 15 MPa.

To analyse the influence of fractured extent on triaxial strength and deformation behavior of red sandstone under complex loading path A, two specimens in Fig. 4.12 are loaded to different post-peak deformation positions (the ratio of the axial deviatoric stress at the unloading point to TCS is different) at $\sigma_3 = 5$ MPa, and then the axial deviatoric stress is unloaded to zero. Specimen RS-J17# dropped to 97.24 MPa after peak stress, and the ratio of the axial deviatoric stress at the unloading point to TCS was approximately 95.7%. However, Specimen RS-J3# dropped to 59.45 MPa after peak stress, and the ratio of the axial deviatoric stress at the unloading point to TCS was approximately 60%. The ratio of the axial deviatoric stress at the unloading point to TCS was smaller, which indicated that the fractured extent of the specimen was also more obvious. Thus, triaxial experimental results of two specimens with different post-peak deformation positions under complex loading path A are also different, as shown in Fig. 4.13.

Figure 4.13 illustrates axial deviatoric stress-axial strain curves of red sandstone under complex loading path A. For comparison, the axial deviatoric stress-axial strain curves of red sandstone under simple loading path ($\sigma_3 = 5$ MPa) are also plotted in Fig. 4.13. In accordance with Fig. 4.13, the corresponding strength and deformation values of red sandstone under complex loading A are listed in Table 4.4. Based on listed data in Table 4.4, the influence of confining pressure on elastic modulus of red sandstone under complex loading path A can be investigated, as shown in Fig. 4.14. For comparison, elastic moduli of red sandstone under simple loading path are also plotted in Fig. 4.14.

From Fig. 4.14, it is very clear that elastic modulus of red sandstone under complex loading path A is lower than that under simple loading path, which results mainly from larger stress drop after peak strength (see Fig. 4.13). With the increase of confining pressure, elastic modulus of red sandstone also increases nonlinearly, but the increasing extent varies with the fractured extent of the specimen. The elastic modulus of specimen RS-J3# under complex loading path A is most sensitive to the confining pressure, and increases from 12.64 to 21.5 GPa as σ_3 increases from 5 to 35 MPa. While elastic modulus of specimen RS-J17# under complex loading path A

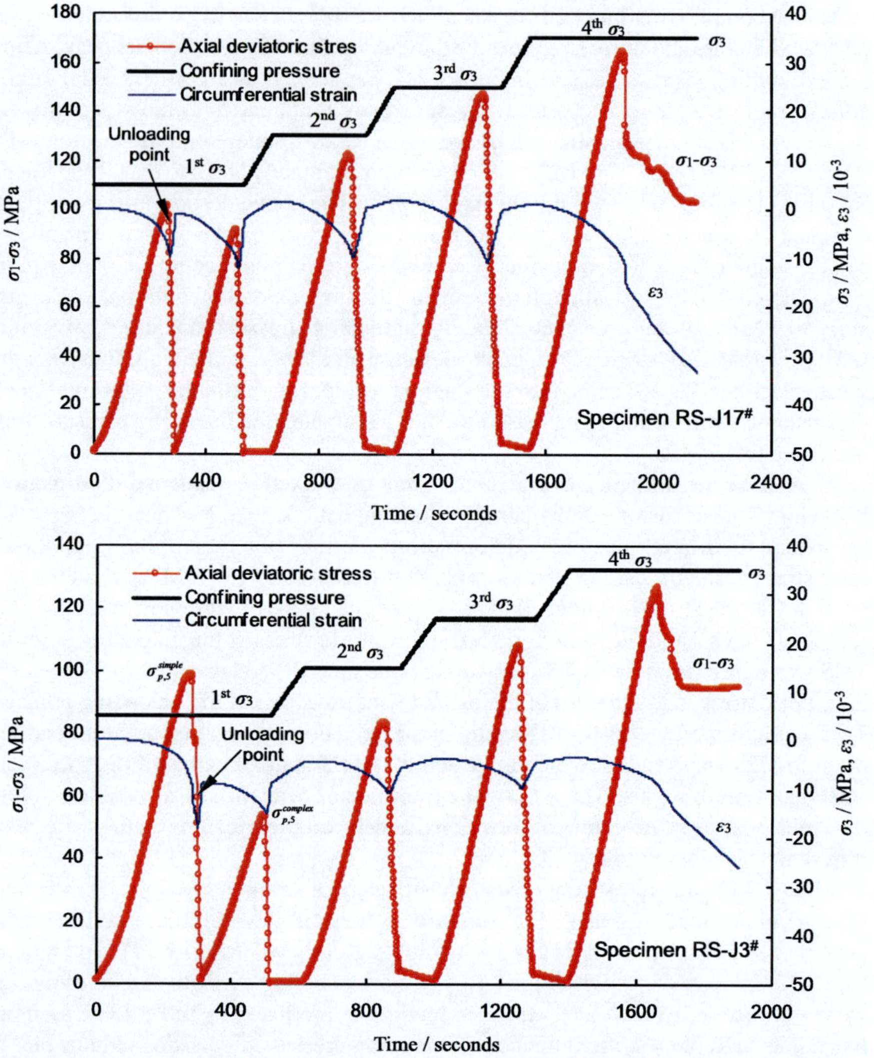

Fig. 4.12 Experimental results on change of axial deviatoric, confining pressure, circumferential strain of red sandstone with the increase of time under complex loading path A

does not depend distinctly on the confining pressure, and only increases from 18.07 to 21.45 GPa as σ_3 increases from 5 to 35 MPa. Under complex loading path A, the difference of elastic modulus between specimen RS-J17# and RS-J3# also decreases gradually with the increase of confining pressure, and at $\sigma_3 = 35$ MPa the elastic modulus of two specimens RS-J17# and RS-J3# are approximately equal. The above descriptions also indicate that elastic modulus of red sandstone at higher confining

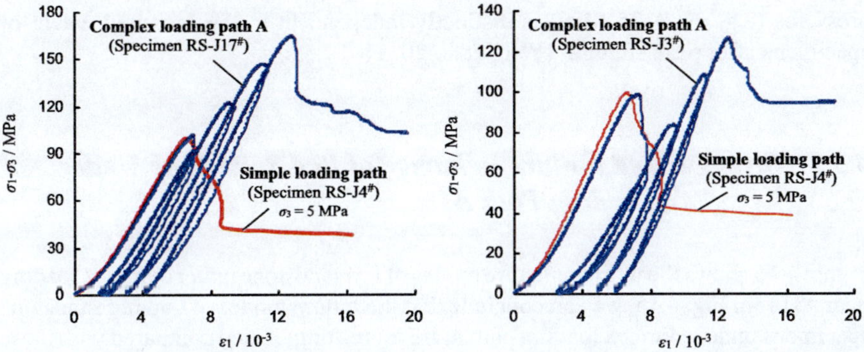

Fig. 4.13 Axial deviatoric stress-axial strain curves of red sandstone under complex loading path A

Table 4.4 Strength and deformation values of red sandstone specimen under complex loading path A

Specimen	Multi-stage	σ_3/MPa	E_S/GPa	σ_{cd}/MPa	σ_{sd}/MPa	σ_p/MPa	σ_S/MPa
RS-J17#	1st	5	18.07	16.5	21.5	92.0	97.0
	2nd	15	19.27	30.6	45.6	122.2	137.2
	3rd	25	20.73	32.7	57.7	147.0	172.0
	4th	35	21.45	39.4	74.4	164.7	199.7
RS-J3#	1st	5	12.64	16.1	21.1	54.0	59.0
	2nd	15	16.26	30.2	45.2	83.7	98.7
	3rd	25	19.42	31.2	56.2	108.4	133.4
	4th	35	21.50	37.3	72.3	126.6	161.6

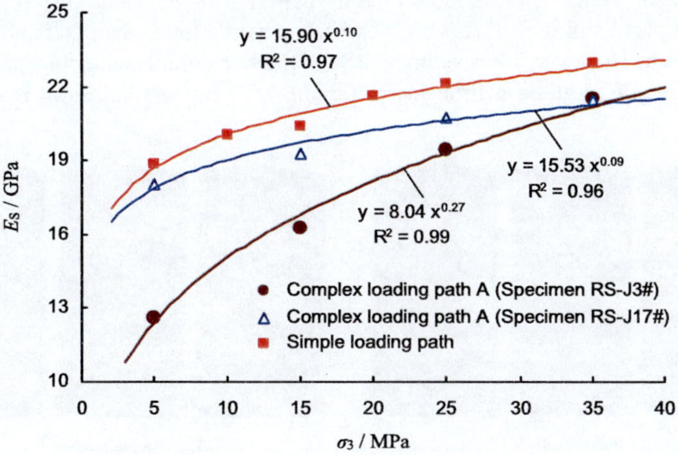

Fig. 4.14 Effect of confining pressure on elastic modulus of red sandstone under complex loading path A

pressures (e.g. $\sigma_3 = 35$ MPa) is distinctly independent to the fractured extent of specimens after peak strength (Yang et al. 2011).

4.3.2 *Strength and Failure Behavior of Red Sandstone Under Complex Loading Path A*

Figure 4.15 shows the ultimate failure modes of red sandstone under complex loading path A. From Fig. 4.15, we can conclude that the failure modes of two red sandstone specimens under complex loading path A are more complicated compared with those under simple loading path (Fig. 4.10f). Two red sandstone specimens under complex loading path A all take on typical shear failure mode with multiple shear fracture planes. At the same time, some lateral tension cracks are also observed in two red sandstone specimens.

Table 4.4 lists the peak strength values of red sandstone under complex loading path A. Based on listed data in Table 4.4, the influence of σ_3 on peak strength σ_S for red sandstone under complex loading path A can be presented in Fig. 4.16. From Fig. 4.16, it can be seen that peak strength of red sandstone under complex loading path A shows distinct linear behavior with the increasing σ_3 in the range of σ_3 from 5 to 35 MPa, which indicates that the peak strength behavior of red sandstone is in a good agreement with the linear Mohr–Coulomb criterion.

In accordance with the linear Mohr–Coulomb criterion, the peak strength parameters of red sandstone under complex loading path A are presented in Table 4.5. Note, under simple loading path, the cohesion and internal friction angle of red sandstone are respectively 22.42 MPa and 37.8°. But for specimen RS-J17#, under complex loading path A, the C value of red sandstone is 22.38 MPa, which is approximately equal to that under simple loading path, but the Φ value of red sandstone under complex loading path A is 33.3°, which is a little lower than that under simple loading path. However, for specimen RS-J3#, under complex loading path A, the internal friction angle is a little lower (about 4.5°) but the cohesion is distinctly

(a) Specimen RS-J17 # (b) Specimen RS-J3 #

Fig. 4.15 Ultimate failure modes of red sandstone under complex loading path A

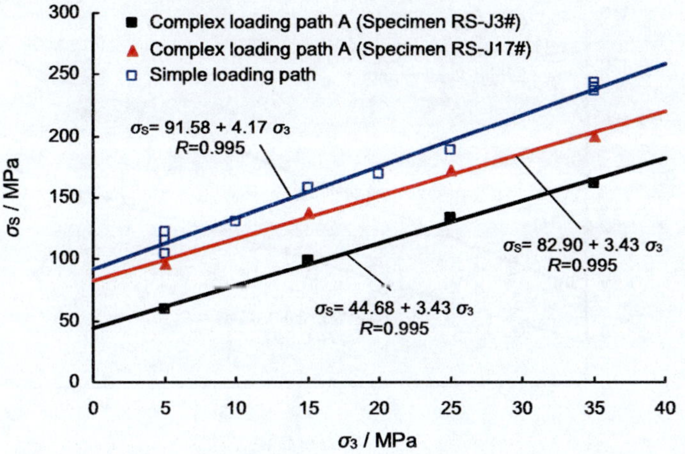

Fig. 4.16 Peak strength of red sandstone under complex loading path A. All the symbols represent the experimental results and the line represents the theoretical value in accordance with Mohr–Coulomb criterion. (*R* is the correlation coefficient of determination of the regression.)

Table 4.5 Peak strength parameters of red sandstone under complex loading path A by the linear Mohr–Coulomb criterion

Complex loading path A	σ_0/MPa	*q*	*C*/MPa	Φ/(°)	*R*
Specimen RS-J17#	82.90	3.43	22.38	33.3	0.995
Specimen RS-J3#	44.68	3.43	12.06	33.3	0.995

lower than that under simple loading path. Moreover under complex loading path A, the cohesion obtained by specimen RS-J3# is obviously lower than that by specimen RS-J17#, whereas the internal friction angle obtained by specimen RS-J3# is equal to that by specimen RS-J17#. Thus, specimens under complex loading path A produce some damage, which can be interpreted as a loss in specimen cohesion. Moreover, it should be noted that the internal friction angle of red sandstone did not depend on the fractured extent of specimens after peak stress.

Figure 4.17 depicts critical damage threshold of red sandstone under complex loading path A (See listed data in Table 4.4), which are confirmed by the volumetric strain of red sandstone under complex loading path A (the axial stress corresponded to the filled circles in Fig. 4.18 is regarded as the critical damage threshold). For comparison, critical damage threshold of red sandstone under simple loading path are also plotted in Fig. 4.17. In accordance with Fig. 4.17, it can be seen that critical damage threshold of red sandstone under complex loading path A are all lower than that under simple loading path, which indicates indirectly that the deformation parameter in Fig. 4.14 is a function of the damage occurring to the specimen and

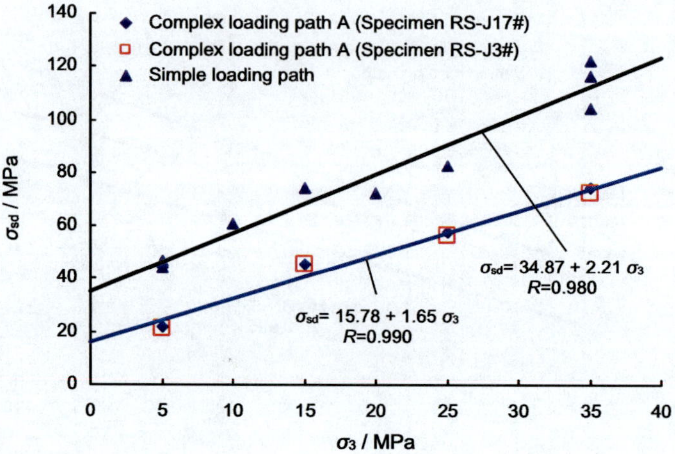

Fig. 4.17 Critical damage threshold of red sandstone under complex loading path A. All the symbols represent the experimental results and the line represents the theoretical value in accordance with Mohr–Coulomb criterion. (*R* is the correlation coefficient of determination of the regression.)

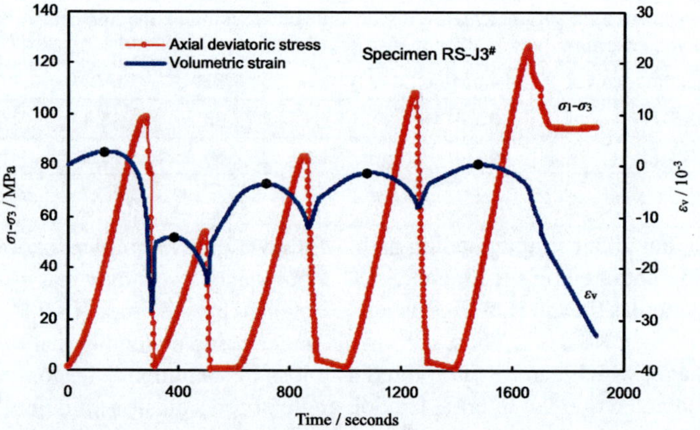

Fig. 4.18 Typical change of volumetric strain of red sandstone with the increase of time under complex loading path A. Crack damage threshold of red sandstone under complex loading path A can be confirmed by the volumetric strain curve

may reflect the creation and re-opening of fractures under repeated loading. Furthermore, critical damage threshold of red sandstone under complex loading path A also depends on the confining pressure.

From Fig. 4.17, we can conclude that critical damage threshold of red sandstone under complex loading path A increases linearly with the confining pressure. Moreover, the difference of critical damage threshold of red sandstone between simple and complex loading path A also increases gradually with the increasing σ_3, which indicates that it is easier to create and re-open of cracks under complex loading path A. However, it should be noted that critical damage threshold of red sandstone under complex loading path A is distinctly independent to the fractured extent of specimens after peak strength. Using the regression analysis for the critical damage threshold, the values of C and Φ of red sandstone under complex loading A were respectively 6.14 MPa and 14.2° for specimen RS-J17#, which were all approximately equal to those for specimen RS-J3#. But, these values were all lower than the values of C and Φ under simple loading path.

4.4 Strength and Deformation Behavior of Red Sandstone Under Complex Loading Path B

4.4.1 Deformation Behavior of Red Sandstone Under Complex Loading Path B

In order to investigate the influence of fractured extent on triaxial strength and deformation behavior of red sandstone under complex loading path B, two specimens in Fig. 4.19 are carried out multi-stage triaxial test. Specimen RS-J16# was unloaded to zero while the axial deviatoric stress was loaded to 200.74 MPa (approximately 97.6% TCS) before peak stress at $\sigma_3 = 35$ MPa. Whereas specimen RS-J15# dropped to the stage of residual strength, and the ratio of the axial deviatoric stress at the unloading point to TCS was approximately 42.5%.

Figure 4.20 depicts the axial deviatoric stress-axial strain curves of red sandstone under complex loading path B. For comparison, the axial deviatoric stress-axial strain curves of red sandstone under simple loading path ($\sigma_3 = 35$ MPa) are also plotted in Fig. 4.20. In accordance with Fig. 4.20, the corresponding strength and deformation values of red sandstone under complex loading B are listed in Table 4.6. Based on listed data in Table 4.6, the influence of confining pressure on the elastic modulus of red sandstone under complex loading path B can be investigated, as shown in Fig. 4.21.

From Fig. 4.21, we can see that under complex loading path B, elastic modulus of specimen RS-J16# at $\sigma_3 = 25$ and 35 MPa are higher than that under simple loading path, which mainly results from unloading the axial deviatoric stress prior to peak strength at $\sigma_3 = 35$ MPa (see Fig. 4.20). Based on the axial stress-axial strain curve before unloading, we can obtain that elastic modulus of specimen RS-J16# at $\sigma_3 = 35$ MPa is 23.63 GPa, which is approximately equal to that under simple loading path (See Table 4.2). But elastic modulus of specimen RS-J16# at $\sigma_3 = 5$ and 15 MPa are lower than that under simple loading path, which attributes to the influence of loading history. However, under complex loading path B, elastic moduli of specimen RS-J15# under four confining pressures are all lower than that under simple loading path. For specimen RS-J16# under complex loading path B, the elastic modulus decreases from 24.87 to 15.71 GPa as σ_3 decreases from 35 to 5 MPa; and

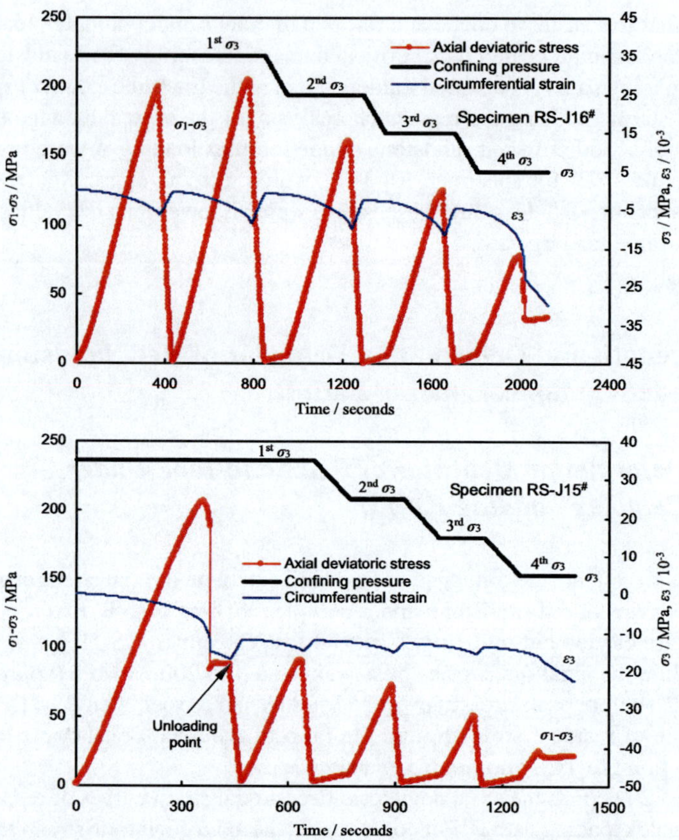

Fig. 4.19 Experimental results on change of axial deviatoric, confining pressure, circumferential strain of red sandstone with the increase of time under complex loading path B

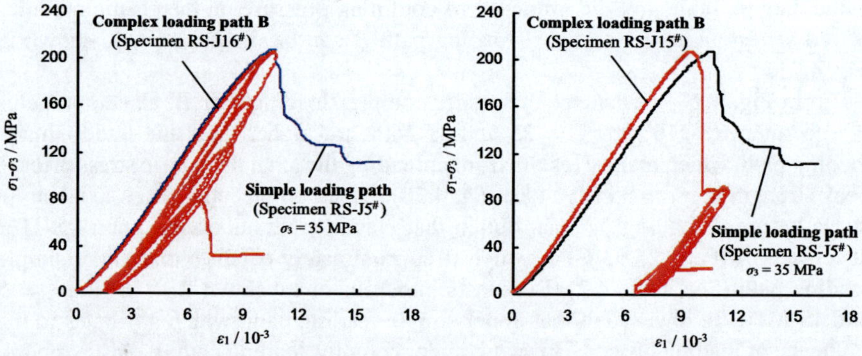

Fig. 4.20 Axial deviatoric stress-axial strain curves of red sandstone under complex loading path B

Table 4.6 Strength and deformation values of red sandstone specimen under complex loading path B

Specimen	Multi-stage	σ_3/MPa	E_S/GPa	σ_{cd}/MPa	σ_{sd}/MPa	σ_p/MPa	σ_S/MPa
RS-J16#	1st	35	24.87	61.1	96.1	205.3	240.3
	2nd	25	22.72	41.3	66.3	161.5	186.5
	3rd	15	20.12	25.3	40.3	124.9	139.9
	4th	5	15.71	7.2	12.2	77.7	82.7
RS-J15#	1st	35	22.44	36.0	71.0	91.4	126.4
	2nd	25	21.34	17.3	42.3	73.1	98.1
	3rd	15	18.28	13.5	28.5	51.3	66.3
	4th	5	13.08	6.3	11.3	25.4	30.4

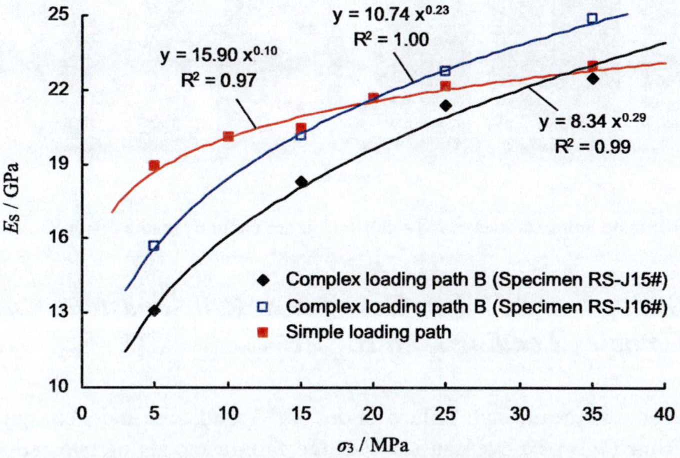

Fig. 4.21 Effect of confining pressure on elastic modulus of red sandstone under complex loading path B

elastic modulus of specimen RS-J15# under complex loading path B decreases from 22.44 to 13.08 GPa as σ_3 decreases from 35 to 5 MPa. Therefore, elastic modulus of red sandstone under complex loading path B is more sensitive to the confining pressure than under simple loading path. In sum, elastic modulus of red sandstone at lower confining pressures (e.g. $\sigma_3 = 5$ MPa) is distinctly dependent to the fractured extent of specimens after peak strength.

(a) Specimen RS-J16#

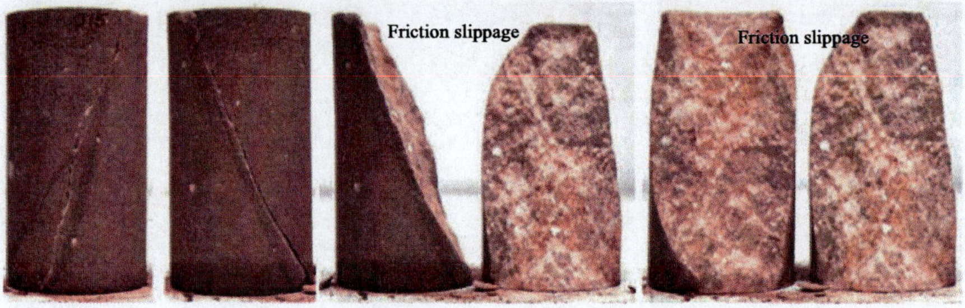

(b) Specimen RS-J15#

Fig. 4.22 Ultimate failure modes of red sandstone under complex loading path B

4.4.2 Strength and Failure Behavior of Red Sandstone Under Complex Loading Path B

Figure 4.22 shows the ultimate failure modes of red sandstone under complex loading path B. From Fig. 4.22, we can see that the failure modes of two red sandstone specimens under complex loading path B are distinctly different. Specimen RS-J16# took on a typical tension failure mode with some axial and lateral tension cracks, and this failure mode is approximately same with that under simple loading path ($\sigma_3 = 5$ MPa), which indicated that the ultimate failure mode of this specimen was the result of occurring the complete failure at the 4th confining pressure under complex loading path B. However, for specimen RS-J15#, due to being unloaded at the stage of residual strength, therefore, the failure mode of specimen was approximate with that under simple loading path ($\sigma_3 = 35$ MPa). Specimen RS-J15# took on the shear failure with a single plane, and the fracture surface was very smooth. At the 1st confining pressure, the red sandstone took on the friction slippage between two blocks. With the decrease of confining pressure, two fractured blocks also began to support lower axial stress by friction. From Fig. 4.20, it can be seen that axial supporting capacity of red sandstone specimen RS-J15# tends towards ideal plasticity at the final confining pressure ($\sigma_3 = 5$ MPa).

Table 4.6 lists the peak strength values of red sandstone under complex loading path B. Based on listed data in Table 6, the influence of σ_3 on peak strength σ_S for

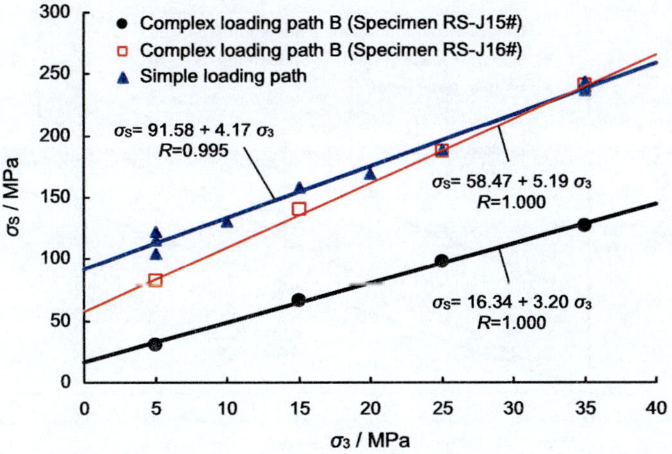

Fig. 4.23 Peak strength of red sandstone under complex loading path B. All the symbols represent the experimental results and the line represents the theoretical value in accordance with Mohr–Coulomb criterion. (*R* is the correlation coefficient of determination of the regression.)

red sandstone under complex loading path B can be presented in Fig. 4.23. From Fig. 4.23, it can be seen that peak strength of red sandstone under complex loading path B shows distinct linear behavior with the increasing σ_3 in the range of σ_3 from 5 to 35 MPa, which indicates that the peak strength behavior of red sandstone is in a good agreement with the linear Mohr–Coulomb criterion.

In accordance with the linear Mohr–Coulomb criterion, the peak strength parameters of red sandstone under complex loading path B are presented in Table 4.7. For specimen RS-J15#, under complex loading path B, the C and Φ values of red sandstone are respectively 4.57 MPa and 31.6°, which are all lower than that under simple loading path. However, for specimen RS-J16#, under complex loading path B, the C value is 12.83 MPa, which is distinctly lower than that under simple loading path, whereas the Φ value is 42.6°, which is a little higher than that under simple loading path.

The peak strength difference of red sandstone (specimen RS-J16#) between simple and complex loading path has an obvious increasing tendency with the decrease of confining pressure, which mainly results from the influence of loading history. However, the peak strength difference of red sandstone (specimen RS-J15#) between

Table 4.7 Peak strength parameters of red sandstone under complex loading path B by the linear Mohr–Coulomb criterion

Complex loading path B	σ_0/MPa	q	C/MPa	Φ/(°)	R
Specimen RS-J16#	58.47	5.19	12.83	42.6	1.000
Specimen RS-J15#	16.34	3.20	4.57	31.6	1.000

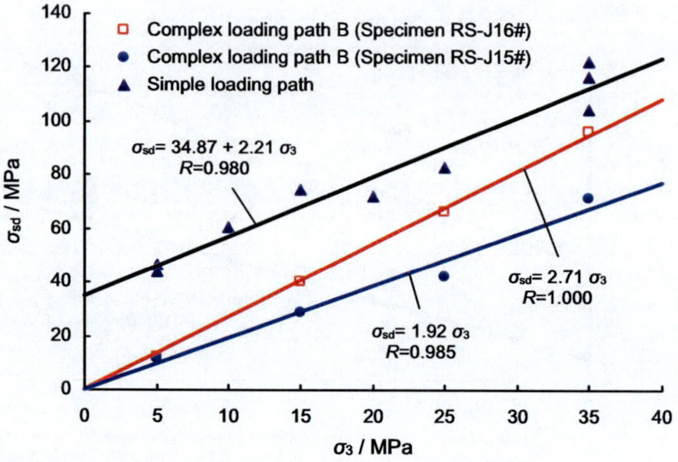

Fig. 4.24 Critical damage threshold of red sandstone under complex loading path B. All the symbols represent the experimental results and the line represents the theoretical value in accordance with Mohr–Coulomb criterion. (*R* is the correlation coefficient of determination of the regression.)

simple and complex loading path has a little decreasing tendency with the decrease of confining pressure, which possibly results from the increase of specimen stiffness.

Figure 4.24 depicts critical damage threshold of red sandstone under complex loading path B (See listed data in Table 4.6), which are confirmed by the volumetric strain of red sandstone under complex loading path B. In accordance with Fig. 4.24, it can be seen that critical damage threshold of red sandstone under complex loading path B are all lower than that under simple loading path Moreover, critical damage threshold of red sandstone under complex loading path B also depends on the confining pressure, which has a positive linear relation with the confining pressure. Notably, the critical damage threshold of red sandstone under complex loading path B is distinctly dependent to the fractured extent of specimens after peak stress, which is different from that under complex loading path A (See Fig. 4.17). For specimen RS-J16$^{\#}$, the critical damage threshold of red sandstone under complex loading path B is more sensitive to the confining pressure than that under simple loading path. However for specimen RS-J15$^{\#}$, the critical damage threshold of red sandstone under complex loading path B has an approximately same sensitivity to the confining pressure as that under simple loading path.

4.5 Discussion on Confirming Peak Strength Parameters of Red Sandstone with One Specimen

Throughout the comparison of strength characteristics of red sandstone between complex loading path A and B, it can be seen that the cohesion and internal friction angle of red sandstone under complex loading path B have obvious difference with those under simple loading path. Moreover, it is more difficult to evaluate for the difference. However, under complex loading path A, the internal friction angle and critical damage threshold of red sandstone under complex loading path A do not depend on the fractured extent of specimens after peak stress. Only a loss in specimen cohesion can be found under complex loading path A. Therefore, we recommend predicting the peak strength of rock under simple loading path by adopting complex loading path A not complex loading path B. If this cohesion drop is calibrated to triaxial test under simple loading path, the peak strength under complex loading path A can also be used to develop a failure criterion on basis of linear Mohr–Coulomb criterion, as illustrated by Eq. (4.2).

$$\sigma_S = \frac{h^* \times 2C\cos\phi + \sigma_3(1+\sin\phi)}{1-\sin\phi} \tag{4.2}$$

where h^* is defined as a cohesion loss factor of rock material, which can be calculated by Eq. (4.3).

$$h^* = \left(\frac{\sigma_{p,5}^{simple} + 5}{\sigma_{p,5}^{complex} + 5}\right)^2 \tag{4.3}$$

where, $\sigma_{p,5}^{simple}$ and $\sigma_{p,5}^{complex}$ respectively, refer to triaxial strength of red sandstone under simple loading path and complex loading path A at $\sigma_3 = 5$ MPa (See Fig. 4.12).

Next, we take specimen RS-J3# as an example to make an analysis. By Eq. (4.3) and triaxial experimental results of specimen RS-J3#, the following equation can be obtained.

$$h^* = \left(\frac{\sigma_{p,5}^{simple} + 5}{\sigma_{p,5}^{complex} + 5}\right)^2 = \left(\frac{98.7+5}{54+5}\right)^2 \approx 3.091 \tag{4.4}$$

In accordance with the above revised method, we make a comparison between revised Mohr–Coulomb criterion by using peak strength parameters under complex loading path A (Table 4.5) and the peak strength of red sandstone under simple loading path, which is shown in Fig. 4.25. From Fig. 4.25, it can be seen that for specimen RS-J17#, revised Mohr–Coulomb criterion is very much closer to the peak strength of red sandstone under simple loading path, which reveals that the above revised method Eq. (4.3) for peak strength of red sandstone under complex loading path A is approximately reasonable.

However for specimen RS-J3#, obtained peak strengths by revised Mohr–Coulomb criterion are all higher than peak strengths of red sandstone under simple loading path. Therefore, Eq. (4.3) has to be corrected for the specimens with a larger stress drop after peak strength. Taking into account that the critical damage threshold

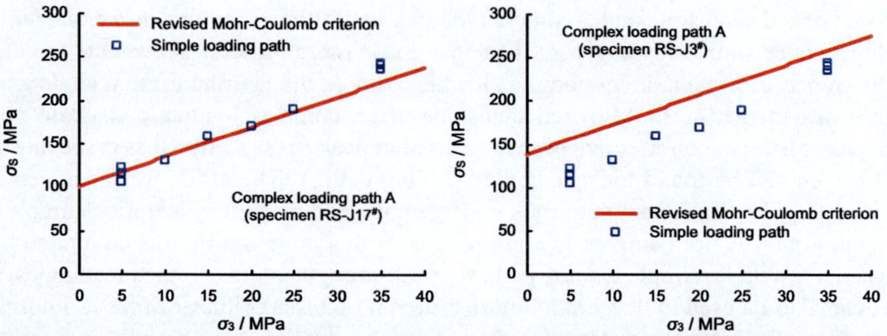

Fig. 4.25 Comparison between revised Mohr–Coulomb criterion (Eq. (4.3)) by using peak strength parameters under complex loading path A and the peak strength of red sandstone under simple loading path. All of the symbols represent triaxial experimental strength of red sandstone under simple loading path

of red sandstone under complex loading path A is independent to the fractured extent of specimens after peak stress, we put forward a new cohesion loss factor by Eq. (4.5).

$$h^* = \frac{\sigma_{cd,5}^{simple} + 5}{\sigma_{cd,5}^{complex} + 5} \tag{4.5}$$

where, $\sigma_{cd,5}^{simple}$ and $\sigma_{cd,5}^{complex}$ respectively, refer to the critical damage threshold of red sandstone under simple loading path and complex loading path A at $\sigma_3 = 5$ MPa.

According to Eqs. (4.5) and (4.2), we also make a comparison between revised Mohr–Coulomb criterion by using peak strength parameters under complex loading path A (Table 4.5) and the peak strength of red sandstone under simple loading path, which is shown in Fig. 4.26. From Fig. 4.26, it can be seen that for specimen RS-J3#, revised Mohr–Coulomb criterion is very much closer to the peak strength of red sandstone under simple loading path. However, for specimen RS-J17#, obtained peak strengths by revised Mohr–Coulomb criterion are distinctly higher than peak strengths of red sandstone under simple loading path.

However, it needs to be noted that the suggested loading path (complex loading path A) for estimation of rock strength under different confining pressures just by using one specimen are very useful for estimation of residual strength of rock. But, this method may be not accurate enough for the estimation of the peak strength of rock under different confining pressures. Because a rock specimen is never able to provide the same amount of cohesion component of its peak strength after it is driven to failure stress for the first time due to the influence of loading history (Yang et al. 2011). Figure 4.27 shows AE hypo-central locations during the load testing on Westerly granite (a kind of typical brittle rock) and its stress–displacement curve. The figure indicates that the failure plane is not likely to develop when the specimen is loaded to peak strength or when the specimen has a smaller stress drop (about 4–5%

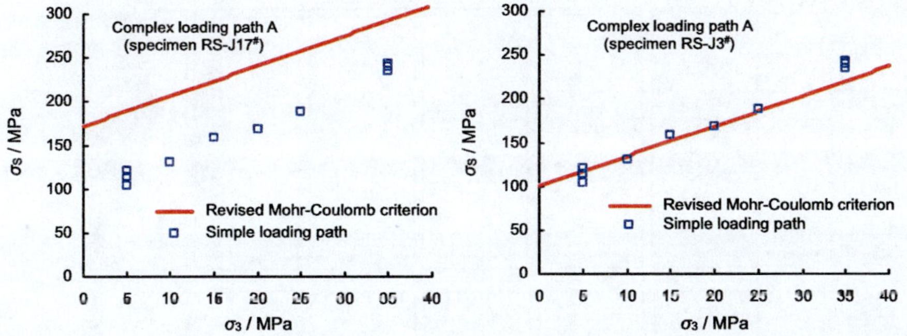

Fig. 4.26 Comparison between revised Mohr–Coulomb criterion (Eq. (4.5)) by using peak strength parameters under complex loading path A and the peak strength of red sandstone under simple loading path. All of the symbols represent triaxial experimental strength of red sandstone under simple loading path

TCS) after peak strength. This observation supports the validity of complex loading path A if the specimen has a smaller stress drop (about 4–5% TCS). Therefore, even though for a kind of brittle rock material, we can conclude that a rock specimen can also provide approximate amount of cohesion component of its peak strength after being driven to failure stress (a smaller stress drop, about 4–5% TCS after the peak strength) for the first time. Thus, for a kind of rock that has been evaluated as a linear Mohr–Coulomb material, in order to predict peak strength of red sandstone under simple loading path using obtained triaxial strength under complex loading path A with only one specimen, the following suggested method can be used to confirm more precisely the peak strength.

For specimen that has a smaller stress drop (about 4–5% TCS) after peak strength, Eqs. (4.2)–(4.3) are validated for predicting the peak strength of red sandstone under simple loading path. Whereas, for specimen that has a larger stress drop (about 40– 45% TCS) after peak strength, the Eqs. (4.2)–(4.3) are not validated due to the loss of an amount of cohesion component of its peak strength after being driven to failure stress. Therefore, we take into account that the critical damage threshold of red sandstone under complex loading path A does not depend on the fractured extent of specimens after peak strength. Equations (4.2) and (4.5) are put forward to predict the peak strength of red sandstone under simple loading path, which is also very applicable for specimens with a larger stress drop (about 40–45% TCS) after peak strength.

In the previous research (Yang et al. 2011), complex loading path A **(i.e. Path II by stepping up the confining pressure, shown in** Fig. 1.2 in Chap. 1) has been suggested to confirm the triaxial strengths of marble under different confining pressures by only one specimen, which has been testified that it is very applicable for a kind of rock with obvious plastic and ductile deformation (e.g. marble, chalk, mudstone, etc.). But in this research, the tested sandstone possesses obvious brittle characteristics even though at the higher confining pressures (e.g. 35 MPa), which is

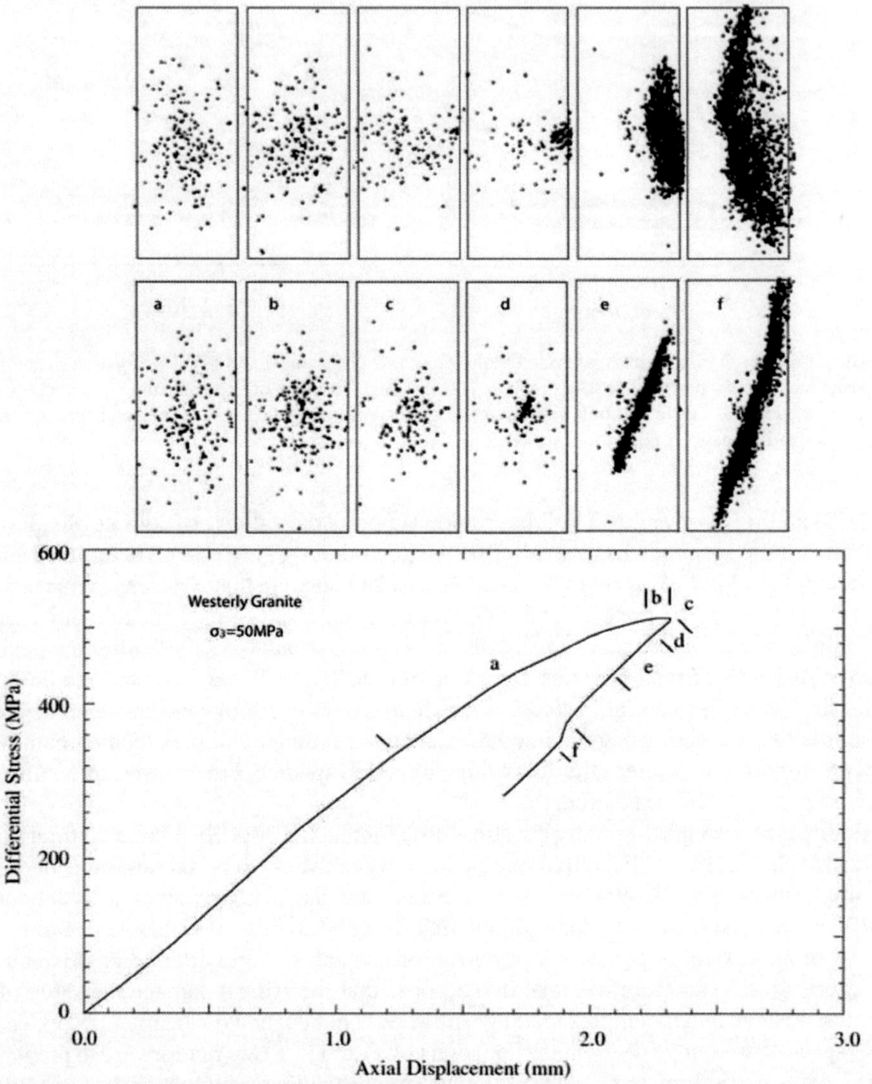

Fig. 4.27 Two-dimensional AE hypo-central locations during fault formation of initially intact Westerly granite at $\sigma_3 = 50$ MPa (Lockner 1995)

different from the deformation behavior of marble (Yang et al. 2011). But the validity of the suggested stress paths for brittle rocks has also been proved by the previous descriptions. However, taking into account the difficulty of performing complex loading path A on more brittle rock material (e.g. granite, basalt, etc.), circumferential deformation control can also be adopted for the complex loading path A, which was a kind of effective testing method to obtain peak strength of brittle rocks under different confining pressures with only one specimen (Youn and Tonon 2010). Of course, due to the difference of the grain shape and size heterogeneity of all kinds of rock types, more experiments for different rock types under the suggested loading path still need to be further strengthened in the future.

4.6 Conclusions

In this chapter, a series of triaxial compression experiment was carried out for red sandstone. Compared the experimental results of red sandstone between simple and complex loading paths, the strength and deformation behavior of tested red sandstone can be summarized as follows.

(1) Under simple loading path, elastic modulus of tested red sandstone increases nonlinearly with increasing confining pressure. The peak strength, residual strength and critical damage threshold of red sandstone all depend on the confining pressure, which are in good agreement with the linear Mohr–Coulomb criterion. The peak cohesion of red sandstone is approximately 54.4% higher than the residual cohesion. However, the peak internal friction angle of red sandstone is only 16.9% higher than the residual internal friction angle. The sensitivity of the critical damage threshold on σ_3 was clearly lower than that for σ_S.

(2) To analyse the re-fractured mechanical behavior of red sandstone with different post-peak stress drops, two complex loading paths are put forward to obtain the strength and deformation parameters of red sandstone. The differences of strength and deformation parameters of red sandstone between simple and complex loading paths are evaluated detailed. The peak strength and critical damage threshold of red sandstone under complex loading path A and B are all in good agreement with the linear Mohr–Coulomb criterion. It is suggested and recommended to predict the peak strength of rock under simple loading path by adopting complex loading path A not complex loading path B.

(3) The difference of critical damage threshold of red sandstone between simple and complex loading path A also increases gradually with the increasing σ_3. Under complex loading path A, the peak internal friction angle is a little lower but the peak cohesion is distinctly lower than that under simple loading path. Specimens under complex loading path A produce some damage, which can be interpreted as a loss in specimen cohesion. However, the internal friction angle and critical damage threshold of red sandstone under complex loading path A do not depend on the fractured extent of specimens after peak stress.

(4) For a kind of rock that has been evaluated as a linear Mohr–Coulomb material, in order to predict peak strength of red sandstone under simple loading path using obtained triaxial strength under complex loading path A with only one specimen, the following suggested method can be used to confirm more precisely the peak strength. For specimen that has a smaller stress drop (about

4–5% TCS) after peak strength, Eqs. (4.2)–(4.3) is adapted to predict the peak strength of red sandstone under simple loading path; however for specimen that has a larger stress drop (about 40–45% TCS) after peak strength, Eqs. (4.2) and (4.5) is adapted to predict the peak strength of red sandstone under simple loading path.

References

Fairhurst CE, Hudson JA (1999) Draft ISRM suggested method for the complete stress–strain curve for the intact rock in uniaxial compression. Int J Rock Mech Min Sci 36(3):279–289

Heap MJ, Baud P, Meredith PG, Bell AF, Main IG (2009) Time-dependent brittle creep in Darley Dale sandstone. J Geophys Res. https://doi.org/10.1029/2008JB006212

Lockner DA (1995) Rock failure. Rock Physics and Phase Relations: A Handbook of Physical Constants. American Geophysical Union, Washington, pp 127–147

Wong TF, David C, Zhu W, 1997. The transition from brittle faulting to cataclastic flow in porous sandstones: Mechanical deformation. J Geophys Res 102((B2)):3009–3025.

Yang SQ, Jing HW, Li YS, Han LJ (2011) Experimental investigation on mechanical behavior of coarse marble under six different loading paths. Exp Mech 51(3):315–334

Yang SQ, Jing HW, Wang SY (2012) Experimental investigation on the strength, deformability, failure behavior and acoustic emission locations of red sandstone under triaxial compression. Rock Mech Rock Eng 45(4):583–606

Yang SQ, Jing HW (2013) Evaluation on strength and deformation behavior of red sandstone under simple and complex loading paths. Eng Geol 164:1–17

Youn H, Tonon F (2010) Multi-stage triaxial test on brittle rock. Int J Rock Mech Min Sci 47(4):678–684

Chapter 5
Mechanical, Acoustic, and Fracture Behaviors of Yellow Sandstone Specimens Under Triaxial Monotonic and Cyclic Loading

In actual rock engineering, the rock mass often experiences cyclic loading owing to the strong blasting and excavation disturbances (Zhu et al. 2010). The mechanical behavior of rock under cyclic loading is notably different from that under monotonic loading (Yang et al. 2015). To better understand the fracture mechanical behavior of rock, a series of triaxial monotonic and cyclic tests is conducted on yellow sandstone in this chapter (Yang et al. 2020). Based on the triaxial monotonic experimental results, the influence of confining pressure on the mechanical parameters of the sandstone is first analyzed, and the influence of the cycle number on the deformation and crack damage parameters of the sandstone is also investigated. During the loading process, real-time AE and ultrasonic velocity monitoring techniques are used to explore the acoustic behavior of yellow sandstone under 3D stress, and the effect of the cycle number on the AE behaviors is explored. Finally, the fractured sandstone specimens are further analyzed by using X-ray micro CT scanning. By analyzing cross-sections and 3-D reconstructed images of the CT scans, the fracture mechanism of sandstone is revealed.

5.1 Experimental Methodology

5.1.1 Yellow Sandstone Material

The testing object was yellow sandstone obtained from Zunyi City, Guizhou Province, China. The mercury intrusion porosimetry (MIP) results indicated 16.20% porosity in the tested sandstone (Huang et al. 2019). The optical microscopic observation and scanning electron microscopic (SEM) images of the micro structure indicated that this sandstone is a medium grained material with some initial micro cracks and pores (Fig. 5.1). The mineral composition of this sandstone is 95.8% quartz with minor quantities of other minerals.

S.-Q. Yang, *Mechanical Behavior and Damage Fracture Mechanism of Deep Rocks*,
https://doi.org/10.1007/978-981-16-7739-7_5

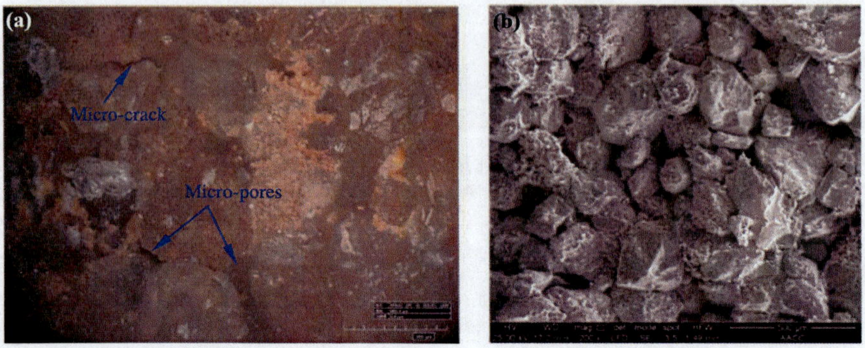

Fig. 5.1 Micro-structure of tested yellow sandstone. **a** Optical microscopic observation. **b** SEM image (Huang et al. 2018)

All intact cylinder specimens were cored from the same block of yellow sandstone material to a diameter of 50 mm and a length of approximately 100 mm, which resulted in a length: diameter ratio of 2:1 for mechanical tests in accordance with the suggested by ISRM (International Society for Rock Mechanics; Fairhurst and Hudson (1999)). The yellow sandstone blocks were first cored, and both ends of each specimen were then carefully ground to produce flat parallel surfaces to ensure uniform load distribution. Before the test, the sandstone was oven-dried for 24 h under 40 °C to avoid possible thermal cracking (Rathnaweera et al. 2016).

5.1.2 Testing System Description

All triaxial compression tests were performed by using a GCTS RTX–4000 high pressure and high temperature triaxial rock testing system at China University of Mining and Technology, China. The testing system (Yang et al. 2018) consists of a hydraulic station, a microprocessor, a loading frame, a triaxial cell, and a confining pressure and pore pressure supercharger (Fig. 5.2a). The maximum axial loading capacity of the servo-controlled system is 4000 kN, the maximum confining pressure is 140 MPa, and the maximum pore pressure is 140 MPa. During experimentation, the axial deformation was measured by using two axial linear variable differential transducers (LVDTs) with a range of 6 mm (Fig. 5.2b). The circumferential deformation was measured using an LVDT attached to a chain wrapped tightly around the specimen (Fig. 5.2b), for which the maximum displacement capacity was 6 mm. The radial LVDT was located in the central part of the specimen to eliminate the effects of friction at the ends (Mogi 1966; Hawkes and Mellor 1970).

It is widely accepted that the AE technique is predominantly related to the release of elastic energy within rock materials (Yang et al. 2012). AE events are often monitored to detect the onset and the evolution of stress-induced cracks. Therefore, in this

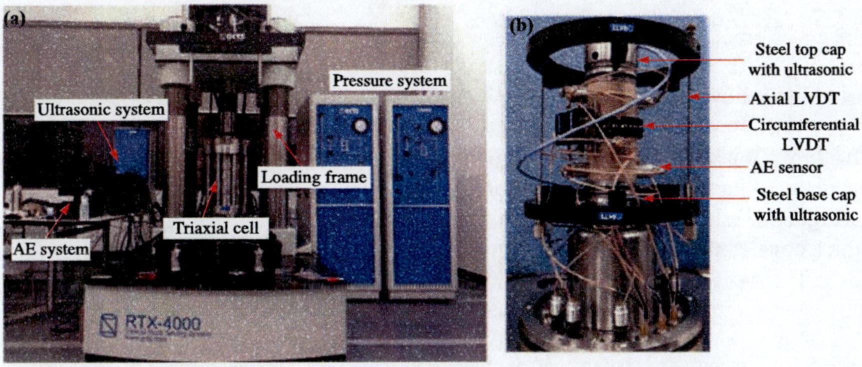

Fig. 5.2 GCTS RTX–4000 rock triaxial testing system. **a** Overview of the testing system (Yang et al. 2018). **b** Details of strain, ultrasonic and AE measurement

study, a micro-II AE measuring system (Physical Acoustic Corporation) was used to investigate the AE evolution of sandstone (Fig. 5.2a).

To further analyze the damage in the sandstone under stress, compressional wave velocity measurements were conducted along the axial stress direction every 5 MPa using an ultrasonic pulse transmission technique. In this test, the transmitter and receiver were positioned on the opposite end surfaces of the specimens. An electric pulse was then generated with a pulse generator, and the waveforms were recorded by using a digital oscilloscope (Fig. 5.3a). The first-cycle wave is stable and renewable under the same transducer and same contact between the transducers and specimens, because it rarely contaminated by other waves arriving late, this wave is easily identified (Wang et al. 2016). After the measurements, the P-wave velocity can be calculated according to length of the specimen and the travel time (or the first arrival time), which was identified from the waveform (Fig. 5.3b).

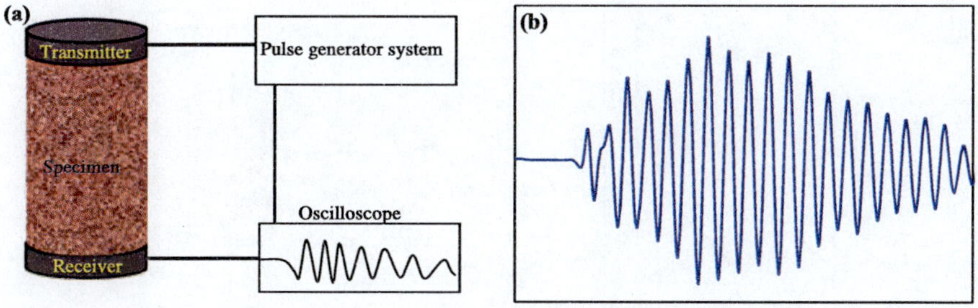

Fig. 5.3 Schematic of ultrasonic measurement process. **a** Schematic of testing system. **b** Identification of the first arrived time

5.1.3 Experimental Procedure

All of the sandstone specimens were loaded under conventional compression testing ($\sigma_2 = \sigma_3$). To obtain the accurate results as well as the best comparison, all experiments were performed under dry conditions at room temperature. Table 5.1 shows the detailed geometry and testing conditions of the sandstone specimens.

Figure 5.4 shows the triaxial monotonic loading path in the principal stress space, which consisted of two below steps. First, the confining pressure was increased to the desired value at a constant rate of 4.0 MPa/min to ensure that the specimen was under

Table 5.1 Geometry and testing content of tested yellow sandstone specimens in this research

Specimen	D/mm	H/mm	m/g	ρ/(kg/m^3)	V_{p0}/(m/s)	σ_3/MPa	Test content
A-214#	49.82	95.51	402.0	2160	2563	0	TCS, P-wave and AE
A-208#	49.81	96.69	401.3	2131	2520	0	TCS, P-wave and AE
A-221#	49.21	99.10	400.8	2128	2538	10	TCS, P-wave and AE
A-95#	50.01	99.82	405.0	2067	2540	10	TCS, P-wave and AE
A-215#	49.82	95.53	396.6	2131	2534	20	TCS, P-wave and AE
A-206#	49.92	96.98	405.6	2138	2520	20	TCS, P-wave and AE
A-218#	49.83	95.54	401.1	2154	2536	30	TCS, P-wave and AE
A-92#	50.25	95.17	399.0	2115	2513	30	TCS, P-wave and AE
A-220#	49.83	95.51	396.8	2131	2350	40	TCS, P-wave and AE
A-222#	49.81	95.54	398.5	2142	2680	60	TCS, P-wave and AE
A-1-C#	49.77	94.99	399.6	2162	–	10	TCS, AE, cyclic loading
A-94-C#	49.88	95.01	400.3	2156	–	30	TCS, AE, cyclic loading
A-95-F#	49.82	94.71	396.1	2145	–	10	TCS, fatigue loading

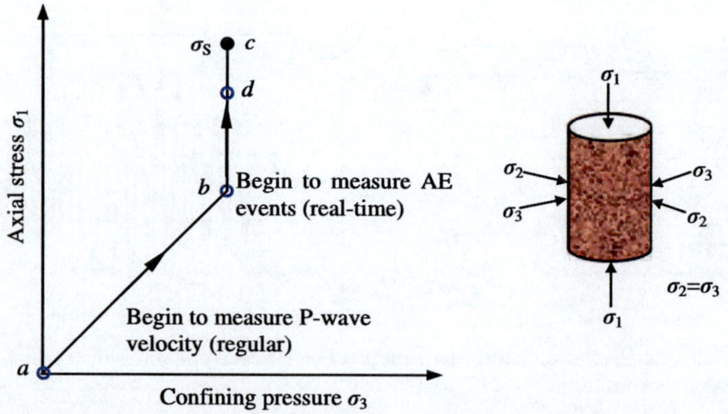

Fig. 5.4 Triaxial monotonic loading path presented in the principal stress space (σ_3, σ_1). Modified from Yang et al. (2012)

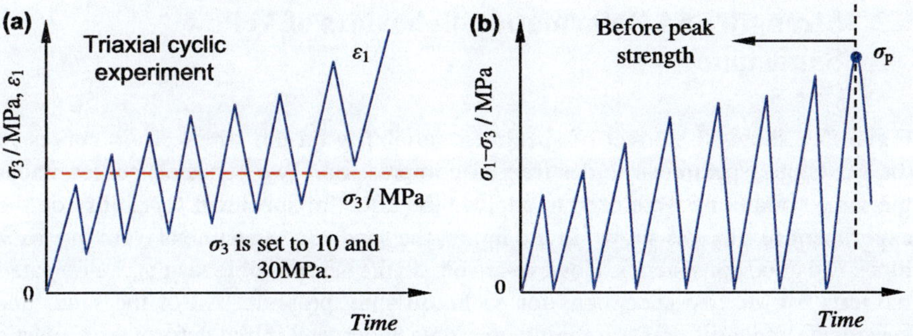

Fig. 5.5 Triaxial cyclic loading path applied in the present research. **a** Variation of confining pressure and axial strain with time. **b** Variation of axial deviatoric stress with time. Modified from Yang et al. (2015)

uniform hydrostatic stress (path *ab* in Fig. 5.4). At point a, the GCTS RTX–4000 and ultrasonic measuring system were operated simultaneously. In the second step, axial deviatoric stress was applied to the surface of the specimen at a constant axial displacement rate of 0.04 mm/min until failure occurred (path $bc \rightarrow cd$). At point b, the AE events were measured in real-time during the entire rock deformation and failure process. To investigate the effect of depth (confining pressure) on the strength failure behavior of reservoir rock, the sandstone specimens were tested under various confining pressures. These different confining pressure conditions were selected to simulate the reservoir depth effect, and six different confining pressure values, i.e., 0, 10, 20, 30, 40, and 60 MPa were selected for the present experimental study. The corresponding maximum deep burial depth was 2400 m assuming a reservoir rock density of 2500 kg/m^3.

The triaxial cyclic loading experiments (Fig. 5.5) were conducted in three steps. First, hydrostatic pressure was applied to the specimens at a rate of 4.0 MPa/min until the desired value was reached. Second, the specimens were loaded to the first displacement value, corresponding to the first deviatoric stress level, at a controlled axial displacement rate of 0.04 mm/min. The first deviatoric stress level was then unloaded to zero at a controlled axial stress of 0.75 MPa/s. In each subsequent cycle, the second deviatoric stress level was increased, and the specimens again were unloaded to zero. In the final step, the stress-cycling was continued until the specimens eventually failed. It should be noted that in our study, it is difficult to conduct further cyclic loading owing to the obvious brittleness of the tested sandstone after the peak strength was reached, which is highly similar to the results reported in previous studies (Heap et al. 2009). Therefore, in order to explore the effect of fatigue loading on the strength and deformation behavior of the sandstone, more than 1000 cycles were conducted on particular specimen.

5.2 Strength and Deformation Behaviors of Yellow Sandstone

Figure 5.6 shows the effect of specimen variability on the stress–strain curves of the sandstone specimens under triaxial compression. Two repeated curves under the same conditions were used to analyze the effect of specimen variability on the experiment results. As shown in the figure, the sandstone specimens from one rock block had good consistency. However, to obtain the best possible results, we repeated the tests for the two specimens for each confining pressure. All of the sandstone specimens under different confining pressures underwent initial deformation, elastic deformation, nonlinear deformation, stress softening and residual strength stages.

5.2.1 *Strength and Deformation Parameters Under Triaxial Monotonic Loading Path*

Figure 5.7a shows the axial deviatoric stress-axial strain curves of the sandstone under the triaxial monotonic loading path, and Fig. 5.7b identifies the critical strength and deformation parameters according to the curves (Yang et al. 2015). The volumetric strain was calculated from the sum of the axial strain and twice the radial strain. Strength parameters σ_{p} and σ_{cr} represent the peak deviatoric stress and the residual deviatoric stress, respectively. Crack damage stress σ_{cd} is the crack damage threshold of the specimen (Fairhurst and Hudson 1999; Heap et al. 2009; Wong et al. 1997), i.e., the corresponding axial deviatoric stress point at which the volumetric deformation of the specimen switches from compaction-dominated to dilatancy-dominated. The elastic modulus (E_{S}) of the sandstone is the average slope of the approximately straight-line portion (*AB*, i.e., from the elastic initial to the elastic limit) of the axial deviatoric stress-axial strain curve before the peak strength was reached. The deformation modulus (E_{50}) is defined as the slope between the original point and the stress at half of the peak strength. Poisson's ratio (μ) refers to the average value of the ratio of radial strain to axial strain at the stage of elastic deformation. Peak axial strain $\varepsilon_{1\mathrm{c}}$ is the corresponding axial strain at peak strength. $\varepsilon_{1\mathrm{d}}$, $\varepsilon_{3\mathrm{d}}$, and $\varepsilon_{\mathrm{vd}}$ are the corresponding axial strain, circumferential stain and volumetric strain at the crack damage threshold. On the basis of the stress–strain results obtained from the experimental data, the mechanical results of the sandstone specimens with respect to various confining stresses are listed in Table 5.2.

Figure 5.8 shows the effect of confining pressure on the triaxial compressive strength, crack damage stress, and residual strength of the sandstone specimens. These parameters clearly increased linearly with the confining pressure, which can be evaluated by the liner Mohr–Coulomb criterion (Eq. (5.1)):

$$\sigma_S = \sigma_c + q\sigma_3 = \frac{2C\cos\phi + \sigma_3(1+\sin\phi)}{1-\sin\phi} \tag{5.1}$$

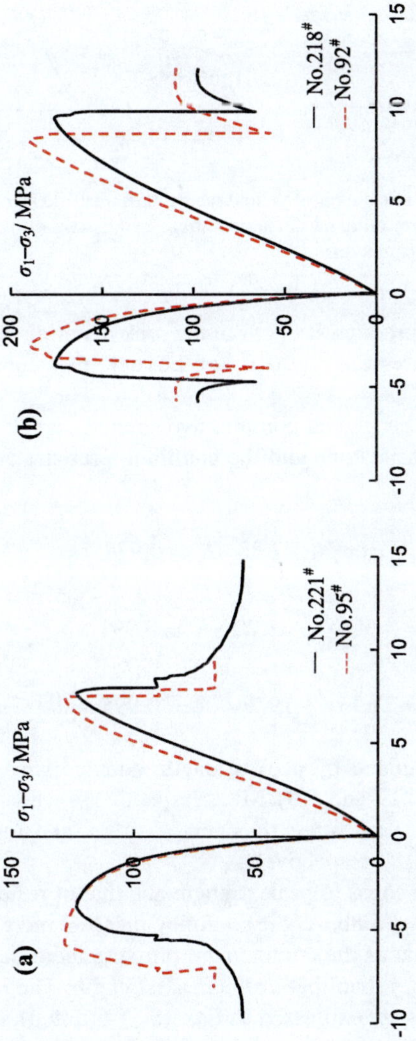

Fig. 5.6 Effect of specimen variability on the stress-strain curves of yellow sandstone specimens under triaxial compression. **a** Two specimens under confining pressure of 10 MPa. **b** Two specimens under confining pressure of 30 MPa

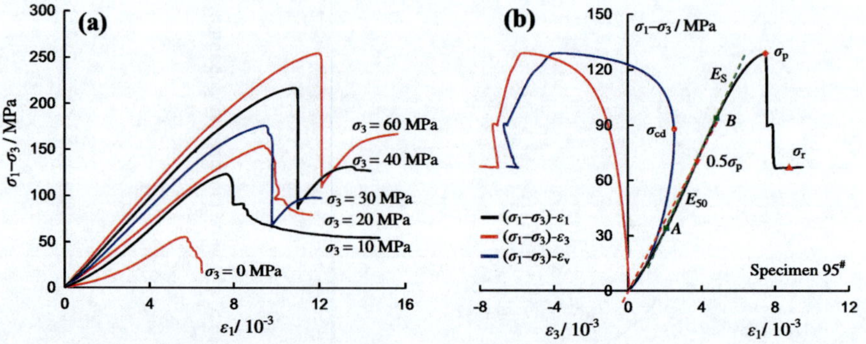

Fig. 5.7 Axial deviatoric stress–strain curves of yellow sandstone under triaxial monotonic loading path. **a** Effect of confining pressure on axial deviatoric stress-axial strain curve. **b** Definition of critical strength and deformation parameters

where σ_c is the uniaxial compressive strength of the rock material, q is an influence coefficient of the confining pressure on the σ_S of the rock, C is cohesion, and φ is the internal friction angle.

The relationships among the triaxial compressive strength, crack damage stress, and residual strength of the sandstone and the confining pressure are presented as Eqs. (5.2)–(5.4), respectively.

$$\sigma_S = 4.36\sigma_3 + 77.33 \quad R = 0.974 \tag{5.2}$$

$$\sigma_{Sd} = 4.50\sigma_3 + 43.22 \quad R = 0.991 \tag{5.3}$$

$$\sigma_{Sr} = 3.69\sigma_3 + 19.24 \quad R = 0.995 \tag{5.4}$$

The cohesion values calculated by peak strength, damage stress, and residual strength were 18.52 MPa, 10.19 and 5.01 MPa, respectively, whereas the internal friction angles calculated by peak strength, damage stress, and residual strength were 38.82°, 39.50°, and 35.01°, respectively.

The ratio of crack damage stress to peak strength and that of residual strength to peak strength are plotted in Fig. 5.8b. As the confining pressure increased, the ratios increased, which indicates that as the confinement pressure increased, the increase in residual strength was greater than that of the peak strength. The relationships of ratios and confining pressures are expressed as Eqs. (5.5) and (5.6), respectively.

$$\sigma_{Sd}/\sigma_S = 0.0054\sigma_3 + 0.66 \quad R = 0.946 \tag{5.5}$$

$$\sigma_{Sr}/\sigma_S = 0.0042\sigma_3 + 0.48 \quad R = 0.941 \tag{5.6}$$

Table 5.2 Strength and deformation parameters of yellow sandstone specimens under triaxial monotonic loading path

Specimen	σ_3/MPa	σ_p/MPa	$\varepsilon_{1c}/10^{-3}$	E_S/GPa	E_{50}/GPa	μ	σ_{cd}/MPa	$\varepsilon_{1d}/10^{-3}$	$\varepsilon_{3d}/10^{-3}$	$\varepsilon_{vd}/10^{-3}$	σ_r/MP
A-214#	0	46.18	5.016	11.29	7.92	0.366	30.53	3.523	−0.863	1.796	-
A-208#	0	55.05	5.707	12.83	7.85	0.350	35.63	4.081	−0.876	2.329	-
A-221#	10	123.57	7.648	20.52	15.75	0.295	85.27	5.053	−0.887	3.280	53.42
A-95#	10	128.71	7.440	22.32	18.45	0.318	84.55	4.387	−0.923	2.540	66.93
A-215#	20	153.70	9.402	19.31	17.79	0.263	110.78	6.152	−1.371	3.410	78.95
A-206#	20	167.15	8.723	22.68	21.51	0.249	134.29	6.108	−1.276	3.556	84.16
A-218#	30	176.13	9.437	22.86	21.61	0.255	141.95	6.630	−1.445	3.740	97.48
A-092#	30	190.37	8.295	26.63	23.88	0.221	158.01	6.521	−1.375	3.771	109.91
A-220#	40	217.08	10.833	25.01	24.40	0.236	166.99	7.060	−1.528	4.003	127.31
A-222#	60	254.58	11.976	26.13	26.19	0.186	253.29	11.717	−2.865	5.987	166.51

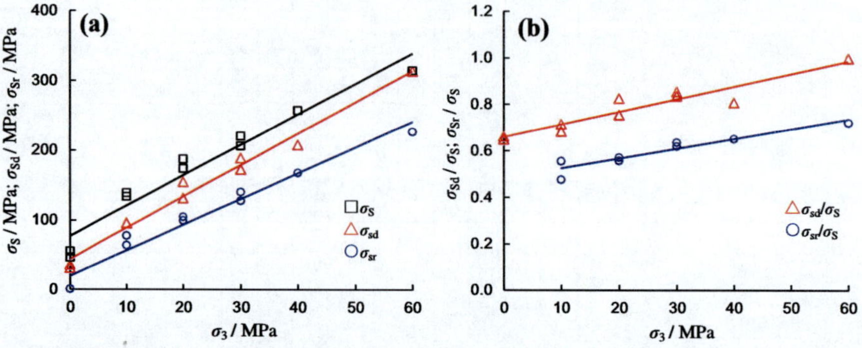

Fig. 5.8 Effect of confining pressure on strength parameters of yellow sandstone specimens. **a** Triaxial compressive strength and crack damage threshold. **b** Ratio of critical stress to triaxial compressive strength. *Note* $\sigma_{\mathrm{sd}} = \sigma_{\mathrm{cd}} + \sigma_3$; $\sigma_{\mathrm{S}} = \sigma_{\mathrm{p}} + \sigma_3$; $\sigma_{\mathrm{Sr}} = \sigma_{\mathrm{r}} + \sigma_3$

Figure 5.9 shows the influence of confining pressure on the Young's modulus, including elastic and deformation moduli. As the confining pressure increased, the Young's modulus increased nonlinearly (Fig. 5.9a). When the confining pressure increased from 0 to 30 MPa, the increments in the elastic and deformation moduli were 105.2 and 188.5%, respectively. However, when the confining pressure increased from 30 to 60 MPa, the increments were 5.6 and 15.1%, respectively. To compare the elastic and deformation moduli, the ratio of E_{50} to E_{S} was plotted against the confining pressure, as shown in Fig. 5.9b. As the confining pressure increased from 0 to 60 MPa, the ratio of E_{50} to E_{S} varied from 0.66 to 1.00. The ratio of E_{50} to E_{S} demonstrated that it increased with confining pressure, which resulted in distinct nonlinear behavior in the tested sandstone specimens at the initial stage under relatively low confining pressure (Fig. 5.7a).

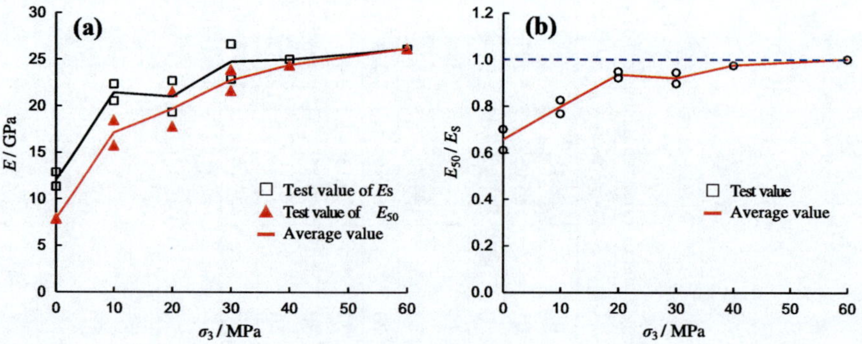

Fig. 5.9 Effect of confining pressure on deformation parameters of yellow sandstone specimens. **a** Elastic modulus and deformation modulus. **b** Ratio of deformation modulus to elastic modulus

Figure 5.10 shows that the Poisson ratios of the sandstone specimens decreased with an increase in the confining pressure. When the confining pressure changed in the range of 0 to 60 MPa, the Poisson ratios decreased from 0.358 to 0.164 (average value). The relationship between the Poisson ratios of the sandstone specimen and the confining pressure was fitted using Eq. (5.7).

$$\mu = -0.0033\sigma_3 + 0.3409 \quad R = 0.960 \tag{5.7}$$

Figure 5.11 shows the critical strain values of the sandstone under different confining pressures. The axial peak strain of the specimen increased with an increase in confining pressure (Fig. 5.11a). The axial strain and volumetric strain for the crack damage increased, whereas the circumferential strain decreased with confining pressure linearly (Fig. 5.11b). The relationships between the strain values and confining pressure are shown in Eqs. (5.8)–(5.11).

$$\varepsilon_{1c} = 0.1072\sigma_3 + 6.0896 \quad R = 0.940 \tag{5.8}$$

$$\varepsilon_{1d} = 0.1179\sigma_3 + 3.5301 \quad R = 0.960 \tag{5.9}$$

$$\varepsilon_{3d} = 0.0586\sigma_3 + 2.1521 \quad R = 0.959 \tag{5.10}$$

$$\varepsilon_{vd} = -0.0296\sigma_3 - 0.689 \quad R = 0.931 \tag{5.11}$$

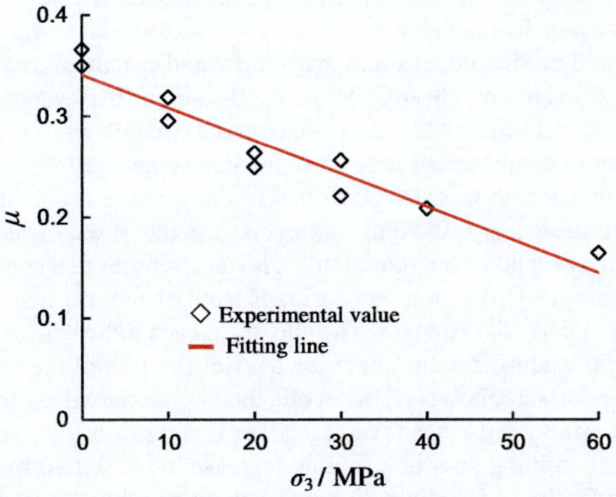

Fig. 5.10 Effect of confining pressure on Poisson's ratio of yellow sandstone specimens under triaxial compression

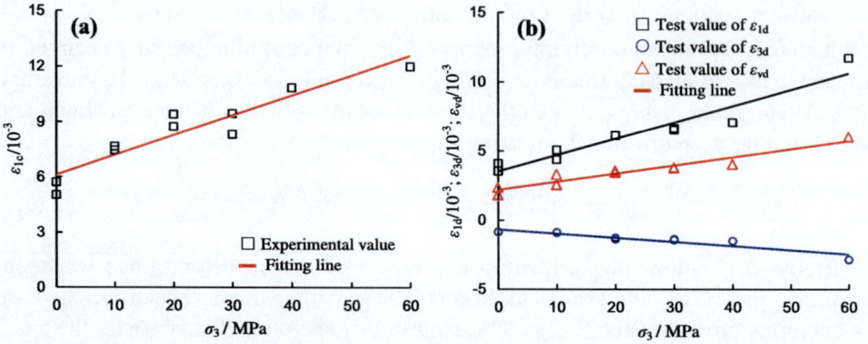

Fig. 5.11 Effect of confining pressure on critical strain of yellow sandstone specimens. **a** Peak axial strain. **b** Axial strain, circumferential strain and volumetric strain at crack damage point

5.2.2 *Strength and Deformation Behavior Under Triaxial Cyclic Loading Path*

Figure 5.12 shows typical triaxial cyclic experimental results of the sandstone at $\sigma_3 = 10$ and 30 MPa, and Fig. 5.13 shows the triaxial fatigue stress–strain curves under 10 MPa of confining pressure. In the figures, the triaxial monotonic experimental curves of the sandstone are also given for comparison purposes. As shown in Fig. 5.12, no cyclic loadings were obtained after the peak stress, which is attributed mainly to the sudden brittle failure of the tested sandstone material after the peak strength.

Figure 5.14a presents a comparison of peak strength between monotonic and cyclic loading based on the triaxial experimental results of the sandstone under monotonic and cyclic loading (Fig. 5.12). No obvious difference was noted between the monotonic and cyclic triaxial strength for the tested specimens. At $\sigma_3 = 10$ MPa, the monotonic triaxial strength (σ_p) of the sandstone, or the average value of two tested specimens, was 126.14 MPa; the cyclic triaxial strength was 123.80 MPa. At $\sigma_3 = 30$ MPa, the monotonic triaxial strength of the sandstone was 183.25 MPa, whereas the cyclic triaxial strength was 192.60 MPa. The comparison revealed that under the lower confining pressures (10 MPa), the cyclic triaxial strength of the sandstone was approximately equal to the monotonic triaxial strength; however, under higher confining pressures (30 MPa), the cyclic triaxial strength of the sandstone was higher than the monotonic triaxial strength. The above analysis indicates that for the tested sandstone, cyclic loading did not affect the triaxial strength of the sandstone under lower confining pressure; however, the cyclic loading increased the triaxial strength under higher confining pressures. The reason for this result is discussed below.

Under lower confining pressures and an increase in axial deformation, the local yield created difficulty in discriminating mineral grains other than the 95.8% quartz grains in the sandstone. Therefore, the cyclic triaxial strength of yellow sandstone is approximately equal to the monotonic triaxial strength at lower confining pressures. However, although an increase in confining pressure led to a gradual decrease in the

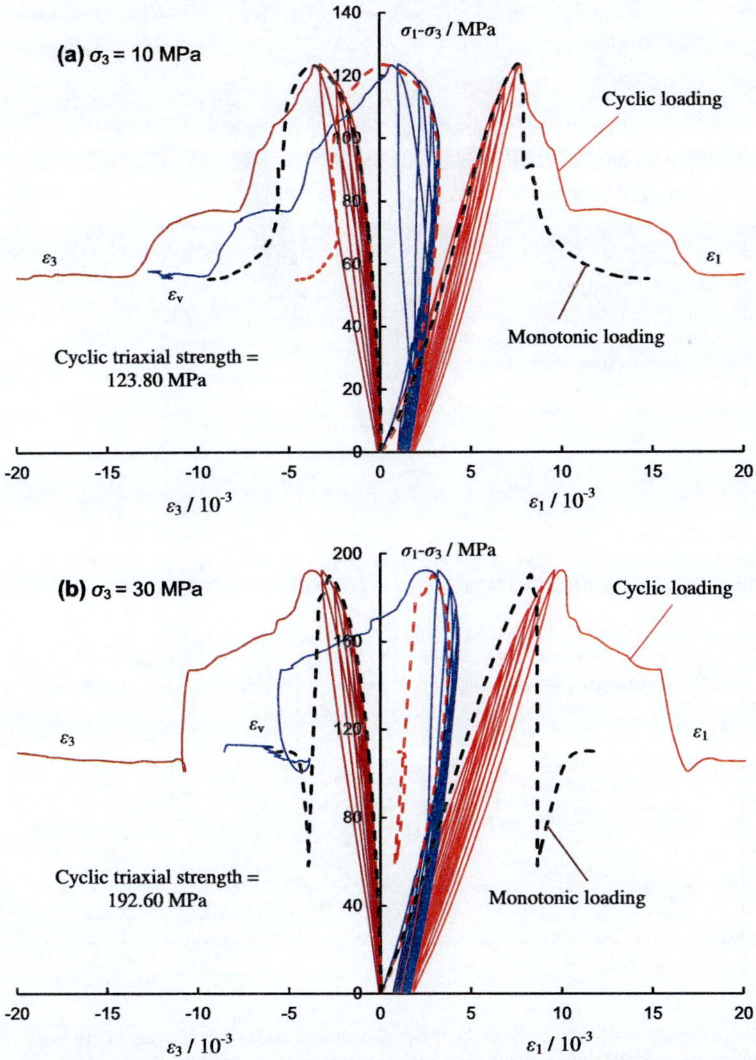

Fig. 5.12 Comparison of triaxial monotonic and cyclic stress-strain curves of yellow sandstone. **a** $\sigma_3 = 10$ MPa. **b** $\sigma_3 = 30$ MPa

sandstone porosity, local yield occurred with increasing deformation under the higher confining pressures. After local yield is reached, extruded and fractured material in the rock can fill the internal voids sandstone and strengthen the supporting structure owing to the greater porosity, which results in higher triaxial strength under the cyclic loading path.

The crack damage threshold of rock specimens (Fairhurst and Hudson 1999; Heap et al. 2009; Wong et al. 1997) can be confirmed by the corresponding axial

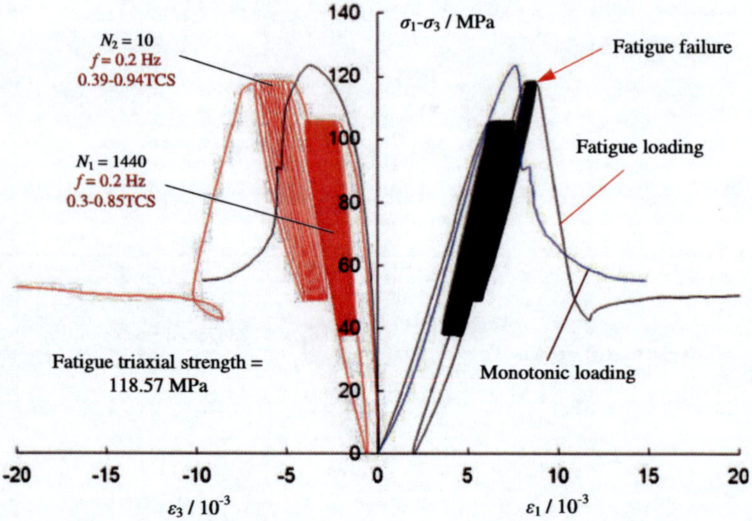

Fig. 5.13 Comparison of triaxial monotonic and fatigue stress–strain curves of yellow sandstone under the confining pressure of 10 MPa

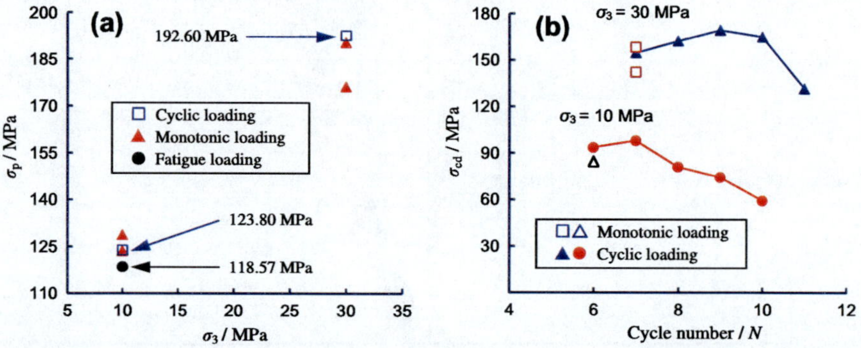

Fig. 5.14 Effect of confining pressure on the peak strength and effect of cycle number on the crack damage threshold of yellow sandstone specimens under triaxial cyclic loading path. **a** Comparison of monotonic triaxial strength, cyclic triaxial strength, fatigue triaxial strength. **b** Relation between crack damage threshold and cycle number

deviatoric stress where the volumetric deformation of the specimen switches from compaction-dominated to dilatancy-dominated (Yang et al. 2012). In accordance with the relation between deviatoric stress and volumetric strain, as shown in Fig. 5.12, the values of the crack damage threshold (σ_{cd}) of the sandstone specimens under triaxial cyclic loading were confirmed, as shown in Fig. 5.14b. The σ_{cd} of the sandstone under monotonic loading is also listed in the figure for comparison. As shown in the figure, when the cycle number was less than 6–7, the specimen did not switch from compaction-dominated to dilatancy-dominated; thus, the σ_{cd} of the specimen could not be obtained. At the same confining pressure, the first σ_{cd} of the sandstone specimen under cyclic loading is approximately equal to that under monotonic loading, which indicates that the discreteness of the tested specimens was extremely small in the present research. With an increase in the cycle number, the crack damage threshold of the specimens increased. For example, at $\sigma_3 = 30$ MPa, the σ_{cd} value of the sandstone increased from 154.5 to 169.3 MPa because it was more difficult for the closed cracks to re-open after unloading with the increase in cycle number in the early cycles. Afterward, the σ_{cd} value decreased from 169.3 to 131.5 MPa, which can be explained as follows (Yang et al. 2015). The closed cracks after unloading re-opened more easily with the increase in the cycle number; however, some new cracks were initiated in the specimen, which propagated and coalesced with the increase in the cycle number. Furthermore, the σ_{cd} value of the sandstone under higher confining pressure was generally greater than that under lower confining pressure. The confining pressure did not affect the evolution characteristics of the crack damage threshold.

Figure 5.15 shows the relationship between the cycle number and strain for the specimens under different confining pressures. An increase in the cycle number led to linear decreases in the elastic and plastic strains for all specimens. Furthermore, at the same cycle number, the plastic strain of the sandstone at $\sigma_3 = 10$ MPa was higher than that at $\sigma_3 = 30$ MPa. In addition, the elastic strain in all specimens was higher than the plastic strain at the same confining pressure. Furthermore, an increase in the cycle number resulted in a gradual increase in the difference between the elastic and plastic strains. The elastic strain was higher than the plastic strain under the same confining pressure because the accumulated damage is low at the pre-peak stage for brittle rock. The above conclusion is clearly different from that reported for marble specimens with high plastic deformation (Yang et al. 2016).

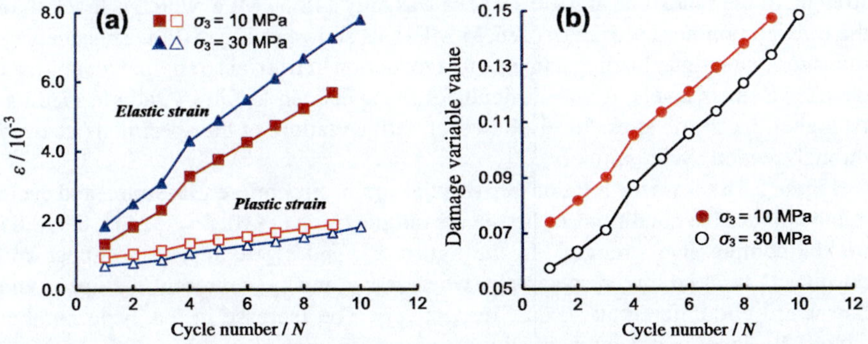

Fig. 5.15 Effect of cycle number on the elastic and plastic strain of yellow sandstone specimens under triaxial cyclic loading path. **a** Effect of cycle number on the elastic and plastic strain; **b** Effect of cycle number on the damage variable

To quantitatively evaluate the damage extent of the sandstone specimens under triaxial cyclic loading, a damage variable is defined in accordance with the irreversible deformation (plastic deformation) for each cycle (Yang et al. 2015), as expressed in Eq. (5.12):

$$D = \frac{(\varepsilon_p)_N}{\sum_{N=1}^{m} (\varepsilon_p)_N} \tag{5.12}$$

where D represents the damage variable of the rock, ε_p represents the plastic strain, N is the cycle number, and m represents the maximum cycle number.

Figure 5.15b shows that the confining pressure had an obvious effect on the damage evolution of the sandstone specimens under different confining pressures. For the same cycle number, the damage variable of the specimens at $\sigma_3 = 10$ MPa was greater than that at $\sigma_3 = 30$ MPa, which means that the confining pressure had a restraint effect on the damage extent of the sandstone. However, at the same confining pressure, the damage variable values of the specimens and the extent of the damage variable increased linearly with an increase in the cycle, which indicates that the damage extent increased with increasing axial deformation.

Figure 5.13 shows the fatigue loading of the sandstone under confining pressure of 10 MPa. At a frequency of 0.2 Hz, the stress amplitude at the first stage of fatigue ranged from 0.30 to 0.85 of triaxial compressive strength of the sandstone under the monotonic loading path. The cyclic time of the first fatigue stage was 2 h; thus, 1440 cycle numbers were completed at the first stage. However, the specimen did not experience failure during the first stage of fatigue. Afterward, at the same frequency of 0.2 Hz, the stress amplitude at the second stage of fatigue ranged from 0.39 to 0.94 of triaxial compressive strength of the sandstone under the monotonic loading path. The specimen experienced fatigue failure after only 10 cycles. Moreover, the fatigue strength of the sandstone at $\sigma_3 = 10$ MPa was only 118.57 MPa, which is lower than the triaxial monotonic strength (126.14 MPa), as shown in Fig. 5.14a. This analysis indicates that fatigue loading can lead to a reduction in triaxial strength. Under lower deviatoric stress levels, it was difficult for the specimen to fail by fatigue loading. At higher deviatoric stress levels, however, fatigue failure of the specimen occurred through several cycle numbers.

Figure 5.16 shows the relationship of axial strain, circumferential strain, and cycle number under the conditions of first-stage fatigue ($N_1 = 1440, f = 0.2$ Hz, 0.3–0.85 triaxial compressive strength). In the figure, ε_{1u} and ε_{1d} represent the upper and downward axial strains, respectively, whereas ε_{3u} and ε_{3d} represent the upper and downward circumferential strains, respectively. The increase in the cycle number caused the upper and downward strains to increase gradually, although the increase rate tended to be stable. This relation is similar to the steady creep of sandstone (Yang and Jiang 2010), which represents the relation of deformation with time. However, the difference between the upper and downward axial strains increased slowly with an increase in the cycle number. However, the difference between the upper and

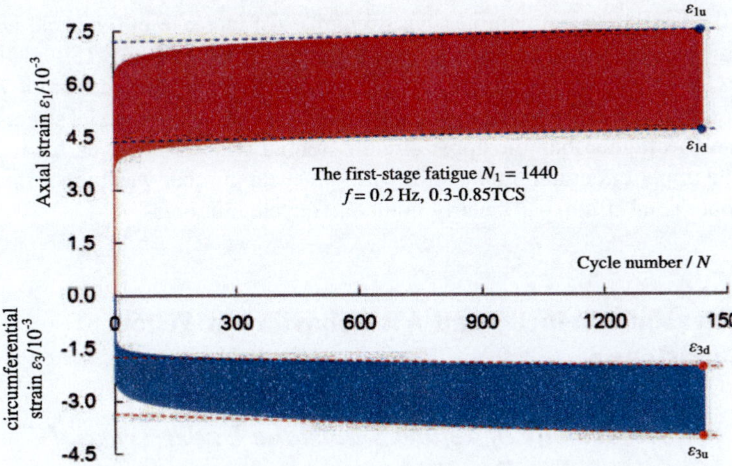

Fig. 5.16 Relationship between axial strain, circumferential strain of yellow sandstone and the cycle number under the condition of the first-stage fatigue. ($N_1 = 1440, f = 0.2$ Hz, 0.3-0.85TCS)

downward circumferential strains increased with increase in the cycle number even though the absolute value of the circumferential strain was lower than that of the axial strain. This indicates that the failure of specimen resulted mainly from the increase in circumferential deformation.

Figure 5.17 shows the relationship of the upper axial strain, upper circumferential strain, and cycle number under the condition of second-stage fatigue ($N_2 = 10, f =$

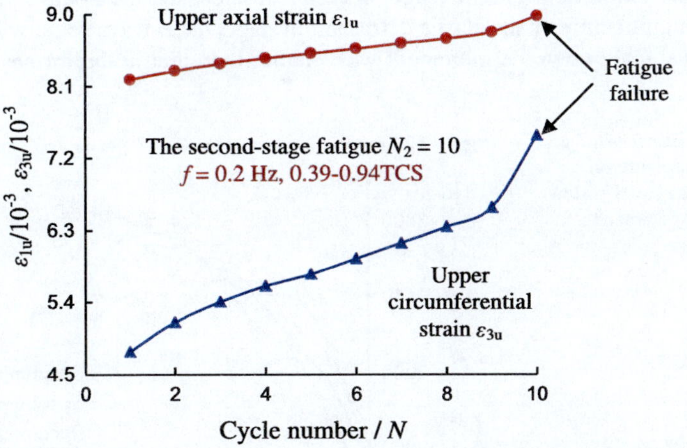

Fig. 5.17 Relationship between the upper axial strain, the upper circumferential strain of yellow sandstone and the cycle number under the condition of the second-stage fatigue. ($N_2 = 10, f = 0.2$ Hz, 0.39-0.94TCS)

0.2 Hz, 0.39–0.94 triaxial compressive strength). An increase in the cycle number caused the upper axial strain of the sandstone to increase from 8.189×10^{-3} to 8.983×10^{-3}; however, the upper circumferential strain increased from 4.789×10^{-3} to 7.484×10^{-3}. Even with the same cycle number, the upper axial strain of the sandstone was higher than the upper circumferential strain. Moreover, the increase rate of the upper circumferential strain with the cycle number was higher than that of the upper axial strain, particularly in the later cycle numbers.

5.3 Ultrasonic Velocity and AE Behaviors of Yellow Sandstone

5.3.1 P-wave Velocity of Yellow Sandstone Under Triaxial Monotonic Loading Path

Figure 5.18 shows that the P-wave velocity of the sandstone increased with an increase in the confining pressure owing to the denser structure of the specimen under higher confining pressure. Taking specimen No. 222 as an example when the confining pressure increased from 0 to 25 MPa, the P-wave velocity increased from 2680 to 3776 m/s; when the former increased from 25 to 60 MPa, the latter increased from 3776 to 4015 m/s. This implies that the increments in P-wave velocity at the lower range of confining pressure were larger than those at the higher range.

Figure 5.19 shows that the P-wave velocity exhibited different characteristics in different stress levels during the entire deformation process under triaxial compression. In the initial deformation stage, crack closure caused the P-wave velocity to increase significantly. In the elastic deformation stage, the P-wave velocity increased continuously, although the increment was smaller than that in the former stage. In

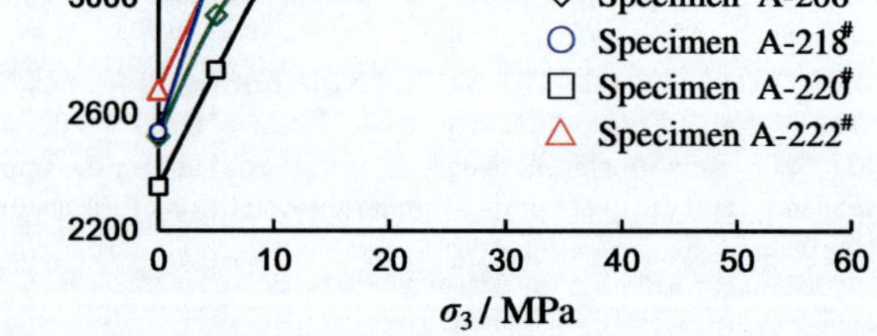

Fig. 5.18 Effect of confining pressure on P-wave velocity of yellow sandstone specimens

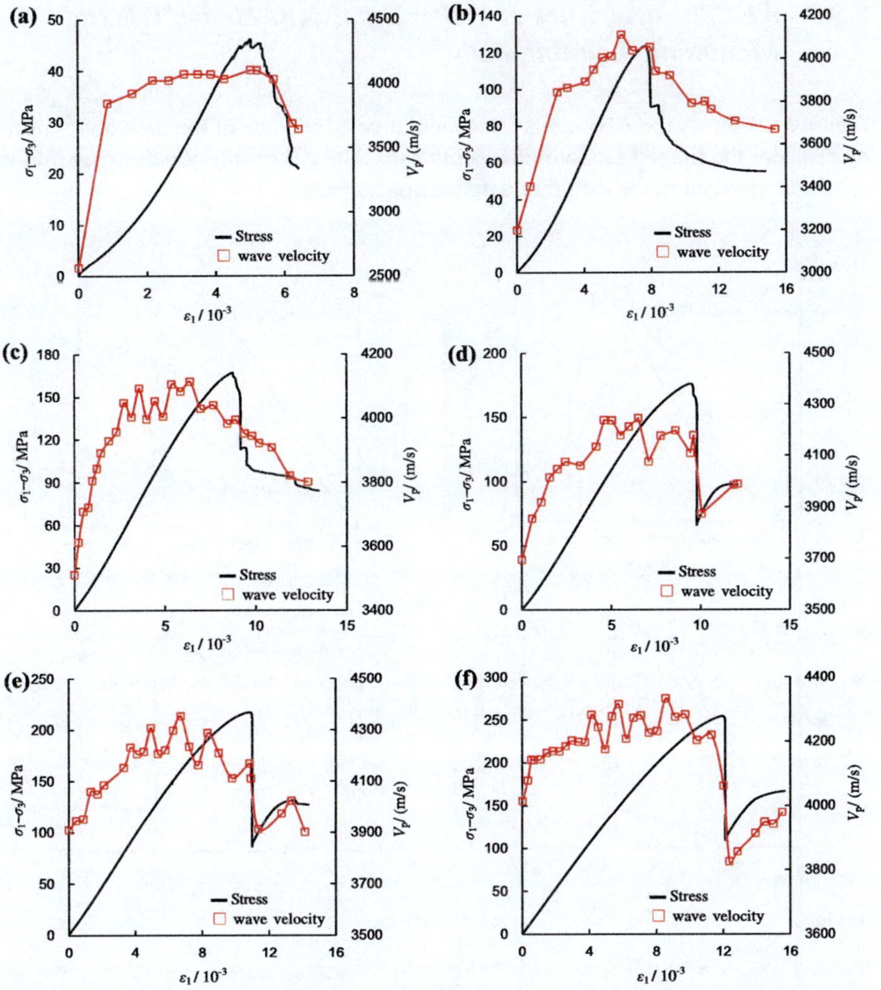

Fig. 5.19 P-wave velocity against axial strain during complete failure process with respect to various confining pressures. **a** Under a confining pressure of 0 MPa. **b** Under a confining pressure of 10 MPa. **c** Under a confining pressure of 20 MPa. **d** Under confining pressure of 30 MPa. **e** Under a confining pressure of 40 MPa. **f** Under a confining pressure of 60 MPa

the nonlinear deformation stage, the P-wave velocity increase fluctuated because of the crack initiation and propagation. After the failure, the P-wave velocity dropped as a result of macro cracking. However, in the residual strength stage, the P-wave velocity was almost unaffected by the increase in deformation. This phenomenon indicates that the P-wave velocity evolution can reflect the crack development in rock specimens.

5.3.2 AE Characteristics of Yellow Sandstone Under Triaxial Monotonic Loading Path

Figure 5.20 shows the AE counts and cumulative AE counts of the sandstone specimens under the triaxial monotonic loading path. The AE counts of yellow sandstone specimen evolved in the different deformation stages.

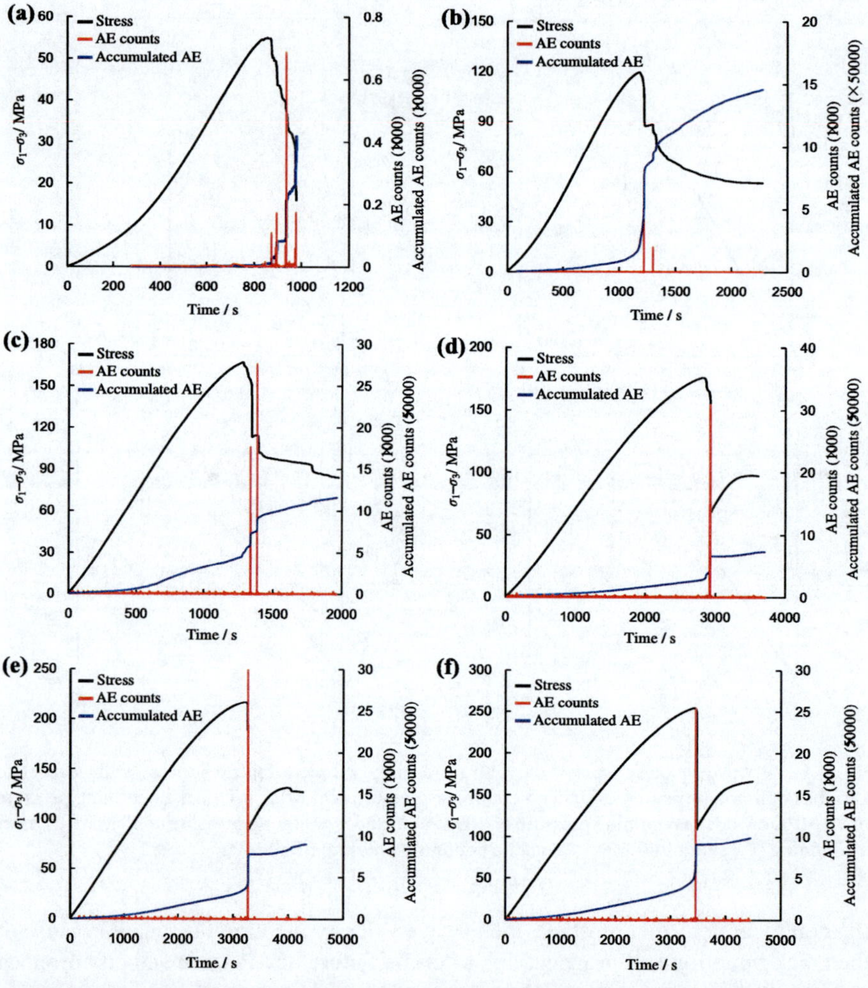

Fig. 5.20 AE counts and accumulated AE counts of yellow sandstone specimens under triaxial monotonic loading path. **a** Under a confining pressure of 0 MPa. **b** Under a confining pressure of 10 MPa. **c** Under a confining pressure of 20 MPa. **d** Under a confining pressure of 30 MPa. **e** Under a confining pressure of 40 MPa. **f** Under a confining pressure of 60 MPa

The evolution process of the AE counts is also closely related to the stress levels, which were divided into four phases (Huang et al. 2018). Under different confining pressures, the overall characteristics of the AE counts were similar. However, the local characteristics of the counts were influenced by the applied confining pressure. When the confining pressure was 0 MPa, multiple large AE counts were noted in the post-peak stage. However, only two large AE counts were observed after peak strength was reached under a confining pressure of 10 MPa. This indicates that more cracks were initiated under uniaxial compression compared with that under triaxial compression. When the confining pressure increased to 20 MPa, the shapes of the AE count curve were similar to those of 10 MPa. When the confining pressures were up to 30, 40, and 60 MPa, the AE behaviors of the specimens were different from the relatively lower confining pressures mentioned previously. Only one large AE count after peak strength was reached was observed under higher confining pressure. The number of large AE counts was dependent on the number of cracks in the specimen, which will be discussed in the following section.

Furthermore, the cumulative AE counts versus time under various confining pressures are also displayed in Fig. 5.20. The cumulative AE count curve of the sandstone under uniaxial compression (Fig. 5.20a) was highly similar to that previously reported for a granite specimen (Yang et al. 2017). The main difference in the curve of the cumulative AE count between uniaxial and triaxial compression was observed in the residual strength stage. The cumulative curves of specimens with different confining pressures had similar shapes. At low stress levels, the curve increased slowly. At higher stress levels before the peak strength was reached, however, the curve increased faster, followed by a sharp increase after the failure occurrence.

5.3.3 AE Characteristics of Yellow Sandstone Under Triaxial Cyclic Loading Path

In this research, two yellow sandstone specimens under different confining pressures were analyzed, with similar results obtained. For the analysis, the AE events considered were localized throughout the specimens. Figure 5.21 shows the AE counts and cumulative AE counts of the sandstone under the triaxial cyclic loading path, where the numbers in the figure denote the cycle numbers. To more clearly show the AE characteristics of triaxial cyclic loading, parts of Fig. 5.21 are enlarged in Fig. 5.22 and Fig. 5.23.

The curve of the cumulative AE events over time increased stepwise until failure stages were reached. Near the point of peak stress, new cracks formed and expanded to produce macro-fracture, which caused the AE event activity to increase significantly. The cumulative AE counts–time curve showed sharp growth, as displayed in Fig. 5.21.

Figures 5.22 and 5.23 show that in the first cycle, from the initial loading point to the unloading point, AE events in the sandstone specimens were active throughout

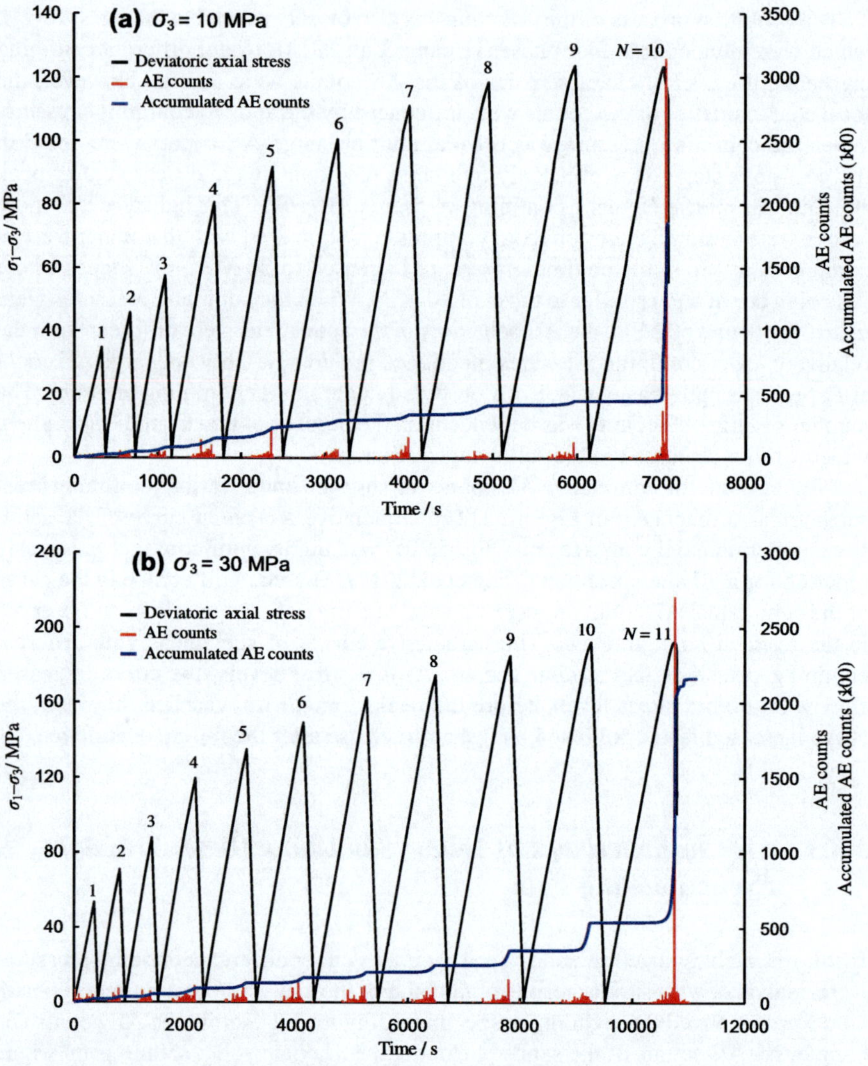

Fig. 5.21 AE counts and accumulated AE accounts of yellow sandstone under triaxial cyclic loading path. The denoted numbers in the figure are the cycle number

the process and were induced mainly by the initial microcrack closure and pore consolidation. From the unloading point to the point of zero stress, no AE events were observed in the specimens. In the second loading cycle, almost no AE events occurred before the deviatoric stress reached a certain threshold that was evidently lower than the previous maximum deviatoric stress. In addition, numerous AE events were detected. This phenomenon, defined as the Felicity effect, occurs when AE

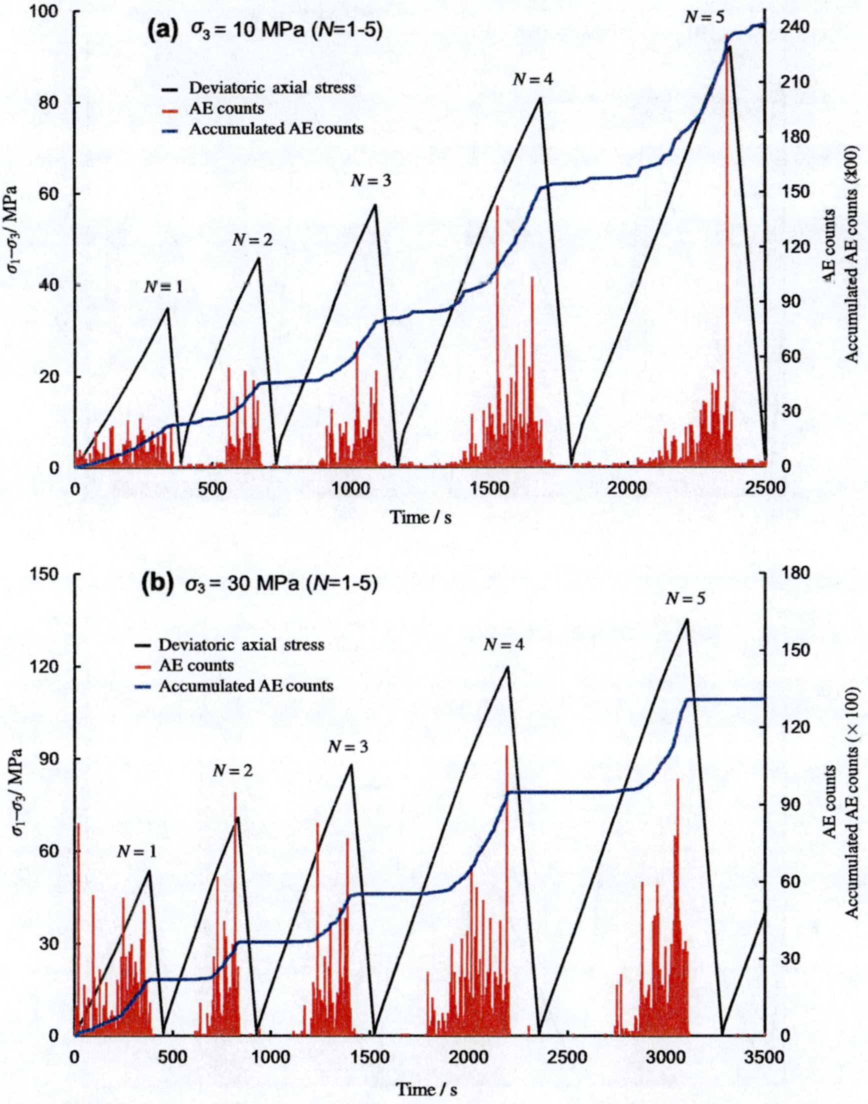

Fig. 5.22 Local magnification (N=1-5) on AE counts and accumulated AE accounts of yellow sandstone under triaxial cyclic loading path

events begin to significantly increase before the current stress reaches the previous maximum stress during the cyclic loading process. The Felicity ratio is an important quantitative evaluation index and is calculated by

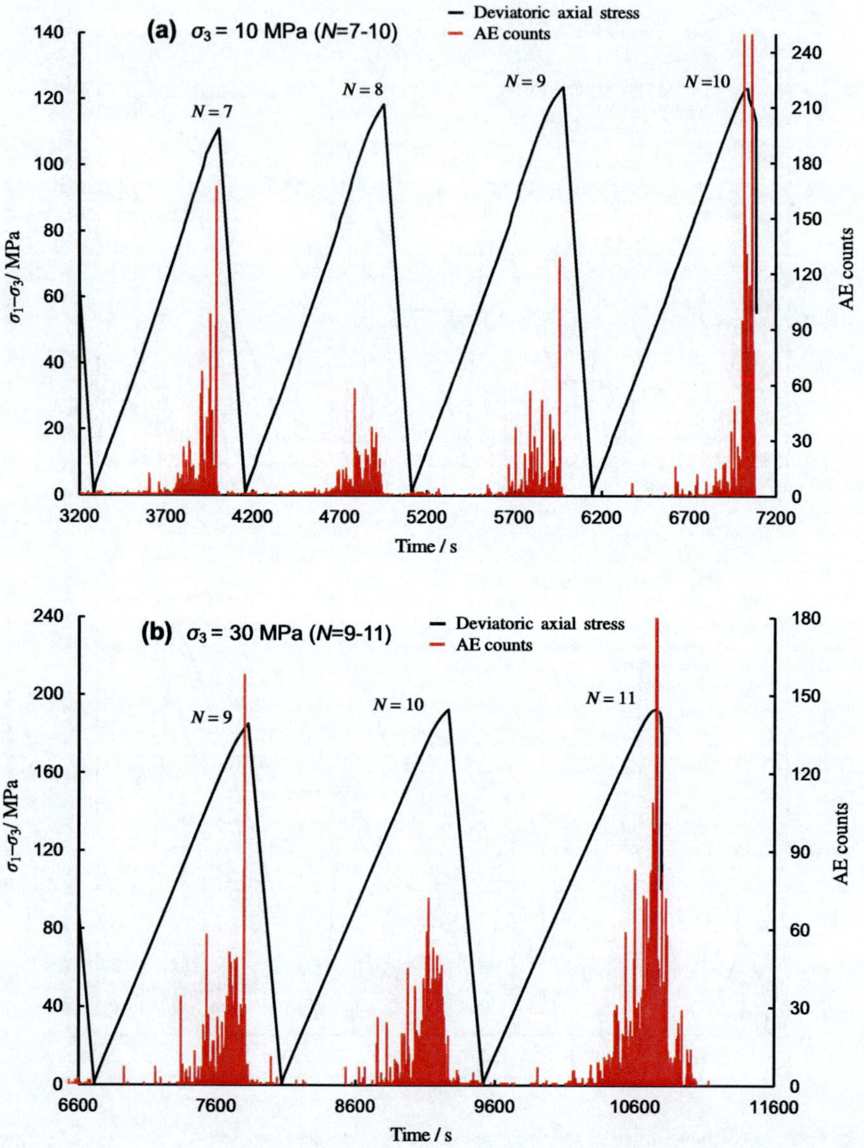

Fig. 5.23 Local magnification (N=7-11) on AE counts and accumulated AE accounts of yellow sandstone under triaxial cyclic loading path

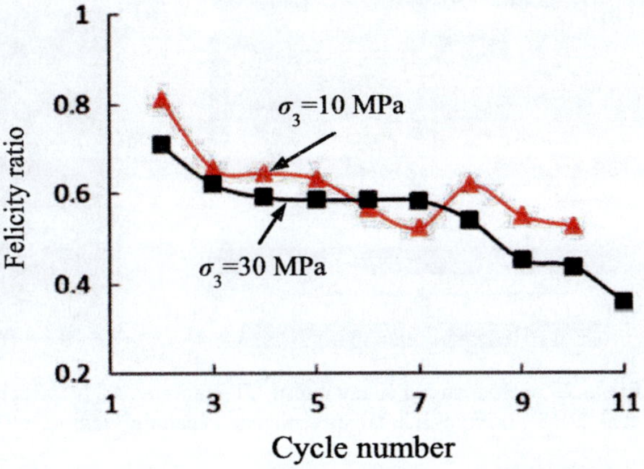

Fig. 5.24 Variation of Felicity ratio with cycle number

$$FR_i = \frac{P_{i+1}}{P_{i\,\max}} \tag{5.13}$$

where FR_i is the Felicity ratio in the ith cycle, P_{i+1} is the current stress level when significant AE events are detected in the $i + 1$ loading stage, and P_{imax} is the maximum stress level reached in the ith cycle. If $FR_i \geq 1.0$, the acoustic emission phenomenon is referred to as the Kaiser effect. Conversely, if $FR_i < 1.0$, it is known the Felicity effect. A smaller Felicity ratio represents a higher severity of initial damage or structural defects of the rock material, and the variation in the ratio reflects the complete evolution of the rock damage.

The changes in the rock Felicity ratio with increasing loading–unloading cycles are shown in Fig. 5.24. The Felicity effect was all obvious for the sandstone under the triaxial cyclic loading path, where the cycle's peak stress increased from cycle to cycle. As the cyclic number increased, the Felicity ratio of the sandstone decreased gradually under two different confining pressures. This indicates that the damage in the sandstone was exacerbated with an increase in cyclic stress.

5.4 Fracture Propagation Behavior of Yellow Sandstone

5.4.1 X-rary Micro-CT Scanning Procedure

To investigate the internal fracture characteristics of the sandstone specimens, advanced X-ray CT scanning was applied after the specimen failure. Figure 5.25 shows a schematic diagram of the X-ray micro-CT scanning system. Owing to the limitation of the scanning space, a height of 68 mm was regarded as the scanning region. In this study, the CT system employed was a Nanotom 160 high-resolution

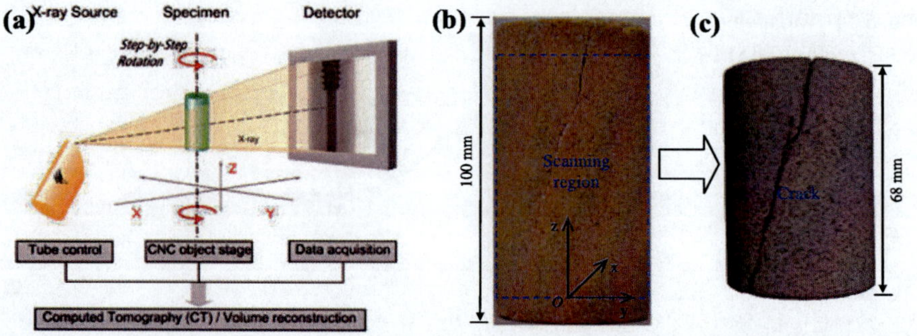

Fig. 5.25 Schematic of X-ray micro-CT scanning. **a** The principle of X-ray CT observation (Yang et al. 2015). **b** Experimental specimen and scanning region. **c** The CT scanning result

Table 5.3 X-ray CT scanning and reconstruction parameters in this study

Items parameters	Items parameters
Spatial resolution	30 μm
Source voltage	140 kV
Source current	80 μA
Number of rows	columns 2048 × 2048
Object to source distance	200 mm
Camera to source distance	500 mm
Exposure	1000 ms
Rotation step	0.18°
Duration (per slice)	4 s
Slice width	height 1730 by 1730 pixels
Smoothing	2
Image format	32 bit float

micro-CT with a spatial resolution of 30 mm. Details of the CT scanning procedure are given in Yang et al. (2016). Table 5.3 lists the parameters used in the CT observation.

5.4.2 *Surface Crack Characteristics of Yellow Sandstone*

Figure 5.26 shows a comparison of surface failure modes between actual photographs and the CT scanning results of the sandstone specimens under the triaxial monotonic loading path. In Fig. 5.26b, black regions with lower CT numbers represent cracks; other regions with higher CT numbers indicate no surface failure. The X-ray CT scanning surface images were highly similar to those of the actual surface cracks,

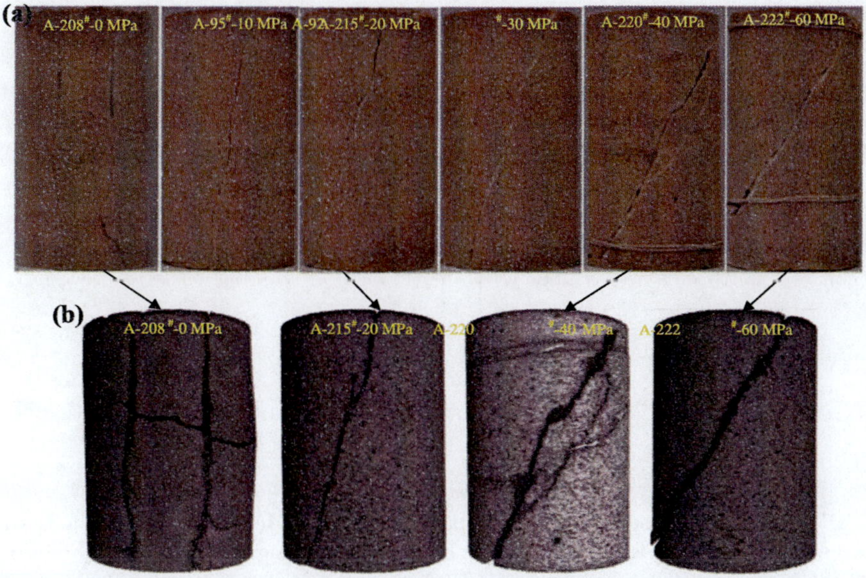

Fig. 5.26 Ultimate failure modes of yellow sandstone specimens under triaxial monotonic loading path. **a** Actual surface crack photographs. **b** X-ray CT scanning surface images

which demonstrates that X-ray micro-CT scanning is effective for exploring the internal damage in the yellow sandstone material.

Figure 5.27 shows ultimate failure modes of the sandstone specimens under the triaxial cyclic loading path. A comparison of Figs. 5.26 and 5.27 revealed that under triaxial cyclic loading, the system of cracks in the sandstone was more complicated than that under triaxial monotonic loading at the same confining pressure (Yang et al. 2015). However, the specimen exhibited typical shear fracture characteristics under the fatigue loading path, which did not lead to the production of new cracks occurring under the triaxial cyclic loading path.

Fig. 5.27 Ultimate failure modes of yellow sandstone specimens under triaxial cyclic loading path

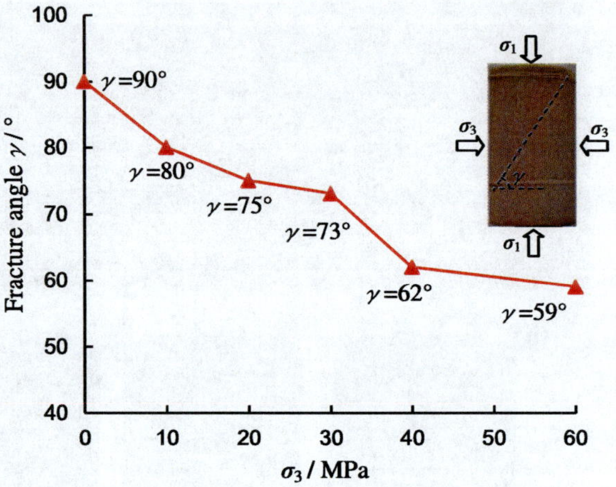

Fig. 5.28 Effect of confining pressure on the fracture angle of yellow sandstone specimens under triaxial monotonic loading path

For the triaxial monotonic loading path, the ultimate failure modes of the sandstone specimens were different under different confining pressures. As shown in Fig. 5.26, several axial tensile cracks and horizontal cracks occurred under uniaxial compression, whereas a single shear crack occurred under triaxial compression. The failure content under unconfined compression was higher than that under confined compression. It is clear that the shear crack crossed each specimen under triaxial compression. Facture angle γ was defined as the angle between the shear plane and the minor principal stress (i.e., confining pressure σ_3) direction. The influence of confining pressure on the fracture angle is plotted in Fig. 5.28. As the confining pressure increased from 10 to 60 MPa, the fracture angle decreased from 80° to 59°, which is consistent that reported in previous research (Wasantha and Darlington 2013).

5.4.3 *Internal Crack Evolution of Yellow Sandstone*

The internal crack patterns were investigated by analyzing cross-sections under different confining pressures. In this study, the front vertical cross-section images were captured at 0.5 mm intervals, and the scanning sequence was from front to back. The horizontal cross-section images were captured at 0.03 mm intervals, and the scanning sequence was from up to down; each image was 0.03 mm thick.

Figure 5.29 shows the vertical cross-section images of the specimens under triaxial compression. When the depth x was 10 mm, under triaxial compression, several cracks were observed in the cross-section. However, under triaxial compression, the fracture system was simple, and an inclined crack was observed in the cross-section. Furthermore, the width of the crack under uniaxial compression was larger than that under triaxial compression. The failure patterns in these cross-section when $x =$

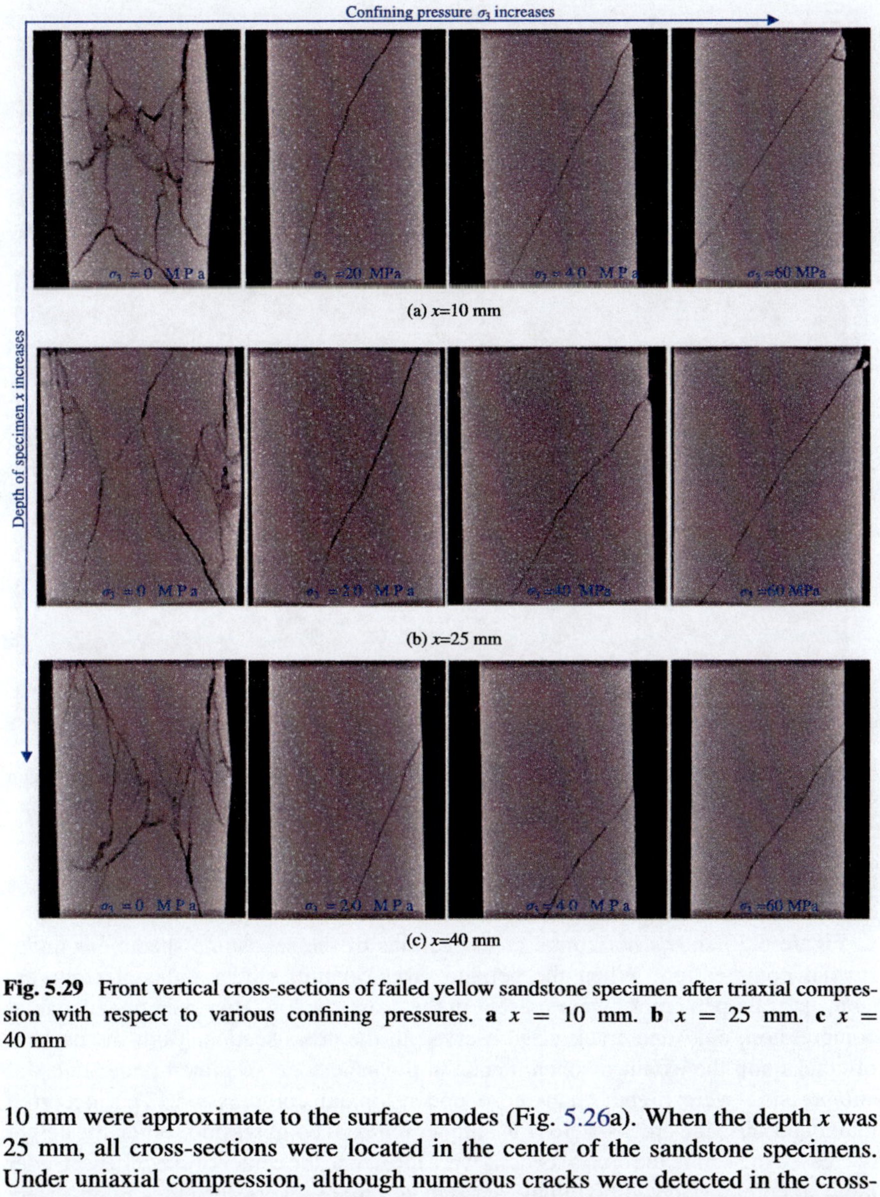

Fig. 5.29 Front vertical cross-sections of failed yellow sandstone specimen after triaxial compression with respect to various confining pressures. **a** $x = 10$ mm. **b** $x = 25$ mm. **c** $x = 40$ mm

10 mm were approximate to the surface modes (Fig. 5.26a). When the depth x was 25 mm, all cross-sections were located in the center of the sandstone specimens. Under uniaxial compression, although numerous cracks were detected in the cross-section, the directions of crack propagation differed significantly from those when $x = 10$ mm. In these cross-sections, the cracks did not propagate along the loading direction, which indicates that the cracks changed during their propagation. Under triaxial compression, there was no notable change in the cracks. When the depth x increased to 40 mm, several new cracks were initiated in the specimen under triaxial

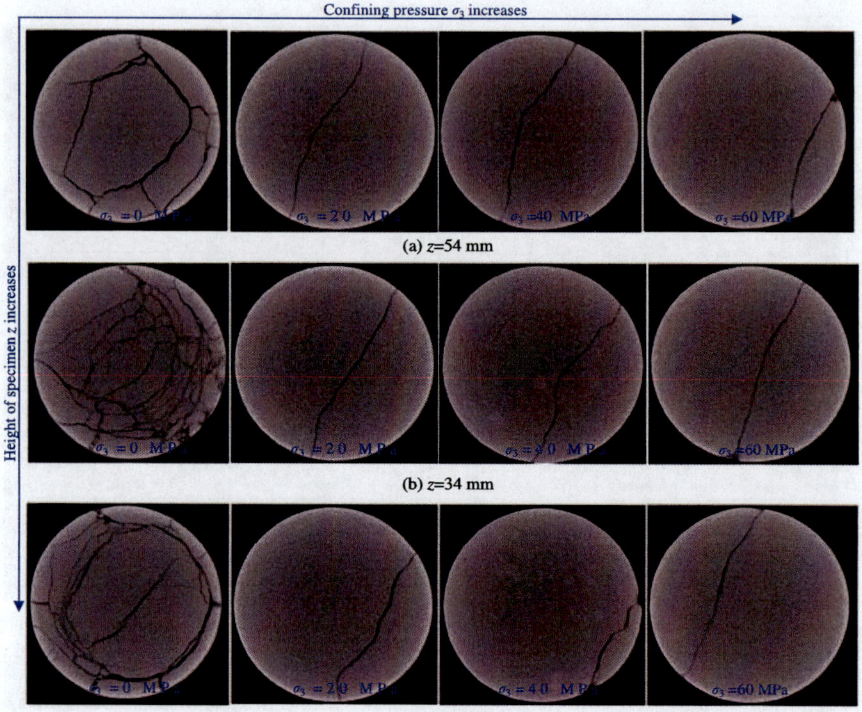

Fig. 5.30 Horizontal cross-sections of failed yellow sandstone specimen after triaxial compression with respect to various confining pressures. **a** $z = 44$ mm. **b** $z = 34$ mm. **c** $z = 14$ mm

compression. This implies that the damage degree of the sandstone specimen is closely related to the position of the cross-section.

Figure 5.30 shows horizontal cross-sections of the sandstone specimens under triaxial compression. When the height z was 54 mm, under uniaxial compression, several open cracks were present in the cross-section. However, under triaxial compression, only one crack was observed in the cross-section. Both the number of cracks and the extent of open cracks in the sandstone specimen under uniaxial compression were higher than those under triaxial compression. This occurred mainly because the cracks were tensileunder uniaxial compression. When the height decreased to 54 mm, the cross-sections were located in the center of specimens. Under uniaxial compression, although the damage was more severe, the crack widths were smaller compared with that when z was 54 mm in the specimen. When the height reached 14 mm, the cracks were circumferential, which caused the rock to separate from the central part under uniaxial compression. However, under triaxial compression, the cracks occurred as rough lines crossing the specimens, indicating that the crack was a curved line in 3D. The above analysis reveals the importance of using X-ray CT scanning to study the internal fracture behavior.

Figure 5.31 shows the 3D patterns of fractures and fracture planes in the sandstone specimens under different confining pressures. The images in Fig. 5.31a are the 3D CT reconstructed results of the sandstone specimens after failure with respect to different confining pressures. In these images, the fracture regions with lower CT numbers are black; other regions with higher CT numbers are transparent. Based on these 3D reconstructed images of the failed specimens, the crack patterns were clearly distinguished in the opaque rock. Axial and horizontal tensile cracks were observed under uniaxial compression; however, these cracks might not have caused the specimen failure. The internal tensile cracks mainly resulted in the specimen failure, which is verified by the internal fracture photograph in Fig. 5.31b. Under triaxial compression, shear cracks were clearly visible in the specimens, and the surfaces of the shear cracks were highly rough. Unlike the clean surfaces of tensile

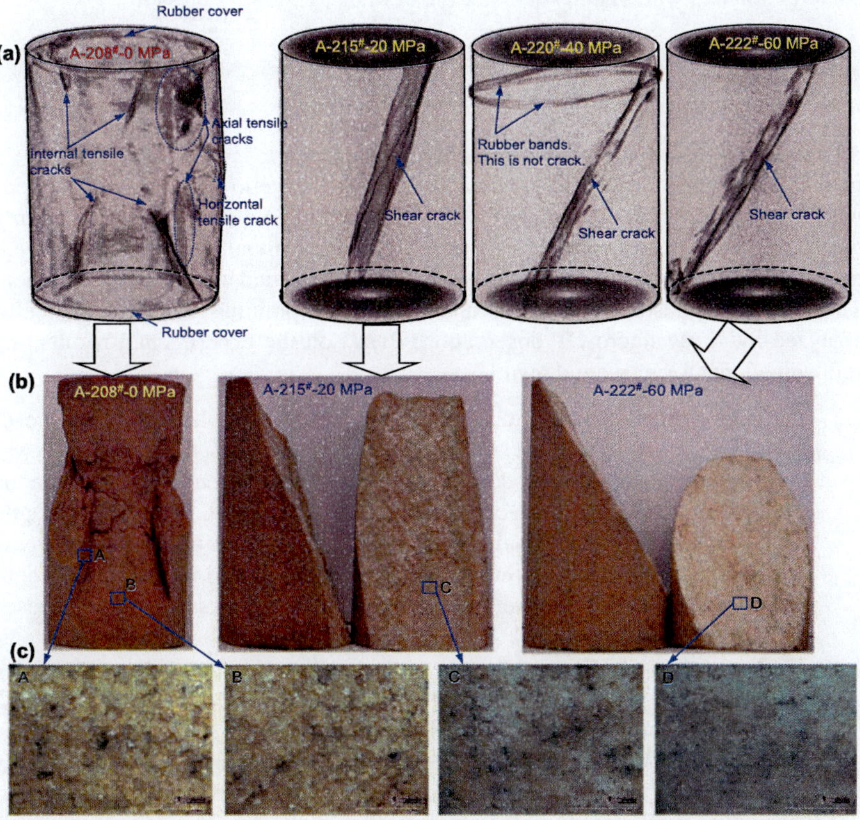

Fig. 5.31 Internal cracks of failed yellow sandstone specimens under different confining pressures. **a** Three-dimensional CT images. **b** Actual photograph of fractures. **c** Micro structures of tensile and shear cracks

cracks under uniaxial compression, powder was observed on the surfaces of the shear cracks under triaxial compression (Fig. 5.31b).

The microstructures of the tensile and shear cracks were observed by using an advanced microscope (model KH-8700E). Some of the representative results are plotted in Fig. 5.31c, which shows local amplification of the locations marked in Fig. 5.31b. These images show axial tensile cracking under uniaxial compression, internal tensile cracking under uniaxial compression, shear cracking under a confining pressure of 20 MPa, and shear cracking under a confining pressure of 60 MPa. The micro-observations for the cracks under uniaxial compression revealed that all grains on the surface remained intact, which indicates that the two cracks are tensile cracks. However, the micro-observations for the cracks under triaxial compression differed significantly from the mode under uniaxial compression. The surfaces of the cracks were smoother compared with those under uniaxial compression and were much smoother under higher confining pressure.

5.5 Conclusions

In this chapter, the strength and deformation behaviors of yellow sandstone specimens were investigated by applying a series of triaxial monotonic and cyclic tests. During the loading process, real-time AE and ultrasonic velocity monitoring techniques were used to explore the acoustic behaviors of the sandstone under triaxial compression. The fractured sandstone specimens under different confining pressures were also analyzed by X-ray micro-CT observation. Based on the experimental results, the following conclusions were drawn.

(1) According to the triaxial monotonic experimental results, the influence of confining pressure on the mechanical parameters of the sandstone was analyzed. The strength and deformation parameters depended strongly on the confining pressure. As the confining pressure increased, the peak strength, crack damage stress, and residual strength of the sandstone under the monotonic loading path all increased linearly, which was expressed in the linear Mohr–Coulomb criterion. However, the Poisson ratio of the sandstone decreased linearly.

(2) Under lower confining pressures, cyclic loading did not affect the triaxial strength of the sandstone; however, the loading increased the triaxial strength under higher confining pressures. Fatigue loading can lead to a reduction in triaxial strength. Under lower deviatoric stress levels, the specimen was highly resistant to failure by fatigue loading. Under higher deviatoric stress levels, however, fatigue failure of the specimen can occur even through several cycle numbers. The influence of the cycle number on the deformation and crack damage parameters of the sandstone was also investigated.

(3) The AE count and P-wave velocity of the sandstone under triaxial monotonic loading path were directly dependent on the stress level. Generally, the cumulative evolution of the AE count included few counts and linear, exponential, and stable increments, whereas the P-wave velocity included rapid increments, stable increments, unstable decrements, and constant values during the entire stress–strain process. Under the triaxial cyclic loading path, the Kaiser and Felicity effects were highly significant during the entire loading and unloading deformation.

(4) The internal fracture characteristics of the sandstone specimens under different confining pressures were investigated by analyzing the cross-section and 3D reconstructed images of CT scans, which revealed the fracture mechanism of the sandstone. Two failure patterns were observed in the specimens: axial splitting under uniaxial compression and single shear failure under triaxial compression. The fracture angle to the horizontal direction decreased with an increase in the confining pressure. The cracking modes in the CT scan cross-sections changed with the depth and height of the slices.

References

Fairhurst CE, Hudson JA (1999) Draft ISRM suggested method for the complete stress-strain curve for intact rock in uniaxial compression. Int J Rock Mech Min Sci 36:279–289

Hawkes I, Mellor M (1970) Uniaxial testing in rock mechanics laboratories. Eng Geol 4(3):179–285

Heap MJ, Baud P, Meredith PG, et al. (2009) Time-dependent brittle creep in Darley Dale sandstone. J Geophys Res Solid Earth 114(B7)

Huang YH, Yang SQ, Hall MR et al (2018) The effects of NaCl concentration and confining pressure on mechanical and acoustic behaviors of brine-saturated sandstone. Energies 11:385. https://doi.org/10.3390/en11020385

Huang YH, Yang SQ, Tian WL (2019) Crack coalescence behavior of sandstone specimen containing two pre-existing flaws under different confining pressures. Theoret Appl Fract Mech 99:118–130

Mogi K (1966) Some precise measurements of fracture strength of rocks under uniform compressive stress. Rock Mech. Eng. Geol. 4:41–55

Rathnaweera TD, Ranjith PG, Perera MSA (2016) Experimental investigation of geochemical and mineralogical effects of CO2 sequestration on flow characteristics of reservoir rock in deep saline aquifers. Sci Rep. https://doi.org/10.1038/srep19362

Wang Y, Li X, Zheng B (2016) Experimental study on mechanical properties of clay soil under compression by ultrasonic test. Eur J Environ Civ Eng. https://doi.org/10.1080/19648189.2016.1217791

Wasantha PLP, Darlington WJ, Ranjith PG (2013) Characterization of mechanical behaviour of saturated sandstone using a newly developed triaxial apparatus. Exp Mech 53(5):871–882

Wong TF, David C, Zhu W (1997) The transition from brittle faulting to cataclastic flow in porous sandstones: Mechanical deformation. J Geophys Res: Solid Earth 102(B2):3009–3025

Yang SQ, Jiang YZ (2010) Triaxial mechanical creep behavior of sandstone. Min Sci Technol 20(3):339–349

Yang SQ, Jing HW, Wang SY (2012) Experimental investigation on the strength, deformability, failure behavior and acoustic emission locations of red sandstone under triaxial compression. Rock Mech Rock Eng 45(4):583–606

Yang SQ, Ranjith PG, Huang YH et al (2015) Experimental investigation on mechanical damage characteristics of sandstone under triaxial cyclic loading. Geophys J Int 201(2):662–682

Yang SQ, Ju Y, Gao F et al (2016) Strength, deformability and X-ray micro-CT observations of deeply buried marble under different confining pressures. Rock Mech Rock Eng 49(11):4227–4244

Yang SQ, Ranjith PG, Jing HW, Tian WL, Ju Y (2017) An experimental investigation on thermal damage and failure mechanical behavior of granite after exposure to different high temperature treatments. Geothermics 65:180–197

Yang SQ, Huang YH, Ranjith PG (2018) Failure mechanical and acoustic behavior of brine saturated-sandstone containing two pre-existing flaws under different confining pressures. Eng Fract Mech 193:108–121

Yang SQ, Huang YH, Tang JZ (2020) Mechanical, acoustic, and fracture behaviors of yellow sandstone specimens under triaxial monotonic and cyclic loading. Int J Rock Mech Min Sci 130:104268

Zhu WC, Li ZH, Zhu L et al (2010) Numerical simulation on rockburst of underground opening triggered by dynamic disturbance. Tunn Undergr Space Technol 25(5):587–599

Chapter 6
Triaxial Strength and Deformation Failure Behavior of Coarse Marble Under Six Different Loading Paths

In the past, some experiments on the influence of loading path on mechanical behavior of rock have been carried out (Lee et al. 1999; Xiong and Zhou 2006; Yang et al. 2011), but the strength, deformation and failure behaviors of rock under various loading paths have not been fully investigated, and the fundamental mechanisms of loading path on the triaxial mechanical behavior of rock are still not fully understood. Besides, real engineering rock mass experienced by loading path is more complex, which makes that mechanical behavior of rock under complex loading paths has been more and more important. Re-fracture mechanical behavior of fractured rock has been one of the most important scientific questions (Yang et al. 2008, 2009; Han and Yang 2009). However the re-fracture mechanical behavior of fractured rock is not clear under different loading paths, so it is necessary to explore the mechanical behavior of rock under different loading paths, which is helpful to the effective control of underground engineering rock unstable failure.

Therefore in this chapter, in order to investigate the influence of loading path on mechanical behavior of rock, a number of triaxial compression experiments were carried out for the coarse marble specimens under six different loading paths by the rock mechanics servo-controlled testing system. This research is focused on to investigate the influence of loading path on the strength parameters (cohesion and internal friction angle) of coarse marble material, to analyze the re-fracture strength and deformation behavior of flawed and damaged rock specimen under different loading paths, and to discuss the failure mode of coarse marble under different loading paths.

S.-Q. Yang, *Mechanical Behavior and Damage Fracture Mechanism of Deep Rocks*,
https://doi.org/10.1007/978-981-16-7739-7_6

6.1 Experimental Methodology

6.1.1 Coarse Marble Material and Specimen Preparation

In this research, the coarse marble material located the midland ground of China was chosen for the experimental object, which was the same as that in literature (Yang et al. 2008). The coarse marble has a crystalline and blocky structure, which is macroscopically very homogeneous with average unit weight about 2700 kg/m^3. The minerals in the coarse marble specimens are mainly calcite, dolomite and magnesite, and the chemical components are mainly Ca and Mg. The crystal size range of the coarse marble is 4.0 − 6.0 mm (5.0 mm in average). The average crystal size of the coarse marble was determined by optical microscope and SEM.

In order to investigate the mechanical behavior of rock under various loading path, a number of coarse marble specimens were prepared to carry out triaxial experiment. When drilling, we machined the coarse marble specimens along the same direction in order to avoid the influence of anisotropy on the experimental results of coarse marble. At the same time, machined coarse marble specimens were observed and carefully selected to preserve specimens suitable for testing. In order to obtain the exact results as well as the best comparison, all the experiments were carried out in natural and dry conditions.

According to the method suggested by the ISRM (Fairhurst and Hudson 1999), the length to diameter ratio of tested specimen should be in the range of 2.0 to 3.0 in order to minimize the influence of the end friction effects on the testing results. Therefore, the size of all tested coarse marble specimens is cylindrical with 50 mm in diameter and 100 mm in length. As a result, all tested specimen with the length to diameter ratio of 2.0 ensure a uniform stress state within the specimens. The mechanical behaviors of specimens under different loading paths were also determined according to the method suggested by the ISRM.

6.1.2 Testing Procedure for Six Different Loading Paths

All compression experiments for coarse marble specimens were carried out on a rock servo-controlled system with the maximum loading capacity of 2000KN and the maximum displacement capacity of 25 mm. This servo-controlled system can test specimens in load or displacement control while the data from LVDT (Linear Variable Displacement Transducer) are recorded and analyzed in real-time.

In this research, we measured the axial force with the loading capacity of 1000KN and the axial deformation with the displacement capacity of 5 mm. During the whole uniaxial and triaxial compression experiments, loads and deformations of the tested coarse marble specimens were recorded simultaneously. Moreover, two rigid steel cylinders (50 mm in diameter and 20 mm in length) were placed between the loading frame and rock specimen, which decreased distinctly the influence of the end friction

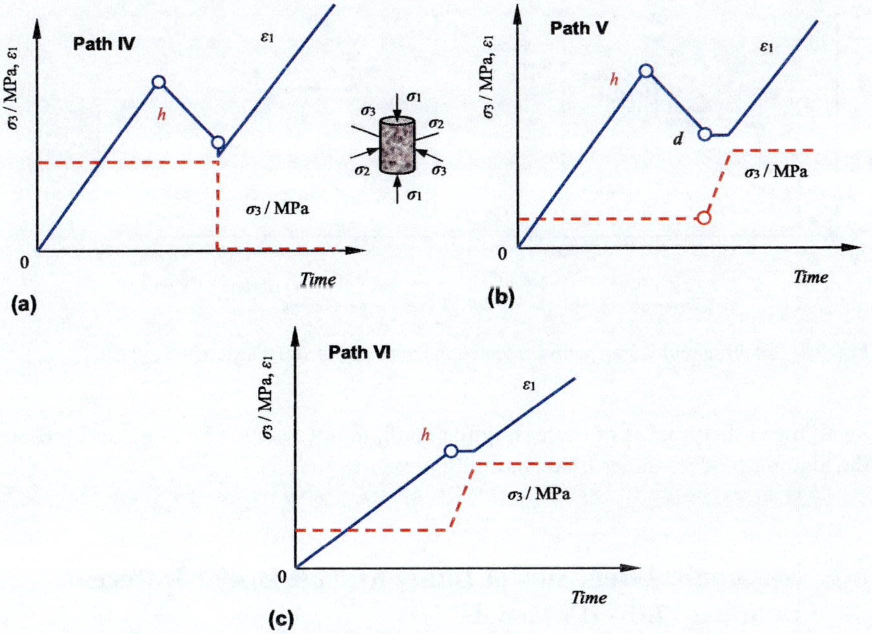

Fig. 6.1 Three different loading paths in the present study

effects on the testing results of marble specimens with the length to diameter ratio −2.0 (Yang et al. 2005a, 2005b).

In this study, to investigate the effect of the loading path on the strength and deformation failure behavior of coarse marble under conventional triaxial compression, six different loading paths were chosen. It should be noted that three loading paths (Path I, II and III) had been presented in Fig. 1.2 (Chapter 1), and another three loading paths was described in Fig. 6.1. For Path I, σ_3 is set to 0, 5, 10, 20 and 30 MPa.

Under six different loading paths, the confining pressure is first applied to the specimen at a constant rate of 0.5 MPa/s, which ensures that the specimen is under uniform hydrostatic stresses ($\sigma_1 = \sigma_2 = \sigma_3$). And then, the axial deviatoric stress begins to impose the surface of rock specimen at the 0.002 mm/s (2×10^{-5}/s) or 0.127 MPa/s control rate. After above procedure, some different loading paths are designed to investigate the effect of loading path on mechanical behavior of coarse marble, which can be seen from Fig. 6.1.

The above six loading paths in the principal stress space (σ_3, σ_1) are further presented in Fig. 6.2, in which *oa* and *of* all represent hydrostatic stress ($\sigma_1 = \sigma_2 = \sigma_3$). *ab* → *bc*, *fg* → *gh* and *ab* → *bh* all represent the loading process under initial confining pressure; *cd* shows the process of confining pressure reduction while keeping axial strain constant; *hi* and *hd* all mean the process of unloading the axial deviatoric stress; *io* represents the process of unloading the confining pressure; *de* and

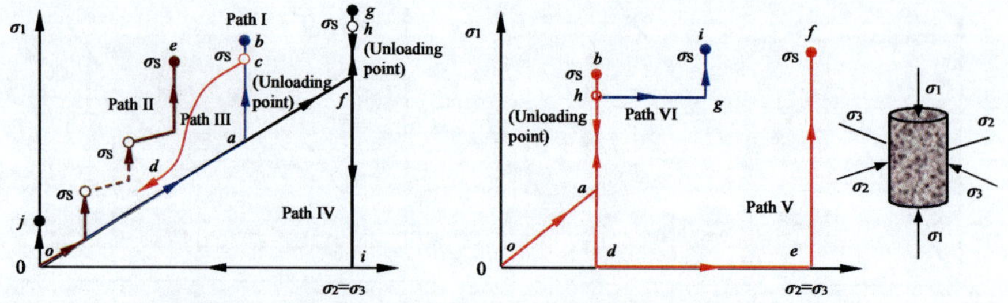

Fig. 6.2 Six different loading paths presented in the principal stress space (σ_3, σ_1)

hg all mean the process of increasing the confining pressure; *ef* and *gi* all represent the loading process under final confining pressure.

6.2 Mechanical Behavior of Intact Marble Under Different Loading Paths (Paths I-II)

Typical axial stress-axial strain curves for intact coarse marble under Path I are shown in Fig. 6.3, in which the denoted number is the confining stress, MPa. In accordance with Fig. 6.3, we can conclude that the yielding stress (the stress value that begins to depart the linear elastic phase in the axial stress-axial strain curve) and TCS (triaxial compressive strength) of specimen increase gradually with the increase of confining pressure. For intact coarse marble as shown in Fig. 6.3, the post-peak deformation

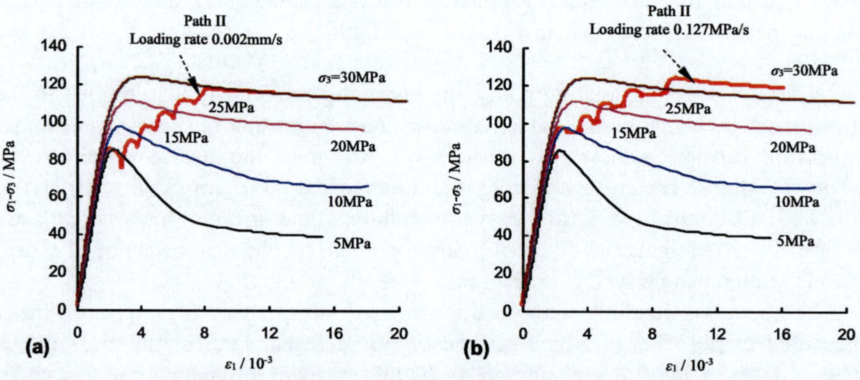

Fig. 6.3 Typical triaxial stress–strain curves of intact coarse marble under Path I and II (The bold line is the experimental results of coarse marble under Path II; however other experimental results are all obtained under Path I at the loading rate of 0.002 mm/s)

changes gradually from brittleness to ductility. When σ_3 equals approximately to − 30 MPa, the deformation of the specimen comes out distinct yielding platform and the plastic flow takes place, i.e. the post-peak deformation behavior of the specimen changes into ideal plasticity (Yang et al. 2008). Moreover, in the reference (Yang et al. 2008), the mechanism on the brittle-ductile transition of intact coarse marble specimen has also been made a detailed description.

Taking into account larger plastic and ductile deformation behavior of coarse marble after peak strength, a method of loading path for single specimen by Path II is put forward to confirm the peak strength under different confining pressures with only one specimen. From Fig. 6.2, it is very clear that there is an obvious difference between Path I and II, which can be described as follows. Under Path I, one specimen can only confirm a TCS and many specimens must be tested to construct the relation between the peak strength and the confining pressure. However under Path II, single specimen can be used to obtain a series of TCS. In Fig. 6.3, the axial stress-axial strain curves of coarse marble under Path II are also shown at two different loading rates, i.e. 0.002 mm/s (2×10^{-5}/s) and 0.127 MPa/s. From Fig. 6.3, it can be seen that Path II may be used to confirm a series of peak strengths of rock with different confining pressures.

In accordance with the peak strength of intact coarse marble under Path I and II, we can analyze the effect of two loading paths (Path I and II) on the strength behavior by adopting linear Mohr–Coulomb criterion, which can be expressed in terms of the maximum axial supporting capability σ_S and the minimum principal stress σ_3 (i.e. confining pressure) (Yang et al. 2008), i.e. $\sigma_S = \sigma_0 + q\sigma_3$, where σ_0 is usually regarded as UCS (uniaxial compressive strength). The σ_0 and q are related to the cohesion C and the internal friction angle φ, which can be expressed in the following forms, respectively.

$$\sigma_0 = 2C\cos\varphi/(1 - \sin\varphi) \tag{6.1}$$

$$q = (1 + \sin\varphi)/(1 - \sin\varphi) \tag{6.2}$$

In accordance with linear Mohr–Coulomb criterion, the influence of σ_3 on σ_S for intact coarse marble specimen under Path I and II are presented in Fig. 6.4. The line $\sigma_S = 73.68 + 2.77\sigma_3$ is regressed according to the strengths of rock obtained under Path I, which there is a good linear regression coefficient of $R = 0.985$. It is very clear that the strength of coarse marble specimen under Path II has an inappreciable difference with that under Path I, which proves that Path II is very feasible and reasonable to confirm the peak strength of rock under different confining pressures by only one specimen. From Fig. 6.5, we can see out that the intact coarse marble takes on typical shear failure mode even though in uniaxial compression. Therefore, it can be believed that experimental results of intact coarse marble specimen have a smaller deviation from the linear Mohr–Coulomb criterion, which testifies that linear Mohr–Coulomb criterion is suitable for intact coarse marble in this research. The intact coarse marble under Path I all lead to typical shear failure mode with

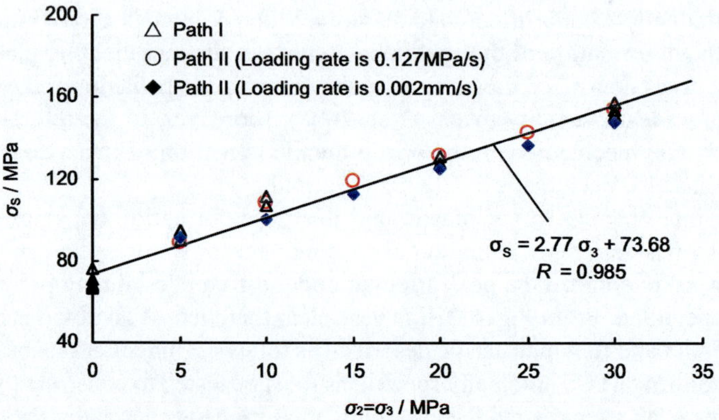

Fig. 6.4 Strength analysis of intact coarse marble under Path I and II. All the symbols represent the experimental results and the line represents the theoretical value in accordance with Mohr–Coulomb criterion. (*R* is the correlation coefficient of determination of the regression.)

Fig. 6.5 Failure modes of intact coarse marble under different confining pressures (Path I)

a single fracture surface (Fig. 6.5) and the fracture surface is very smooth, which results from the friction of macroscopic crack surfaces. However the failure mode of coarse marble specimen under Path II is similar to that ($\sigma_3 = 30$ MPa) under Path I, as shown in Fig. 6.5.

Table 6.1 Effect of loading path on strength parameters of intact coarse marble in accordance with linear Mohr–Coulomb criterion

Loading path	Loading rate	σ_0/MPa	q	C/MPa	φ/(°)	R
Path I	0.002 mm/s	73.68	2.77	22.14	28.0	0.985
Path II	0.002 mm/s	78.35	2.32	25.72	23.4	1.000
	0.127 MPa/s	79.28	2.55	24.82	25.9	0.995

Note All the experiments in Table 6.1 were all carried out in January, 2007

Although there are not obvious differences between peak strengths obtained under Path I and II, we can still see out the diversity of peak strengths under Path II at two different loading rates, which is further presented in Table 6.1. In terms of Table 6.1, it is very clear that the values of σ_0 and q are 78.35 MPa and 2.32 at a control rate of 0.002 mm/s, while the values of σ_0 and q are 79.28 MPa and 2.55 at a control rate of 0.127 MPa/s. According to Eq. (6.1) and (6.2), it can be seen that, for intact coarse marble specimen, the values of C and φ, respectively, equal to 22.14 MPa and 28.0° for Path I; $C = 25.72$ MPa, $\varphi = 23.4°$ for Path II (Loading rate −0.002 mm/s); and $C = 24.82$ MPa, $\varphi = 25.9°$ for Path II (Loading rate − 0.127 MPa/s), which further illustrates the feasibility and accuracy to confirm the cohesion and internal friction angle of rock by loading Path II with only one specimen. It needs to be noticed that suggested rock specimen number is not less than five in the ISRM for triaxial compression experiment (Fairhurst and Hudson 1999). However, due to a limited number of cores drilled on-site, especially from weak, stratified and fractured rock mass, moreover testing items are more; it is very difficult to repeat all the experiments in the laboratory. Therefore for a kind of rock that has obvious plastic and ductile deformation behavior (e.g. marble, chalk, mudstone), the Path II are suggested to determine the cohesion and internal friction angle of rock with only one specimen, which can eliminate the effect of heterogeneity on the peak strength of rock. But it is not very easy to grasp the control point (A and B shown in Fig. 6.6) that the

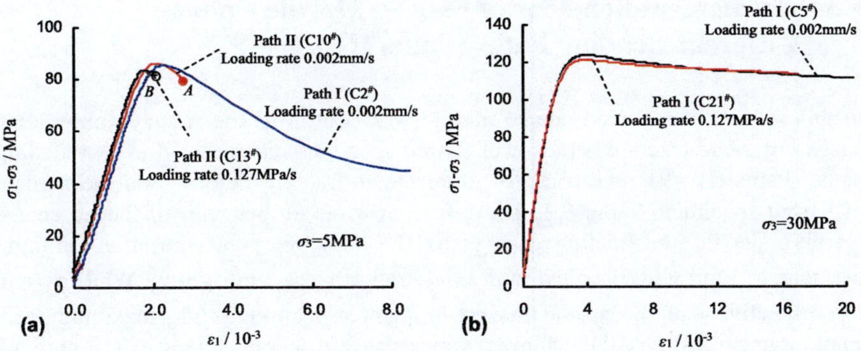

Fig. 6.6 Axial stress-axial strain curves of intact coarse marble under Path I and II ($\sigma_3 = 5$ MPa and 30 MPa)

axial deviatoric stress is stopped immediately when the specimen reaches post-peak stress, which will be explained as follows.

Figure 6.6 shows the axial stress-axial strain curves of intact coarse marble under Path I and II. At $\sigma_3 = 5$ MPa, the strength and deformation behavior of three different specimens have a good consistency. The maximum difference of peak strength is only 2.55 MPa, which can be seen from Fig. 6.6. However, from Fig. 6.6, we can conclude that the difference of peak strength between two different loading rates besides $\sigma_3 = 5$ MPa all reach 6.0 MPa (Fig. 6.4), which results from the following possibilities:

(1) Heterogeneity of rock material, which reflects the difference among different specimens;
(2) Effect of loading rate on the peak strength;
(3) The difference among damage specimens loaded to post-peak stress (i.e. control point *A* and *B*).

From Fig. 6.6a and b, it is easy to see out that the smaller effect of the above first and second reason on the peak strength of intact coarse marble, which the maximum influence is not more than 2.55 MPa. Therefore the third reason is very significant on the peak strength of intact coarse marble. At the loading rate of 0.127 MPa/s, the specimen C13# was loaded up to *B*-point; while at the loading rate of 0.002 mm/s, the specimen C10# was loaded up to *A*-point. Therefore the difference between C10# and C13# specimens that is applied respectively to control point *A* and *B* are very distinct, the C10# specimen has more damage than the C13# specimen after loading to control point at $\sigma_3 = 5$ MPa, which results in the difference of peak strength as shown in Fig. 6.6. Moreover the difference of peak strength at two loading rates is almost in-dependent to the confining pressure higher than 5 MPa. Therefore it is very important to grasp accurately the control point (*A* and *B* shown in Fig. 6.6) for confirming the strength parameters by the loading path of stepping up the confining pressure with only one specimen (Path II).

6.3 Mechanical Behavior of Flawed Marble Under Different Loading Paths (Paths III-VI)

In this section, the flawed coarse marble was chosen as the research object. By analyzing the re-fracture behavior of flawed specimen under four different loading paths (Paths III –VI), we tried to reveal the re-fracture deformation mechanism of flawed specimen. Table 6.2 lists re-fracture strength behavior of flawed coarse marble under different loading paths (Paths III–VI). σ_3^i represents the initial confining pressure; σ_S^i represents the maximum axial supporting capability at σ_3^i. While σ_3^f and σ_S^f, respectively, represents the final confining pressure and the final maximum axial supporting capability. All the above parameters were denoted by Fig. 6.7. Figure 6.7 shows experimental results on change of the confining pressure, axial deviatoric

Table 6.2 Initial and re-fracture strength behavior of flawed coarse marble under different loading paths (Paths III-VI)

Specimen	D/mm	L/mm	Loading path	Flawed extent/δ	σ_3^i/MPa	$\sigma_S^i - \sigma_3^i$/MPa	σ_S^i/MPa	σ_3^f/MPa	$\sigma_S^f - \sigma_3^f$/MPa	σ_S^f/MPa
C7#	49.7	100.8	III	0.265	30	135.97	165.97	Reducing confining pressure after peak strength		
C14#	49.7	100.1	IV	0.311	30	133.19	163.19	0	49.94	49.94
C27#	49.7	100.2	IV	0.327	30	132.22	162.22	0	38.37	38.37
C15#	49.7	99.4	V	0.697	3	59.44	62.44	30	124.52	154.52
C22#	49.7	99.9	V	0.658	5	70.14	75.14	30	124.70	154.70
C19#	49.7	99.5	V	0.526	10	97.95	107.95	30	129.16	159.16
C16#	49.7	99.8	VI	0.424	5	80.69	85.69	10	71.29	81.29
C24#	49.7	99.8	VI	0.643	5	70.04	75.04	20	101.52	121.52
C31#	49.7	99.8	VI	0.518	5	80.97	85.97	20	103.21	123.21
C26#	49.7	101.0	VI	0.614	5	68.04	73.04	30	122.91	152.91

Note All the experiments in Table 6.2 were all carried out in July, 2009

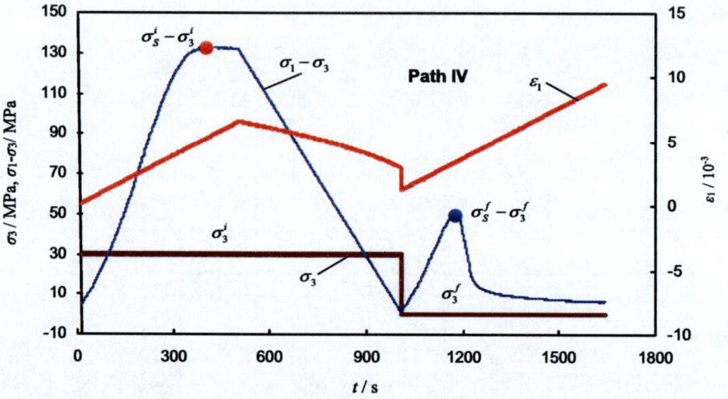

Fig. 6.7 Experimental results on change of confining pressure, axial deviatoric stress, axial strain of coarse marble with the increase of time under Path IV, which is approximately same compared with Fig. 7.1a

stress, axial strain of coarse marble with the increase of time under Path IV, which is approximately same compared with Fig. 6.1a.

Figure 6.8 shows triaxial stress–strain curves of flawed coarse marble at initial confining pressures with different loading paths. Although in this research all the coarse marble specimens under Paths I –VI were machined from the same block along the same direction, the experimental coarse marble still shows two kinds of different mechanical behaviors due to the difference of experimental date. In order to distinguish the difference, under Paths I –II, tested specimens were all called "intact specimen"; while under Paths III –VI, test specimens were all called "flawed specimen". It needs to notice that "flawed specimen" in this research is distinctly different from that with pre-existing cracks in (Yang et al. 2008, 2009).

From Fig. 6.8, we can conclude that the flawed specimens fail with smaller Young's moduli and larger failure strains compared with intact specimens at the same confining pressure. At lower confining pressures ($\leq$10 MPa), an interest experiment phenomenon for flawed coarse marble was observed as follows. All the axial stress-axial strain curves of flawed specimens show a turning point (an abrupt change of slope) at the deviatoric stress of about 35 MPa, coincident with local yield in the coarse marble. For flawed specimens, except for C15#, C22#, C24# and C26# specimens (Table 6.2 and Fig. 6.8), the local yield extent of other flawed specimens are all approximately equal. From Fig. 6.8a, it can be seen that local yield extent of C24# and C26# specimens are larger than that of C16# and C31# specimens, which results in smaller Young's moduli, larger failure strains and lower strengths of C24# and C26# specimens. Moreover C24# and C26# specimens have more plasticity after peak stress than that of C16# and C31# specimens. Under highest confining pressure of 30 MPa, the axial stress-axial strain curve of flawed specimen C27# does not show a turning point at the deviatoric stress of 35 MPa, which results from the occurrence of local yield under hydrostatic stress of 30 MPa. The above analysis reflects the effect of confining pressure on the local yield, while, the confining pressure does not

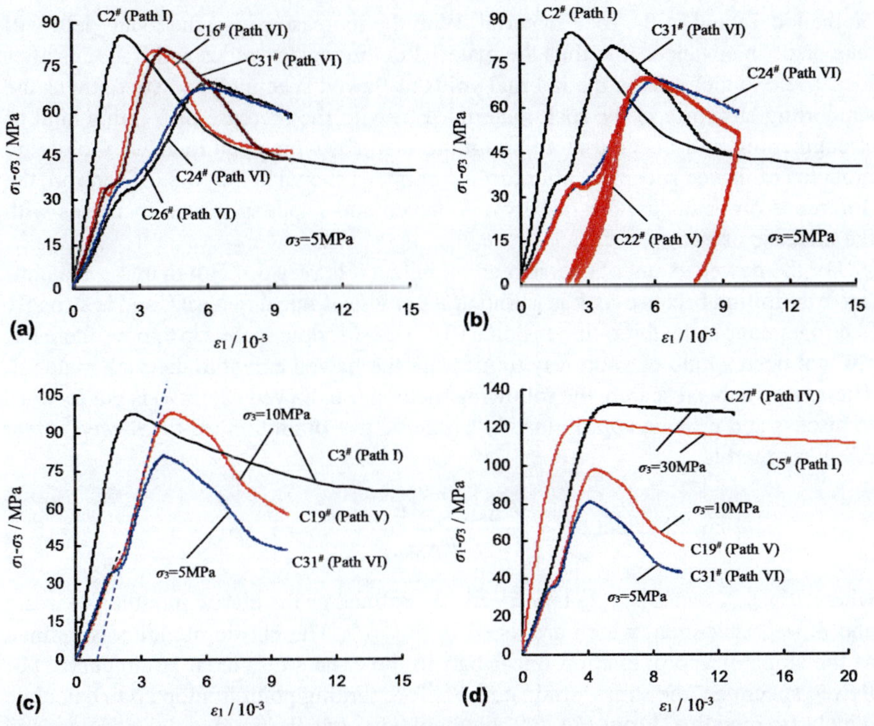

Fig. 6.8 Triaxial stress–strain curves of flawed coarse marble at initial confining pressures with different loading paths. Under Paths I–II, tested specimens were all "intact specimen"; while under Paths III –VI, test specimens were all "flawed specimen"

have a distinct influence on the Young's modulus after turning point, which can be seen from Fig. 6.8d. For flawed specimen, except for four specimens (C15#, C22#, C24# and C26#), we can conclude that the Young's modulus of flawed specimen is in-dependent to the confining pressure; if not taking into account the effect of turning point on the stress–strain curve, the Young's moduli of other flawed specimens are approximately equal (Fig. 6.8d). From the stress–strain curve of C22# and C24# specimens, we can see that the flawed specimen C22# have larger plastic deformation after unloading, but it is very clear that two cycle loading–unloading do not increase the peak strength of flawed specimen, while increase the slope of axial stress-axial strain curve after turning point.

From Table 6.2 and Fig. 6.8, the following conclusions can also be found out for all the flawed specimens. Under lower confining pressures (e.g. 5 MPa), the peak strength of flawed specimen is lower than that of intact specimen. Under medium confining pressures (e.g. 10 MPa), the peak strength of flawed specimen is approximately equal to that of intact specimen. However under higher confining pressures (e.g. 30 MPa), the peak strength of flawed specimen is higher than that of intact specimen. The above conclusion can be explained as follows. Confining pressure has a key effect on the local yield in the rock material. With the increase of σ_3, the local yield will take place in advance. And thus the material in the rock extrudes and fractures after local yield, which fills in the internal voids of flawed specimen and strengthens the supporting structure in the rock material. But with the increase of σ_3, this kind of strengthening will become more and more distinct, which can increase the elastic modulus of flawed specimen after turning point, as shown in Fig. 6.8c. Therefore the difference on peak strength of between flawed and intact specimen increases with the increase of σ_3.

For the flawed extent of coarse marble, in fact it is very difficult to make a quantitative definition because rock is a kind of natural geological material, and is strongly heterogeneous at the three-dimensional (3-D) mesoscopic scale. Up to now, there has still not been a kind of good way to measure the flawed extent in the rock material. Therefore, in this research, the following method (i.e. flawed extent δ) is put forward to discuss and provide approximately a quantitative definition for the flawed extent of coarse marble.

$$\text{Flawed extent}: \delta = \frac{E_{Sintact} - E_{Sflawed}}{E_{Sintact}} = 1 - \frac{E_{Sflawed}}{E_{Sintact}} \tag{6.3}$$

where, $E_{Sintact}$ and $E_{Sflawed}$ is respectively defined as the elastic modulus of intact and flawed specimen, which are listed in Fig. 6.9. The elastic modulus is defined as the slope of approximation linear part in the axial stress-axial strain curve. For flawed specimen, the stress–strain curve before turning point is adopted to calculate the elastic modulus. From Fig. 6.9, on the whole, the elastic modulus of intact and flawed coarse marble all increase with the confining pressure. In accordance with Eq. (6.3), the values of δ are listed in Table 6.2 and the relation between δ and σ_3 is plotted in Fig. 6.9b. From Fig. 6.9b, we can see that the flawed extent δ is not a constant, which varies with heterogeneity of rock material and confining pressure. The average value of δ of four specimens (C15#, C22#, C24# and C26#) is −0.653, and the dispersion extent is about 12.7%. The average value of δ of three specimens (C19#, C31# and C16#) is −0.489, and the dispersion extent is about 20.9%. However, at the

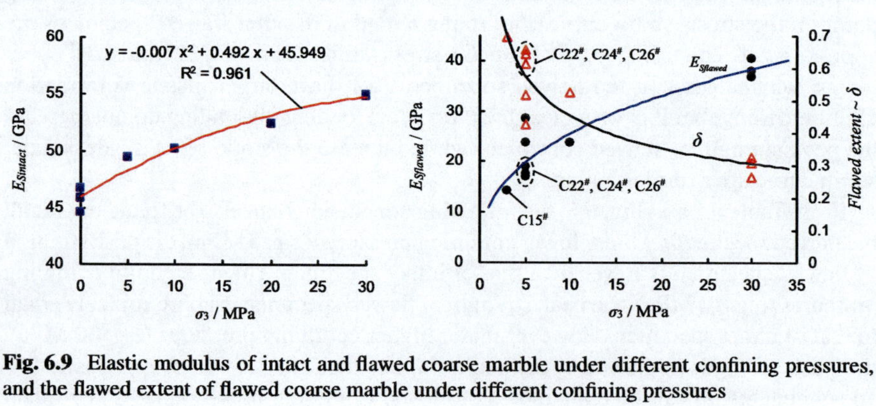

Fig. 6.9 Elastic modulus of intact and flawed coarse marble under different confining pressures, and the flawed extent of flawed coarse marble under different confining pressures

confining pressure of 30 MPa, the average value of δ of three specimens (C7#, C14# and C27#) is 0.301, and the dispersion extent is about 20.6%. The above description shows that the flawed extent δ is not a static parameter but a dynamic parameter, which decreases with the increase of confining pressure as shown in Fig. 6.9b.

6.3.1 Mechanical Behavior of Flawed Marble Under Path III

Triaxial experimental results of flawed coarse marble under Path III are shown in Fig. 6.10. From Fig. 6.10a, it can be seen that the confining pressure is reduced after applying to post-peak stress for the specimen under Path III, which is different from the confining pressure reduction before peak stress (Wang et al. 2008; Zhang et al. 2007). In Fig. 6.10b, the relation between σ_1 and σ_3 is shown. In accordance with linear Mohr–Coulomb criterion, we can obtain the strength parameters of flawed coarse marble, which is listed in Table 6.3.

In accordance with Table 6.3, it can be seen that the values of C and φ are respectively 22.14 MPa and 28.0° for intact coarse marble; while the values of C and φ are respectively 20.76 MPa and 30.4° for flawed coarse marble by different specimens except for four specimens (C15#, C22#, C24# and C26#), which shows only −10% difference between intact and flawed coarse marble. However the values of C and φ are respectively 19.95 MPa and 31.2° for flawed coarse marble under

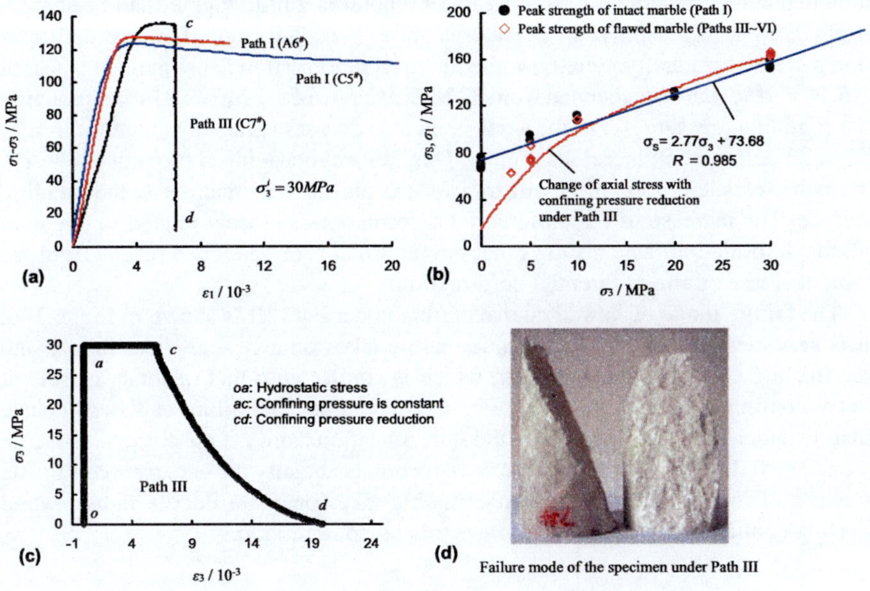

Fig. 6.10 Triaxial experiment results of flawed coarse marble under Path III

Table 6.3 Strength parameters of flawed coarse marble in accordance with linear Mohr–Coulomb criterion

Specimen	Loading path	σ_0/MPa	q	R	C/MPa	φ/(°)	Note
Intact coarse marble	I	73.68	2.77	0.985	22.14	28.0	With UCS
Flawed coarse marble	III –VI	72.52	3.05	0.995	20.76	30.4	Different specimens expect for four specimens (C15#, C22#, C24# and C26#)
Flawed coarse marble	III	70.81	3.15	0.995	19.95	31.2	One specimen (10 MPa $\leq \sigma_3 \leq$ 30 MPa)

Path III (10 MPa $\leq \sigma_3 \leq$ 30 MPa), which is approximately equal with that for flawed coarse marble by different specimens. The above conclusion testifies the feasibility and accuracy to confirm the cohesion and internal friction angle of rock by Path III with only one specimen.

Figure 6.10c presents the relation between the confining pressure and circumferential strain of flawed coarse marble under Path III. During increasing the hydrostatic stress − *oa*, the circumferential deformation equals to zero approximately. And then the σ_3 is maintained a constant level, the axial deviatoric stress is increased until post-peak stress "*c* point" at a constant displacement rate of 0.002 mm/s. It needs to notice that corresponding axial strain of *c* point is a little higher than peak axial strain. During increasing axial deviatoric stress − *ac*, the circumferential deformation increases gradually. Finally, while the axial deformation is maintained constant -6.44×10^{-3}, the σ_3 is reduced from 30 MPa at a rate of 0.1 MPa/s. During reducing the confining pressure − *cd*, the axial stress also decrease gradually, but the circumferential deformation increases rapidly. Due to local behavior in the process − *cd*, the initial elastic deformation will transfer the plastic deformation on the yielding surface. The increase of circumferential deformation is closely related to the axial plastic deformation, which shows that the failure of rock specimen results from the rapid increase of circumferential deformation.

The failure mode of flawed coarse marble under Path III is shown in Fig. 6.10d. It is very clear that the flawed coarse marble takes on a typical shear failure and the fracture surface is very smooth, which is similar with that of intact marble at lower confining pressure shown in Fig. 6.5. Therefore the failure of flawed coarse marble under Path III is essentially the state transition from the plastic state to brittle state. The failure of confining pressure reduction is actually the occurrence of brittle fracture of rock. However the coarse marble takes on more ductile failure while loading continuously for the rock under triaxial compression.

6.3.2 Re-Fracture Mechanical Behavior of Flawed Marble Under Path IV and V

Figure 6.11 shows axial stress-axial strain curves of flawed coarse marble under Path IV, and the corresponding re-fracture strength and deformation parameters are presented in Table 6.4. The ~ $-E_S$ is defined as elastic modulus, i.e. the slope of approximation linear part in the axial stress-axial strain curve. The $-\varepsilon_{1c}$ represents the axial strain value at rupture in terms of axial stress-axial strain curve of flawed coarse marble under final confining pressure ($\sigma_3 = 0$ MPa). The E_{50} represents the deformation modulus, i.e. the slope between the original point and the axial deviatoric stress at an half of TCS.

From Fig. 6.11, it can be seen that the mechanical behavior of two flawed coarse marble specimens at $\sigma_3^i = 30$ MPa have better consistency before unloading, which can be used to investigate the effect of post-peak damage extent on the re-fracture mechanical behavior of flawed coarse marble in uniaxial compression. At $\sigma_3^i =$

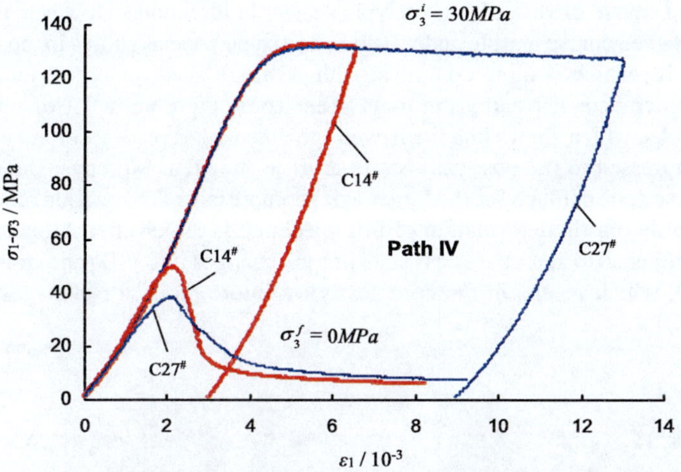

Fig. 6.11 Re-fracture axial stress-axial strain curves of flawed coarse marble under Path IV

Table 6.4 Re-fracture strength and deformation parameters of flawed coarse marble under Path IV

Specimen	Loading path	σ_3^f/MPa	σ_S^f/MPa	$\varepsilon_{1c}/10^{-3}$	E_S/GPa	E_{50}/GPa	Note
C35#	I	0	67.21	1.841	45.82	37.37	Intact specimen
C14#	IV	0	49.94	2.118	27.14	25.24	Damage specimen
C27#	IV	0	38.37	2.074	20.28	21.53	Damage specimen

30 MPa, the C14# specimen is unloaded after the post-peak axial strain -6.55×10^{-3}, while the C27# specimen is unloaded after the post-peak axial strain -13.03×10^{-3}. From the unloading curve, we can see that the C27# specimen has larger plastic deformation than C14# specimen, which results from the larger compressed post-peak axial deformation of C27# specimen.

In accordance with Table 6.4, we can see out that the peak strength and Young's modulus of damage specimens (compressed until post-peak stress at $\sigma_3^i = 30$ MPa) are all lower compared with that of intact specimen; moreover the peak strength and Young's modulus of damage specimens C27# are lower for the larger compressed post-peak plastic deformation of coarse marble, which results from more serious damage of C27# specimen with larger compressed deformation. However the peak axial strains of two specimens are approximately equal, which is not dependent to the flawed or damage extent in the rock. Besides, in final uniaxial compression, the C27# specimen takes on more plasticity after peak strength than C14# specimen.

Under Path IV, final failure modes of flawed coarse marble are shown in Fig. 6.12. In order to compare, the failure mode of intact specimen under uniaxial compression (Path I) is also presented in Fig. 6.12. From Fig. 6.12, it can be found that the intact and flawed coarse marble all take on typical shear failure mode. However, the failure mode of flawed coarse marble under Path IV is a shear zone as shown in the ellipticity of Fig. 6.12, which is more complicated than that (a shear plane) of intact coarse marble under Path I. Nearby the main shear zone, there are a lot of minute and dense cracks, which shows that the flawed specimen takes place a certain of damage after compressed to the post-peak stress at $\sigma_3^i = 30$ MPa. Moreover the specimen has more serious damage for the larger axial compressive deformation, and thus the irrecoverable plastic deformation of this specimen is larger after unloading. After the σ_3 is adjusted to zero, the stress state of rock changes into 1-D (one-dimensional) from 3-D, which results in the complexity of failure mode for damage specimen (Fig. 6.12).

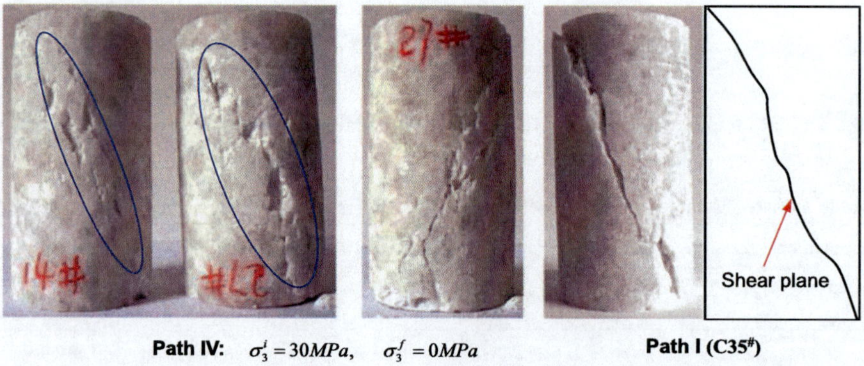

Fig. 6.12 Final failure modes of flawed coarse marble under Path IV. The failure mode of intact specimen under uniaxial compression (Path I) is also shown in this figure

Under Path V, Fig. 6.13 shows axial stress-axial strain curves of flawed coarse marble under initial and final confining pressures. The corresponding re-fracture strength and deformation parameters of flawed coarse marble under different loading paths are listed in Table 6.5. It is very clear that under initial different confining pressures three specimens are loaded to the same post-peak strain -9.8×10^{-3} and then are unloaded to zero at a rate of 2×10^{-5}/s. From Fig. 6.13, it can be seen that the post-peak deformation of fractured rock changes from softening to hardening with the increase of σ_3, which tends towards the stable state from unstable state after post-peak stress. At $\sigma_3^f = 30$ MPa, three damage specimens compressed until post-peak stress under initial confining pressures all have distinct yielding platform, which takes on ideal plasticity. Moreover we can conclude that the relation between axial deviatoric stress and deformation of fractured coarse marble is almost in-dependent to the heterogeneous difference among three flawed coarse marble specimens and the previous loading history, which shows that the maximum supporting capacity and elastic modulus of fractured coarse marble is not related to the fractured extent. The friction among crystal grains determines the strength behavior of damage coarse marble. From Fig. 6.14, the friction slippage is clearly visible, which further validates the fractured coarse marble can support higher axial loads than its material strength by friction. The total deformation in fractured coarse marble tends to uniform and then appears ideal plasticity.

In accordance with Table 6.5, compared with the flawed specimen, the damage specimen failed with lower peak strength, a little smaller elastic moduli, and larger failure strains. However the deformation modulus of damage specimen is a little

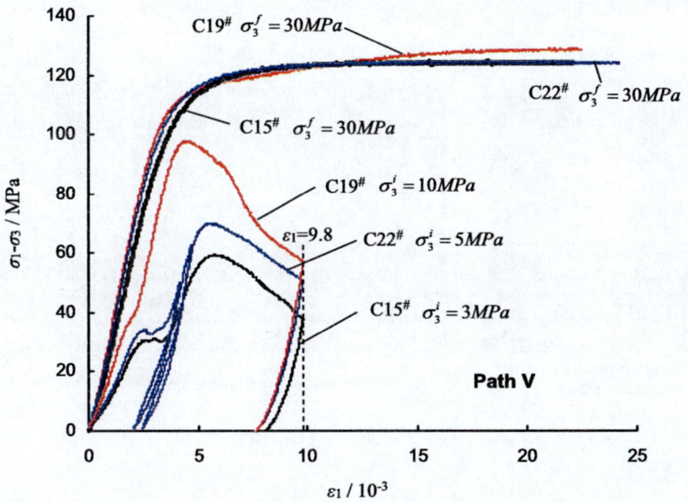

Fig. 6.13 Re-fracture axial stress-axial strain curves of flawed coarse marble under Path V. The damage coarse marble was defined to the flawed specimen compressed until post-peak stress, and then which is unloaded in order to investigate the re-fracture behavior of flawed coarse marble under higher confining pressure of 30 MPa

Table 6.5 Re-fracture strength and deformation parameters of flawed coarse marble under different loading paths

Specimen	Loading path	σ_3/MPa	σ_S/MPa	$\varepsilon_{1c}/10^{-3}$	E_S/GPa	E_{50}/GPa	Note
C7[#]	III	30	165.97	5.246	40.39	30.91	Flawed specimen
C27[#]	IV	30	162.22	6.015	37.00	27.64	Flawed specimen
C14[#]	IV	30	163.19	5.203	37.88	27.99	Flawed specimen
C15[#]	V	30	154.52	15.55	31.97	29.55	Damage specimen
C22[#]	V	30	154.70	16.91	36.07	33.19	Damage specimen
C19[#]	V	30	159.16	22.00	36.68	35.30	Damage specimen

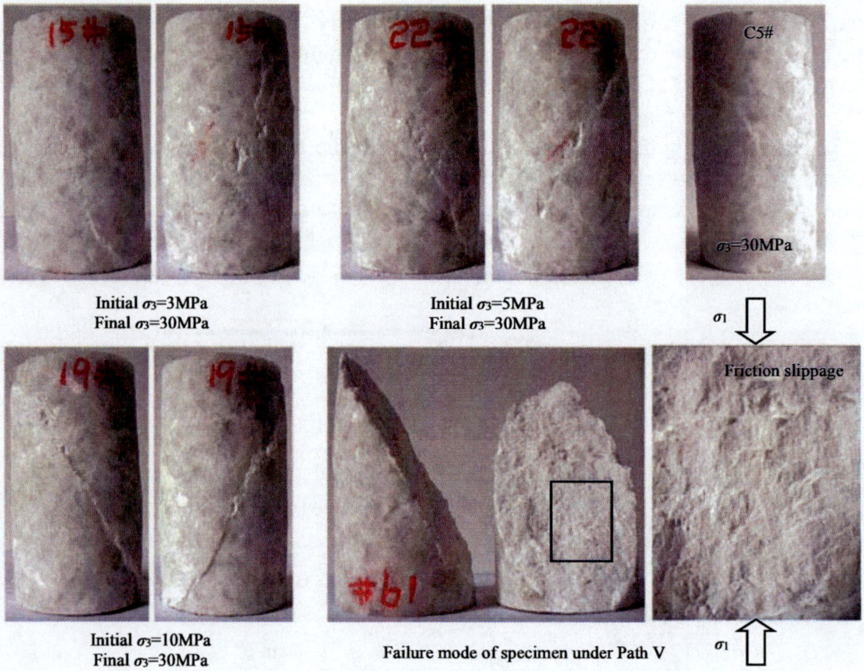

Fig. 6.14 Failure modes of flawed coarse marble under Path V

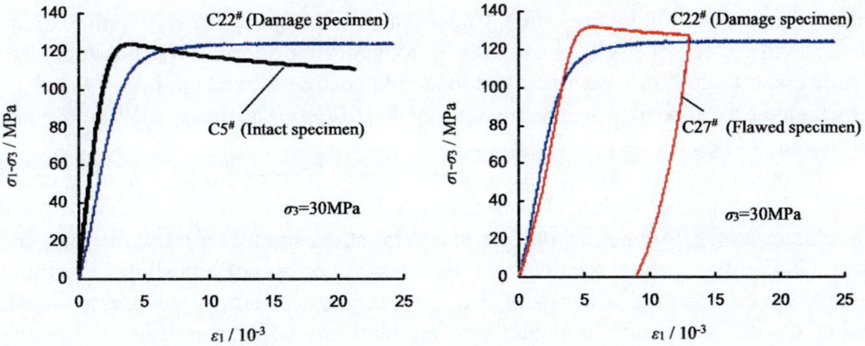

Fig. 6.15 Comparison on axial stress-axial strain curves between intact specimen, damage specimen and flawed specimen

higher than that of flawed specimen, which results from more increase of initial stiffness for damage specimen than flawed specimen. For damage specimen, under higher confining pressures ($\sigma_3 = 30$ MPa), the axial supporting capacity and elastic modulus of damage coarse marble is not related to the loading path, while the deformation modulus and peak strain of damage specimen depend on the difference of initial confining pressure. If the initial confining pressure is lower, the peak axial strain and deformation modulus is also smaller.

Figure 6.14 shows the failure modes of flawed coarse marble under Path V. From Fig. 6.14, we can conclude that all flawed coarse marble takes on the shear failure with a single plane, and the fracture surface is very smooth. After peak stress, the flawed coarse marble takes on the friction slippage between two blocks. With the increase of confining pressure from σ_3^i to σ_3^f, two fractured blocks begin to close and support higher axial stress by friction. Figure 6.15 shows the comparison on axial stress-axial strain curves between intact specimen, flawed specimen and damage specimen. It is very clear that axial supporting capacity of damage specimen tends towards ideal plasticity, while intact and flawed specimen firstly comes out a peak stress, and then with the increase of axial deformation, the axial supporting capacity of intact and flawed specimen decreases slowly towards the residual strength. Moreover damage specimen has more non-linear plastic deformation after yielding stress than intact or flawed specimen.

6.3.3 Re-Fracture Mechanical Behavior of Flawed Marble Under Path VI

In order to investigate the weakening of peak strength of fractured rock and the difference of strength parameters obtained by various loading paths, the re-fracture mechanical experiments of flawed coarse marble under Path VI were carried out under different confining pressures. Under Path VI, the specimen was applied to the same axial strain -9.8×10^{-3} and then the axial deviatoric stress was not unloaded; while under Path V, the specimen was loaded to post-peak axial strain -9.8×10^{-3} and then the axial deviatoric stress was unloaded to zero. The detailed difference can be found in Figs. 6.1, 6.2.

Figure 6.16 shows triaxial stress–strain curves of flawed coarse marble under Path VI. In Fig. 6.16, when the specimen is loaded until post-peak control point *h*, a larger stress drop comes out, i.e. the axial stress reduces a lot while the axial strain is constant, which results from the stress relaxation of adjusting the parameter to increase the σ_3. In terms of Fig. 6.16, damage specimen compressed under initial confining pressure has better yielding platform, which shows ideal plasticity even though under lower confining pressure of 10 MPa. The re-fracture deformation

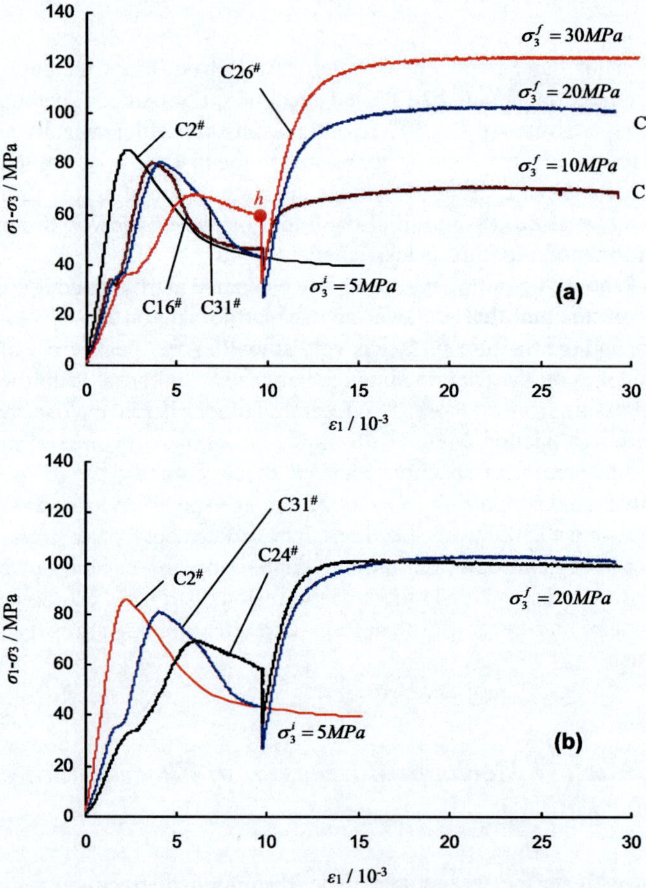

Fig. 6.16 Triaxial stress–strain curves of flawed coarse marble under Path VI

behavior is distinctly different from the softening characteristics of flawed coarse marble as shown in Fig. 6.8 at $\sigma_3 = 10$ MPa.

The above analysis shows that it is necessary to take into account whether rock engineering is the fractured rock mass or not when carrying out the numerical simulation for engineering rock mass in practice. We can not make the numerical analysis by a simple strain-softening model of rock. Only by looking for the constitutive model adapted to different stress states of rock mass, the calculated results would be able to approach the rock engineering practice. The fractured rock mass is in essence a non-continuous medium, but which is closer to the behavior of a continuous medium under higher confining pressures. Therefore it may be more convenient to investigate the problem of non-continuous medium by adopting the method of continuous media under higher triaxial stress states.

In Fig. 6.16a, the axial stress-axial strain curve of the specimen C26# has great difference compared with other two specimens at $\sigma_3^i = 5$ MPa, which causes the following problem. Whether the re-fracture mechanical test results of damage specimens compressed for flawed specimens with different flawed extents can be used to investigate the weakening of peak strength of fractured rock or not ? To explain this above problem, the experimental results of two flawed specimen at the same confining pressure are presented in Fig. 6.16b. From Fig. 6.16b, it can be seen that the axial supporting capacity of two damage specimens at $\sigma_3^f = 20$ MPa are approximately equal with the difference only 1.69 MPa although the strength and deformation behavior of two flawed specimens are distinctly different at $\sigma_3^i = 5$ MPa. Therefore, the experimental results shown in Fig. 6.16 can be used to analyze the strength parameters of flawed and damage coarse marble obtained by different loading paths, which is presented in Fig. 6.17.

From Fig. 6.17, it is very clear that the sensitivity of the peak strength of damage specimen on the confining pressure is higher than that of flawed specimen on the confining pressure. Flawed specimen under initial lower confining pressure was compressed until after post-peak stress, which formed the damage specimen. At this time, the macroscopic failure surface has come out in the damage specimen. Therefore under higher confining pressure, the damage specimen supports the axial deviatoric stress only by friction among fracture surfaces. In accordance with Fig. 6.17, it is clear that σ_0 equals to 72.52 and 48.11 MPa for flawed and damage coarse marble, respectively. However q is 3.05 and 3.58 for flawed and damage coarse marble, respectively. According to Eqs. (6.1) and (6.2), the values of cohesion (C) and internal friction angle (φ) calculated for intact, flawed and damage specimens under different loading paths are presented in Table 6.6. From Table 6.6, it can be seen that the value of C for flawed coarse marble equal to 20.76 MPa, while the value of φ are 30.4°. However, the value of C of damage coarse marble equals to 12.71 MPa, while the value of φ is 34.3°. Therefore, the value of C of damage specimen is lower than that of flawed specimen, but the value of φ of damage specimen is higher than that of flawed specimen.

According to Fig. 6.17, the influence of loading path on peak strength of coarse marble can be concluded. With the increase of σ_3, the difference of peak strength between flawed and damage specimen tends to decrease. At $\sigma_3 = 10$ MPa, the peak

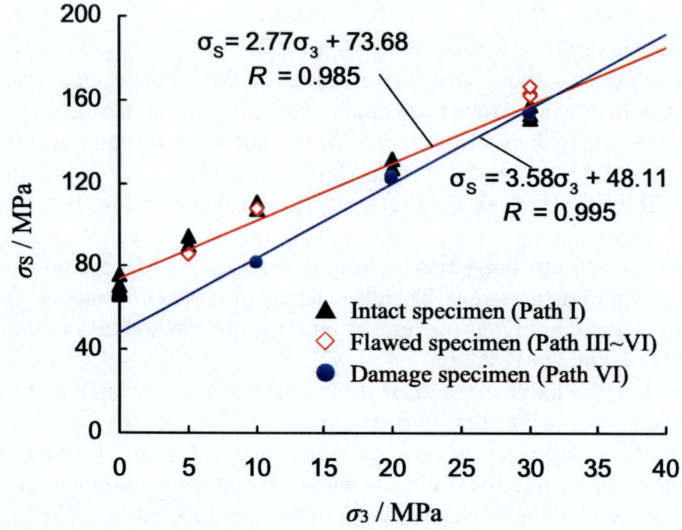

Fig. 6.17 Relation between peak strength and the confining pressure of flawed and damage coarse marble under different loading paths

Table 6.6 Re-fracture strength parameters of flawed and damage coarse marble under different loading paths

Specimen	Loading path	σ_0/MPa	q	R	C/MPa	φ/(°)	Note
Flawed coarse marble	III –VI	72.52	3.05	0.995	20.76	30.4	Different specimens
Damage coarse marble	VI	48.11	3.58	0.995	12.71	34.3	N/A

strength of damage specimen is lower −26.66 MPa than that of flawed specimen. However at $\sigma_3 = 30$ MPa, the peak strength of flawed and damage specimen is approximately equal, which shows no significant effect of loading path on the flawed coarse marble under this confining pressure. Moreover from Fig. 6.17, we can predict approximately that beyond $\sigma_3 = 30$ MPa, the peak strength of flawed and damage coarse marble may be independent of the loading path.

Figure 6.18 shows failure modes of flawed coarse marble under Path VI. At $\sigma_3^f =$ 10 MPa, the flawed coarse marble leads to typical shear failure mode with a single fracture surface and the fracture angle is about 63° with the direction of σ_3. At $\sigma_3^f =$ 20 MPa, two flawed coarse marble all lead to typical shear failure mode with a single fracture surface, but the fracture angle of specimen C24# is about 55°, which is a littler lower than that of the specimen C31# (60°). Moreover despite the above two specimens has single shear failure mode, the specimen C24# is more complicated than the specimen C31#, which possesses double slippage planes in the specimen C24#. At $\sigma_3^i = 5$ MPa, the specimen C24# has taken on shear failure after being loaded

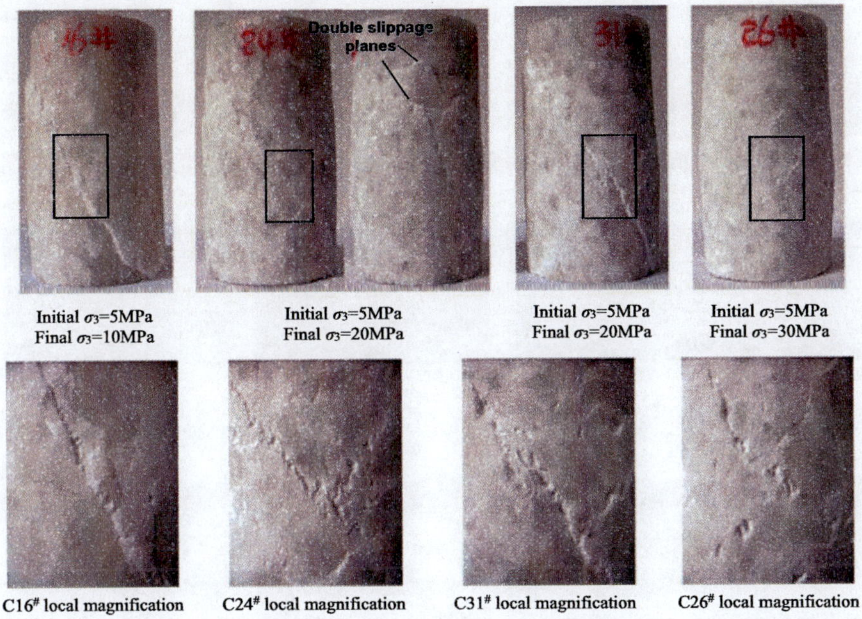

Fig. 6.18 Failure modes of flawed coarse marble under Path VI

to post-peak stress. At the time, when the confining pressure is adjusted to 20 MPa, the increase of axial deformation makes the internal material in the specimen happen the yielding fracture from lower strength to higher strength, which results in new failure surface. Therefore two surfaces in the specimen C24# take place yielding step by step. However at $\sigma_3^f = 30$ MPa, the flawed coarse marble shows typical conjugate shear failure and takes on drum. From the local magnification of C26#, we can see that the specimen possesses conjugate shear failure characteristics and the fracture angle is approximately 45 –50°.

Figure 6.19 shows typical friction slippage of flawed coarse marble under higher confining pressures. From Fig. 6.19, we can see that the friction slippage is very visible. Compared with that of the specimen C19# shown in Fig. 6.14, the two specimens in Fig. 6.19 lead to shear failure mode with a single fracture surface. But the two specimens in Fig. 6.19 takes on a little cone shear failure, not a complete plane shear failure of the specimen C19# shown in Fig. 6.14. Moreover the two specimens in Fig. 6.19 take on the failure behavior with step shape. Although the final confining pressure of the specimens C16# and C31# are different, the fracture surface and slippage behavior of two specimens are almost completely similar, which further validates that the fractured specimen supports the axial stress by friction.

The effect of two loading paths (Path IV and VI) on the complete stress–strain curves of flawed coarse marble is presented in Fig. 6.20. The circumferential deformation was measured by circumferential LVDT. From Fig. 6.20, it can be seen that two damage coarse marble specimens compressed until post-peak stress all have

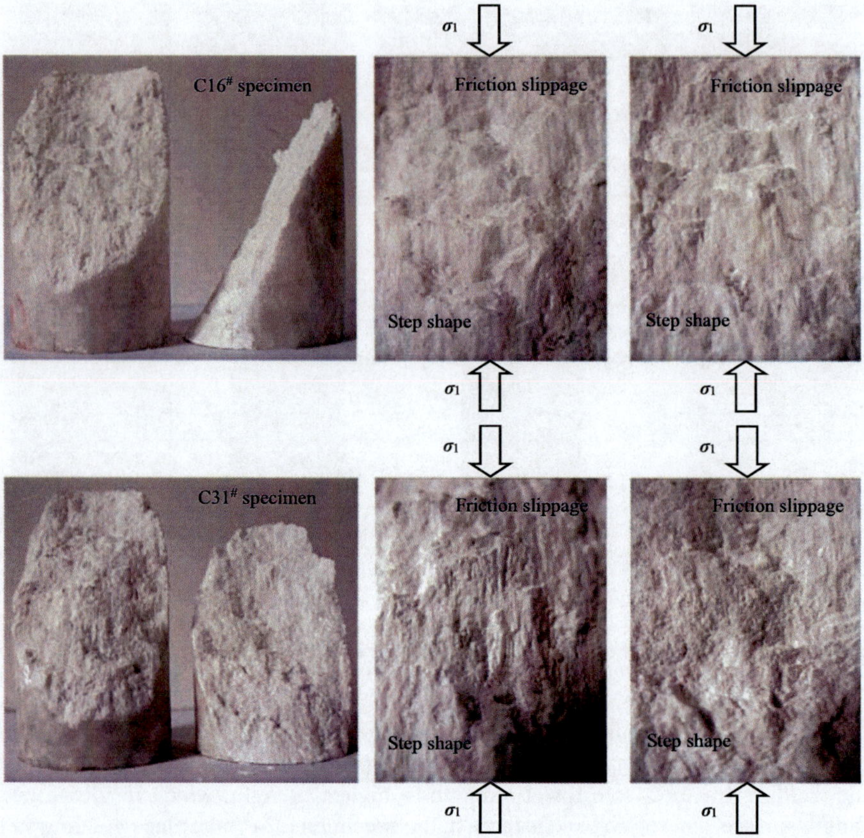

Fig. 6.19 Typical friction slippage of flawed coarse marble under higher confining pressures

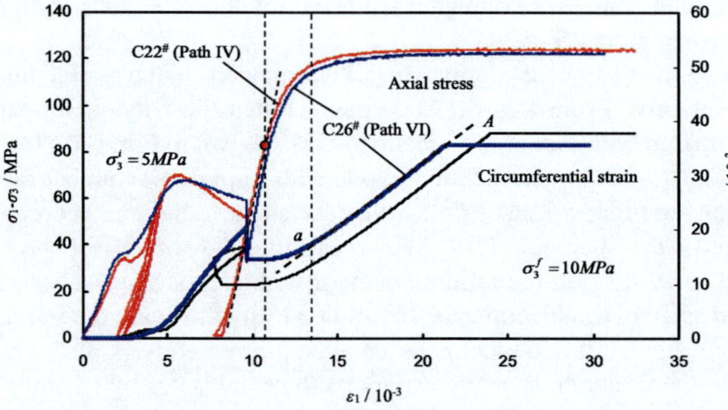

Fig. 6.20 Effect of loading path on the complete stress–strain curves of flawed coarse marble

distinct circumferential plastic strain. If not taking into account initial circumferential plastic strain, we can see that the circumferential strain of damage specimen is smaller at the stages of crack closure and elastic deformation (i.e. before the round dot), which is almost inappreciable compared that after peak strength. At the stage of deformation localization, the circumferential strain of damage specimen increases rapidly with the axial strain, which results from that the circumferential strain of damage specimen after yielding stress reflect sensitively the yielding and softening process of internal material in the rock. However after *a* point, with the increase of axial strain, the circumferential strain of damage specimen increases linearly, which can be explained as follows. The damage specimen supports the axial stress only by friction. Moreover we can conclude that the increasing coefficient k ($k = \varepsilon_3/\varepsilon_1$) of circumferential strain with axial strain is completely equal, which shows no effect of loading path on the coefficient k of damage specimen, and further testifies the supporting by friction slippage. It needs to be noticed that the circumferential strain of damage specimen after *A* point is parallel with the axial strain, which results from the circumferential deformation has reached the maximum measurement range of circumferential LVDT, therefore the circumferential strain after *A* point can not reflect actually circumferential deformation behavior of flawed coarse marble.

6.4 Conclusions

In this chapter, an experimental investigation on the strength, deformation and failure behavior of coarse marble under six different loading paths was carried out. Based on the experimental results of coarse marble specimens under various loading paths, the following conclusions can be drawn.

Loading path has a significant effect on the strength, deformation and failure behavior of coarse marble under triaxial compression. In present study, tested coarse marble is a kind of typical plastic and ductile rock material and takes on typically shear failure mode even though in uniaxial compression. With the increase of confining pressure, the post-peak deformation of coarse marble changes gradually from brittleness to ductility. Flawed specimens fail with smaller Young's moduli and larger failure strains compared with intact specimens at the same confining pressure. Under lower confining pressures ($\leq$10 MPa), all the axial stress-axial strain curves of flawed specimens show a turning point (an abrupt change of slope) at the deviatoric stress of about 35 MPa, coincident with local yield in the coarse marble. But with the increase of σ_3, this turning point at the deviatoric stress of about 35 MPa will decrease step by step.

Confining pressure has different effect on the strength difference between intact and flawed specimen. Under lower confining pressures (e.g. 5 MPa), the peak strength of flawed specimen is lower than that of intact specimen. Under medium confining pressures (e.g. 10 MPa), the peak strength of flawed specimen is approximately equal to that of intact specimen. However under higher confining pressures (e.g. 30 MPa), the peak strength of flawed specimen is higher than that of intact specimen. The above conclusion can be explained as follows. With the increase of σ_3, the local yield will take place in advance. And thus the material in the rock extrudes and fractures after local yield, which fills in the internal voids of flawed specimen and strengthens the supporting structure in the rock material. But with the increase of σ_3, this kind of strengthening will become more and more distinct, which can increase the elastic modulus of flawed specimen after turning point, as shown in Fig. 6.8c. Therefore the difference on peak strength of between flawed and intact specimen increases with the increase of σ_3.

Three loading paths (Paths I-III) are put forward to confirm the strength parameters (cohesion and internal friction angle) of coarse marble in accordance with linear Mohr–Coulomb criterion. Compared among the strength parameters, two loading paths (i.e. Path II by stepping up the confining pressure and Path III by reducing the confining pressure after peak strength) are suggested to confirm the triaxial strengths of rock under different confining pressures by only one specimen, which is very applicable for a kind of rock that has obvious plastic and ductile deformation behavior (e.g. marble, chalk, mudstone, etc.).

In order to investigate re-fracture mechanical behavior of rock material, three loading paths (Paths IV-VI) are also put forward for flawed coarse marble. The peak strength and deformation failure mode of flawed coarse marble are found depending on the loading path (Paths IV-VI). Under lower confining pressures, the peak strength and Young's modulus of damage specimens (compressed until post-peak stress under higher confining pressure) are all lower compared with that of flawed specimen; moreover mechanical parameters of damage specimens are lower for the larger compressed post-peak plastic deformation of coarse marble. However under higher confining pressures (e.g. $\sigma_3 = 30$ MPa), the axial supporting capacity and elastic modulus of damage coarse marble (compressed until post-peak stress under lower confining pressure) is not related to the loading path, while the deformation modulus and peak strain of damage specimen depend on the difference of initial confining pressure and post-peak plastic deformation. The friction among crystal grains determines the strength behavior of flawed coarse marble under different loading paths.

In the range of tested final confining pressures, the sensitivity of the peak strength of damage specimen on the confining pressure is higher than that of flawed specimen on the confining pressure, which results in that the value of cohesion of damage specimen is lower than that of flawed specimen, but the value of internal friction angle of damage specimen is higher than that of flawed specimen.

References

Fairhurst CE, Hudson JA (1999) Draft ISRM suggested method for the complete stress–strain curve for the intact rock in uniaxial compression. Int J Rock Mech Min Sci 36:279–289

Han LJ, Yang MJ (2009) Re-fracture process and mechanical characteristics of cracked rock samples. Int J Rock Mech Min Sci 46:738–747

Lee DH, Juang CH, Chen JW et al (1999) Stress paths and mechanical behavior of a sandstone in hollow cylinder tests. Int J Rock Mech Min Sci 36:857–870

Wang B, Zhu JB, Wu AQ et al (2008) Experimental study on mechanical properties of Jinping marble under loading and unloading stress paths. Chin J Rock Mech Eng 27(10):2138–2145

Xiong SH, Zhou HM (2006) Deformation properties of rock mass of TGP permanent shiplock slopes under complex stress paths. Chin J Rock Mech Eng 25(Suppl. 2):3636–3641 (in Chinese)

Yang SQ, Su CD, Xu WY (2005) Experimental investigation on strength and deformation properties of marble under conventional triaxial compression. Rock Soil Mechs 26(3):475–478 (in Chinese)

Yang SQ, Su CD, Xu WY (2005) Experimental and theoretical study of size effect of rock material. Eng Mech 22(4):112–118 (in Chinese)

Yang SQ, Jiang YZ, Xu WY, Chen XQ (2008) Experimental investigation on strength and failure behavior of pre-cracked marble under conventional triaxial compression. Int J Solids Struct 45(17):4796–4819

Yang SQ, Dai YH, Han LJ et al (2009) Experimental study on mechanical behavior of brittle marble samples containing different flaws under uniaxial compression. Eng Fract Mech 76(12):1833–1845

Yang SQ, Jing HW, Li YS, Han LJ (2011) Experimental investigation on mechanical behavior of coarse marble under six different loading paths. Exp Mech 51(3):315–334

Zhang HB, Song XG, Huang MS, Shang QS (2007) Research on failure features of rocks under different stress unloading path. J Shandong Univer (Eng Sci) 37(6):83–86

Chapter 7
Deformation Failure Characteristics of Crystalline Marble Under Triaxial Cyclic Loading

Marble, a type of metamorphic rock, is ubiquitous in all kinds of rock engineering. Yang et al. (2011) designed three complex loading paths to investigate the re-fracture mechanical behavior of flawed coarse marble. They found that the peak strength and deformation failure mode of the flawed coarse marble depended on the loading paths. They then analyzed the effects of complex loading paths on strength, deformation and failure behavior of the flawed coarse marble under different confining pressures. Pei et al. (2016) carried out uniaxial compression experiments with acoustic emission (AE) monitoring on marble specimens containing natural fractures. They analyzed the characteristics of the AE events and spatial evolution of fractures in marble specimens containing various types of natural fractures, which revealed the crack propagation processes by the spatial evolutions of the AE events. Migliazza et al. (2011) compared the results of static and cyclic experimental tests on Carrara marble. They evaluated fatigue strength and estimated fatigue crack growth rate. Qiu et al. (2014) proposed an incrementally cyclic loading–unloading press test to quantify stress-induced micro-fracturing and fracturing in marble under a condition of reductions in confining pressure. The experimental results demonstrated pre-peak damage and deformation characteristics of the marble specimens could be easily quantified by irreversible strains. Two damage stages, namely, a linear steady stage and a nonlinear unsteady stage, which were, respectively, represented using linear steady rate and nonlinear unsteady rates of damage evolution, occurred along with increasing unloading damage. In accordance with short-term triaxial experimental results for marble under cyclic loading, Yang et al. (2015b) investigated the elastic, plastic and strength behaviors of marble. The experimental results showed that for the same confining pressures, the elastic modulus of the marble remains constant at a lower axial deviatoric level but decreased slowly after yielding strength. However, the plastic modulus of the marble decreased rapidly with increasing axial deviatoric stress at the same confining pressure. Moreover, the elastic and plastic moduli of the tested marble were independent of confining pressure.

Therefore, in this chapter, we report the results of a series of triaxial monotonic and cyclic compression tests on marble under different confining pressures (Yang et al.

S.-Q. Yang, *Mechanical Behavior and Damage Fracture Mechanism of Deep Rocks*,
https://doi.org/10.1007/978-981-16-7739-7_7

2017). Based on the experimental results of the marble under triaxial monotonic loading, the effect of confining pressure on the strength and deformation failure behavior of the marble specimens is first compared. Using the experimental results of the marble under triaxial cyclic loading, the influence of cycle number on the strain damage behavior and elastic modulus of the marble is analyzed. Finally, on the basis of the experimental results of the marble under complex cyclic loading, the strength and deformation parameters of the marble specimens with various damage extents are evaluated.

7.1 Crystalline Marble and Testing Procedure

7.1.1 Crystalline Marble Material and Specimen Preparation

The material used throughout this research was marble collected from Guilin city in Guangxi province, China. The marble material has a connected porosity of 0.57% and an average unit weight of approximately 2680 kg/m^3. According to X-ray diffraction (XRD) test results, the marble material consisted mainly (98.6%) of carbonate minerals (calcite) and fewer clay minerals (1.4%). The marble had a crystalline and blocky structure (Fig. 7.1), which is macroscopically very homogenous. In this experiment, all the specimens were cored from the same block of material to an actual diameter of 50 mm and approximate 100 mm in length. All the experiments were performed on dry specimens at room temperature. Table 7.1 lists the tested marble specimens and conditions in this research.

The triaxial experiments on the marble specimens were all carried out with rock servo-controlled triaxial equipment (Fig. 7.2) (Yang et al. 2014). The equipment included a loading system, constant-stability pressure equipment, a hydraulic pressure transfer system, a pressure chamber, a hydraulic pressure system, and an automatic data collection system. The most important part of the equipment was the self-equilibrium triaxial pressure chamber system, which was comprised of three high precision pumps that controlled axial pressure (P1), confining pressure (P2), and pore pressure (P3, P4). The maximum capacities of P2 and pore pressure (P3 = P4) were 60 MPa. However, the maximum capacity of P1 was as high as 400 MPa for a normal cylindrical specimen with a diameter of 50 mm. When testing, the axial displacement was measured using two linear variable differential transformers (LVDTs) that were fixed between the bottom and top surfaces of the specimen inside the triaxial cell.

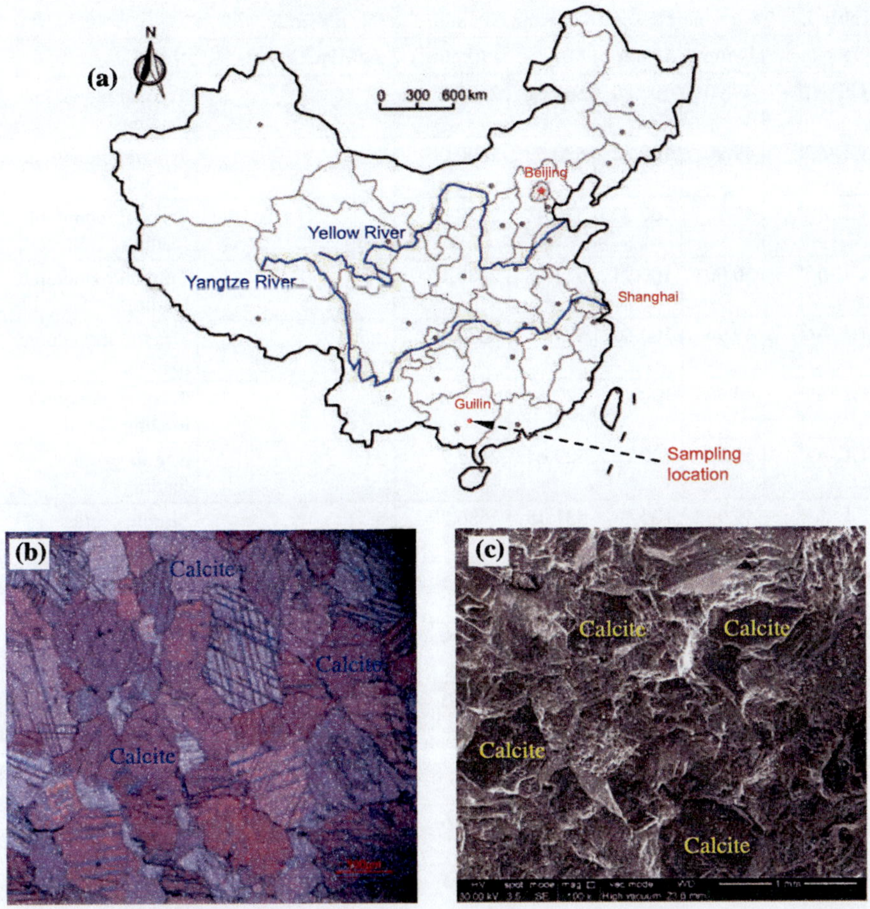

Fig. 7.1 Microscopic structure of marble used in this research. **a** Sampling location; **b** optical microscopy, and **c** SEM photograph

7.1.2 Two Types of Triaxial Tests

To investigate the evolution of the mechanical damage of the marble material, we performed two types of triaxial tests: (1) monotonic, and (2) cyclic.

The triaxial monotonic experiments (Fig. 7.3) (i.e., the conventional triaxial experiment) were carried out under different confining pressures (σ_3) of 2.5, 5.0, 7.5 and 10 MPa and consisted of the following two steps. First, the confining pressure was increased to the desired value at a constant rate of 0.1 MPa/s to ensure that the specimen was under uniform hydrostatic stresses. The deviatoric stress (σ_1–σ_3) was then applied to the surface of the specimen at a constant axial displacement rate of

Table 7.1 Tested marble specimens and conditions in this research

Specimen	D/mm	L/mm	M/g	ρ/(kg/m^3)	σ_3/MPa	Note
GL-A1#	49.90	100.80	530.81	2692.7	0	Triaxial monotonic loading
GL-A3#	49.90	100.82	530.39	2690.0	0	Triaxial monotonic loading
GL-A4#	49.98	100.90	530.48	2679.8	5	Triaxial monotonic loading
GL-A5#	50.00	100.94	531.18	2680.1	10	Triaxial monotonic loading
GL-A6#	49.96	100.68	527.06	2670.4	2.5	Triaxial monotonic loading
GL-A7#	49.98	100.76	529.64	2679.2	7.5	Triaxial monotonic loading
GL-A8#	50.00	100.70	529.64	2687.7	2.5	Simple cyclic loading
GL-A9#	49.96	100.88	531.35	2686.8	5	Simple cyclic loading
GL-A10#	49.98	100.38	528.93	2685.8	7.5	Simple cyclic loading
GL-A11#	49.98	100.38	527.39	2677.9	10	Simple cyclic loading
GL-A12#	50.00	100.06	527.56	2685.2	10–2.5–5.0–7.5–10	Complex cyclic loading
GL-A13#	49.96	100.88	528.06	2670.2	10–2.5–5.0–7.5–10	Complex cyclic loading
GL-A14#	49.98	100.40	527.63	2678.6	10–2.5–5.0–7.5–10	Complex cyclic loading
GL-A16#	50.00	100.40	533.32	2705.4	10–2.5–5.0–7.5–10	Complex cyclic loading
GL-A17#	50.00	100.80	529.86	2677.1	10–2.5–5.0–7.5–10	Complex cyclic loading

Note *D* diameter; *L* length; ρ density; *M* mass; σ_3 confining pressure

0.02 mm/s until failure. In Fig. 7.3, ε_1 represents the axial strain, and σ_1 represents major principal stress.

The triaxial cyclic experiments can be divided into two groups. The first group is triaxial cyclic loading, which was conducted using the following three steps. First, the hydrostatic pressure was applied to the specimens at a rate of 0.1 MPa/s until the desired values were reached. Second, the specimens were loaded to the first displacement value (corresponding to the first deviatoric stress level) at a controlled axial displacement rate of 0.02 mm/s, and the first deviatoric stress level was then unloaded to zero at a rate of 0.4 MPa/s through controlling axial force. In each subsequent cycle the second deviatoric stress level was increased, and the specimens were

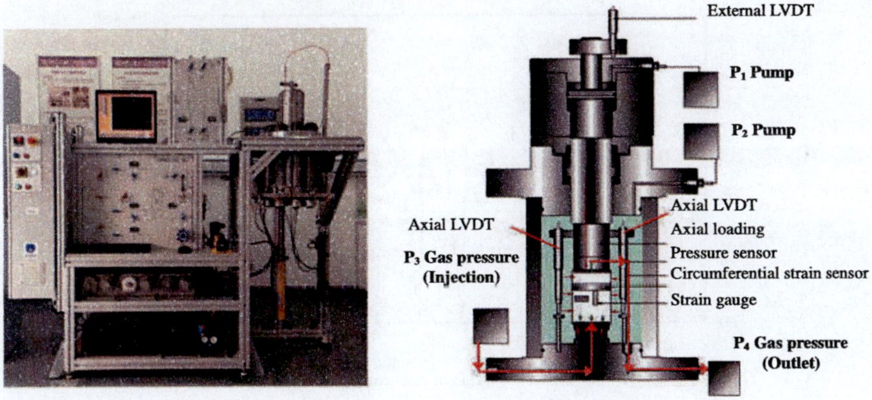

Fig. 7.2 Rock servo-controlled triaxial equipment (modified from Yang et al. 2014)

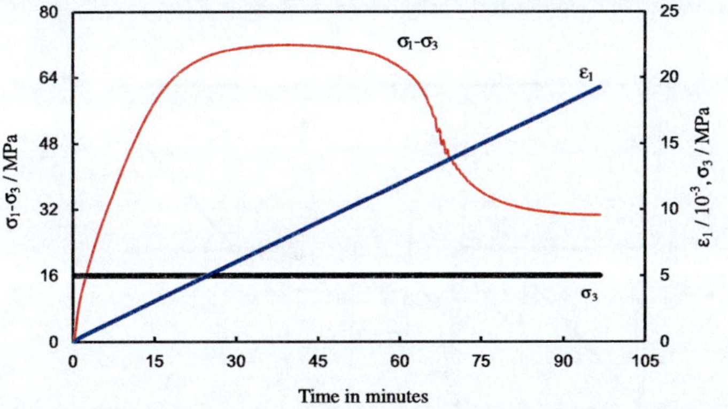

Fig. 7.3 A typical experimental result of triaxial monotonic loading path on marble specimen. The blue line represents the axial strain, the black line represents the confining pressure and the red line represents the axial deviatoric stress

again unloaded to zero. Finally, the stress-cycling was continued in that way until the specimens eventually failed. It should be noted that in our study, after peak strength, multiple cyclic loadings were continued to analyze the mechanical behaviors of the fractured rock material, a procedure unlike that used in previous studies (Heap et al. 2009; Trippetta et al. 2013). Figure 7.4 shows a typical experimental result of triaxial cyclic loading path for a marble specimen that failed at the 11th loading cycle.

The second group is complex cyclic loading (Yang and Jing 2013), which was conducted with the following steps. First, a hydrostatic pressure of 10 MPa was applied to the specimens at a rate of 0.1 MPa/s. Second, the specimens were loaded to the first displacement value, i.e., point *A* shown in Fig. 7.5 (Note: this displacement value of point *A* can be designed before or after peak strength) at a controlled axial

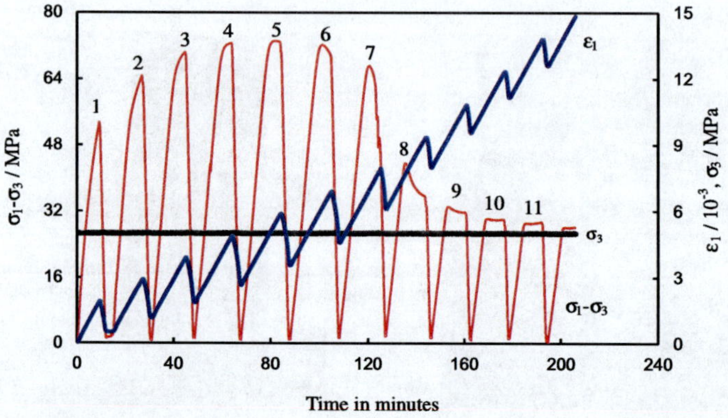

Fig. 7.4 A typical experimental result of simple cyclic loading path from increasing amplitude, cyclic stressing triaxial experiment on marble specimen. The numbers in this figure represent the cycle number. The blue line represents the axial strain, the black line represents the confining pressure and the red line represents the axial deviatoric stress

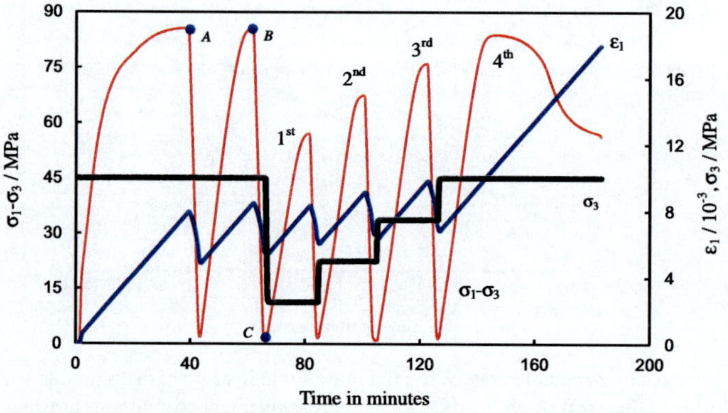

Fig. 7.5 A typical experimental result of complex cyclic loading path from increasing amplitude, cyclic stressing triaxial experiment on marble specimen. The blue line represents the axial strain, the black line represents the confining pressure and the red line represents the axial deviatoric stress

displacement rate of 0.02 mm/s, and the deviatoric stress level was then unloaded to zero at a rate of 0.4 MPa/s through controlling axial force. Maintaining a constant 10 MPa confining pressure, the specimens were reloaded to point *B* at a controlled axial displacement rate of 0.02 mm/s. Afterwards, the deviatoric stress level was unloaded to zero (point *C*) at a controlled axial force of 0.4 MPa/s.

At that time, the 10 MPa confining pressure was reduced to 2.5 MPa (1st stage) at a rate of 0.1 MPa/s. The deviatoric stress was loaded at a constant axial displacement control rate of 0.02 mm/s until the specimen reached post failure, at which point, the deviatoric stress dropped by ~2% from its peak. The confining pressure was then hydrostatically increased to the 2nd stage, i.e., 5.0 MPa. After that, the preceding steps were repeated at the third confining pressure (3rd stage), 7.5 MPa. At the fourth confining pressure (4th stage) of 10 MPa, the deviatoric stress was loaded to the surface of the specimens at a constant axial displacement control rate of 0.02 mm/s until failure occurred. Thus, we were able to obtain a series of peak strengths of the damaged specimens under different confining pressures. Figure 7.5 shows a typical experimental result of complex cyclic loading path for a marble specimen.

7.2 Experimental Results of Crystalline Marble Under Triaxial Monotonic Loading

The deviatoric stress-axial strain curves for the marble under different confining pressures are presented in Fig. 7.6a. According to Fig. 7.6a, the post-peak deformation of the marble changed gradually from strain-softening to strain-hardening with increasing confining pressure. When σ_3 equals ~10 MPa, the deformation of the marble specimen shows a distinct yielding platform and plastic flow occurs, i.e., the post-peak deformation behavior of the specimen changes into ideal plasticity. The above brittle-ductile transition of the marble is also approximate to the previous findings for Jinping marble (Yang et al. 2016).

On the basis of the peak strength values shown in Fig. 7.6a, the influence of σ_3 on the peak strength of the marble can be presented in Fig. 7.6b. Figure 7.6b indicates that the peak strength of the marble showed distinctly linear behavior with increasing σ_3, in good agreement with the linear Mohr–Coulomb criterion, i.e., Eq. (7.1):

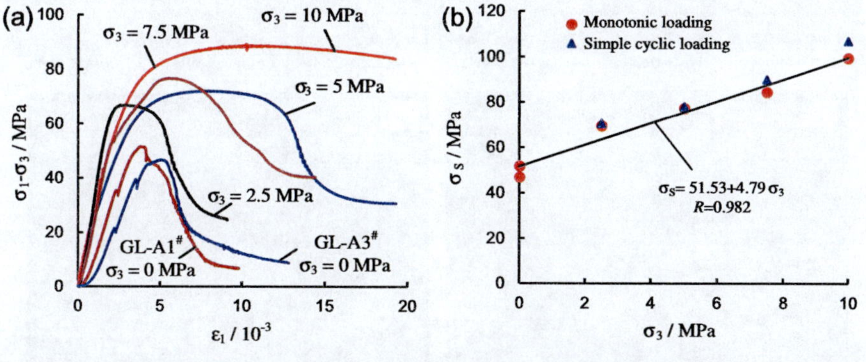

Fig. 7.6 Experimental results of marble specimens under triaxial monotonic loading. **a** Deviatoric stress-axial strain curves of marble specimen under different confining pressures; **b** comparison of monotonic and simple cyclic triaxial strength of marble specimen under different confining pressures

$$\sigma_S = \sigma_0 + q\sigma_3 = \frac{2C\cos\varphi + \sigma_3(1+\sin\varphi)}{1-\sin\varphi} \quad (7.1)$$

where σ_S is the maximum axial supporting capacity of the rock, σ_0 is usually regarded as the uniaxial compressive strength (UCS) of rock material, and q is an influence coefficient of the confining pressure on σ_S of the rock. The σ_0 and q are related to the cohesion C and internal friction angle φ of the rock material.

By using the linear Mohr–Coulomb criterion, the peak strength parameters of the marble under triaxial monotonic loading can be calculated. We determined that the cohesion of the marble was 11.77 MPa, and the internal friction angle was 40.9°.

Figure 7.7 shows the ultimate failure modes of the marble specimens under triaxial monotonic loading. From Fig. 7.7, we can see that the marble specimens experienced some axial splitting tensile cracks, which validates the brittle failure of the marble material under uniaxial compression and was also approximately that of previous findings on Jinping marble (Yang et al. 2016). Furthermore, specimens A1# and A3# exhibited some differences, even though both displaying the tensile fracture modes, which resulted from the influence of rock heterogeneity on the fracture characteristics. However, at confining pressures of 2.5–7.5 MPa, the marble specimens mainly

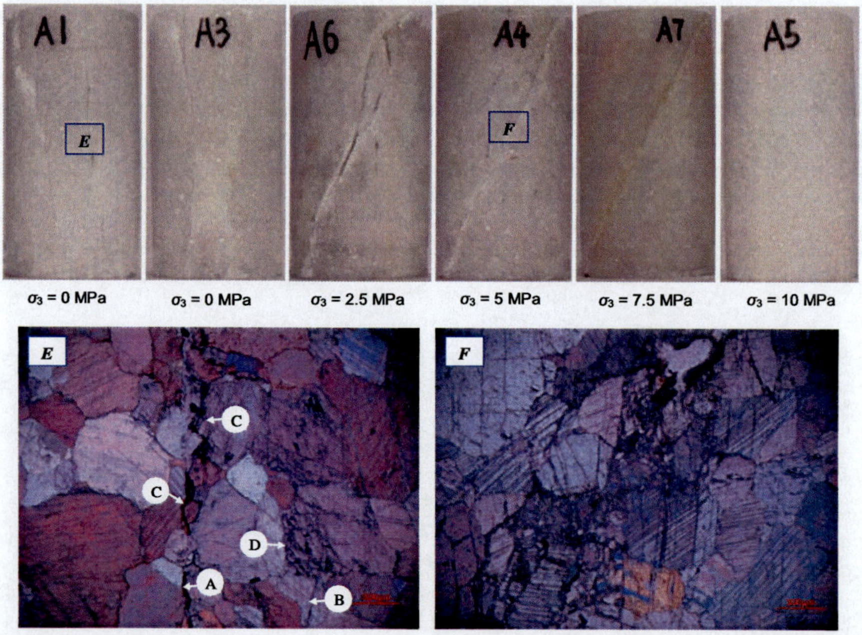

Fig. 7.7 Ultimate failure mode and photomicrographs of marble specimen under triaxial monotonic loading. Type A and B mechanisms were favored by the distinctive polygonal shapes of the crystal grains in the marble, the Type C mechanism is associated with gliding along twin lamellae, and the Type D mechanism operates in grains that failed mainly due to compressive stress rather than tensile stress concentrations (Cheng et al. 2016)

failed by shear localization along an inclined macroscopic shear band with a single shear zone across each specimen. The shear fracture angle was approximately 65°, which was approximately equal to 45° + $\varphi/2$. Furthermore, the fracture angles did not depend on confining pressure at $\sigma_3 = 2.5$–7.5 MPa. However, at $\sigma_3 = 10$ MPa, no obvious shear fracture surfaces could be observed by the unaided eye, which is also evidenced by the axial deviatoric stress-axial strain curve shown in Fig. 7.6a.

To analyze the deformation mechanism of the marble specimen after triaxial monotonic loading failure, an optical microscope is used to observe the microscopic characteristics of failed marble specimen. The previous research for Carrara marble (Cheng et al. 2016) has verified that the optical observation is favored for identifying grain boundaries and cracks because the different grains will show a different degree of grayish color and open cracks will remain black under optical microscope. Two square regions shown in Fig. 7.7 will be chosen to perform the microscopic examination, since a lot of serious damage has occurred in these regions. To avoid any possible additional damage associated with saw cutting, a dye epoxy was used to fill the flaws and newly developed microcracks and strengthened the failed specimens. A large amount of polishing was carried out to form a better thin section.

Figure 7.7 also shows microscopic observation results of a marble specimen after triaxial monotonic loading failure with respect to $\sigma_3 = 0$ and 5 MPa. Compared with the microscopic observation results of the uncompressed marble specimen shown in Fig. 7.1a, it is clear that the local damage failure of the marble specimen after triaxial monotonic loading obviously differed, depending on the confining pressure. A further analysis can be illustrated as follows. It can be seen that under uniaxial compression, tensile cracks were observed along the direction of the major principal stress. In the previous research (Cheng et al. 2016), four principal nucleation mechanisms for transgranular cracks were observed in Carrara marble subjected to uniaxial compression loading. Type A and B mechanisms were favored by the distinctive polygonal shapes of the crystal grains in the marble. The local tensile stress concentration in those two mechanisms is attributed to grain sliding and divergent normal contact force, respectively. The Type C mechanism is associated with gliding along twin lamellae, and the Type D mechanism operates in grains that failed mainly due to compressive stress rather than tensile stress concentrations. The above four mechanisms were also observed for our marble specimens under uniaxial compression, as shown in Fig. 7.7. However, at confining pressures of 5 MPa, the nucleation mechanism in the marble differed from that under uniaxial compression. It is clear that at $\sigma_3 = 5$ MPa, shear slippage was very obvious; furthermore, different regions showed various damage extents.

7.3 Triaxial Experimental Results of Crystalline Marble Under Simple Cyclic Loading

Figure 7.8 shows typical triaxial experimental results for the marble specimens under simple cyclic loading at $\sigma_3 = 2.5$, 5, 7.5 and 10 MPa. It should be noted that even after the peak stress, the specimens still continued to undergo cyclic loading until the residual strength, which occurred easily due to ductile failure of the marble material after peak strength.

Based on the triaxial experimental results of the marble under monotonic and simple cyclic loading shown respectively in Figs. 7.8 and 7.6a, the comparison of the peak strengths under monotonic and simple cyclic loading is presented in Fig. 7.6b. From Fig. 7.6b, we can see that the cyclic triaxial strength of the marble was approximately equal to the monotonic triaxial strength, which indicates that in the range of the tested confining pressures, cyclic loading did not affect the triaxial strength of the marble compared with monotonic loading.

In accordance with Fig. 7.8, we confirmed the elastic and plastic deformations (irreversible deformation) of marble specimens for each loading–unloading cycle.

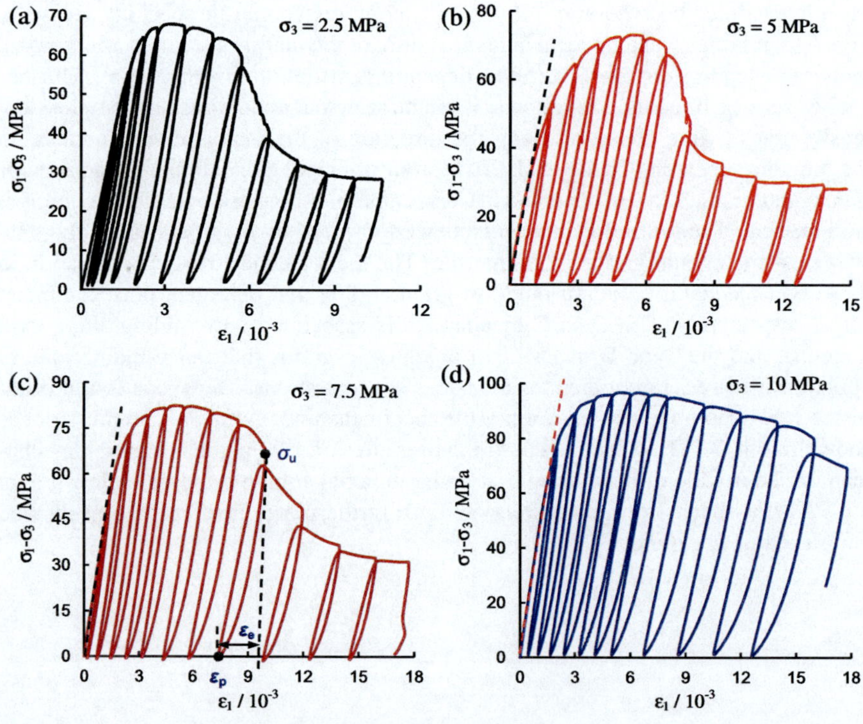

Fig. 7.8 Triaxial experimental results of marble specimens under simple cyclic loading in accordance with the path shown in Fig.7.4

The ε_e, ε_p and σ_u are defined the elastic strain, plastic strain and unloading deviatoric stress level of rock, respectively, as depicted in Fig. 7.8c.

7.3.1 Effect of Cyclic Number and Unloading Stress Level on the Strain Behavior

Figure 7.9 illustrates the relationship between unloading stress level and the strains of the marble specimens under different confining pressures. From Fig. 7.9a, it can be seen that with the increase in unloading stress level, the elastic deformation of the marble also increased gradually before peak strength; furthermore, that variance did not depend obviously on confining pressure. However, after peak strength, the relationship between unloading stress level and the strains of the marble specimens depended on the confining pressure. At $\sigma_3 = 5.0$ MPa, after the peak strength, the elastic deformations of the marble specimens decreased with the decreases in the unloading stress levels. However, at $\sigma_3 = 2.5$, 7.5 and 10 MPa, after the peak strength, the elastic deformations of the marble specimens first increased gradually and then decreased with the decreasing unloading stress level.

Figure 7.9b illustrates the effects of unloading stress levels on the plastic deformation of the marble under different confining pressures, which were obviously unlike those of the elastic deformation of the marble shown in Fig. 7.9a. For the same confining pressure, and at early several cyclic stress levels, the elastic deformation of the marble exceeded the plastic deformation, but afterwards, the plastic deformation of the marble was higher than the elastic deformation, which was a result of the increased number of cycles, whereby the specimens experienced increasingly greater crack damage, which resulted in an increasing amount of irreversible deformation. It should be noted that the plastic deformation of the marble increased slowly as

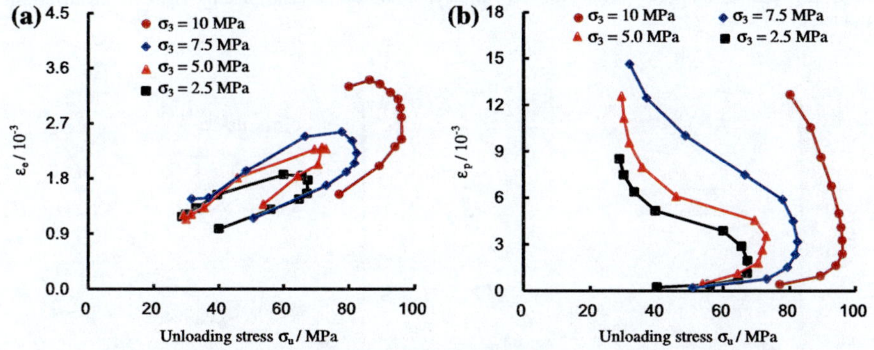

Fig. 7.9 Relationship between unloading stress level and strains of marble under different confining pressures. **a** Relationship between elastic strain and unloading stress level; **b** relationship between plastic strain and unloading stress level

unloading stress level increased; whereas after the peak strength, the plastic deformation of marble increased dramatically, despite the reduction in unloading stress level. The rate of plastic deformation increased with applied confining pressure.

Figure 7.10 shows the relationship between cycle number and strains for the marble specimens under different confining pressures. From Fig. 7.10a, we can see that with the increase in cycle number, the elastic strains for the marble specimens at first increased and then decreased, and all had maximum values. However, the cycle numbers corresponding to the maximum elastic strain values of the marble specimens differed for various confining pressures. From Fig. 7.10b, it can be seen that the plastic strains of the marble specimens increased nonlinearly with cycle number.

Figure 7.11 shows the comparison of elastic and plastic strains of marble at the same confining pressure. From Fig. 7.11, we can see that in lower cycle number, the elastic strain of the marble specimens was slightly higher than the plastic strain for the same confining pressure. However, in higher cycle number, the elastic strain

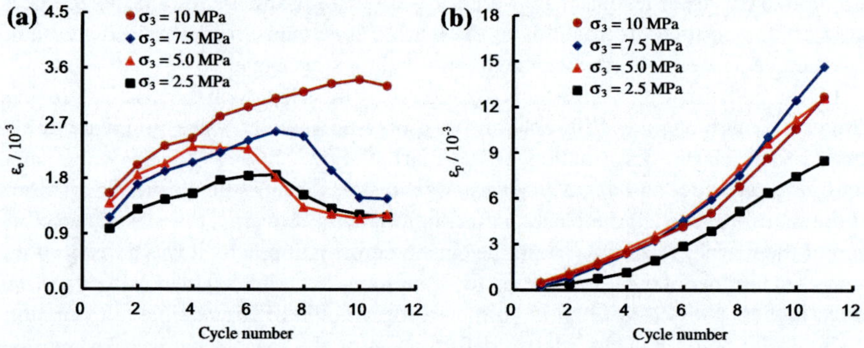

Fig. 7.10 Relationship between cycle number and strains of marble under different confining pressures. **a** Relationship between elastic strain and cycle number; **b** relationship between plastic strain and cycle number

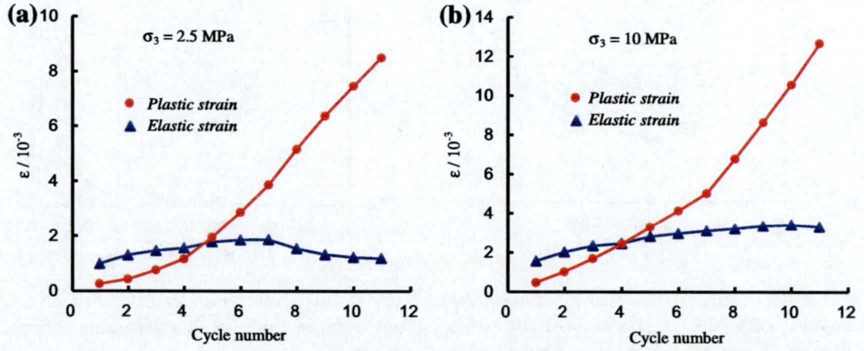

Fig. 7.11 Comparison of elastic and plastic strains of marble at the same confining pressure

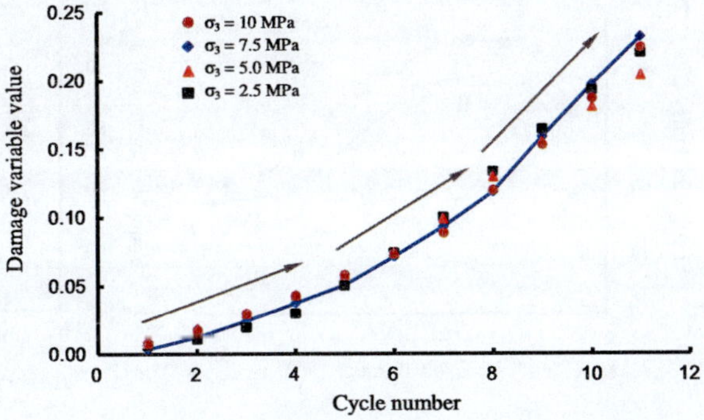

Fig. 7.12 Relationship between cycle number and damage variable value of marble under different confining pressures

of the marble specimen was obviously lower than the plastic strain for the same confining pressure. Furthermore, with the increase of cycle number, the difference between the elastic and plastic strain increased gradually.

To quantitatively evaluate the damage extent of the marble specimens under triaxial cyclic loading, a damage variable is defined in accordance with the irreversible deformation (plastic deformation) for each cycle (Yang et al. 2015a), and is expressed in Eq. (7.2):

$$D = \frac{(\varepsilon_p)_N}{\sum_{N=1}^{m} (\varepsilon_p)_N} \tag{7.2}$$

where, D represents the damage variable of rock. ε_{p} represents the irreversible deformation of rock, and N is the cycle number. The letter m represents the maximum cycle number.

Figure 7.12 shows the damage evolution of the marble specimens under different confining pressures. From Fig. 7.12, it is clear that the confining pressure had almost no effect on the damage evolution of the marble specimens. However, at the same confining pressure, the damage variable values of the marble specimens increased with increasing cycle number. Furthermore, the extent of the damage variable increased with increasing cycle number.

7.3.2 *Evolution of Elastic Modulus of Crystalline Marble with Cyclic Number*

The evolution of elastic modulus of marble can be characterized as having four stages (Fig. 7.13): (a) Stage I: material strengthening; (b) Stage II: material degradation; (c) Stage III: shear failure; and (d) Stage IV: structure slippage. At $\sigma_3 = 2.5$, 5.0 and 7.5 MPa, the above four stages were very obvious. During Stage I, the elastic modulus (E) of the marble increased obviously, which indicates that the marble became stiffer

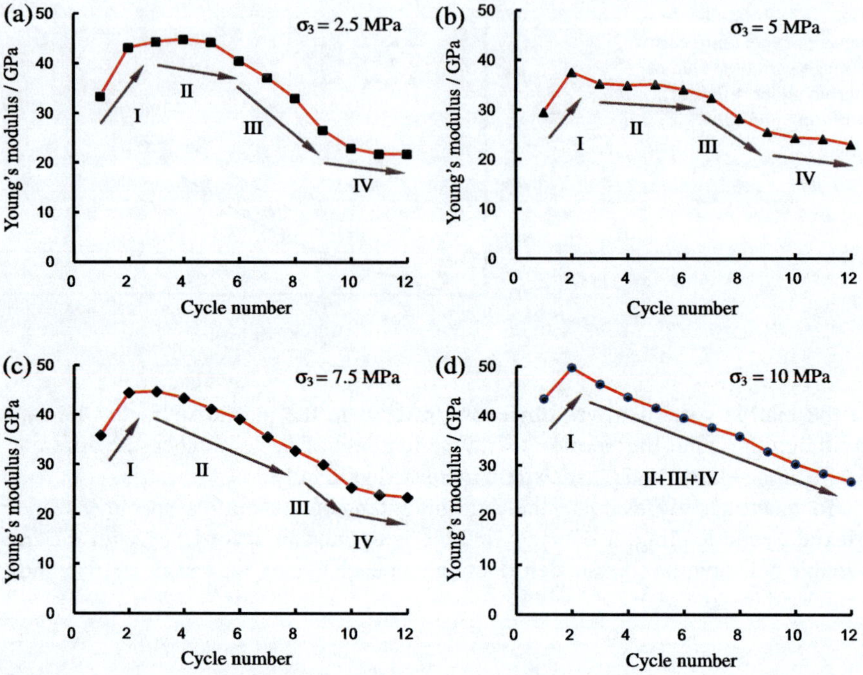

Fig. 7.13 Evolution of elastic modulus of marble specimens with the increase of cycle number under various confining pressures

and more difficult to deform. In our research, the unloading stress levels of the marble specimens for the first cycle were located at the stage of yielding before peak strength. The reason for the evolution can be presented as follows. With the increase in cycling stress, some mesoscopic elements with lower strengths and stiffnesses first reached their maximum supporting capacities, yield softening, and produced plastic deformation as the specimens yielded (Yang et al. 2008). Some extruded and fractured elements then filled the internal voids of the marble specimens. On the other hand, pre-existing pores and fissures could also produce closure. The above two factors strengthened the supporting structure of the marble material, which resulted in the increase of *E*. During Stage II, we observe a significant decrease in *E*. At $\sigma_3 =$ 2.5 MPa, the elastic modulus decreased from 44.73 to 32.95 GPa during increasing amplitude cycling stressing to failure. At $\sigma_3 = 5.0$ MPa, the elastic modulus decreased from 37.48 to 32.33 GPa during increasing amplitude cycling stressing to failure. At $\sigma_3 = 7.5$ MPa, the elastic modulus decreased from 44.63 to 32.73 GPa during increasing amplitude cycling stressing to failure. The decrease in *E* can be attributed to a progressive degradation in the stiffness of the rock due to the increase in crack damage with increasing cycle number. At higher stress levels, each cycle applied to a specimen will produce an additional increment of crack damage, resulting in an increasing amount of irreversible deformation. During Stage III, it is very clear that the elastic modulus of marble specimen decreased. At $\sigma_3 = 2.5$ MPa, the elastic modulus decreased significantly from 32.95 to 22.82 GPa during this stage. At $\sigma_3 = 5.0$ MPa, the elastic modulus decreased from 32.33 to 24.26 GPa during this stage. At $\sigma_3 = 7.5$ MPa, the elastic modulus decreased from 32.73 to 25.33 GPa during this stage. The variance in E for the marble specimens during this stage can be explained as follows. During this stage, larger macroscopic cracks will have formed, resulting in the failure of the structure of the rock material. Therefore, the increase in the number of macroscopic cracks reduces axial stiffness. During Stage IV, the elastic modulus decreased slightly from 22.82 to 21.66 GPa at $\sigma_3 = 2.5$ MPa. The elastic modulus changed from 24.26 to 22.88 GPa at $\sigma_3 = 5.0$ MPa. At $\sigma_3 = 7.5$ MPa, the elastic modulus changes from 25.33 to 23.35 GPa. This indicates that the elastic modulus of the marble is dependent on cycle number during this stage, which results from the following reason. Due to the constriction of confining pressure, the specimens had entered the stage of residual strength, and mainly supported the axial capacity by friction slippage in the macroscopic failure plane, which resulted in no obvious variance in the E of the marble specimens.

However at $\sigma_3 = 10$ MPa, the above Stages II, III and IV could only be distinguished with great difficulty, as shown in Fig. 7.13d. This can be explained as follows. Under the confining pressure, after the first cycle, with the increase of axial deformation, the mesoscopic material elements with low strength and stiffness first reached their maximum supporting capacities, experienced yield softening and produced the plastic deformation. Afterwards, if we want the specimen to fail macroscopically, the axial stress must be increased continuously. Thus, when the macroscopic stress–strain response of the tested marble specimens reached the peak stress, the mesoscopic material elements with higher strengths and stiffnesses in the specimens also reached their maximum supporting limits, yielding failure and initiating plastic deformation. Therefore, the total deformations in the specimens would tend to be uniform, and the plastic deformations of the specimens would also increase with cycle number, as shown in Fig. 7.10d, which results in distinct yielding platform in the macroscopic stress–strain curves near the peak stress, as shown in Fig. 7.8d. Therefore, the three stages II, III and IV are very difficult to distinguish.

Figure 7.14 depicts the ultimate failure mode of the marble specimens under simple cyclic loading. From Fig. 7.14, it can be seen that although all of the entire marble specimens experienced single shear fracture, the shear fracture angles (the angles between the shear fracture surfaces and the horizontal directions) depended on the confining pressure. At $\sigma_3 = 2.5$ MPa, the shear fracture angles of marble was approximately 68°; at $\sigma_3 = 5.0$ MPa, the shear fracture angle of marble was approximately 65°; at $\sigma_3 = 7.5$ MPa, the shear fracture angle of marble was approximately 60°; and at $\sigma_3 = 10$ MPa, the shear fracture angle of marble was approximately 57°. The above analysis indicates that the shear fracture angle of marble under triaxial cyclic loading decreased gradually with increasing confining pressure.

Figure 7.15 shows the microscopic observation results of marble specimen after simple cyclic loading failure with respect to $\sigma_3 = 2.5$ and 10 MPa. Compared with the microscopic observation results of marble specimen under triaxial monotonic loading failure shown in Fig. 7.7, it is clear that the damage to the marble specimen

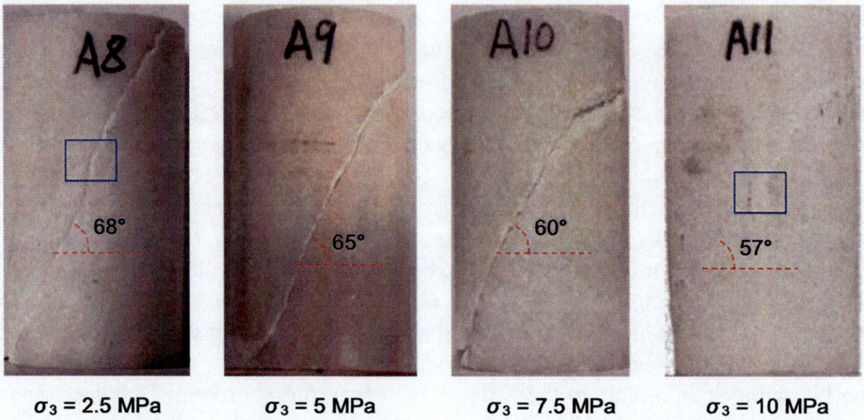

Fig. 7.14 Ultimate failure mode of marble specimen under simple cyclic loading

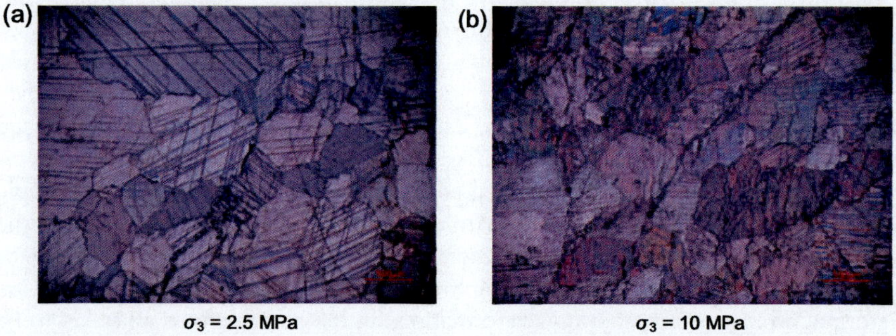

Fig. 7.15 Photomicrographs of marble specimen under simple cyclic loading

after triaxial cyclic loading failure was more serious, which agrees very well with the previous research results on sandstone specimen (Yang et al. 2015a). Furthermore, it should be noted that under higher confining pressures, near the main shear crack, the mineral grains also experienced more serious damage due to the following reason (Yang et al. 2016). Under lower confining pressures, the specimen failed mainly along the shear fracture plane, which was dominated by grain boundary cracks or transgranular cracks, whereas in other regions, most of the mineral grains were intact and no obvious grain boundary cracks and transgranular cracks were observed. However, under higher confining pressures, the mineral grains with low strengths and stiffnesses first reached maximum supporting capacity, and experienced yield softening (Yang et al. 2008). Afterwards, if we want to cause the specimen to fail macroscopically, the axial deformation must be increased continuously. Thus, when the macroscopic stress–strain response of the tested marble reaches the peak strength, the mesoscopic grains with higher strengths and stiffnesses in the specimen will also reach their maximum supporting limits and experience yielding failure. Therefore, from Fig. 7.15b, we can observe a large amount of damage and failure of the mineral grains besides one main shear failure plane in the crystalline marble at $\sigma_3 = 10$ MPa.

7.4 Triaxial Experimental Results of Crystalline Marble Under Complex Cyclic Loading

In accordance with the path shown in Fig. 7.5, Fig. 7.16 illustrates the experimental results of marble specimens under complex cyclic loading. It should be noted that in Fig. 7.16, five specimens experienced different axial strain values. For the same specimen, the stress level at point *A* was approximately equal to that at point *B*, but their axial strain values are different. The marble specimen with larger axial strain at point *A* indicates that the marble had a larger damage extent. In Fig. 7.16, the points *A*, *B* and *C* correspond to those shown in Fig. 7.5. After point *C*, the specimens had become damaged, and the axial deviatoric stress-axial strain curve of the damaged marble specimens at four different confining pressures can be also obtained. Furthermore, it can be seen from Fig. 7.16 that the five specimens had very small differences, which can be used to investigate the mechanical behavior

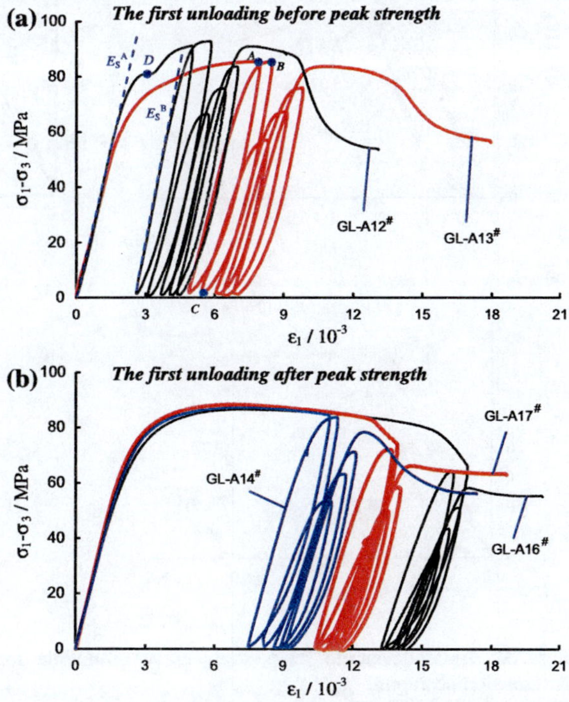

Fig. 7.16 Experimental results of marble specimens under complex cyclic loading in accordance with the path shown in Fig. 8.5

of the marble compressed different deformation positions, as shown in Fig. 7.17. In Fig. 7.17, each axial deviatoric stress-axial strain curve of marble specimens was redrawn by adjusting the axial strain shown in Fig. 7.16 to the zero point. The corresponding elastic moduli and peak strengths of the damaged marble specimens are listed in Tables 7.2 and 7.3, respectively. We can therefore analyze the effects of damaged extent on the peak strength and elastic modulus of the marble in detail.

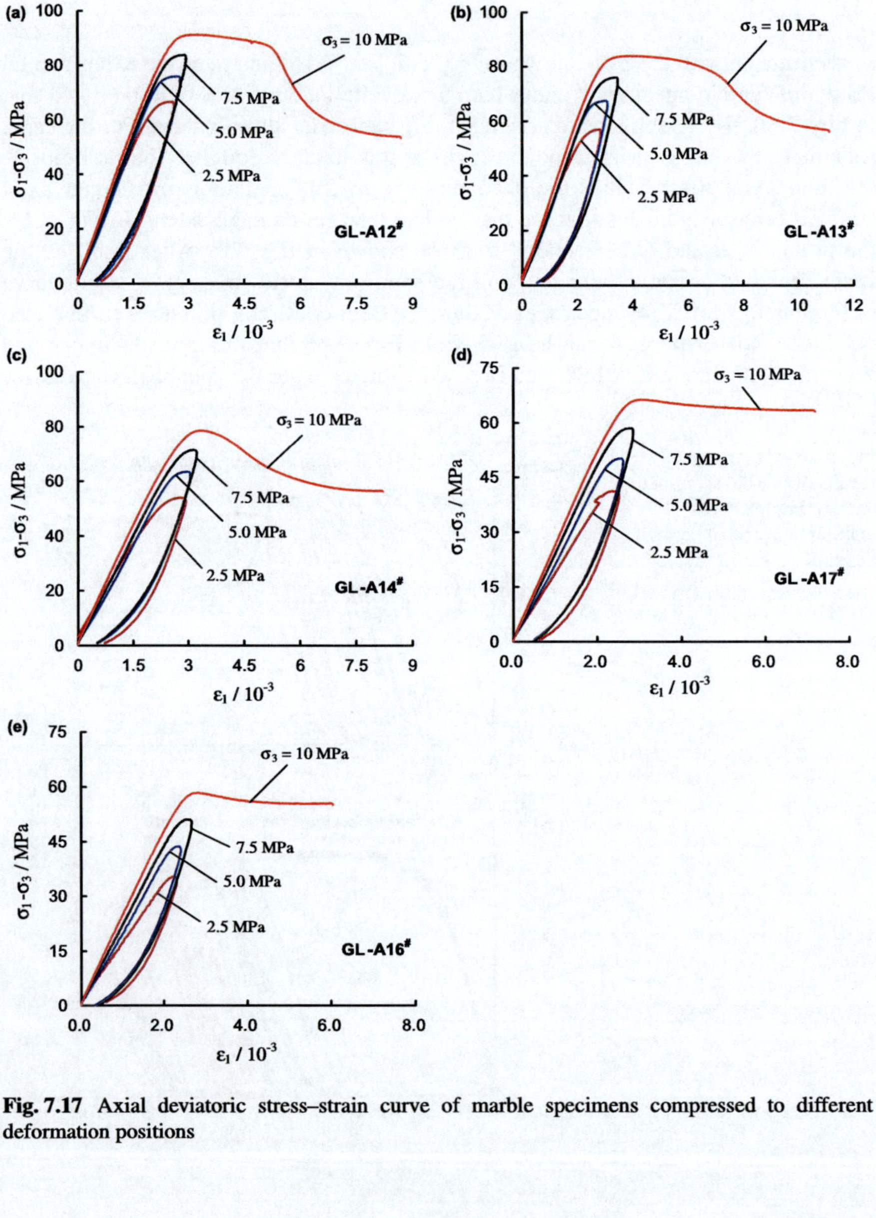

Fig. 7.17 Axial deviatoric stress–strain curve of marble specimens compressed to different deformation positions

7.4.1 *Elastic Modulus of Crystalline Marble Under Complex Cyclic Loading*

Table 7.2 lists the elastic moduli of the marble specimens under complex cyclic loading. Based on the experimental data listed in Table 7.2, we can detailed investigate the variations in the elastic moduli of the marble specimens under complex cyclic loading, as shown in Figs. 7.18 and 7.19.

From Fig. 7.18, it can be seen that at $\sigma_3 = 10$ MPa, the Es^B values of marble decreased gradually from 43.37 to 24.51 GPa as ε_{1A} increases from 5.035 × 10^{-3} to 16.96 × 10^{-3}. The variation in the Es^{4th} values of the marble with the axial deformation at point *A* was similar to that of Es^B, which also decreased from 37.27 to 23.76 GPa in the variation range of the same axial strain. This can be interpreted as follows. The marble specimen with larger compressed axial deformation suffered more damages in comparison with that of the specimen with less compressed axial deformation, which resulted in decreasing of the Es^B or Es^{4th} values of the marble specimen with increasing compressed deformation. However, the Es^A values of the marble first decreased sharply from 40.25 to 34.85 GPa as ε_{1A} increased from 5.035 ×

Table 7.2 Elastic modulus of marble under complex cyclic loading (unit: GPa)

Specimen	$\varepsilon_{1A}/10^{-3}$	$E_S{}^A$/GPa (σ_3 = 10 MPa)	$E_S{}^B$/GPa (σ_3 = 10 MPa)	$E_S{}^{1st}$/GPa (σ_3 = 2.5 MPa)	$E_S{}^{2nd}$/GPa (σ_3 = 5.0 MPa)	$E_S{}^{3rd}$/GPa (σ_3 = 7.5 MPa)	$E_S{}^{4th}$/GPa (σ_3 = 10 MPa)
GL-A12#	5.035	40.25	43.37	34.20	35.70	36.77	37.27
GL-A13#	7.958	34.85	33.60	25.13	29.06	29.15	30.12
GL-A14#	10.98	33.99	30.82	22.52	26.19	27.10	28.28
GL-A17#	13.92	36.12	27.98	18.72	22.82	24.57	26.06
GL-A16#	16.96	31.37	24.51	16.53	20.37	22.07	23.76

Note E_S represents the elastic modulus of rock, which is confirmed according to the slope of elastic deformation stage. The elastic modulus of marble at point *A* and *B* are denoted in Fig. 7.16. The four stages of marble under four different confining pressures are denoted in Fig. 7.5

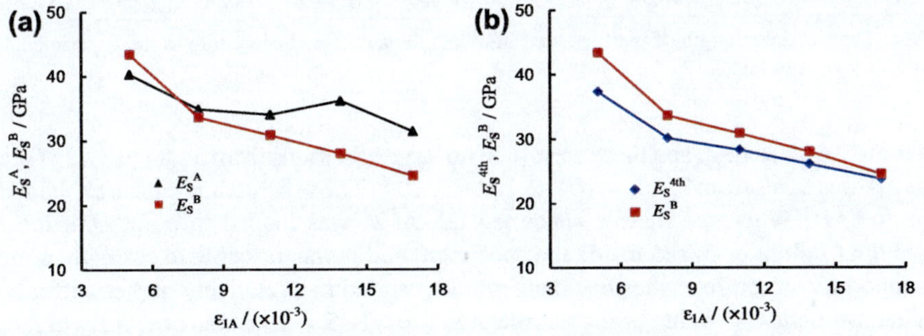

Fig. 7.18 Elastic modulus of marble specimen at various deformation positions (σ_3 = 10 MPa)

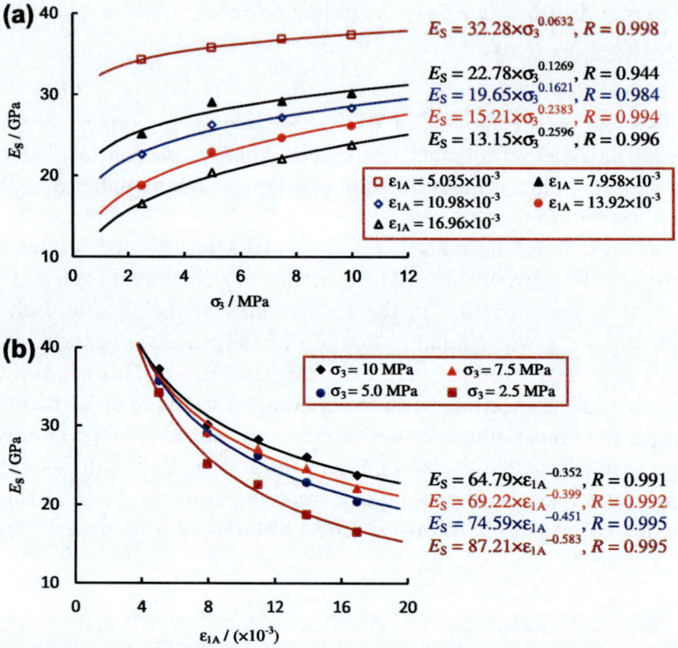

Fig. 7.19 Effect of confining pressure and axial strain at point *A* on elastic modulus of marble specimen under complex cyclic loading

Table 7.3 Triaxial strength of marble under complex cyclic loading (unit: MPa)

Specimen	$\varepsilon_{1A}/10^{-3}$	$\varepsilon_{1B}/10^{-3}$	$\sigma_3 =$ 2.5 MPa	$\sigma_3 =$ 5 MPa	$\sigma_3 =$ 7.5 MPa	$\sigma_3 =$ 10 MPa	*C*/MPa	*Φ*/°
GL-A12#	5.035	5.842	66.53	75.89	83.76	91.06	14.28	38.3
GL-A13#	7.958	8.444	57.05	67.55	76.09	83.90	11.45	39.8
GL-A14#	10.98	11.27	53.91	63.41	71.44	78.59	11.19	38.4
GL-A17#	13.92	13.51	41.26	50.22	58.61	66.38	7.95	38.8
GL-A16#	16.96	16.38	35.36	43.67	51.11	58.28	6.97	37.2

Note The triaxial strength of marble listed in this table is obtained according to the experimental results shown in Fig. 7.17

10^{-3} to 7.958×10^{-3}, and there were then no large variations from 34.85 to 31.37 GPa as ε_{1A} increases from 7.958×10^{-3} to 16.96×10^{-3}. Those behaviors can be explained as follows. When the marble specimen GL-A12# was loaded to point *D*, a local yielding failure occurred inside the specimen, and it was possible to strengthen the supporting structure of the specimen, which resulted in a relatively higher stiffness. Therefore, the Es^A value of the marble at $\varepsilon_{1A} = 5.035 \times 10^{-3}$ was obviously higher than those for the other four various axial strain values. However, the Es^A values of the marble at the other four various axial strain values were approximately equal, which shows the good consistency of the four tested specimens.

However, it should be noted that in all cases, the Es^{4th} values of the marble were lower than the Es^{B} values even though the confining pressures were all 10 MPa, but the difference between Es^{4th} and Es^{B} decreased with the increase in axial deformation at point *A*, as shown in Fig. 7.18b. The above description indicates that the loading history after point *C* showing in Fig. 7.16 greatly affected the elastic modulus of the marble and the effect decreased as the compressed axial deformation increased. When the marble was compressed to 16.96×10^{-3} at point *A*, the specimen almost supported the axial capacity by friction slippage, as shown in Fig. 7.16b. Therefore, the loading history had almost no any effect on the elastic modulus of the marble, which can be inferred from approximate elastic modulus of the marble at $\varepsilon_{1A} = 16.96 \times 10^{-3}$.

Figure 7.19 shows the effects of confining pressure and axial strain at point *A* on the elastic modulus of marble specimen under complex cyclic loading. From Fig. 7.19a, it can be seen that for the same axial strain at point *A*, the elastic modulus of damaged marble specimen increased nonlinearly with the increase in confining pressure, which can be expressed by the equation: $E_S = a \times {\sigma_3}^b$. *a* and *b* are both material parameters, which were determined according to the experimental results. Furthermore, the elastic modulus of marble increased more and more slowly with increasing confining pressure. However, from Fig. 7.19b, it is clear that for the same confining pressure, the elastic modulus of the damaged marble specimen decreased nonlinearly with the increase in axial strain at point *A*, which can be expressed by the equation: $E_S = c \times {\varepsilon_{1A}}^{-d}$. *c* and *d* are both material parameters, which were determined according to the experimental results. Furthermore, the elastic modulus of the marble decreased more and more slowly with increasing axial strain at point *A*.

7.4.2 Peak Strength of Crystalline Marble Under Complex Cyclic Loading

Figure 7.20 shows peak strength behavior of the marble specimens compressed to different deformation positions, which were obtained according to the experimental results shown in Fig. 7.17. From Fig. 7.20, we can see that the peak strength of marble changed with different axial deformation positions at point *A*. At the same confining pressure, the peak strength of marble specimen decreased with the increase of axial strain at point *A*. For example, at $\sigma_3 = 2.5$ MPa, the peak strength of marble specimen decreased from 66.53 to 35.36 MPa as the axial strain at point *A* increased from 5.035 $\times 10^{-3}$ to 16.96×10^{-3}, which was a result of the gradual increase in damage extent with the axial deformation. For the same strain at point *A*, the peak strength of marble specimen increased with the confining pressure, which could be well expressed by the linear Mohr–Coulomb criterion. By using Eq. (7.1), we confirmed the cohesions

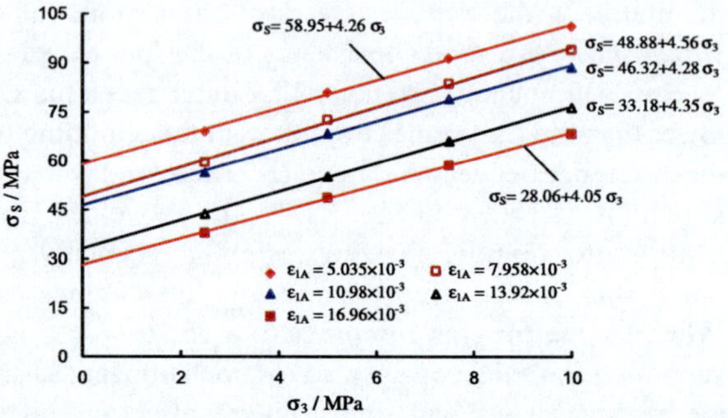

Fig. 7.20 Peak strength behavior of marble specimens compressed to different deformation positions

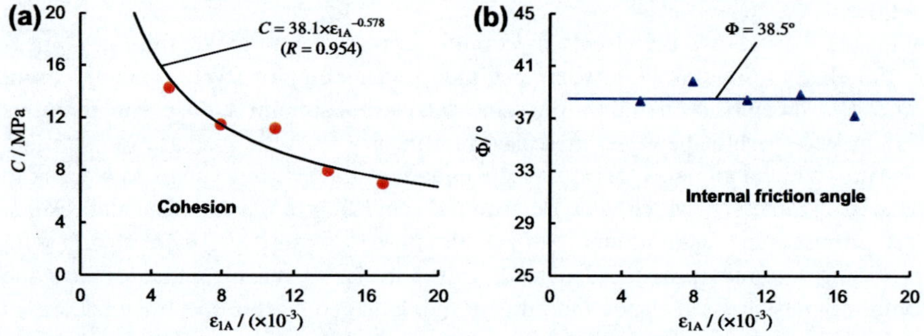

Fig. 7.21 Effect of axial strain at point *A* on the strength parameters of marble specimen under complex cyclic loading

and internal friction angles of the marble specimens for different axial strain values, all of which were listed in Table 7.3. In accordance with the cohesions and internal friction angles (Table 7.3), the effect of the axial strain at point *A* on the strength parameters of marble specimen under complex cyclic loading can be analyzed.

Figure 7.21 plots the relationship between axial strain at point *A* and the strength parameters of marble specimen under complex cyclic loading. In the case of the variation in cohesion with increasing axial strain at point *A*, the cohesion of marble decreased gradually from 14.28 to 6.97 MPa with increasing axial strain at point *A*. However, the attenuating extent was different, and the cohesion of the marble decreased rapidly at the beginning, and was then followed by a slow attenuation, which can be well expressed by the power function relation $C = 38.1 \times \varepsilon_{1A}{}^{-0.578}$, for which the correlation coefficient of the nonlinear regression was approximately 0.954, as evidenced by Fig. 7.21a. However, when compared with the cohesion, the internal friction angle of the marble was relatively insignificant with the axial strain (point *A*) region considered here, as shown in Fig. 7.21b. The internal friction angle of marble at various axial strain values ranged from 37.2° to 39.8° and the average value was approximately 38.5°. Therefore, complex cyclic loading caused some damage to the marble specimens, which can be interpreted as losses in specimens' cohesions.

Compared with the cohesion of the marble under triaxial monotonic loading (11.77 MPa), only the cohesion of the marble at $\varepsilon_{1A} = 5.035 \times 10^{-3}$ was relatively higher (14.28 MPa), primarily because the tested marble specimen GL-A12# had a relatively higher supporting capacity than that used under triaxial monotonic loading, as shown in Fig. 7.16a. Afterwards, it can be seen that cohesion of marble specimen at $\varepsilon_{1A} = 7.958 \times 10^{-3}$ and 10.98×10^{-3} were all approximately equal to those under triaxial monotonic loading because the specimen experienced minor damage after being loaded to the corresponding strain at point *A*. However, the cohesions of the marble specimen at $\varepsilon_{1A} = 13.92 \times 10^{-3}$ and 16.96×10^{-3} were all obviously lower than those under triaxial monotonic loading due to the serious damage experienced by the specimen when being loaded to the corresponding strain at point *A*. However, the internal friction angles of the marble specimens under complex cyclic loading were all approximately equal to those under triaxial monotonic loading (approximately 40.9°), which means that the damage extent had an insignificant effect on the internal friction angle of the marble, as evidenced by Fig. 7.21b.

To describe the peak strength characteristics of the marble specimens under complex cyclic loading and various deformation positions, a revised strength criterion for damaged rocks is put forward on the basis of the linear Mohr–Coulomb criterion:

$$\sigma_S^* = \frac{2C^* \cos k\varphi_0 + \sigma_3(1 + \sin k\varphi_0)}{1 - \sin k\varphi_0} == \frac{2mC_0\varepsilon_{1A}^{-n} \cos k\varphi_0 + \sigma_3(1 + \sin k\varphi_0)}{1 - \sin k\varphi_0} \tag{7.3}$$

where C^* is the cohesion of damaged marble, which depends on various damage extents, and is equal to $mC_0\varepsilon_{1A}{}^{-n}$, where, C_0 is the cohesion of undamaged marble, which was confirmed in the triaxial monotonic loading test in this research to be 11.77 MPa, φ_0 is the internal friction angle of undamaged marble, which was 40.9° in this research, and *k* is a coefficient, obtained by the ratio of the average value of damaged marbles to φ_0 and was equal to 0.94.

Figure 7.22 illustrates the ultimate failure modes of the marble specimens under complex cyclic loading. From Fig. 7.22, it is clear that all the marble specimens

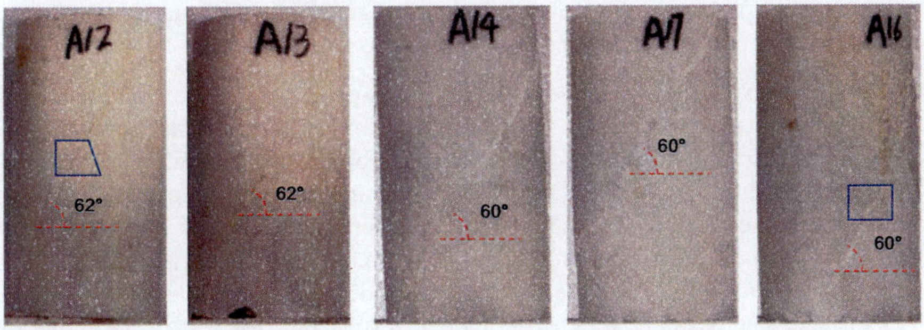

Fig. 7.22 Ultimate failure modes of marble specimens under complex cyclic loading

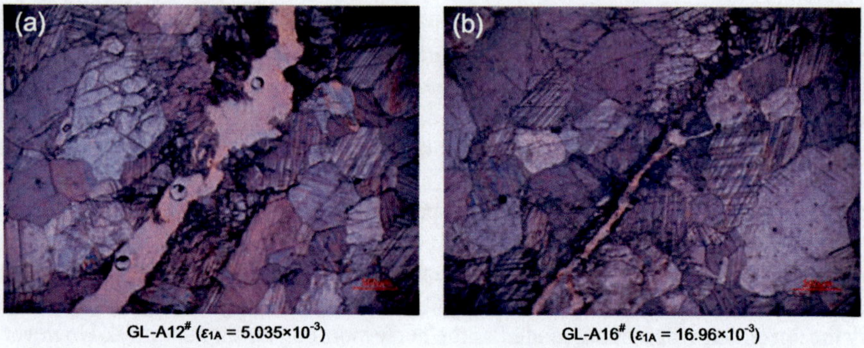

Fig. 7.23 Photomicrographs of marble specimen under complex cyclic loading

under complex cyclic loading exhibited typical shear failure modes with a single shear fracture plane. Furthermore, the angles of the shear fracture planes were all approximately 60°.

Figure 7.23 shows the microscopic observation results of marble specimen after complex cyclic loading failure with respect to different axial deformation value. Compared with the microscopic observation results of marble specimen under simple cyclic loading failure shown in Fig. 7.15, it is clear that the damage of marble specimen after complex cyclic loading failure was more serious, which indicates that complex loading path can lead to more obvious damage on the rock specimen. From Fig. 7.23, it can be seen that main shear failure is very obvious, but nearby the main shear failure, many grains suffered serious damage.

7.5 Conclusions

This chapter explored the mechanical damage characteristics of crystalline marble using a series of triaxial monotonic and cyclic experiments. Photomicrographs of marble specimens after triaxial monotonic and cyclic loading are simultaneously captured to deeply reveal the deformation failure mechanism of marble. Based on the experimental results, the following conclusions can be drawn:

1. Under triaxial monotonic loading, the post-peak deformation of the tested marble changed gradually from strain-softening to strain-hardening as σ_3 increased from 0 to 10 MPa. Under uniaxial compression, the marble specimen showed brittle failure characteristics, with some axial splitting tensile cracks; in the range of $\sigma_3 = 2.5$–7.5 MPa, the marble specimen took on single shear fracture characteristics, with a larger fracture angle of approximately 65°, but at $\sigma_3 = 10$ MPa, the marble specimen did not have an obvious shear fracture surface. The triaxial cyclic experimental results indicate that in the range of tested confining pressures, the triaxial strengths of the marble specimens under

cyclic loading were approximately equal to those under monotonic loading. Although there was only one single shear fracture, the fracture angle of marble under triaxial cyclic loading decreased gradually with increasing σ_3.

2. On the basis of the experimental results of the marble under triaxial cyclic loading, it can be seen that at the same confining pressure, the plastic deformation of the marble increased slowly with the increase in unloading stress level, whereas after the peak strength, the plastic deformation of marble increased dramatically even with the reduction in unloading stress level. With the increase in cycle number, the elastic strain of marble at first all increased and then decreased, having a maximum value, but the plastic strain of marble increased nonlinearly. To quantitatively evaluate the damage extent of the marble specimen under triaxial cyclic loading, a damage variable was defined in accordance with the irreversible deformation for each cycle. Furthermore, the evolution characteristics of Young's modulus of marble under triaxial cyclic loading were evaluated.
3. Under complex cyclic loading, the peak strength of marble changed with different axial deformation positions at point *A*. At the same confining pressure, the peak strength of marble specimen decreased with the increase of axial strain at point *A*, which was a result of the gradual increase of damage extent with the axial deformation. For the same axial strain at point *A*, the peak strength of marble specimen increased with the confining pressure, which could be well expressed by the linear Mohr–Coulomb criterion. The cohesion of marble attenuated following a power function relation with the increase of axial strain at point *A*, whereas the internal friction angle of marble under complex cyclic loading path was not dependent on axial strain at point *A*. For the damaged specimen, the elastic modulus of marble increased nonlinearly with the confining pressure. To describe the peak strength characteristics of marble specimens under complex cyclic loading with various deformation positions, a revised strength criterion for damaged rocks was put forward on the basis of the linear Mohr–Coulomb criterion.

References

Cheng Y, Wong LNY, Maruvanchery V (2016) Transgranular crack nucleation in Carrara marble of brittle failure. Rock Mech Rock Eng 49:3069–3082

Heap MJ, Vinciguerra S, Meredith PG (2009) The evolution of elastic moduli with increasing crack damage during cyclic stressing of a basalt from Mt Etna Volcano. Tectonophysics 47:153–160

Migliazza M, Ferrero AM, Spagnoli A (2011) Experimental investigation on crack propagation in Carrara marble subjected to cyclic loads. Int J Rock Mech Min Sci 48:1038–1044

Pei JL, Fei WP, Liu JF (2016) Spatial evolution and fractal characteristics of natural fractures in marbles under uniaxial compression loading based on the source location technology of acoustic emission. Environ Earth Sci 75:828

Qiu SL, Feng XT, Xiao JQ, Zhang CQ (2014) An experimental study on the pre-peak unloading damage evolution of marble. Rock Mech Rock Eng 47:401–419

Trippetta F, Collettini C, Meredith PG, Vinciguerra S (2013) Evolution of the elastic moduli of seismogenic Triassic Evaporites subjected to cyclic stressing. Tectonophysics 592:67–79

Yang SQ, Jing HW (2013) Evaluation on strength and deformation behavior of red sandstone under simple and complex loading paths. Eng Geol 154:1–17

Yang SQ, Jiang YZ, Xu WY, Chen XQ (2008) Experimental investigation on strength and failure behavior of pre-cracked marble under conventional triaxial compression. Int J Solids Struct 45:4796–4819

Yang SQ, Jing HW, Li YS, Han LJ (2011) Experimental investigation on mechanical behavior of coarse marble under six different loading paths. Exp Mech 51:315–334

Yang SQ, Jing HW, Cheng L (2014) Influences of pore pressure on short-term and creep mechanical behavior of red sandstone. Eng Geol 179:10–23

Yang SQ, Ranjith PG, Huang YH, Yin PF, Jing HW, Gui YL, Yu QL (2015a) Experimental investigation on mechanical damage characteristics of sandstone under triaxial cyclic loading. Geophys J Int 201(2):662–682

Yang SQ, Xu P, Ranjith PG, Chen GF, Jing HW (2015b) Evaluation of creep mechanical behavior of deep-buried marble under triaxial cyclic loading. Arab J Geosci 8:6567–6582

Yang SQ, Ju Y, Gao F, Gui YL (2016) Strength, deformability and X-ray micro-CT observations of deeply buried marble under different confining pressures. Rock Mech Rock Eng 49(11):4227–4244

Yang SQ, Tian WL, Ranjith PG (2017) Experimental investigation on deformation failure characteristics of crystalline marble under triaxial cyclic loading. Rock Mech Rock Eng 50(11):2871–2889

Chapter 8
Strength, Deformability and X-ray Micro CT Observations of Deeply-Buried Marble Under Different Confining Pressures

The Jinping II Hydropower Project is located in the upriver of the Yalong River in Sichuan Province, south-western China. As described by Wu et al. (2010) and Feng et al (2015), Jinping II mainly consists of three auxiliary tunnels and a sluice dam on the western side of the Jinping Mountain, four headrace tunnels through the mountain and an underground powerhouse complex on the eastern side of the mountain (Fig. 8.1). The total installed capacity is 4800 MW. Four headrace tunnels with a diameter of 12–13 m and a total length of 16.67 km are constructed at depths of 1500–2000 m, and the maximum buried depth is up to 2525 m. The average principal stress of the rock mass ranges from 35 to 50 MPa, and the maximum vertical principal stress is as high as 60 MPa, according to the measured geo-stress data and back analysis of the geo-stress field (Jiang et al. 2010). The main stratum excavated for the underground tunnels is Triassic marble. The uniaxial compressive strength of the marble in the project region ranges from 80 to 120 MPa (Li et al. 2012a, b), which indicates that it is a medium to hard rock material. Therefore, the experimental studies of the strength, deformability and failure behavior of the marble under higher geo-stress conditions are significant to understand the stability of the Jinping II Hydropower Project (Aliha 2014; Yang et al. 2015a).

Li et al. (2013) conducted conventional triaxial compression and reducing confining pressure tests for marble specimens, and analyzed the influence of two loading paths on the mechanical behavior of the marble. The experimental results show that in the loading and unloading test, peak strength, lateral strain, axial strain, and plastic deformation increase significantly as the confining pressure increases. The fracture mode is mainly a single shear fracture for the triaxial compression test and post-peak test, and the angle between the failure surface and the ends of the rock material becomes smaller with the increase of confining pressure. By conducting triaxial compression loading–unloading tests for marble specimens under different confining pressures and unloading rates, Huang and Li (2014) found that the magnitude of the initial confining pressure and unloading rate significantly influenced the rock failure modes and strain energy conversion (accumulation, dissipation and release) during unloading. Yang et al. (2008) completed conventional triaxial

S.-Q. Yang, *Mechanical Behavior and Damage Fracture Mechanism of Deep Rocks*,
https://doi.org/10.1007/978-981-16-7739-7_8

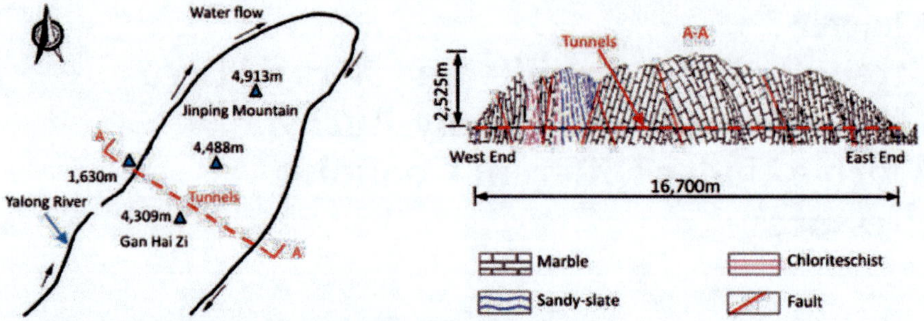

Fig. 8.1 The deeply-buried tunnels of the Jinping II Hydropower Station, China (Feng et al. 2015)

compression experiments on flawed marble specimens. The results show that the peak strength and failure mode of the marble depend not only on the geometry of the flaw, but also on the confining pressure. By proposing a new incrementally cyclic loading–unloading pressure test method, Qiu et al (2014) experimentally analyzed the pre-peak unloading damage evolution characteristics of the marble from the Jinping II Hydropower Project. The results demonstrated that the pre-peak damage and deformation characteristics of marble specimens could be easily quantified by the irreversible strains.

Rock masses commonly have a large number of cracks and flaws (Yang et al. 2009), and investigating their integrity and load bearing capacity via the frame work of fracture mechanics may provide more reliable predictions under real situations (Ayatollahi and Aliha 2007; Aliha et al. 2012, 2013; Funatsu et al. 2014). Next, we briefly describe fracture mechanics and their abilities to accommodate the cracked and flawed rock masses and structures. Ayatollahi and Aliha (2007) proposed a modified maximum tangential stress (MTS) criterion to predict the mixed mode I (crack opening mode) and II (crack sliding mode) fracture toughness data on the basis of the experimental results of semi-disk brittle rock specimens. Akbardoost and Ayatollahi (2014) completed the experiments on the mixed mode fracture behavior of marble using a center cracked circular disk (CCCD) specimen and an edge cracked triangular (ECT) specimen. Then, they proposed a modified MTS criterion to predict the mixed mode fracture resistance of marble obtained from the ECT specimens only using the mode I fracture resistance measured from a CCCD specimen. Based on the experimental results of the fracture toughness of marble under mixed-mode conditions, Chang et al. (2002) analyzed the size effects in terms of the specimen thickness, diameter and notch length on the fracture toughness of marble. By taking into account the geometry and loading conditions of the tested specimen, Aliha et al. (2012, 2013) put forward generalized maximum tangential stress (GMTS) criterion, which is in a good agreement with the experimental results of the Brazilian disc (BD), the semi-circular bend (SCB) and ECT marble specimens.

However, previous experimental studies of the strength and deformation failure behavior of marble using X-ray computer tomography (CT) observations have rarely considered triaxial tests. Generally, X-ray CT is the best non-invasive non-destructive method to characterize complex pore structures, and has been applied to explore the internal damage behavior of some rock materials (Ge et al. 2001; Hirono et al. 2003; Feng et al. 2004; Sufian and Russell 2013; Zhao et al. 2014; Meier et al. 2014; Yang et al. 2015b). Therefore, in this chapter, we report the results of a series of uniaxial and conventional triaxial compression tests on deeply-buried marble under different confining pressures ranging from 0 to 50 MPa (Yang et al. 2016). Based on the experimental results, we first investigate the influence of the confining pressure on the strength and deformation parameters of marble. Then, using an X-ray micro CT scanning system, the internal damage characteristics of the deformed marble specimens are analyzed in detail.

8.1 Deeply-Buried Marble Material and Testing Procedure

Marble, located in the number 3# test branch tunnel of an auxiliary cave in the Jinping II Hydropower Project, was chosen for testing. This marble is buried at a depth of approximately 2270 m. According to X-ray diffraction (XRD) (Fig. 8.2), the marble material mainly consists of carbonate minerals (calcite), with small quantities of quartz and amorphous solids. The marble has a crystalline and blocky structure (Fig. 8.2); that is macroscopically very homogeneous with a connected porosity of 1.09% and an average density of approximately 2710 kg/m^3.

In accordance with the method suggested by the ISRM (Fairhurst and Hudson 1999), the length-to-diameter ratio of the tested specimens should be between 2.0 and 3.0 to minimize the influence of the end friction effects on the results and to ensure a uniform stress state within the specimen. Therefore, all of the tested marble specimens are cylindrical, 50 mm in diameter and 100 mm long. The mechanical behaviors of specimens under triaxial compression were determined following the method suggested by the ISRM (Fairhurst and Hudson 1999).

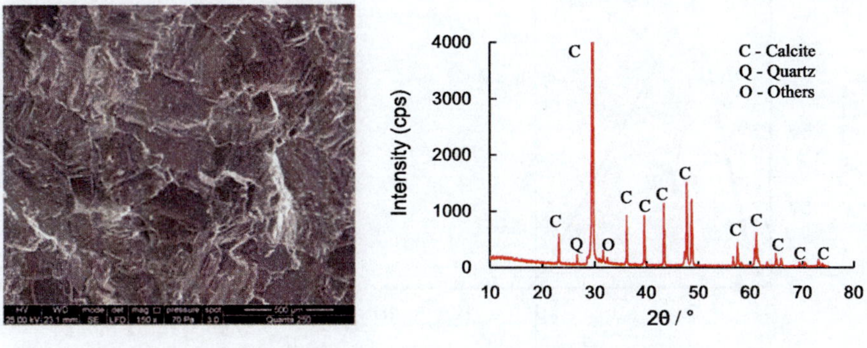

Fig. 8.2 SEM photomicrographs and XRD of the marble in this research

The experiments were conducted on servo-controlled triaxial equipment, which can be used to perform hydrostatic pressure tests, conventional triaxial compression tests under drained or undrained conditions, triaxial seepage tests, triaxial creep tests and chemical corrosion tests. When testing, the axial displacement is measured with two LVDTs (Linear Variable Differential Transformer), which are fixed between the bottom and top surfaces of the specimen inside the triaxial cell. The triaxial compression experiment had the following steps: the confining pressure was first applied to a desired value at a constant rate of 0.5 MPa/s, which ensured that the specimen was under uniform hydrostatic stress. The deviatoric stress was then loaded on the top surface of the specimen at a constant axial stress rate of 0.127 MPa/s until failure occurred. A total of eight specimens have been tested with confining pressures ranging from 0 to 50 MPa, and a total of ten specimens were carried out indirect Brazilian test. All marble specimens that were in the indirect Brazilian test were cylindrical, 50 mm in diameter and 25 mm thickness.

8.2 Strength and Deformation Behavior of Deeply-Buried Marble

8.2.1 Uniaxial Compression and Tensile Failure Behavior of Deeply-Buried Marble

Figure 8.3 presents the axial stress-axial strain curve and ultimate failure mode of marble under uniaxial compression. From Fig. 8.3, the uniaxial compressive strength (UCS) of marble specimen A4# is 99.11 MPa and the UCS of specimen A3# is 96.37 MPa. Furthermore under uniaxial compression, two marble specimens have

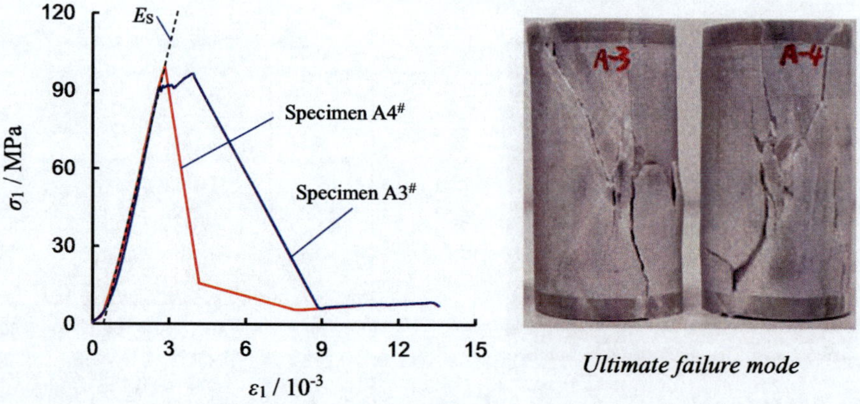

Fig. 8.3 Uniaxial compression test results of marble

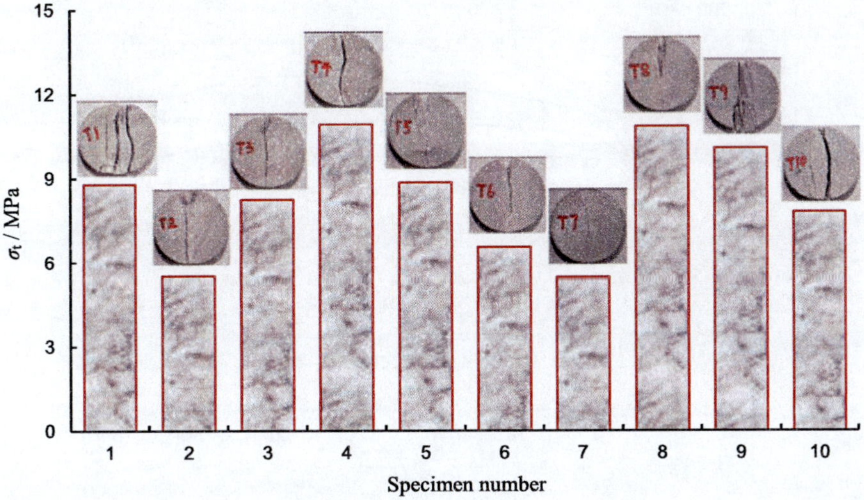

Fig. 8.4 The indirect Brazilian tensile strength of marble

a consistent deformation behavior before peak strength, but marble specimen A4# shows more brittleness than A3# after peak strength. However, from the ultimate failure mode, the marble specimens have noticeable axial splitting tensile cracks, which validates the obviously brittle failure of the marble material under uniaxial compression.

Figure 8.4 illustrates the indirect Brazilian tensile strength (σ_t) of ten marble specimens and the corresponding tensile failure modes. From Fig. 8.4, the tensile strength of the marble specimens ranges from 5.47 to 10.91 MPa, and the average is approximately 8.30 MPa. Thus, we obtain a UCS to σ_t ratio of approximately 11.78. The dispersion coefficient (the ratio of the D-value between the maximum and the minimum to the average) of the tensile strength of the tested marble is approximately 65.5%. The tensile failure mode of the tested marble is basically entirely the central splitting fracture, and main tensile cracks can be observed in the central region of the marble specimen.

8.2.2 Triaxial Strength and Deformation Behavior of Deeply-Buried Marble

The deviatoric stress-axial strain curves for marble under different confining pressures are presented in Fig. 8.5, where, σ_1 and σ_3 represent the major principal stress and the confining pressure, respectively; and ε_1 is the axial strain; and σ_{cr} and σ_p are the residual strength and peak strength, respectively, which are confirmed in accordance with the deviatoric stress-axial strain curves of marble specimens. A

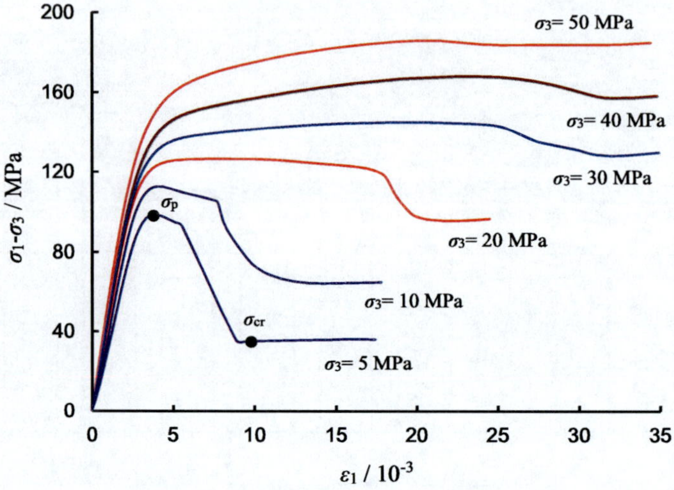

Fig. 8.5 Deviatoric stress-axial strain curves of marble under different confining pressures

typical example at $\sigma_3 = 5$ MPa is shown in Fig. 8.5. In Fig. 8.5, the post-peak deformation changes gradually from strain-softening to strain-hardening with increasing confining pressure. When σ_3 equals ~50 MPa, the deformation of the marble specimen has a distinct yielding platform and plastic flow occurs, i.e., the post-peak deformation behavior of the specimen changes into ideal plasticity.

To analyze the influence of the confining pressure on the elastic modulus (E_S) of marble, we used the average slope of the approximately straight line portion of the deviatoric stress–axial strain curve before the peak strength to calculate E_S, as shown in Fig. 8.3 However, the deformation modulus (E_{50}) is defined as the slope between the origin and the stress at half of the peak strength. The peak axial strain (ε_{1c}) refers to the strain value reaching the peak strength. Table 8.1 lists the detailed deformation values of the marble specimens in this study.

Table 8.1 Strength and deformation parameters of marble under different confining pressures

σ_3/MPa	E_{50}/GPa	E_S/GPa	ε_{1c}/($\times 10^{-3}$)	σ_{cr}/MPa	σ_{sr}/MPa	σ_P/MPa	σ_S/MPa
0 (A3#)	28.47	46.50	3.957	7.57	7.57	96.37	96.37
0 (A4#)	30.15	43.61	2.845	6.30	6.30	99.11	99.11
5	32.46	36.36	3.822	35.57	40.57	98.56	103.56
10	38.63	42.34	4.108	64.11	74.11	112.54	122.54
20	44.47	46.08	8.267	95.77	115.77	126.28	146.28
30	46.41	46.53	19.30	128.51	158.51	144.84	174.84
40	46.76	48.08	23.45	157.36	197.36	167.92	207.92
50	51.75	53.07	20.64	183.89	233.89	185.71	235.71

Note $\sigma_{sr} = \sigma_{cr} + \sigma_3$; $\sigma_S = \sigma_P + \sigma_3$

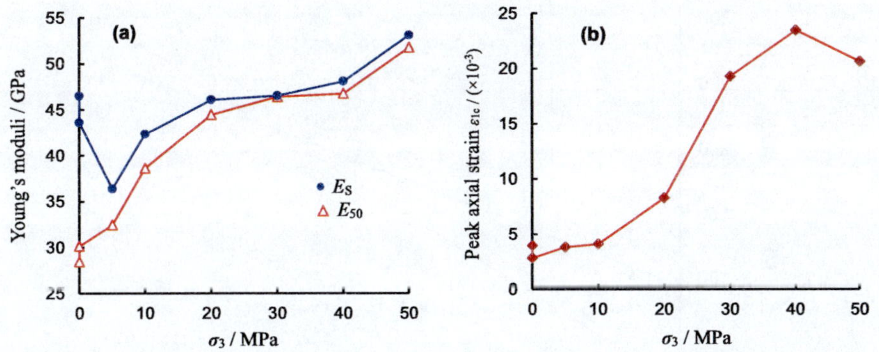

Fig. 8.6 Influence of the confining pressure on the Young's modulus and peak axial strain of marble

Based on the data listed in Table 8.1, Fig. 8.6 shows the influence of the confining pressure on the Young's modulus and peak axial strain of marble specimens in the range of tested confining pressures from 0 to 50 MPa. As Fig. 8.6a illustrates, the Young's modulus increased non-linearly with increasing confining pressure. It should be noted that the elastic modulus of the marble specimen under uniaxial compression ($\sigma_3 = 0$ MPa) is approximately 45.06 MPa, which is higher than that at $\sigma_3 = 5$ MPa, but the deformation modulus at $\sigma_3 = 0$ MPa is lower than that at $\sigma_3 = 5$ MPa. At $\sigma_3 = 5$ MPa, the elastic modulus and deformation modulus were 36.36 and 32.46 GPa, respectively. When σ_3 was increased to 50 MPa, the elastic modulus and the deformation modulus increased nonlinearly to 53.07 and 51.75 GPa, respectively. Furthermore, the difference between the elastic modulus and the deformation modulus tended to decrease as σ_3 increased, which was the result of the less pronounced stage of microcrack closure at higher confining pressures. To describe the relationship between the Young's moduli of the marble and confining pressures, a power function is proposed (Fig. 8.6a). However, from Fig. 8.6b, an increase in the confining pressure led the peak axial strain of marble to increase greatly as σ_3 increased to 40 MPa, whereas, the peak axial strain decreased from 23.45×10^{-3} to 20.64×10^{-3} as σ_3 increased from 40 to 50 MPa.

The Mohr–Coulomb criterion is a common strength criterion, which have been applied widely to rock mechanics and engineering practice. In this section, the Mohr–Coulomb criterion is used to investigate the effect of the confining pressure on the peak strength and residual strength of marble. The linear Mohr–Coulomb criterion can be expressed with the following equation:

$$\sigma_S = \sigma_0 + q\sigma_3 = \frac{2C\cos\phi + \sigma_3(1+\sin\phi)}{1-\sin\phi} \tag{8.1}$$

where, σ_S is the maximum axial supporting capacity of rock; σ_0 is usually regarded as the uniaxial compressive strength (UCS) of rock material; and q is an influence

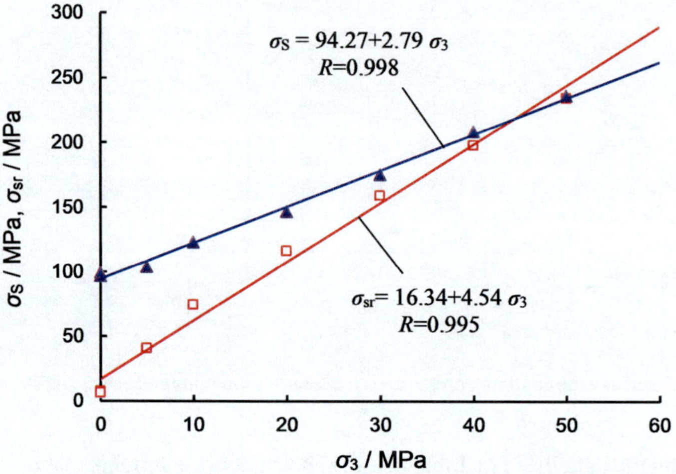

Fig. 8.7 Comparison of the peak and residual strength, and crack damage threshold of marble. The symbols represent the experimental results, and the lines represent the theoretical values following the Mohr–Coulomb or Hoek–Brown criteria (*R* is the correlation coefficient of determination of the regression)

coefficient of confining pressure on the σ_S of rock. The σ_0 and q are related to the cohesion C and the internal friction angle ϕ (Jaeger et al. 2007).

Based on the strength data listed in Table 8.1, the influence of σ_3 on the peak strength of marble is presented in Fig. 8.7. The figure indicates that the peak strength of marble has a distinct linear behavior with an increase in σ_3. Furthermore, the sensitivity of the residual strength on the confining pressure was clearly higher than that for the peak strength. By using the linear Mohr–Coulomb criterion, the peak and residual strength parameters of marble are calculated and listed in Table 8.2. The peak cohesion of marble is 28.22 MPa, which is considerably higher than the residual cohesion of approximately 3.83 MPa. However, the peak internal friction angle of marble is 28.2°, which is significantly lower than the residual internal friction angle of approximately 39.7°.

Table 8.2 Strength parameters of marble by the linear Mohr–Coulomb criterion

Different regression analysis	σ_0/MPa	q	C/MPa	ϕ/(°)	R
Using peak strength (σ_S)	94.27	2.79	28.22	28.2	0.998
Using residual strength (σ_{sr})	16.34	4.54	3.83	39.7	0.995

8.3 X-ray Micro CT Observations of Deep-Buried Marble

In this study, X-ray computed tomography (CT) scanning of the marble specimen was conducted using a Nanotom 160 high-resolution micro-CT at a spatial resolution of 30 μm. The X-ray beam penetrating the specimen was measured using an array of detectors. The X-rays are produced by electrons striking a Mo-W alloy target in an X-ray tube. The electron current was 90 μA, the accelerating voltage was 130 kV, and the scanning time was 4 s. The degree of X-ray attenuation depends on the density and the atomic number of the materials composing the specimens. Materials with a higher density and higher atomic number generally have a higher attenuation of X-rays. The X-ray projection data from various directions are obtained by rotating the specimen 360°, as shown in Fig. 8.8a. We collected two-dimensional images at intervals of 0.18°, and thus 2000 slice images were collected per specimen. A two-dimensional image representing the linear distribution of the X-ray attenuation was reconstructed using Fourier transformation of the projection data. A 3-D data set of the specimen was obtained by stacking consecutive two-dimensional images.

The CT images were obtained from the center portion (a 50 mm diameter and 68 mm long) of each specimen, as shown in Fig. 8.8b, c. The slice images of the horizontal cross-section shown in Fig. 8.9a were captured at 0.03 mm intervals. Therefore, a total of 2266 horizontal slices were obtained during the X-ray micro CT testing of each marble specimen along the specimen axis. For the vertical cross-section shown in Fig. 8.9b, the slice images were captured at 0.5 mm intervals and the scanning sequence was from front to back. The horizontal and vertical slice images can be used to explore the internal damage in the marble material. In Fig. 8.9, z and x represent the height and the depth, respectively. Each horizontal CT slice image was divided into 1730 × 1730 pixels and the spatial voxel resolution was approximately 0.03 mm.

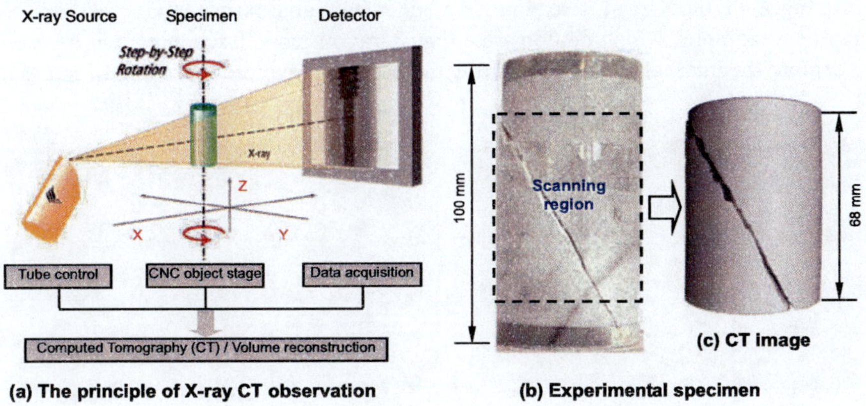

Fig. 8.8 The principle of the X-ray micro CT observation and illustration of the scanning region

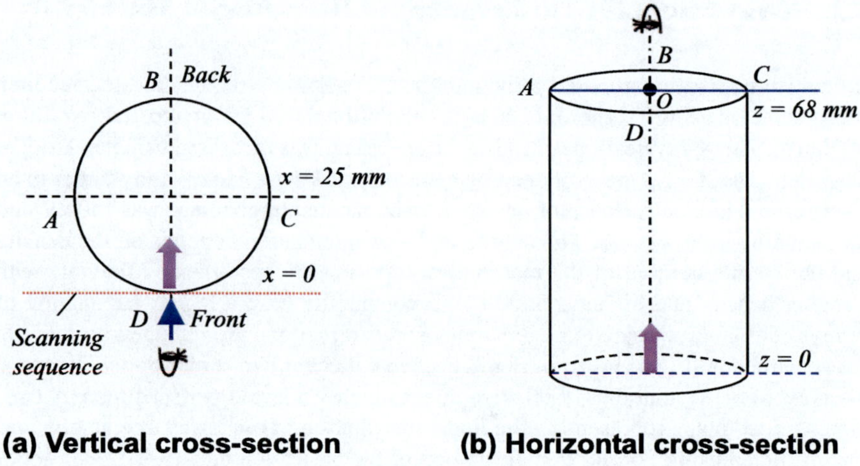

Fig. 8.9 Illustration of the scanning direction on the vertical and horizontal cross-section

Figure 8.10 illustrates the ultimate failure mode of the marble specimen under different confining pressures. From Fig. 8.10, the ultimate failure mode of marble specimens is very approximate, and took on the shear fracture characteristics, which does not change with the confining pressure. Therefore, in this research, we use CT images of marble specimens tested at $\sigma_3 = 0$ (uniaxial compression) and 10 MPa to explore the internal damage mechanism. Figure 8.11 shows the comparison of the X-ray CT scanning surface images and the actual surface crack photographs of marble specimens after uniaxial and triaxial compression failure, where the black regions represent cracks or surface spalling, and other regions indicate no surface failure.

In Fig. 8.11, the X-ray CT scanning surface images approximate the actual surface crack photographs, which demonstrate that X-ray micro-CT scanning can be used to explore the internal damage in marble materials. To explore the extent of internal

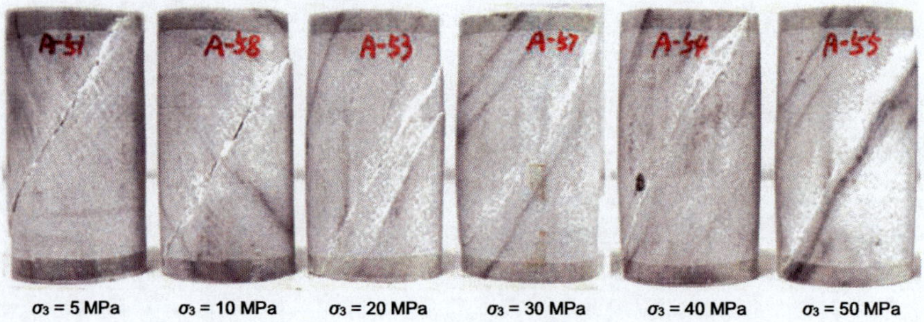

Fig. 8.10 The ultimate failure mode of marble specimen under different confining pressures

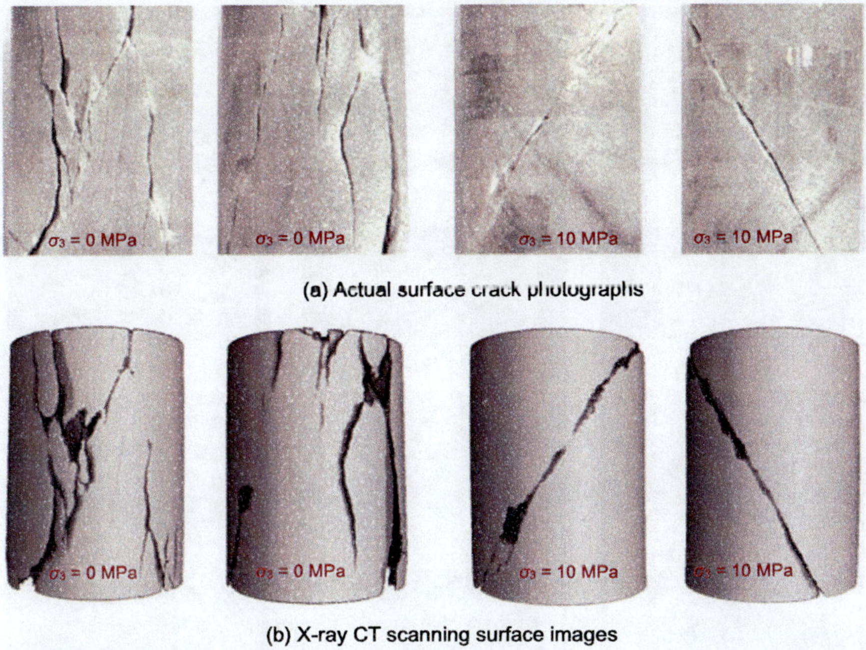

Fig. 8.11 Comparison of the X-ray CT scanning surface images and actual surface crack photographs of marble specimens after uniaxial ($\sigma_3 = 0$ MPa) and triaxial compression failure ($\sigma_3 = 10$ MPa)

damage in specimens after uniaxial and triaxial compression failure, Figs. 8.12 and 8.13 show vertical cross-sections with different depths of marble specimens after uniaxial and triaxial compression failure ($\sigma_3 = 10$ MPa), respectively. Generally, as seen in Figs. 8.12 and 8.13, the system of cracks under triaxial conditions is less complicated than that under uniaxial conditions, which also agrees well with the findings from the macro CT (the resolution was approximately 0.3~1.0 mm) observations of faulting in the Westerly granite under uniaxial and triaxial conditions (Kawakata et al. 1999). Furthermore, the fracture planes under triaxial compression have lower angles (the angle that between the normal to crack plane and the maximum principal stress) than under uniaxial compression.

Figure 8.14 shows 3-D CT images of failed marble specimens. In the 3-D images, the fracture regions are black, and the other regions are transparent. From Fig. 8.14, the marble specimen under uniaxial compression has a typical axial splitting failure mode, which is also identical to the surface fracture shown in Fig. 8.11 and the vertical cross-sections shown in Fig. 8.12. However, the failure mode of the marble specimen after triaxial compression failure is different to that after uniaxial compression failure. At a confining pressure of 10 MPa, the specimen shows a single shear failure mode. By rotating the image of the specimen, the trace of the shear fracture appears as an approximate plane, which indicates that the dip is more uniform throughout the shear

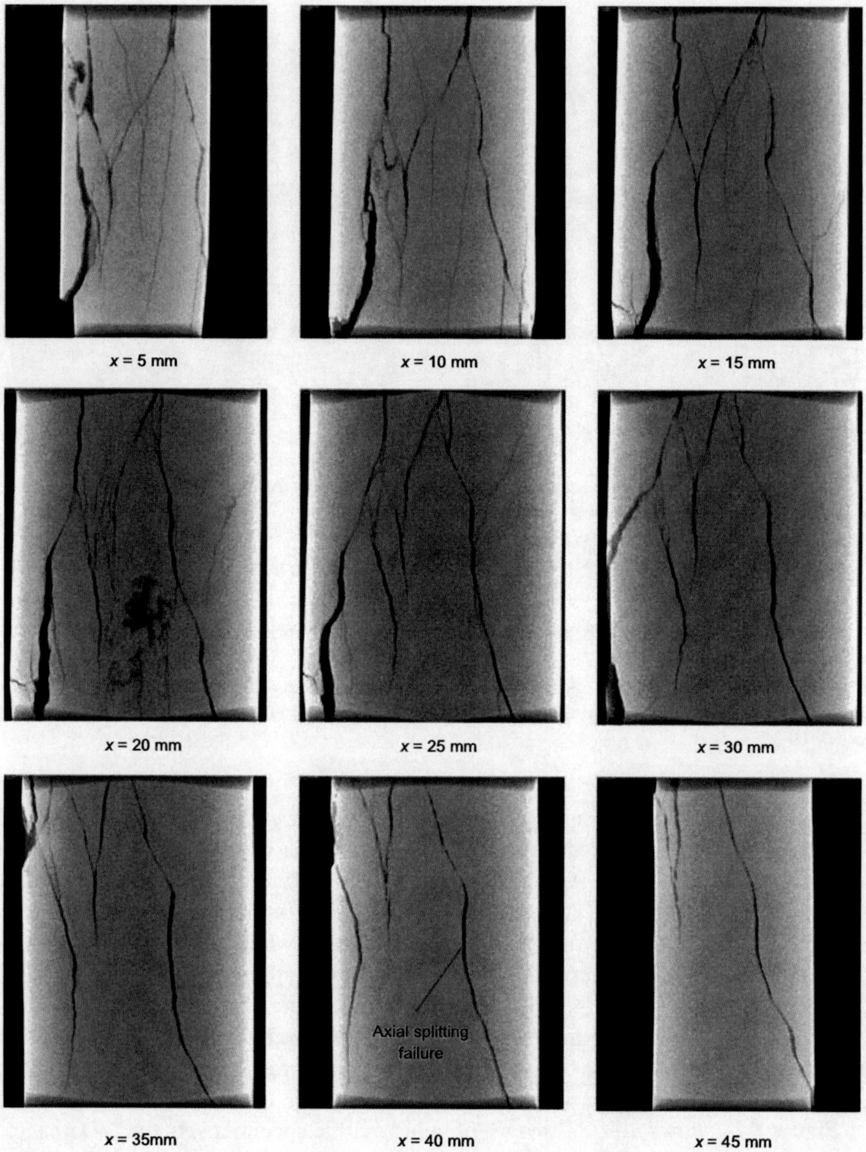

Fig. 8.12 Vertical cross-sections with different depths of marble specimen after uniaxial compression failure

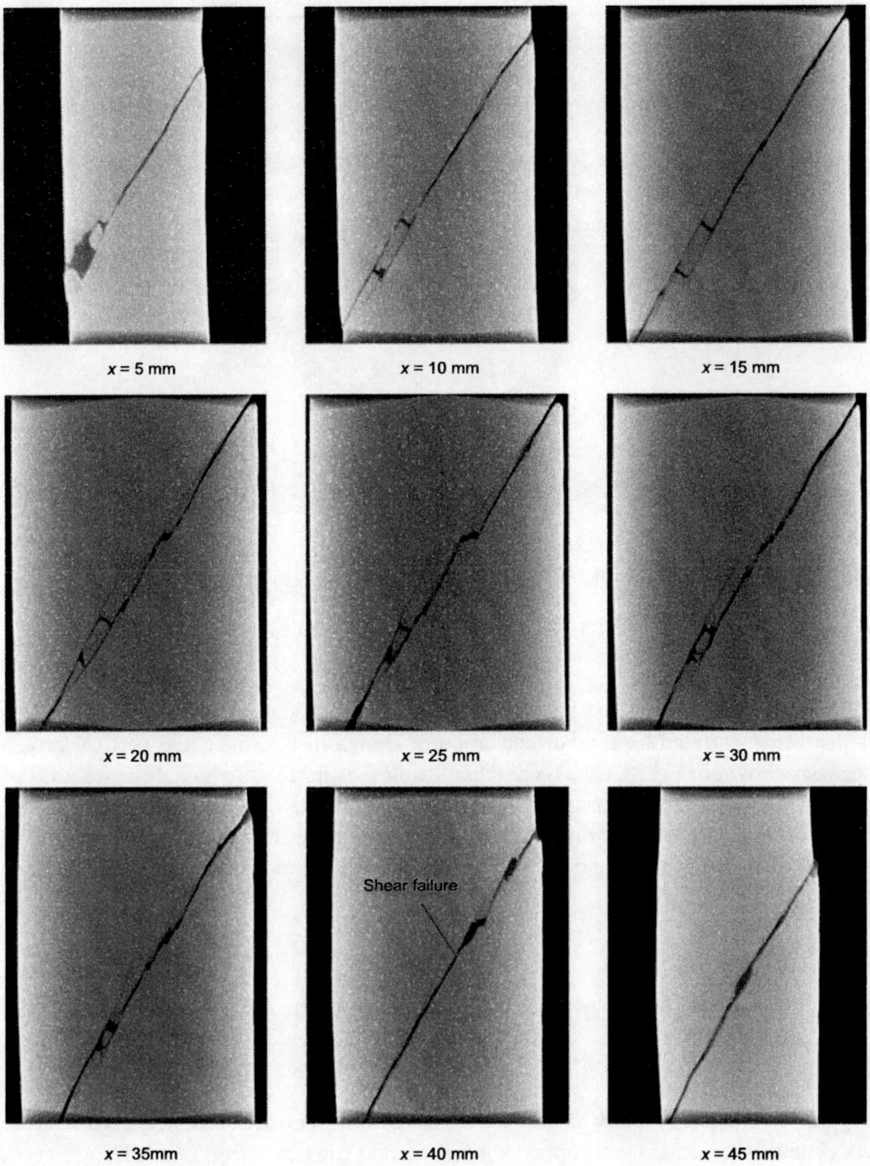

Fig. 8.13 Vertical cross-sections with different depths of marble specimen after triaxial compression failure ($\sigma_3 = 10$ MPa)

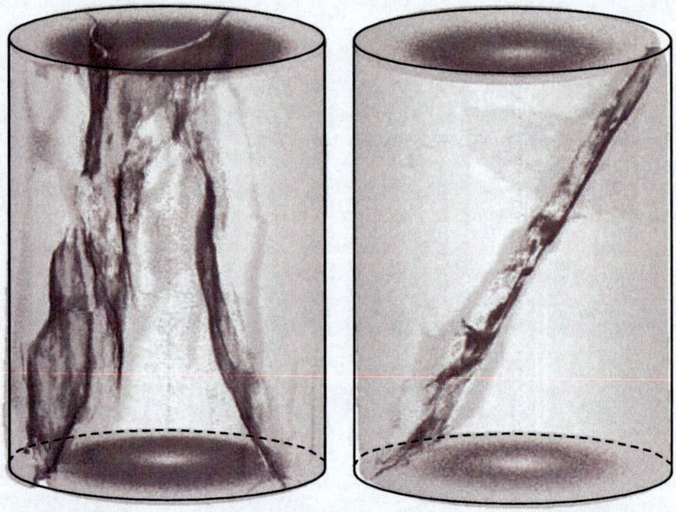

Fig. 8.14 Three-dimensional CT images of failed marble specimen

fracture curve. The single shear fracture curve under a confining pressure of 10 MPa is also demonstrated by the surface fracture shown in Fig. 8.11 and vertical cross-sections shown in Fig. 8.12. The axial splitting tensile fracture and shear fracture of the marble specimen under uniaxial and triaxial compression can be further inferred from microscopic observations on the fracture surface of the failed specimen, as shown in Fig. 8.15. The shear slippage surface is very smooth, but the tensile fracture surface is very rough.

8.4 Discussion on Failure Mechanism of Deep-Buried Marble

Figures 8.16 and 8.17 show horizontal cross-sections with different heights of marble specimens after uniaxial compression failure and triaxial compression failure at $\sigma_3 = 10$ MPa, respectively. From Fig. 8.16, we can analyze the effect of the height (z) on the extent of the internal damage to the marble material after uniaxial compression failure. At $z = 5$–20 mm, the horizontal cross-section is located in the bottom region of the specimen, and a number of cracks are observed, but the system of cracks is very simple, which can also be demonstrated from the bottom region shown in Fig. 8.14. Moreover, the width and length of these cracks are different, indicating various damage extents in the same cross-section. At $z = 20$–50 mm, the internal

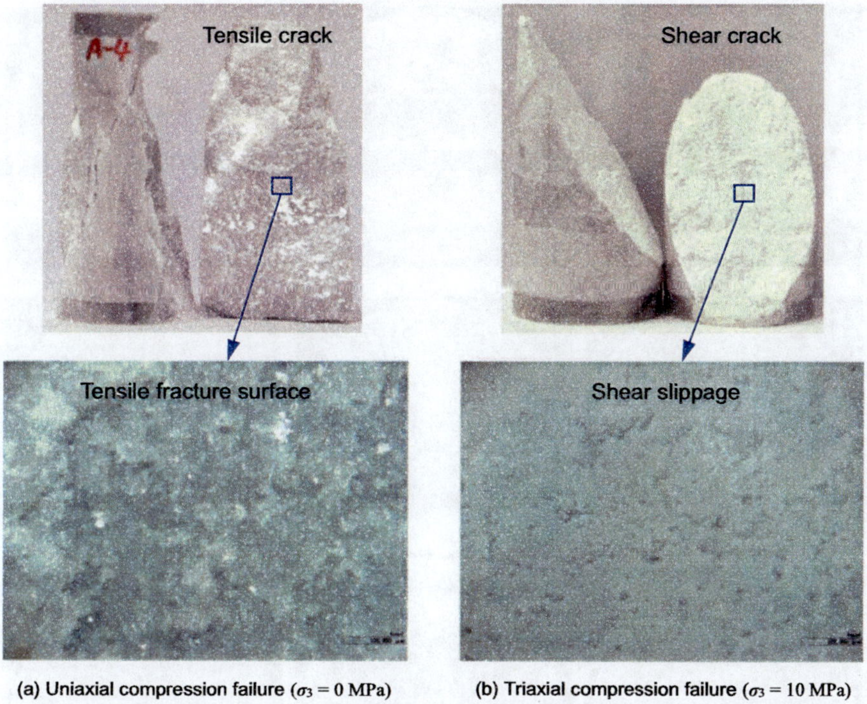

Fig. 8.15 Microscopic observations of the fracture surface of marble specimens under uniaxial and triaxial compression. **a** tensile fracture surface represents the tensile crack; and **b** shear slippage indicates the shear crack

damage to the specimen is more serious. Many cracks are observed and the system of cracks is very complex, which results from crack propagation and coalescence inside the specimen with the increase of deformation. However, at $z = 50$–60 mm, the crack system becomes more simple compared with that at $z = 20$–50 mm. The above statement suggests that the cracks are first initiated in the center of the specimen and then propagate towards the top and bottom regions of the specimen. In addition, with the increase of height, the width of crack 1 first decreases and then increases, and the path of crack 1 is not a line but a curve. Crack 2 moves from the right side to the central region, and the length of crack 3 decreases gradually with the increase of height. The angle between crack 4 and the horizontal direction decreases with the increase of height.

From Fig. 8.17, one main crack cuts through the cylindrical specimen. At $z = 5$–10 mm and 20–60 mm, only one main crack is observed; whereas at $z = 10$–20 mm two cracks were observed. With the increase of height, the main crack moves towards the right side from the left side. The main crack is approximately linear, which also indicates that the shear fracture is an approximate plane. The above statement may

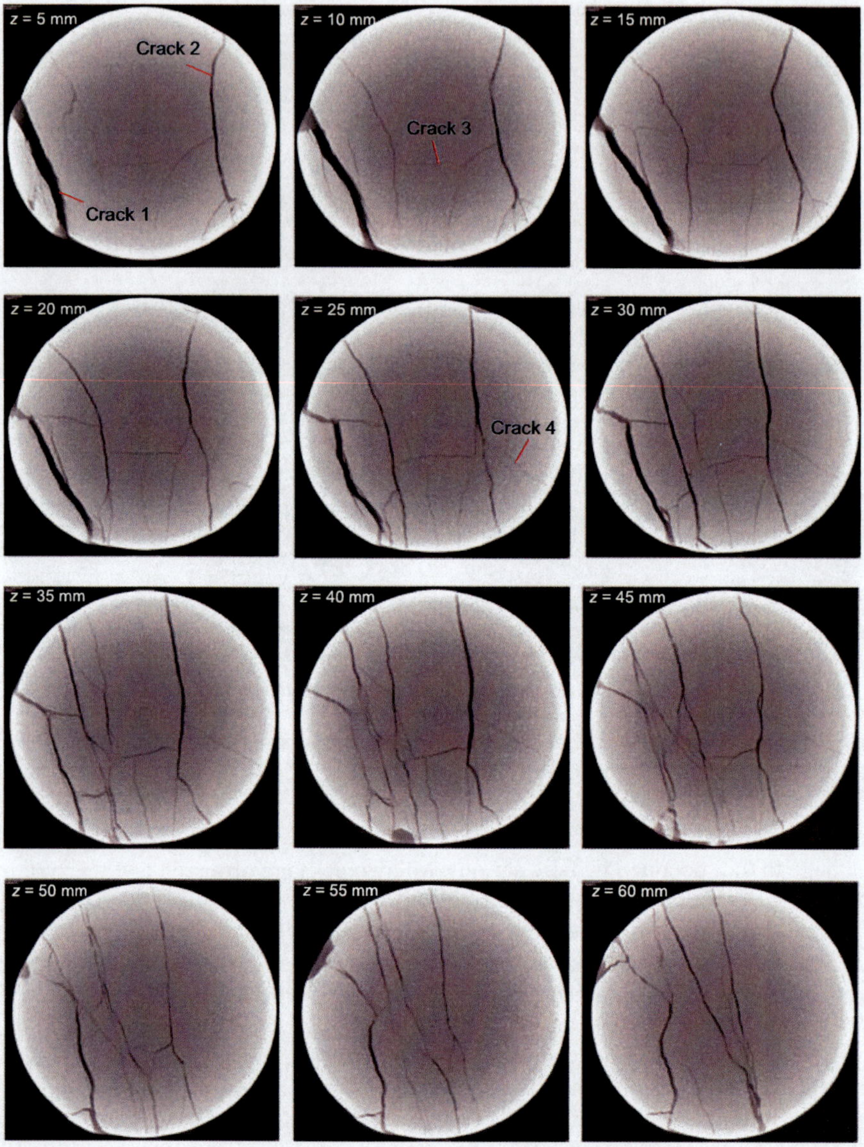

Fig. 8.16 Horizontal cross-sections with different heights of marble specimen after uniaxial compression failure

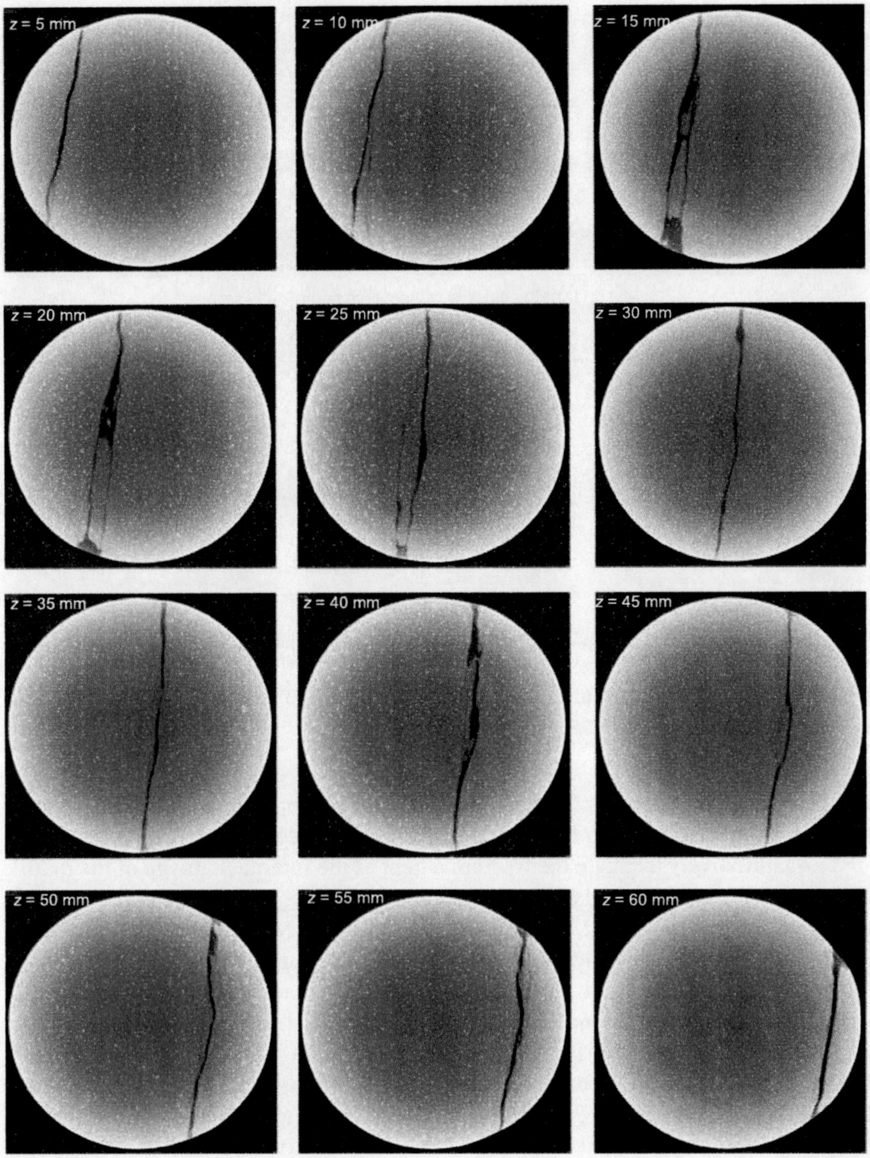

Fig. 8.17 Horizontal cross-sections with different heights of marble specimen after uniaxial compression failure ($\sigma_3 = 10$ MPa)

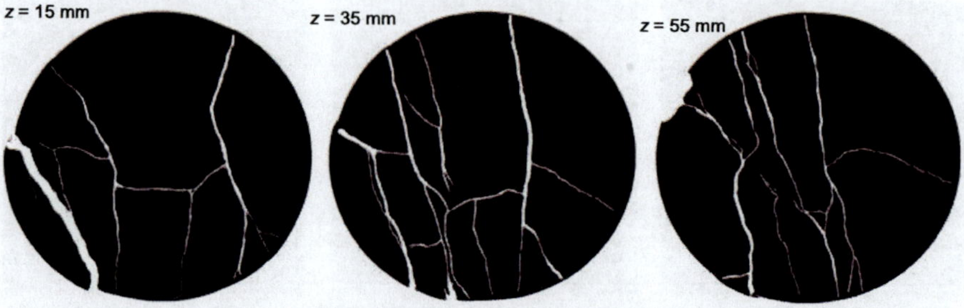

(a) After uniaxial compression failure

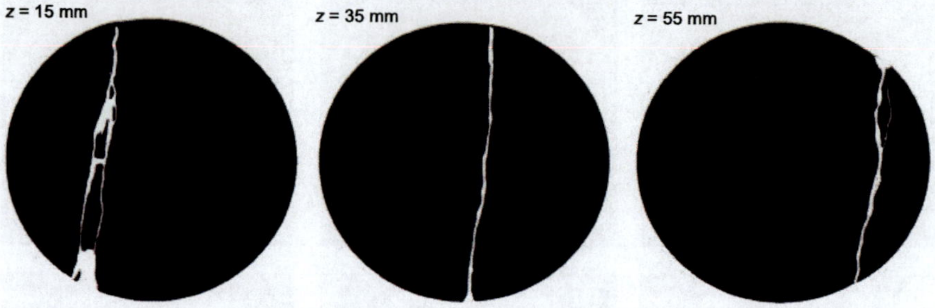

(b) After triaxial compression failure ($\sigma_3 = 10$ MPa)

Fig. 8.18 Typical binarized pictures of horizontal cross-sections of marble specimen after uniaxial compression and triaxial compression failure ($\sigma_3 = 10$ MPa)

indicate that cracks first initiate from the bottom or the top boundary of the specimen and then propagate towards the central region of the specimen.

To quantitatively evaluate the fracture extent of the internal damage of the marble specimen after uniaxial and triaxial compression failure, each horizontal CT image, as shown in Figs. 8.16 and 8.17 was processed using Matlab software (Gui et al. 2012; Yang et al. 2015b) and changed to binarized pictures (Fig. 8.18). In Fig. 8.18, the white regions represent the cracks, whereas the black regions represent intact rocks that did not experience any failure. Generally, the crack system shown in Fig. 8.18 agrees well with the experimental results shown in Figs. 8.16 and 8.17. Based on the binarized pictures for all horizontal cross-sections after uniaxial and triaxial compression failure, we determined the crack area and aperture extent (i.e., the percentage ratio of the crack area to the entire area) for each horizontal cross-section. Thus, we were able to quantitatively investigate the effect of height (z) on the extent of the internal damage of the marble material.

Figure 8.19 shows the evolution of crack area and aperture extent along different heights (z) after uniaxial and triaxial compression failure ($\sigma_3 = 10$ MPa). From Fig. 8.19a, the crack area after uniaxial compression failure increases from 88.88 mm^2 at z = 5 mm to 102.02 mm^2 at z = 15 mm, and then decreases to 93.02 mm^2 at z = 20 mm. Second, the crack area increases to 111.4 mm^2 at z = 30 mm. Then, the crack

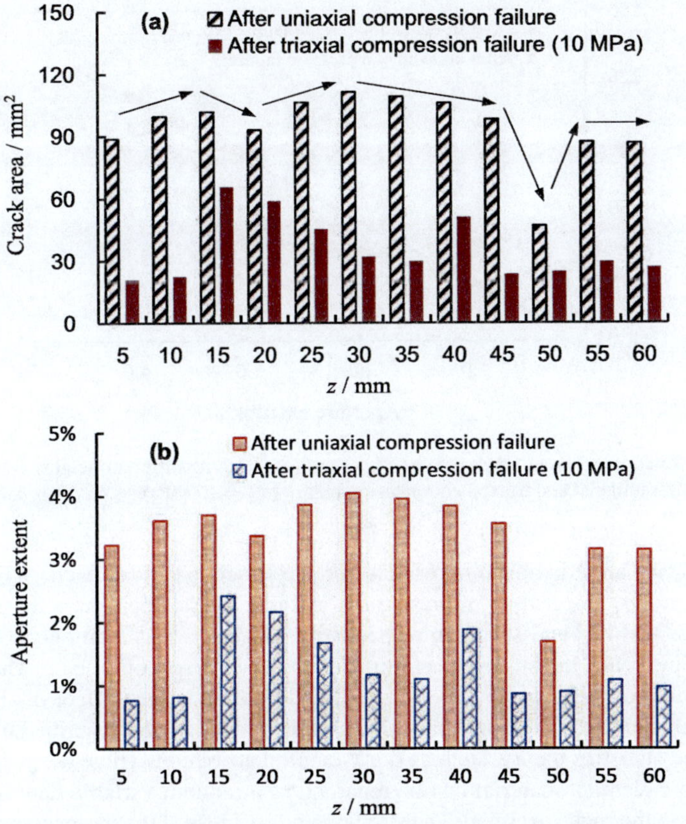

Fig. 8.19 Evolution of the crack area and aperture extent along different heights (z) of marble specimen after uniaxial and triaxial compression failure ($\sigma_3 = 10$ MPa)

area decreases from 111.4 mm^2 at z = 30 mm to 46.86 mm^2 at z = 50 mm. Finally, the crack area increases to 86.64 mm^2 at z = 60 mm. However, after triaxial compression failure ($\sigma_3 = 10$ MPa), the crack area increases gradually from 20.91 mm^2 at z = 5 mm to 65.74 mm^2 at z = 15 mm, and then decreases to 29.77 mm^2 at z = 35 mm. Then, the crack area obviously increases to 51.09 mm^2 at z = 40 mm. In the end, the crack area of marble specimens tends to a stable value with the increase of height. From Fig. 8.19b, we can see that the evolution of the aperture extent approximates to that of the crack area. Figure 8.20 illustrates the relationship between crack area and aperture extent along different heights (z) of marble specimen after uniaxial and triaxial compression failure ($\sigma_3 = 10$ MPa). From Fig. 8.20, it can be seen that the crack area has a good linear relationship with the aperture extent along different heights (z), which does not depend on the confining pressure.

To analyze the brittle-ductile transition mechanism of the marble specimen shown in Fig. 8.5, conceptual failure models (see Figs. 8.21 and 8.22) of the marble material

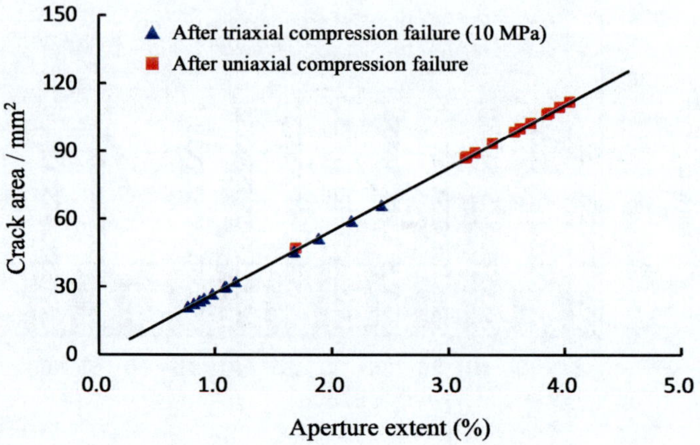

Fig. 8.20 Relation between crack area and aperture extent along different heights (z) of marble specimen after uniaxial and triaxial compression failure ($\sigma_3 = 10$ MPa)

under uniaxial and triaxial compression were proposed in this research. In Figs. 8.21 and 8.22, σ_a, σ_p, σ_b and σ_{cr} represent the yielding stress, peak strength, post-peak drop stress and residual strength, respectively. Marble is very homogeneous at the macroscopic scale, but strongly heterogeneous at the mesoscopic scale. The macroscopic fracture phenomenon may be regarded as the average effect of the mesoscopic failure and the heterogeneous mesoscopic failure can be described by a probability and statistics theory, such as the Weibull distribution. Thus, we propose that mesoscopic element material can be regarded as a random variable and distributes randomly in the rock specimen. The strength and stiffness of the mesoscopic element material in the specimen varies strongly from one element to another.

Under uniaxial compression, the marble specimen is loaded to σ_a with increasing axial deformation, the tensile cracks initiate from the tips of the fissures for larger fissure angles, whereas the tensile cracks initiate from a certain distance to the tips of the fissures for smaller fissure angles (Fig. 8.21a). At this time, a small part of the mesoscopic material elements with low strength and stiffness first reach their maximum supporting capacity, and produces plastic deformation during the yielding of specimen. Afterwards, when the specimen is loaded to peak strength (σ_p), tensile cracks begin to propagate (Fig. 8.21b) in a direction that is approximately parallel to the axial stress, e.g. wing cracks. However, after the macroscopic stress–strain response of the tested marble specimen reaches peak stress, the mesoscopic material elements with high strength and stiffness will unload because they did not reach their maximum supporting limit, and plastic deformation in the specimen is not uniform. Therefore following a decrease in the macroscopic axial deviatoric stress, the plastic deformation that is initiated will further concentrate on those failed mesoscopic material elements with low strength and stiffness, which results in localized deformation. Thus, specimens under lower confining pressures have a softening behavior (Fig. 8.21c). During deformation, some tensile cracks coalesce. When the specimen

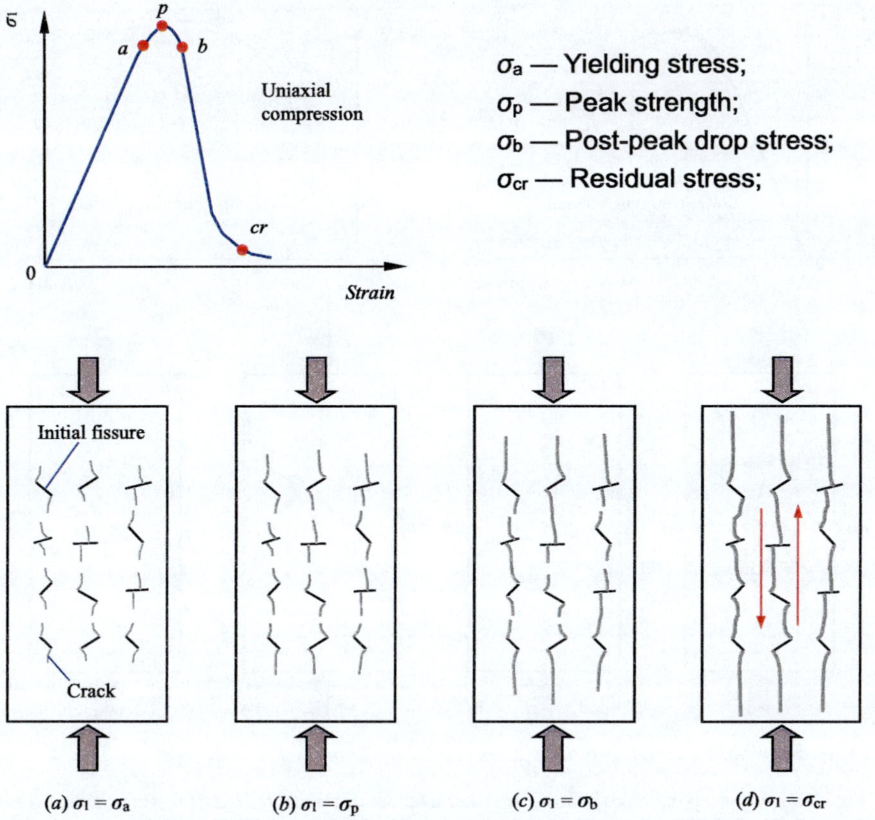

Fig. 8.21 A conceptual model of marble material under uniaxial compression: axial splitting tensile fracture is due to the coalescence of wing cracks starting from pre-existing fissures

reaches the residual stress (σ_{cr}), some axial splitting tensile fractures are observed (Fig. 8.21d).

When the marble specimen is loaded to σ_a under confining pressure with the increase of axial deformation, the cracks are initiated from the tips of fissures (Fig. 8.22a). Most of the mesoscopic material elements with low strength and stiffness first reach their maximum supporting capacity, yield softening and produce the plastic deformation. Then, the axial stress must be increased continuously to allow the specimen to fail macroscopically. When the specimen is loaded to peak strength (σ_p), the cracks begin to propagate (Fig. 8.22b). The direction of crack propagation is different under uniaxial compression, and the cracks will propagate towards the left and right boundaries of the specimen. At this time, the mesoscopic material elements with higher strength and stiffness in the specimen will also reach their maximum supporting limit, yielding failure and initiating plastic deformation. Therefore the total deformation in the specimen tends to uniform and the plastic

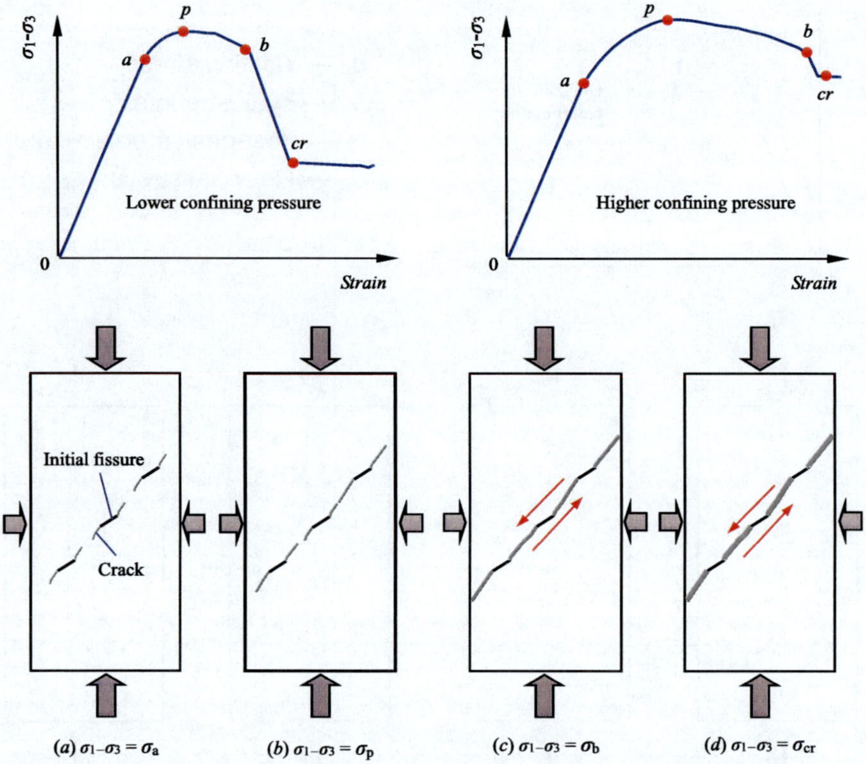

Fig. 8.22 A conceptual model of marble material under different confining pressures: single shear fracture

deformation of the specimen will increases, which results in a distinct yielding platform in the macroscopic stress–strain curves close to the peak stress. Furthermore, the yielding platform of the specimen with higher confining pressures is more obvious. When the stress is loaded to post-peak drop stress (σ_c), the specimen will form the shear fracture due to the crack coalescence (Fig. 8.22c). When the specimen reaches the residual stress (σ_{cr}), the shear fiction slippage will control the axial capacity (Fig. 8.22d).

8.5 Conclusions

1. The Young's modulus of deeply-buried marble increased nonlinearly with increasing confining pressure. The difference between the elastic modulus and the deformation modulus tended to decrease as the confining pressure increased, which was the result of the more pronounced stage of microcrack closure at lower

confining pressures. The peak and residual strengths of marble exhibited a clear linear relationship with the confining pressure, which can be best described by the linear Mohr–Coulomb criterion. The cohesion using the regression analysis for peak strength was higher than that for residual strength, but the internal friction angle using the regression analysis for peak strength was lower than that for residual strength.

2. After uniaxial and triaxial compression failure, the marble specimens were analyzed using an X-ray micro CT scanning system. Many X-ray CT images of the specimens were generated, and 3-D images of the specimens were reconstructed. The surface X-ray CT scanning surface images of the marble specimens were in good agreement with the actual surface crack photographs, which demonstrates that X-ray micro CT scanning can be used to explore the damage of rock material. To quantitatively evaluate the effect of the height on the extent of the internal damage of the marble specimen, the crack area and aperture extent for each horizontal cross-section were calculated by analyzing the binarized pictures, and indicated that height had a distinct influence on the crack area or aperture extent.
3. Under uniaxial compression, marble specimens were dominated by axial splitting tensile cracks. However under different confining pressures, marble specimens were dominated by a single shear crack. Based on the horizontal and vertical cross-sections of marble specimens, the system of crack planes under uniaxial compression was more complicated than that under triaxial compression, which also agrees well with the crack area and aperture extent with respect to the confining pressure. Finally, the brittle-ductile transition mechanism of marble were discussed and interpreted according to the proposed conceptual models.

References

Akbardoost J, Ayatollahi MR (2014) Experimental analysis of mixed mode crack propagation in brittle rocks: the effect of non-singular terms. Eng Fract Mech 129:77–89

Aliha MRM (2014) Indirect tensile test assessments for rock materials using 3-D disc-type specimens. Arab J Geosci 7(11):4757–4766

Aliha MRM, Ayatollahi MR, Akbardoost J (2012) Typical upper bound-lower bound mixed mode fracture resistance envelopes for rock material. Rock Mech Rock Eng 45(1):65–74

Aliha MRM, Hosseinpour GR, Ayatollahi MR (2013) Application of cracked triangular specimen subjected to three-point bending for investigating fracture behavior of rock materials. Rock Mech Rock Eng 46(5):1023–1034

Ayatollahi MR, Aliha MRM (2007) Fracture toughness study for a brittle rock subjected to mixed mode I/II loading. Int J Rock Mech Min Sci 44:617–624

Chang SH, Lee CI, Jeon S (2002) Measurement of rock fracture toughness under modes I and II and mixed-mode conditions by using disc-type specimens. Eng Geol 66:79–97

Fairhurst CE, Hudson JA (1999) Draft ISRM suggested method for the complete stress-strain curve for the intact rock in uniaxial compression. Int J Rock Mech Min Sci 36(3):279–289

Feng XT, Chen SL, Zhou H (2004) Real-time computerized tomography (CT) experiments on sandstone damage evolution during triaxial compression with chemical corrosion. Int J Rock Mech Min Sci 41(2):181–192

Feng GL, Feng XT, Chen BR, Xiao YX (2015) Microseismic sequences associated with rockbursts in the tunnels of the Jinping II hydropower station. Int J Rock Mech Min Sci 80:89–100

Funatsu T, Kuruppu M, Matsui K (2014) Effects of temperature and confining pressure on mixed-mode (I-II) and mode II fracture toughness of Kimachi sandstone. Int J Rock Mech Min Sci 67:1–8

Ge XR, Ren JX, Pu YB, Ma W, Zhu YL (2001) Real-in time CT test of the rock meso-damage propagation law. Sci China (Series E) 44(3):328–336

Gui YL, Zhao GF, Khalili N (2012) Experimental investigation of desiccation of clayey soils. In: 22th Australasian conference on the mechanics of structure and materials (ASMSM22), Dec 11–14, Sydney, Australia

Hirono T, Takahashi M, Nakashima S (2003) In situ visualization of fluid flow image within deformed rock by X-ray CT. Eng Geol 70:37–46

Huang D, Li YR (2014) Conversion of strain energy in triaxial unloading tests on marble. Int J Rock Mech Min Sci 66:160–168

Jaeger JC, Cook NGW, Zimmerman RW (2007) Fundamentals of rock mechanics (4th edn). Blackwell Publishing

Jiang Q, Feng XT, Xiang TB, Su GS (2010) Rockburst characteristics and numerical simulation based on a new energy index: a case study of a tunnel at 2500 m depth. Bull Eng Geol Environ 69:381–388

Kawakata H, Cho A, Kiyama T, Yanagidani T, Kusunose K, Shimada M (1999) Three-dimensional observations of faulting process in Westerly granite under uniaxial and triaxial conditions by X-ray CT scan. Tectonophysics 313:293–305

Li SJ, Feng XT, Li ZH (2012) Evolution of fractures in the excavation damaged zone of a deeply buried tunnel during TBM construction. Int J Rock Mech Min Sci 55(10):125–138

Li SJ, Feng XT, Li ZH, Chen BR, Zhang CQ, Zhou H (2012) In situ monitoring of rockburst nucleation and evolution in the deeply buried tunnels of Jinping II hydropower station. Eng Geol 137(138):85–96

Li XP, Zhao H, Wang B, Xiao TL (2013) Mechanical properties of deep-buried marble material under loading and unloading tests. J Wuhan Univ Technol-Mat 28(3):514–520

Meier T, Rybacki E, Backers T, Dresen G (2014) Influence of bedding angle on borehole stability: a laboratory investigation of transverse isotropic oil shale. Rock Mech Rock Eng. https://doi.org/10.1007/s00603-014-0654-1

Qiu SL, Feng XT, Xiao JQ, Zhang CQ (2014) An experimental study on the pre-peak unloading damage evolution of marble. Rock Mech Rock Eng 47:401–409

Sufian A, Russell AR (2013) Microstructural pore changes and energy dissipation in Gosford sandstone during pre-failure loading using X-ray CT. Int J Rock Mech Min Sci 57:119–131

Wu SY, Shen MB, Wang J (2010) Jinping hydropower project: main technical issues on engineering geology and rock mass. Bull Eng Geol Environ 69:325–332

Yang SQ, Jiang YZ, Xu WY, Chen XQ (2008) Experimental investigation on strength and failure behavior of pre-cracked marble under conventional triaxial compression. Int J Solids Struct 45:4796–4819

Yang SQ, Dai YH, Han LJ, Jin ZQ (2009) Experimental study on mechanical behavior of brittle marble samples containing different flaws under uniaxial compression. Eng Fract Mech 76:1833–1845

Yang SQ, Xu P, Ranjith PG, Chen GF, Jing HW (2015) Evaluation of creep mechanical behavior of deep-buried marble under triaxial cyclic loading. Arab J Geosci 8:6567–6582

Yang SQ, Ranjith PG, Huang YH, Yin PF, Jing HW, Gui YL, Yu QL (2015) Experimental investigation on mechanical damage characteristics of sandstone under triaxial cyclic loading. Geophys J Int 201:662–682

Yang SQ, Ju Y, Gao F, Gui YL (2016) Strength, deformability and X-ray micro-CT observations of deeply buried marble under different confining pressures. Rock Mech Rock Eng 49(11):4227–4244

Zhao GF, Russell AR, Zhao XB, Khalili N (2014) Strain rate dependency of uniaxial tensile strength in Gosford sandstone by the Distinct Lattice Spring Model with X-ray micro CT. Int J Solids Struct 51:1587–1600

Chapter 9
Deformation and Damage Failure Behaviour of Mudstone Specimens Under Single-Stage and Multi-stage Triaxial Compression

As is well-known, mudstone is a typical sedimentary rock, widely distributed in all kinds of underground engineering (nuclear waste disposal engineering, deep mining engineering, etc.). Previously, the strength, deformation and failure behaviour of mudstone have been the focus of a number of experimental studies. Geertsema (2002) indicated that the experimental shear strength of mudstone under saturated conditions is much lower than expected (or published in the reference), which should be paid attention to in the design of dams on mudstone or similar rock types with planar joints. He et al. (2008) used a self-developed soft rock hydrophilic experimental system with SEM (scanning electronic microscopy) and X-ray techniques to analyse the main influential factors of the hydrophilic behaviour of mudstone in a deep well. His experimental results showed that porosity, content and the sort of minerals as well as the clay occurrence are the main factors influencing the hydrophilic behaviour of mudstone. Some conventional triaxial compression (i.e., single-stage triaxial compression) for cylindrical specimens is also adopted to study the strength and deformation failure behaviour of mudstone material. Corkum and Martin (2007) studied the mechanical behaviour of weak mudstone recovered at the Mont Terri rock laboratory under uniaxial and triaxial compression. Their study showed that the stress–strain behaviour was highly nonlinear at low confining pressures. The uniaxial compressive strength (UCS) tests, confirmed by laboratory seismic measurements, showed that a correlation exists between the degree of non-linearity and peak strength. Furthermore triaxial data demonstrated that stiffness increased with confining pressure.

However, in the past, few experiments and numerical simulations were conducted to investigate the mechanical properties of damaged mudstone specimens. Therefore in this chapter, a multi-stage triaxial compression experiment and discrete element modelling were used to investigate the deformation and mechanical failure behaviour of mudstone specimens with different damage extents (Yang et al. 2019). This study has three aims: (1) to calibrate the micro-parameters in PFC^{2D} by single-stage triaxial compression of mudstone, (2) to analyse the influence of the extent of damage on the cohesion and internal friction angle of the mudstone specimen, and (3) to reveal the damage evolution mechanism of the mudstone specimen under triaxial compression.

S.-Q. Yang, *Mechanical Behavior and Damage Fracture Mechanism of Deep Rocks*,
https://doi.org/10.1007/978-981-16-7739-7_9

9.1 Experimental and Numerical Methodology

9.1.1 Mudstone Material and Testing Equipment

The material used throughout this study was mudstone, collected from the Yuncheng coal mine at a depth of 960 m. In the Yuncheng Coal mine, the maximum principal stress is horizontal (33.4 MPa), and the vertical stress is approximately 20 MPa, indicating that the maximum horizontal principal stress is 1.67 times the vertical stress as the consolidation condition. The experimental mudstone had an average unit weight of approximately 2550 kg/m^3. Mudstone is a fine-grained heterogeneous material (the particle size is very fine, which is generally smaller than 0.063 mm) with a connected porosity of 2.55%, as shown in Fig. 9.1. Furthermore, the tested mudstone is isotropic (Vervoort et al. 2014). According to X-ray diffraction (XRD), the minerals in the specimens are mainly kaolinite, quartz, illite/smectite interstratified minerals,

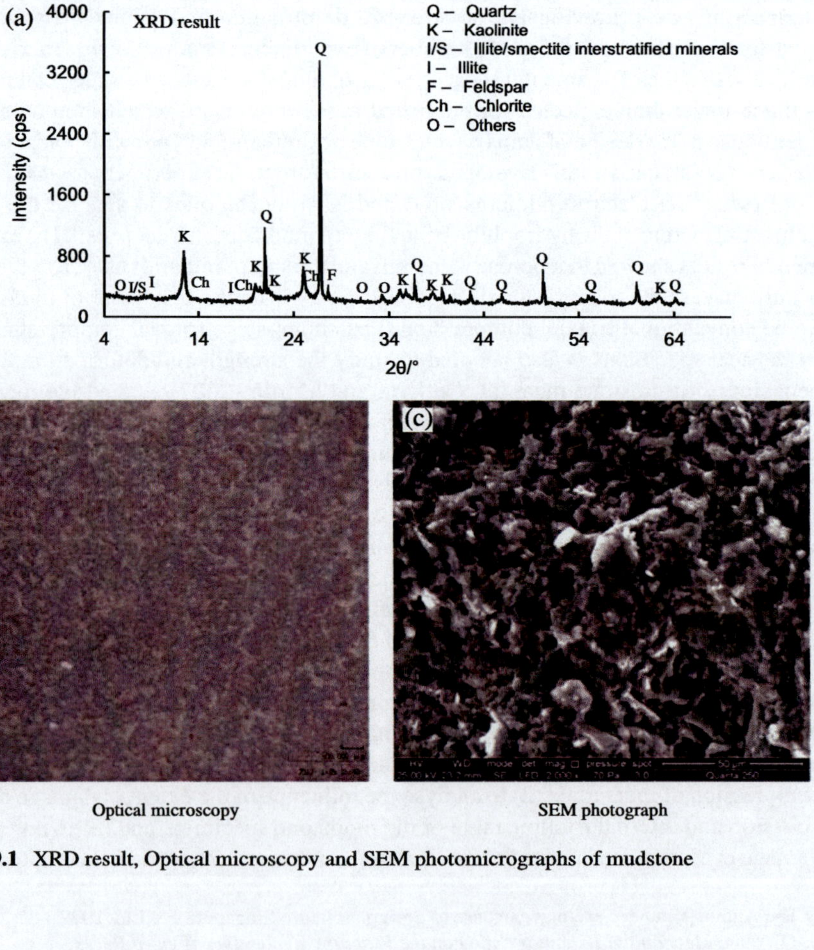

Optical microscopy SEM photograph

Fig. 9.1 XRD result, Optical microscopy and SEM photomicrographs of mudstone

Fig. 9.2 MTS815.02 rock mechanics servo-controlled testing system

illite, feldspar and allochite. The detailed composition of this rock is described as follows: 16% quartz, 64% kaolinite, 7% illite/smectitle interstratified minerals, 5% illite, 2% feldspar, 3% chlorite, and 3% other minerals.

All triaxial experiments for intact mudstone were carried out on cylindrical specimens 50 mm in diameter and 100 mm in length under natural and dry conditions. All compression experiments were carried out on an MTS815.02 rock servo-controlled system with a maximum loading capacity of 2 700 kN and a maximum displacement capacity of 25 mm (Fig. 9.2). This servo-controlled system can test specimens in load or displacement control. In this study, we measured the axial force at a loading capacity of 600 kN and the axial deformation at a displacement capacity of 5 mm. However, the circumferential deformation was measured by an LVDT attached to a chain wrapped tightly around the specimen, during which the maximum displacement capacity can reach 8 mm. During uniaxial and triaxial experiments, the loads and deformations of the tested mudstone specimens were recorded simultaneously. In addition, two rigid steel cylinders (50 mm in diameter and 20 mm in length) were placed between the loading frame and the rock specimen, distinctly decreasing the influence of the end friction effects on the testing results of the mudstone specimens with the length to diameter ratio of ~ 2.0.

9.1.2 Single-Stage and Multi-stage Triaxial Compression

To study the evolution of the strength reduction and mechanical damage of mudstone, we performed two types of triaxial compression experiments: single-stage and multi-stage.

Single-stage or conventional triaxial compression was carried out under different confining pressures (σ_3) of 0, 5, 15 and 25 MPa and consisted of two steps, as shown in Fig. 9.3a and b. First, the confining pressure was increased to the desired value at a constant rate of 0.1 MPa/s to ensure that the specimen was under uniform hydrostatic stress. Then, the deviatoric stress ($\sigma_d = \sigma_1$–σ_3, where σ_1 represents the maximum

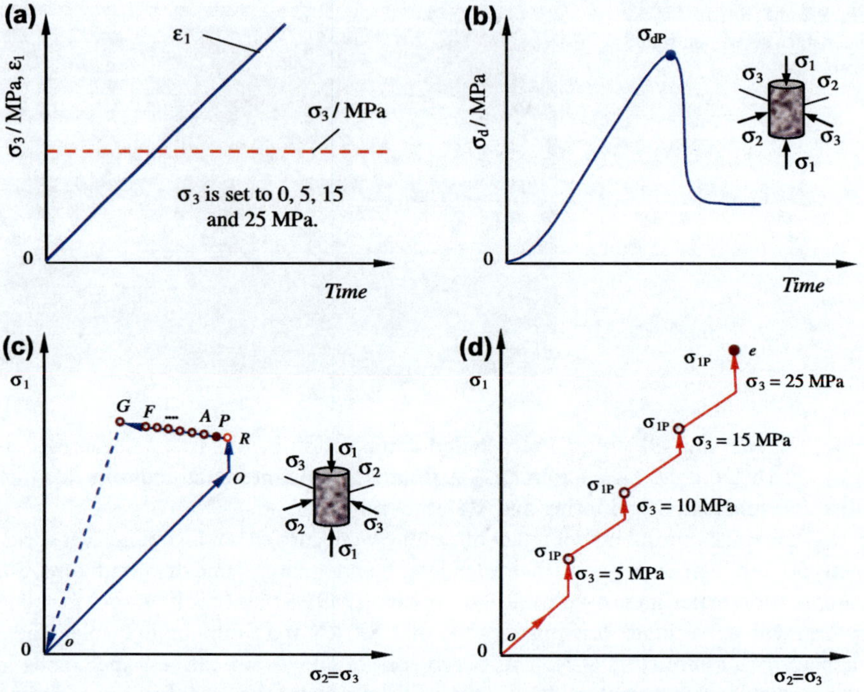

Fig. 9.3 Single-stage and multi-stage triaxial compression applied in the present research. **a** Variation of confining pressure and axial strain with time under single-stage triaxial compression. **b** Variation of axial deviator stress with time. **c** Step I: Prefabricate damage specimen designed by multi-stage triaxial compression. **b** Step II: Confirm the strength parameter of damaged specimen designed by multi-stage triaxial compression

principal stress) was applied to the end surface of the specimen at a constant axial displacement rate of 0.002 mm/s (the axial strain rate is approximately 2.0×10^{-5}/s) until failure. In this study, ε_1 represents the axial strain.

Figure 9.3c and d shows the multi-stage triaxial compression test designed in the present study, which includes two steps: (1) prefabricating damaged specimen (step I), and (2) confirming the strength parameter of the damaged specimen (step II). More details are provided belows.

In step I, hydrostatic pressure was applied to the specimen at a rate of 0.05 MPa/s until $\sigma_1 = \sigma_2 = \sigma_3 = 25$ MPa (i.e., *oO*). The applied hydrostatic pressure (~25 MPa) was in accordance with the actual burial depth (approximately 960 m) in the tunnel in the Yuncheng coal mine. Then, the confining pressure (σ_3) was maintained at 25 MPa, and the deviatoric stress was loaded at a constant axial displacement control rate of 0.002 mm/s until pre-peak was reached at 80% of peak strength (i.e., *OR*), which is regarded as the actual stress state of the surrounding rock mass in a deep tunnel before excavation. Then, the confining pressure is reduced gradually until the axial deviatoric stress is decreased to its post-peak strength point (i.e., *RG*).

While the confining pressure is reduced, the axial loading plate was controlled by displacement rate of 0.0005 mm/s, which is intended to prevent a sudden brittle failure of the damaged specimen due to a reduction in the confining pressure (Yang et al. 2012). This path was designed to gradually decrease the radial stress and gradually increase the circumferential stress during excavation unloading of the deep tunnel. To prefabricate the damaged specimen to different extents, the post-peak strength points were designed with different stress values. Finally, the axial stress and confining pressure applied to the damaged specimen were all unloaded to zero.

Note that in Fig. 9.3c, *R* is the beginning point for reducing the confining pressure and *P* is the peak strength point; σ_{1P}. *A*, *B*, *C*, *D*, *E*, *F* and *G* are designed to be post-peak different stress points. To better describe various damage extents after the post-peak stress is reached, the post-stress reduction ratio (i.e., λ) is defined as the ratio of the difference between σ_{dP} (σ_{dP} represents the peak strength, which is confirmed in accordance with the axial deviatoric stress-axial strain curves of the mudstone specimens. $\sigma_{dP} = \sigma_{1P}$-$\sigma_3$.) and σ_{dU} (i.e., the deviatoric stress unloading at the point).

$$\lambda = \frac{\sigma_{dP} - \sigma_{dU}}{\sigma_{dP}} \tag{9.1}$$

For an intact specimen without damage, λ equals zero. We designed *A*, *B*, *C*, *D*, *E*, *F* and *G* for $\lambda = 0.02$, 0.15, 0.25, 0.4, 0.55, 0.7 and 0.85, respectively. If the λ value is larger, the specimen will exhibit more obvious damage.

After step I, the loading path (step II) for a single specimen (i.e., *oe*) is used to confirm the strength parameter of the damaged specimen. First, hydrostatic pressure was applied to the specimen at a rate of 1.0 MPa/s. Then, the initial confining pressure was maintained at 5 MPa. The deviatoric stress was loaded at a constant axial displacement control rate of 0.002 mm/s until the specimen reached post failure, at which point, the deviatoric stress dropped by ~ 2% from its peak. Then, the confining pressure was hydrostatically increased to the second level, i.e., 10 MPa. Subsequently, the preceding steps were repeated at the third confining pressure, 15 MPa. At the fourth confining pressure of 25 MPa, the deviatoric stress was loaded to the surface of the specimen at a constant axial displacement control rate of 0.002 mm/s until failure took place. Note that, for each confining pressure, a 2% axial stress drop after peak strength could be controlled automatically in the MTS815 experimental loading system. Through step II, we were able to obtain a series of peak strengths of the damaged specimens under different confining pressures, effectively avoiding the great influence of the difference of the strength parameter among damaged specimens. Table 9.1 lists the loading method design for the present study.

Table 9.1 Loading method designed in the present study

Series	σ_3 / MPa	Post-peak stress point	λ	Loading method
1	0	–	0	STC
2	5	–	0	STC
3	15	–	0	STC
4	25	–	0	STC
5	Step I, II	A	0.02	MTC
6	Step I, II	B	0.15	MTC
7	Step I, II	C	0.25	MTC
8	Step I, II	D	0.4	MTC
9	Step I, II	E	0.55	MTC
10	Step I, II	F	0.7	MTC
11	Step I, II	G	0.85	MTC

Note STC = Single-stage triaixal compression; MTC = Multi-stage triaxial compression

9.1.3 Numerical Model and Micro-parameters in PFC

It is known that PFC^{3D} has been exploited for many years; however, PFC^{3D} consumes more computing time compared with PFC^{2D} and cannot obtain a clear failure mode compared with PFC^{2D}. Therefore, many researchers use PFC^{2D} to simulate rock mechanics tests (Cao et al. 2016; Mu et al. 2015; Yang et al. 2014, 2018). To simulate the experiments of intact mudstone, a numerical specimen was constructed by PFC^{2D} at the same scale as the experimental specimen: 50 mm in diameter (*D*) and 100 mm in length (*L*). All numerical specimens had a height-to-width ratio of 2.0 to ensure a uniform stress state within the central part of the specimen. Each numerical intact specimen was discretized into 17 232 particles with 44 907 contacts and 34 515 parallel bonds, as shown in Fig. 9.4. The particle size followed a uniform distribution that ranged from 0.24 mm to 0.36 mm. In Fig. 9.4, the $1^{\#}$ and $2^{\#}$ boundaries are regarded as the loading platen, and thus we can apply the velocity on the specimen by the loading platen. At the same time, we can use the $3^{\#}$ and $4^{\#}$ boundaries to apply the confining pressure on the specimen in order to simulate a conventional triaxial compression experiment.

In the numerical simulation, a vertical displacement was applied to the top of the rock specimens. To ensure that the specimen remained in quasi-static equilibrium throughout the test, a sufficiently low displacement loading rate of 0.05 m/s was applied under a simple loading path. The micro-parameters in this study were confirmed using the following method (Yoon 2007; Lee and Jeon 2011; Yang et al. 2014). First, we established a quantitative correction between the macro- and micro-parameters of mudstone by conducting the sensitivity analysis of the micro-parameters. Then, the micro-parameters were confirmed through trial and error. We

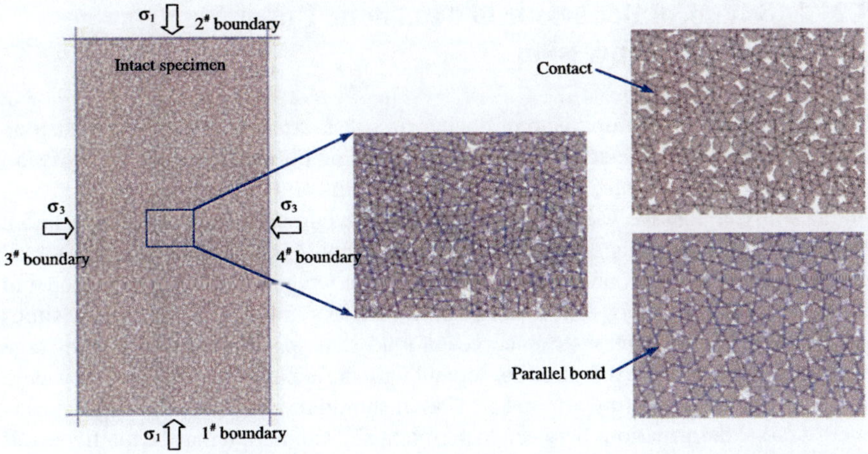

Fig. 9.4 Numerical intact specimen generated in PFC^{2D} in this research. The numerical mudstone specimen has the following dimensions of 50 mm in diameter (D) and 100 mm in length (L)

used the macroscopic behaviour of intact mudstone obtained by experiment to calibrate the micro-parameters. The macroscopic results obtained by numerical simulation after each trial were used to verify the micro-parameters. This process was repeated until the numerical results of intact mudstone aligned with the experimental results. In accordance with the above method, we propose the micro-parameters for the mudstone specimens in this study, which are listed in Table 9.2.

Table 9.2 The micro-parameters for mudstone specimens in this research

Micro-parameters	Values	Remarks
Young's modulus of the particle, E_c (GPa)	7.9	
Young's modulus of the parallel bond, $\overline{E}_c$(GPa)	7.9	
Ratio of normal to shear stiffness of the particle, k_n/k_s	1.4	
Ratio of normal to shear stiffness of the parallel bond, $\overline{k}_n/\overline{k}_s$	1.4	
Particle friction coefficient, μ	0.35	
Parallel-bond normal strength (σ_n), mean (MPa)	32.1	Normal distribution
Parallel-bond normal strength, standard deviation (MPa)	8.5	
Parallel-bond shear strength (τ_n), mean (MPa)	226	Normal distribution
Parallel-bond shear strength, standard deviation (MPa)	90	

9.2 Mechanical Behaviour of Mudstone Under Single-stage Triaxial Compression

Figure 9.5 shows the comparison of deviatoric stress–strain curves of the mudstone specimens under single-stage triaxial compression obtained by experiment and particle flow simulation. ε_1 and ε_3 represent the axial and radial strain, respectively. In this study, a positive value for the strain means compressive deformation and a negative value means dilatational deformation. It should be noted that we only tested one specimen for each confining pressure because we had limited specimens from the same block. From Fig. 9.5, it can be seen that, in general, the numerically simulated axial deviatoric stress–strain curves of mudstone specimens under single-stage triaxial compression are in a good agreement with the experimental curves. However, at the post-peak region ($\sigma_3 = 0$ and 5 MPa), it should be noted that the larger difference of radial deformations between experiment and simulation may be mainly result from the difference of macroscopic crack evolution speed along the radial direction. Under the lower confining pressures, the macroscopic crack evolution speed along

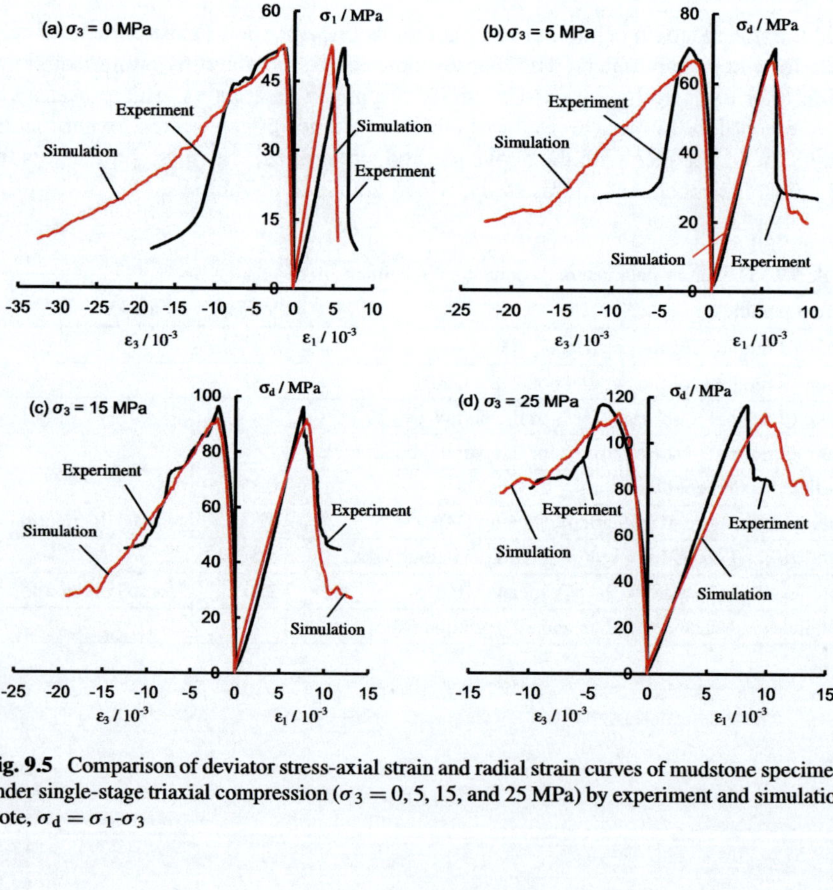

Fig. 9.5 Comparison of deviator stress-axial strain and radial strain curves of mudstone specimens under single-stage triaxial compression ($\sigma_3 = 0$, 5, 15, and 25 MPa) by experiment and simulation. Note, $\sigma_d = \sigma_1 - \sigma_3$

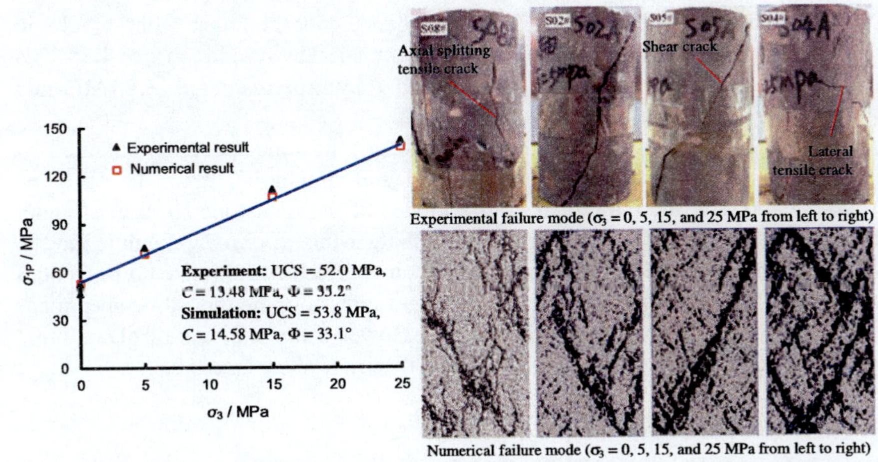

Fig. 9.6 Comparison of peak strength parameters and ultimate failure mode of mudstone specimens obtained by single-stage triaxial compression experiment and simulation. The line represents the theoretical value in accordance with linear Mohr–Coulomb criterion obtained by simulation

the radial direction obtained by numerical simulation is faster than that obtained by experiment due to larger energy dissipation during the numerical simulation. However, under the higher confining pressures ($\sigma_3 = 15$ and 25 MPa), the confining pressure restrains the macroscopic crack evolution speed of specimen obtained by numerical simulation resulting in an approximate radial deformation compared with the experimental results. Moreover, the deviatoric stress–strain curves of mudstone at lower confining pressures do not initially show a non-linear (concave) deformation due to the lower porosity of mudstone.

Figure 9.6 compares the peak strength of intact mudstone specimens under different confining pressures obtained by experiment and by simulation. It can be seen from Fig. 9.6 that the peak strength of mudstone increases linearly with the confining pressure, which can be expressed by the linear Mohr–Coulomb criterion (Mohr 1914; Yang et al. 2011), i.e., Eq. (9.2).

$$\sigma_{1P} = \sigma_c + q\sigma_3 = \frac{2C\cos\varphi + \sigma_3(1+\sin\varphi)}{1-\sin\varphi} \tag{9.2}$$

where $\sigma_{1\mathrm{P}}$ is the maximum supporting capacity and σ_c and q are related to the cohesion C and the internal friction angle Φ of the rock. It should be noted that $\sigma_{1\mathrm{P}}$ equals the sum of σ_{dP} and σ_3.

It can be observed from Fig. 9.6 that the numerical peak strength of intact mudstone under different confining pressures is quantitatively equal to that obtained by experiment. The maximum difference in peak strength between experiment and simulation is at the confining pressure of 15 MPa. At $\sigma_3 = 15$ MPa, the peak strength of intact mudstone by experiment is 111.4 MPa, whereas the numerical peak strength is

106.7 MPa, a difference of 4.7 MPa. In accordance with the linear Mohr–Coulomb criterion, the values of C and Φ of the mudstone material by simulation are 14.58 MPa and 33.1°, respectively, and the values of C and Φ by experiment are 13.48 MPa and 35.2°, respectively. These results differ by only ~ 7.8%.

Figure 9.6 compares the failure modes of the mudstone specimen obtained by single-stage triaxial compression experiment and simulation. It can be observed from Fig. 9.6 that, the numerical failure modes of the mudstone specimens under different confining pressures are very approximate to the experimental failure modes. We can see that the confining pressure significantly affects the ultimate failure modes of the mudstone specimens. Under uniaxial compression, the mudstone specimens mainly demonstrate by axial splitting fracture. However, under the action of confining pressure, the mudstones mainly demonstrate by single shear fracture.

9.3 Mechanical Behaviour of Mudstone Under Multi-stage Triaxial Compression

Figure 9.7 depicts typical complete stress–strain curves of the mudstone specimens under step I obtained by experimentation and numerical simulation. It can be seen from Fig. 9.7a that, the peak strength and deformation behavior of all three specimens were approximately similar, indicating almost no difference among that tested specimens under multi-stage triaxial compression. However, it can be seen from Fig. 9.7b that, in the numerical simulation, the stress and deformability of several specimens before the peak strength reached were all same, except at the post-peak stress points. Thus, we can analyse the influence of various damage extents on the peak strength parameters of mudstone specimens. Furthermore, from Fig. 9.7, we can also see that the numerical result is in a good agreement with the experimental result of mudstone under step I. However, because of the instability of mudstone deformation after its

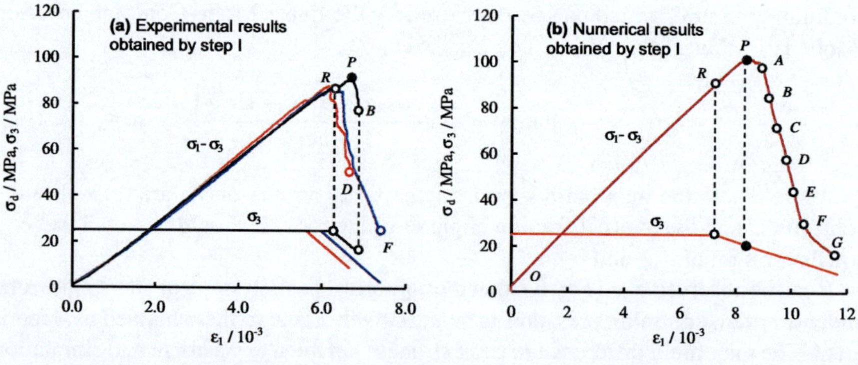

Fig. 9.7 Typical complete stress–strain curves of mudstone specimen under step I shown in Fig. 3c obtained by experiment and numerical simulation

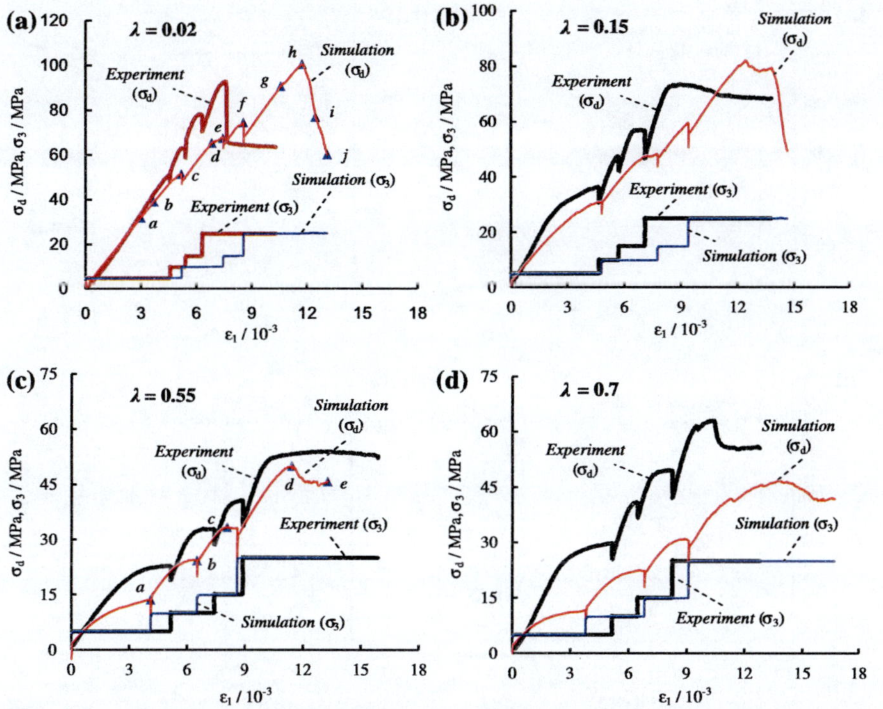

Fig. 9.8 Typical multi-stage triaxial compression curves of mudstone specimen for post-peak different stress points obtained by experiment and numerical simulation

peak strength, the actual post-peak stress points were not exactly with those designed in Table 9.1. Experimental results show that the different post-peak stress points were $\lambda = 0.04$, 0.15, 0.22, 0.43, 0.56, 0.73 and 0.85.

For step II, as shown in Fig. 9.3d, Fig. 9.8 shows typical multi-stage loading curves of the mudstone specimen for different post-peak stress points obtained by experiment and numerical simulation. From Fig. 9.8, we can obtain that the triaxial multi-stage experimental results for four mudstone specimens with different post-peak deformation positions are also different. It can be observed that, in general, the numerical simulated results have some differences with the experimental results of the mudstone specimen under multi-stage triaxial compression. Possible reasons for the difference between experimental and numerical simulation of mudstone specimens are explained as follows: (1) Heterogeneity of rock, which reflects the difference among various specimens. (2) The influence of loading history on the experimental and numerical results. (3) The difference among damaged specimens reaching different post-peak deformation values after peak strength.

Based on the peak strength of the mudstone specimens under multi-stage triaxial compression (Fig. 9.8), the influence of confining pressure on the peak strength obtained by experiment and numerical simulation is presented in Fig. 9.9. From

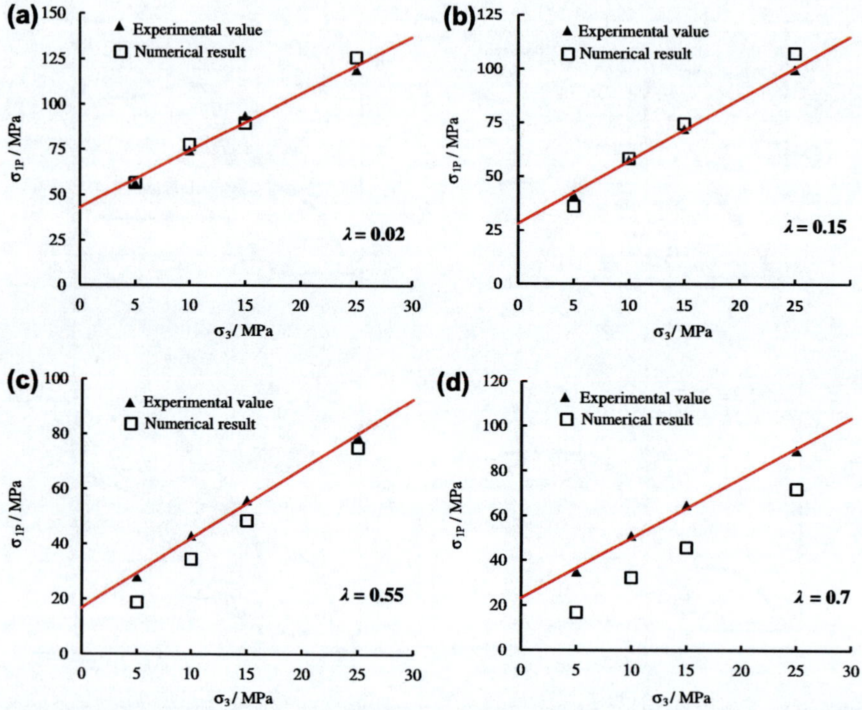

Fig. 9.9 The influence of confining pressure on the peak strength of mudstone specimen for post-peak different stress points obtained by experiment and numerical simulation

Fig. 9.9, it is clear that when the post-peak stress points are located at points A ($\lambda = 0.02$) and B ($\lambda = 0.15$), the difference between the experimental and numerical simulated strength is very small, whereas when the post-peak stress points are located at points E ($\lambda = 0.55$) and F ($\lambda = 0.7$), the difference between the experimental and numerical simulated strength is larger and the experimental strength is clearly higher than the numerical simulated strength.

In accordance with the linear Mohr–Coulomb criterion, Fig. 9.10 shows the peak strength parameters of the mudstone specimens under a complex loading path. It can be observed that, for a damaged specimen, the cohesion and internal friction angle are lower than an intact specimen, where the strength attenuation values are dependent on the extent of the damage.

To describe an attenuation law for strength parameters of mudstone specimens, we proposed the following two equations.

$$C = C_L + \alpha \exp\left(-\frac{\lambda}{\beta}\right) \tag{9.3}$$

$$\varphi = k \exp(-\delta\lambda) \tag{9.4}$$

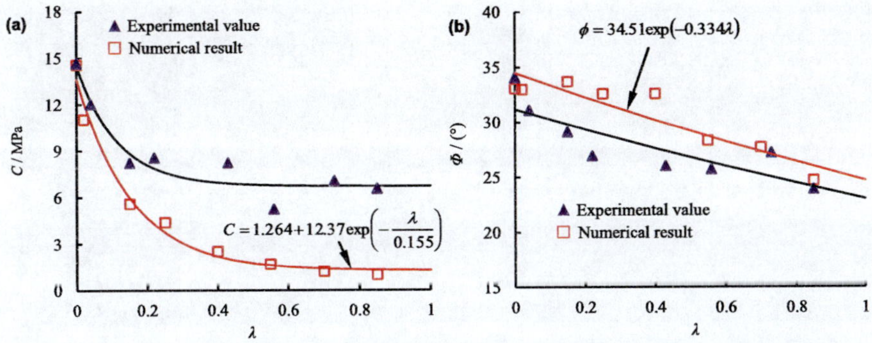

Fig. 9.10 Influence of post-peak different stress point on the cohesion and internal friction angle of mudstone specimen obtained by experiment and numerical simulation. **a** Cohesion; **b** Internal friction angle

where C_L, α, β, k and δ are all fitting parameters of the strength attenuation equations. These parameters can be confirmed using experimental values. Furthermore, the internal friction angle increases with the increase in parameter k for constant δ, but decreases with the increase in parameter δ for constant k. However, the cohesion increases with increasing C_L, α or β for two constant parameters.

From Fig. 9.10, it is clear that the fitting curves of Eqs. (9.3) and (9.4) are in a good agreement with the experimental results. As the post-stress reducing ratio λ increases, the cohesion of mudstone specimen by experiment and numerical simulation decreases and the variance tends to slow down after $\lambda = 0.4$. The internal friction angle of mudstone specimen by experiment and numerical simulation also decreases gradually with the increase of post-peak stress point. Furthermore, it can be seen from Fig. 9.10, that the internal friction angle by experiment is approximately equal to that by numerical simulation; whereas the cohesion by experiment is clearly higher than that by numerical simulation, especially for greater λ.

9.4 Discussion of the Damage Evolution Mechanism of Mudstone

In this section, we analyze the damage evolution mechanism of mudstone under multi-stage triaxial compression, which will be very important for understanding the strength reduction mechanism of the mudstone specimen presented in Fig. 9.10.

Figure 9.11 shows the failure mode of the mudstone specimen at different deformation values under a pre-peak reducing confining pressure by numerical simulation. All the points denoted in Fig. 9.11 correspond to those shown in Fig. 9.7b. In Fig. 9.11, the letter N represents the micro-crack number. As axial deformation increases, when the axial deviatoric stress of the specimen is loaded to point R, N increases from 0 to 1015. Then, the confining pressure applied to the specimen is

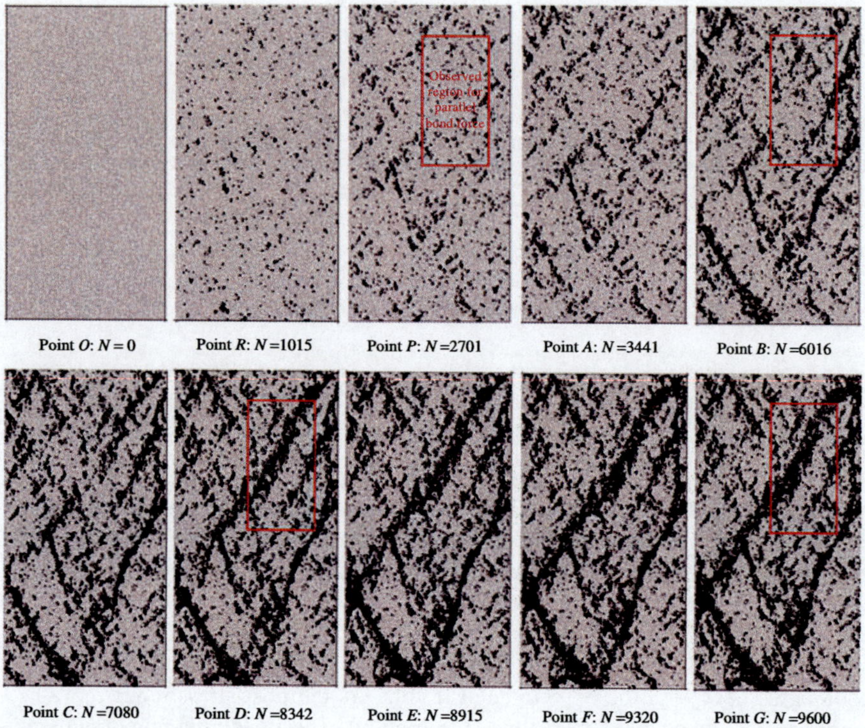

Fig. 9.11 Fracture mode of mudstone specimen at different deformation values under the condition of pre-peak reducing confining pressure by numerical simulation. All the points denoted in this figure correspond to those shown in Fig. 9.7b

reduced from 25 MPa. Until it reduces to 20.05 MPa, at which point the specimen attains its peak strength point *P*, i.e., $\sigma_{1p} = 101.99$ MPa, and *N* increases from 1015 to 2701. After the specimen reaches the peak strength, as the confining pressure is gradually reduced from 20.05 to 8.24 MPa, the axial deviatoric stress also decreases rapidly from 101.99 (point *P*) to 15.23 MPa (Point *G*). While the confining pressure is decreasing, the micro-crack number continues to increase from 2 701 (point *P*) to 9600 (point *G*). However, the rate of increase between point *P* and point *D* is clearly higher than between point *D* and point *G*. At point *D*, the macro-cracks have formed, and after point *D*, even though the confining pressure is reduced gradually, a new macro-crack is very difficult to form.

To evaluate the initial damage extent of the mudstone specimen quantitatively before applying step II as shown in Fig. 9.3c, the following damage variable is applied.

$$D_{\lambda} = \frac{A_{crack}}{A_{total}} \tag{9.5}$$

Table 9.3 Different micro-crack numbers and damage variable at various stress point

Stress point	λ	N	D_λ
P	0	2701	0.100
A	0.02	3441	0.135
B	0.15	6016	0.219
C	0.25	7080	0.249
D	0.4	8342	0.284
E	0.55	8915	0.303
F	0.7	9320	0.311
G	0.85	9600	0.315

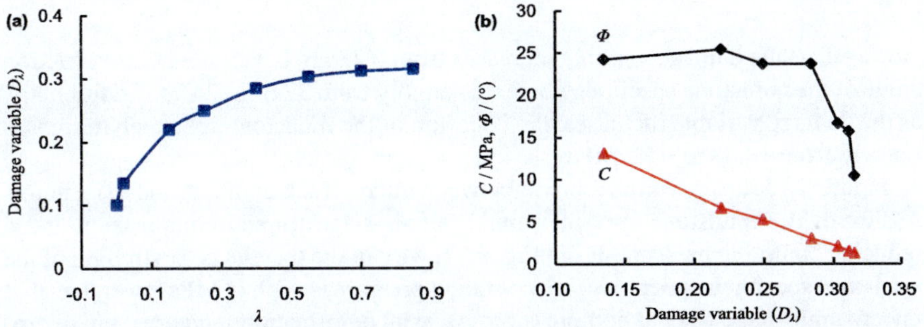

Fig. 9.12 Influence of damage variable on strength parameters of mudstone specimen under multi-stage triaxial compression. **a** Relation between damage variable and post-peak different stress point; **b** Relation between cohesion, internal friction angle and damage variable

where D_λ represents the damage variable, and A_{crack} and A_{total} represent the area of micro-cracks and the total area of the specimen, respectively. Table 9.3 lists different micro-crack numbers and damage variables at various stress points.

Figure 9.12 shows the influence of damage variables on the strength parameters of the mudstone specimen under multi-stage triaxial compression. From Fig. 9.12a, it can be observed that the damage variable increases from 0.100 at $\lambda = 0$ to 0.284 at $\lambda = 0.4$. However, as λ increases from 0.4 to 0.85, the damage variable increases slightly from 0.284 to 0.315. This phenomenon can be explained as follow: When axial stress drops to point D ($\lambda = 0.4$, $D_\lambda = 0.284$), the shear macro-fracture surface has been formed (Fig. 9.11d). Micro-cracks mainly emanate near the shear fracture surface with the loading increase (λ from 0.4 to 0.85), as shown in Fig. 9.11e ~ g, The A_{crack} only increases slightly along the shear macro-fracture surface, therefore the damage variable (D_λ) increases slightly from 0.248 to 0.315. The damage variable varies with λ corresponding to the number of micro-cracks numbers in Fig. 9.11. Figure 9.12b shows the relation between the cohesion and internal friction angle of the mudstone specimen and damage variable. From Fig. 9.12b, we can see that when the damage variable ranges from 0.100 to 0.284, the internal friction angle of the mudstone specimen is approximately constant, varying from only 33.1° to 32.6°.

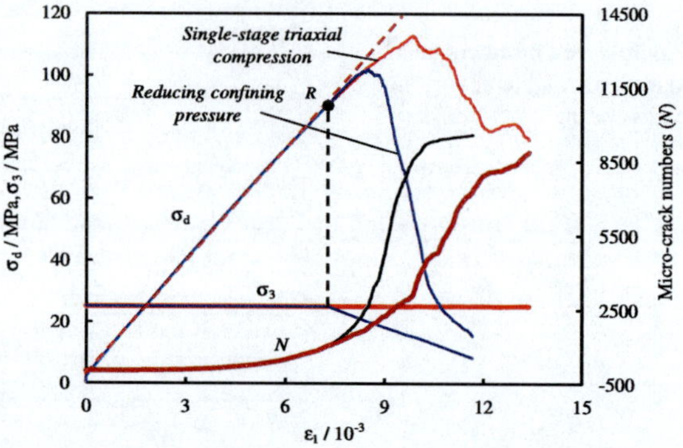

Fig. 9.13 Relation between micro-crack numbers and axial deformation of mudstone specimen under single-stage triaxial compression and reducing confining pressure

However, as the damage variable increases from 0.284 to 0.315, the internal friction angle of the mudstone specimen decreases rapidly from 32.6° to 24.7°. Additionally, as the damage variable increases, the cohesion of the mudstone specimen decreases linearly from 14.58 to 0.99 MPa.

Figure 9.13 illustrates the relation between micro-crack numbers and axial deformation of the mudstone specimen under single-stage triaxial compression and a reducing confining pressure. From Fig. 9.13, we can see that the peak strength of the mudstone specimen under reducing confining pressure is ~ 10.77 MPa lower than that under single-stage triaxial compression. As axial deformation increases, the micro-crack number increases slowly. When the mudstone specimen begins to yield (the stress–strain curve begins to depart from linear behaviour), the rate of change in the micro-crack number (increasing rate of change in the micro-crack number with axial strain) of the mudstone specimens begins to increase. The rate of change in the micro-crack number under a reducing confining pressure is clearly higher than that under single-stage triaxial compression. However, when the mudstone specimen enters the stage of residual strength, the micro-crack number does not increase rapidly due to the macro-cracks having been formed.

Figure 9.14 presents the relation between the micro-crack numbers and the axial deformation of the mudstone specimen under multi-stage triaxial compression. Combined with Fig. 9.8a, Fig. 9.14 allows for detailed analysis of the failure evolution process of mudstone under multi-stage triaxial compression. The initial damage extent was small and the micro-cracks were distributed randomly. Under the first confining pressure of 5 MPa, as axial deformation increased, the micro-crack number remained constant until point *b*. After point *b*, the micro-crack number increased rapidly from 3 550 to 3 988 because of greater damage to the mudstone specimen under higher stress levels. After point *c*, the confining pressure increased to 10 MPa from 5 MPa. Due to the constraint of confining pressure, even though the axial strain increased from 5.21×10^{-3} to 6.84×10^{-3} and the corresponding axial deviatoric stress increased from 51.44 (point *c*) to 65.39 MPa (point *d*), the micro-crack number increased slightly from 3 988 to 4 110. The further increase in axial deformation continued until the specimen was loaded to point e ($\sigma_d = 61.50$ MPa)

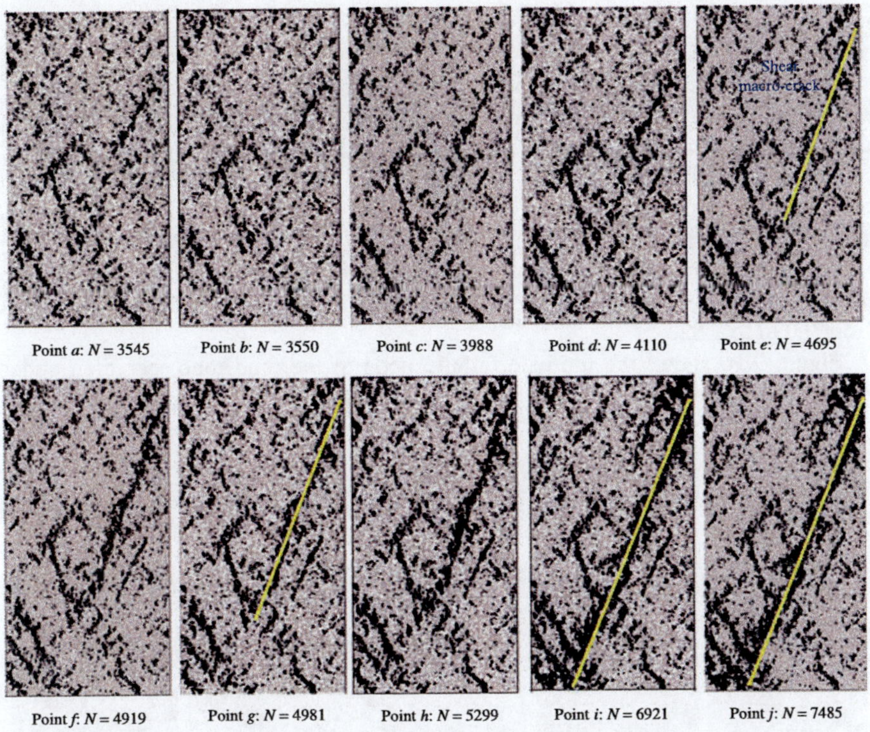

Fig. 9.14 Crack evolution process of mudstone specimen under multi-stage triaxial compression ($\lambda = 0.02$). The points in this figure correspond to those denoted in Fig. 9.8a

and the corresponding micro-crack number increased faster from 4 110 to 4 695. It should be noted that from point *c* to point *e*, a distinct shear macro-crack can be observed in the specimen because many micro-cracks were connected.

Then, when the third confining pressure of 15 MPa was applied to the specimen, the axial supporting capacity began to increase slowly with increasing deformation. However, due to some damage of the supporting structure at the first and second confining pressures of 5 and 10 MPa, respectively, the peak strength at the third confining pressure of 15 MPa only increased to from 67.5 MPa 74.55 MPa. The corresponding micro-crack number of the mudstone specimen increased from 4 695 to 4 919. After point *f*, the confining pressure increased to 25 MPa, the rate of change in the micro-crack number first decreased due to a restrained higher confining pressure, and then increased as the axial deformation increased from 4 919 to 5 299 until reaching the peak strength (point *h*) at $\sigma_3 = 25$ MPa. After the peak strength, the mudstone specimen supported axial capacity by the friction of shear fracture surface, which can be verified by a great number of micro-crack numbers emanating near the main shear macro-crack. As the axial stress decreased, the micro-crack number increased from 5 299 to 7 485, and the increasing rate of change in the micro-crack number was very fast, predicting the friction slippage failure of the mudstone specimen.

However, in accordance with Fig. 9.15 and c, it can be seen that the initial damage extent to the mudstone specimen is large and several shear macro-cracks formed in the specimen, which is clearly different than at $\lambda = 0.02$. This result can be explained by the fact that the macro-cracks in the specimen were formed before loading, so the specimen mainly supported the axial capacity by the friction slippage on its shear failure surface, which can be verified by no obvious change to several shear macro-cracks during the deformation. Furthermore, the micro-crack number of the mudstone specimen at $\lambda = 0.55$ increased less than at $\lambda = 0.02$, which only increased from 9 096 to 9 826.

Figure 9.16 shows the ultimate failure mode of the mudstone specimen under multi-stage triaxial compression obtained by experiment. By comparing Fig. 9.16

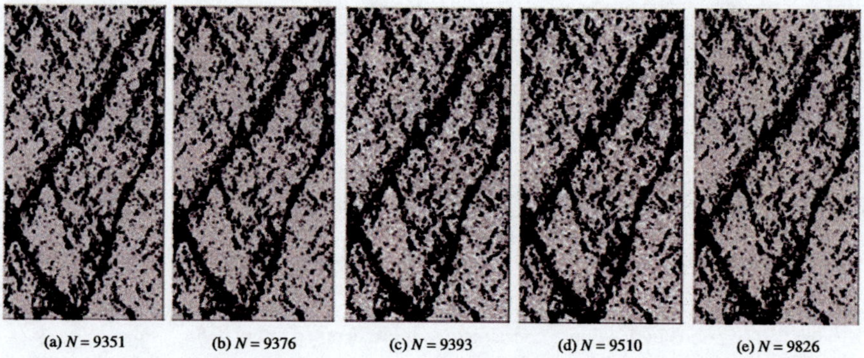

Fig. 9.15 Crack evolution process of mudstone specimen under multi-stage triaxial compression ($\lambda = 0.55$). The points in this figure correspond to those denoted in Fig. 9.8c

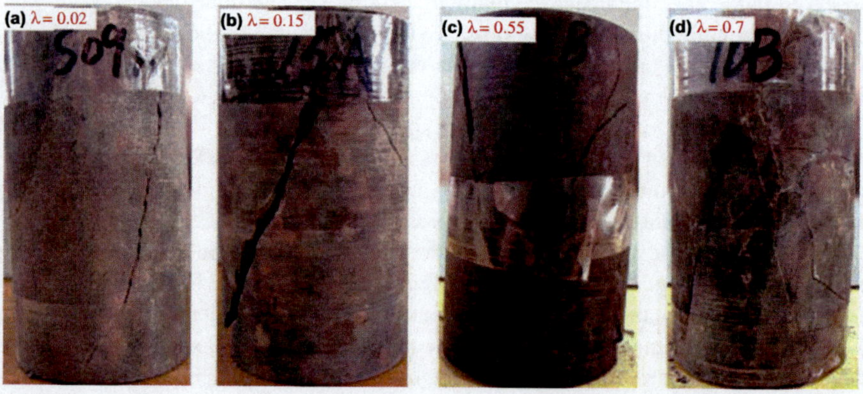

Fig. 9.16 Ultimate failure mode of mudstone specimen under multi-stage triaxial compression obtained by experiment

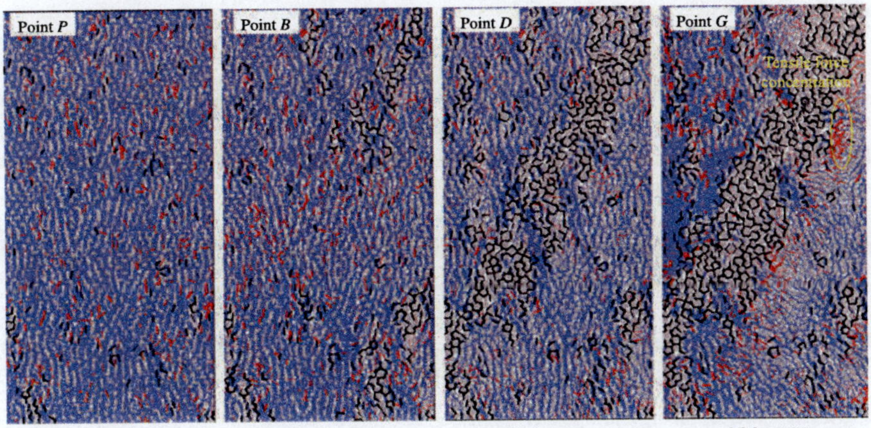

Fig. 9.17 Distribution of parallel bond force in the mudstone specimen. The points *P*, *B*, *D* and *G* in this figure correspond to those shown in Fig. 9.11

with Figs. 9.14 –9.15, it can be concluded that the numerically simulated ultimate failure mode is in a good agreement with the experimental result. When the post-stress points are located at points *A* ($\lambda = 0.02$) and *B* ($\lambda = 0.15$), the mudstone specimen takes on a shear fracture mode with a single failure surface. However, when the post-stress points are located at points *E* ($\lambda = 0.55$) and *F* ($\lambda = 0.7$), the mudstone specimen takes on a multi-shear fracture mode with several failure surfaces. This description indicates that when the damage extent of the mudstone specimen is more serious, the fracture mode is more complicated.

Figure 9.17 shows the distribution of the parallel bond force in the mudstone specimen. Points *P*, *B*, *D* and *G* in Fig. 9.17 correspond to those shown in Fig. 9.11. Additionally, in Fig. 9.17, the parallel bond force is represented by the discrete straight line segments and red represents a tensile force, while blue represents a compressive force. In this figure, the line thickness and orientation correspond to force magnitude and direction, respectively. The black line represents a micro-crack. Figure 9.17 is a local magnification figure of the observed region shown in Fig. 9.11. From Fig. 9.17a, it can be seen that at $\lambda = 0$, the tensile and compressive forces are distributed randomly, and fewer macro-cracks are observed in the specimen. As λ increases, the number of micro-cracks and their length increased significantly, and more macro-crack numbers as well as the length of the macro-crack increased greatly. The corresponding distribution of the parallel bond force also changed greatly. Once the macro-crack was initiated, the tensile and compressive force concentrations were released, as indicated by the blank areas of the figures, which can be observed by comparing Fig. 9.17b and c. Moreover, from Fig. 9.17d, we can see some tensile force concentrations that can be inferred from the red regions shown in the oval region. These tensile force concentrations predict the initiation of some new micro-cracks (Yang et al. 2014).

9.5 Conclusions

To explore the strength reduction and damage mechanical behavior of mudstone from the Yuncheng coal mine in the Shandong province of China at a depth of 960 m, the experiment and PFC numerical simulation were conducted on the mudstone specimens in this chapter. Through compared the mechanical behavior of mudstone under conventional and multi-stage triaxial compression obtained by experiment and numerical simulation, and analyzed the results of experiment and numerical simulation, the following conclusion can be obtained.

PFC not only can be used to simulate the mudstone specimens under conventional triaxial compression with different confining pressures but also can be used to simulate the mudstone specimens under multi-stage triaxial compression.

The peak strength of intact and damaged mudstone increased with confining pressure, which can be well expressed in accordance with the linear Mohr–Coulomb criterion. However, the cohesion and internal friction angle of the intact mudstone specimens are all higher than those of the damaged specimen, but which are dependent on the damage extent. With an increase in damage extent, the strength parameters of mudstone specimen all take on an attenuation law. The exponential attenuation equations of cohesion and internal friction angle with respect to the post-stress reducing ratio were constructed, and they aligned with the experimental and numerical results.

A damage variable is defined by the ratio of the area of micro-cracks to the total area of the specimen. With the increase in the post-stress reducing ratio, the damage variable first increases rapidly and then keeps basically constant after the post-stress reducing ratio reaches 0.4. The rate of change in the micro-crack number increases under a reducing confining pressure is clearly higher than that under single-stage triaxial compression.

References

Cao RH, Cao P, Lin H, Pu CZ, Ou K (2016) Mechanical behavior of brittle rock-like specimens with pre-existing fissures under uniaxial loading: experimental studies and particle mechanics approach. Rock Mech Rock Eng 49(3):1–21

Corkum AG, Martin CD (2007) The mechanical behavior of weak mudstone (Opalinus clay) at low stresses. Int J Rock Mech Min Sci 44:196–209

Geertsema AJ (2002) The shear strength of planar joints in mudstone. Int J Rock Mech Min Sci 39:1045–1049

He MC, Zhou L, Li DJ et al. (2008) Experimental research on hydrophilic characteristics of mudstone in deep well. Chin J Rock Mech Eng 27(6):1113–1120 (In Chinese)

Lee H, Jeon S (2011) An experimental and numerical study of fracture coalescence in pre-cracked specimens under uniaxial compression. Int J Solids Struct 48(6):979–999

Mohr O (1914) Abhandlungen aus dem Gebiete der Technischen Mechanik, 2nd edn. Ernst, Berlin

Mu K, Yu J, Cai YY, Chen X (2015) Acoustic emission of sandstone with hydro-mechanical coupling and PFC-based modelling of energy dissipation. Rock and Soil Mech 36(5):1496–1504

Vervoort A, Min KB, Konietzky H, Cho JW, Debecker B, Dinh QD, Frühwirt T, Tavallali A (2014) Fracturing of transversely isotropic rock under Brazilian test conditions. Int J Rock Mech Min Sci 70:343–352

Yang SQ, Jing HW, Wang SY (2012) Experimental investigation on the strength, deformability, failure behavior and acoustic emission locations of red sandstone under triaxial compression. Rock Mech Rock Eng 45(4):583–606

Yang SQ, Huang YH, Jing HW, Liu XR (2014) Discrete element modeling on fracture coalescence behavior of red sandstone containing two unparallel fissures under uniaxial compression. Eng Geol 178:28–48

Yang SQ, Jing HW, Li YS, Han LJ (2011) Experimental investigation on mechanical behavior of coarse marble under six different loading paths. Exp Mech 51(3):315–334

Yang SQ, Tian WL, Huang YH (2018) Failure mechanical behavior of pre-holed granite specimens after elevated temperature treatment by particle flow code. Geothermics 72:124–137

Yang SQ, Tian WL, Jing HW, Huang YH, Yang XX, Meng B (2019) Deformation and damage failure behavior of mudstone specimens under single-stage and multi-stage triaxial compression. Rock Mech Rock Eng 52:673–689

Yoon J (2007) Application of experimental design and optimization to PFC model calibration in uniaxial compression simulation. Int J Rock Mech Min Sci 44:871–889

Chapter 10
Deformation, Peak Strength and Crack Damage Behavior of Hollow Sandstone Under Conventional Triaxial Compression

During construction, a disturbed or damaged zone may develop around the excavation boundary of a tunnel, cavern, borehole or excavated surface. It is essential to evaluate the stress levels at different stages of the crack evolution of rock and estimate the extent of the disturbed zone. In the past, the experimental studies have mainly focused on conventional solid cylinder specimens. The testing of hollow cylinder specimens in rock engineering is very common and has at least two advantages compared to testing conventional solid cylinder specimens. The first advantage is that tests using hollow cylinders can both evaluate rock behavior and perform physical simulation at reduced scale of rock engineering applications (Alsayed 2002; Labiouse et al. 2014). The second advantage is that tests using hollow cylinders can better simulate stress conditions around deep well bores in petroleum engineering or the safety of deep excavation damage zones in tunnel engineering (Monfared et al. 2011; Haimson and Kovacich 2003). Lee et al. (1999) carried out a series of hollow cylindrical triaxial tests to investigate the potential effect of stress path on the mechanical behavior of sandstone. Alsayed (2002) developed a new multiaxial hollow cylinder test cell and demonstrated its versatility by performing uniaxial, biaxial, triaxial and polyaxial compression tests on hollow sandstone specimens. But, in the past, few experimental studies have analyzed the deformation, peak strength and crack damage behavior of hollow sandstone with various hole diameters.

Therefore, in this chapter, the results of a series of conventional triaxial compression tests using hollow sandstone with various hole diameters are reported (Yang 2016). Based on the experimental results, the influences of confining pressure and hole diameter on the deformation parameters of hollow sandstone are first discussed. Then, the effects of hole diameter and confining pressure on the peak strength parameters of the hollow sandstone are investigated in detail. Finally, the effects of hole diameter and confining pressure on crack damage behavior of the hollow sandstone are analyzed in detail.

S.-Q. Yang, *Mechanical Behavior and Damage Fracture Mechanism of Deep Rocks*,
https://doi.org/10.1007/978-981-16-7739-7_10

10.1 Hollow Sandstone and Testing Procedure

10.1.1 Sandstone Material and Hollow Specimen Preparation

The material used throughout this study was sandstone collected from Rizhao city in Shandong province, China. The sandstone material (Fig. 10.1) is a fine- to medium-grained, feldspathic rock material with a connected porosity of 6.88% and a bulk density of 2410 kg/m^3. According to the results of X-ray diffraction (XRD), the minerals in the sandstone material are quartz, K-feldspar, plagioclase, calcite, dolomite, hematite and clay mineral. The detailed composition of this rock is described as follows: 10.1% quartz, 11.7% K-feldspar, 41.6% plagioclase, 10.7% calcite, 13.1% dolomite, 2.4% hematite and 10.4% clay minerals. In this experiment, all intact specimens were cored from the same block of material to an actual diameter of 49.8 mm and approximate length of 100 mm. Based on intact specimens, hollow specimens with hole diameters (d) of 11, 15 and 26 mm (Fig. 10.2) were machined successfully. The machining method can be described as follows: Internal holes are first drilled in a large rectangular rock block (approximately 600*600*120 mm) using diamond bits with various outer diameters (d = 11, 15 and 26 mm). Then, a

Fig. 10.1 Thin section images of sandstone used in this research. **a** crossed polarization photograph of microstructure, and **b** plane polarization photograph of microstructure

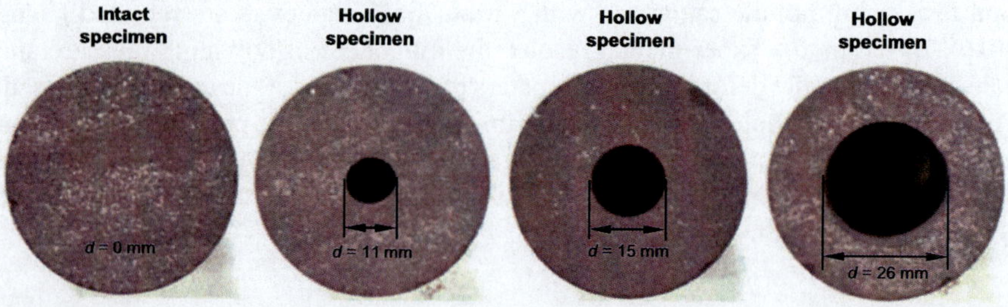

Fig. 10.2 Intact specimen and hollow specimen in the present study

larger diamond bit with an internal diameter of 50 mm is used to machine the hollow specimens. All the experiments were performed on dry specimens at room temperature. The average *P*-wave velocity and *S*-wave velocity of the tested sandstone are 3813 m/s and 2290 m/s, respectively.

10.1.2 Testing Equipment and Procedure

All experiments were performed in a TAW-1000 servo-controlled rock mechanics experimental system (Yang et al. 2015). The maximum loading capacity of the servo-controlled system is 1000 kN, and the maximum confining pressure is 60 MPa. During experimentation, the axial deformation was measured with an axial linear variable differential transducer (LVDT) with a range of 8 mm. The radial deformation was measured using an LVDT wrapped tightly around the specimen, for which the maximum displacement capacity was 4 mm. To eliminate end friction effects (Mogi 1966; Hawkes and Mellor 1970), the axial and radial LVDTs were located in the central part of the specimen in the direction of axial stress (Yang et al. 2015).

Conventional triaxial experiments were carried out under different confining pressures of 0, 8, 16, 24 and 35 MPa. They consisted of the following two steps. First, a rubber membrane was used to seal the specimen and prevent hydraulic oil from penetrating into the rock. Then, the confining pressure was increased to the desired value at a constant rate of 0.5 MPa/s to ensure that the specimen was at a uniform hydrostatic stress. A deviatoric stress was then applied to the specimen at a constant axial displacement rate of 0.08 mm/min (approximately 1.33×10^{-5}/s) until failure, in accordance with the method suggest by the ISRM (International Society for Rock Mechanics) (Fairhurst and Hudson 1999). Note that in this study, the internal pressures of hollow sandstone specimens are all zero.

10.2 Deformation Behavior of Hollow Sandstone

Figure 10.3 illustrates the axial stress–strain curve of intact sandstone under uniaxial compression. As the figure shows, under uniaxial compression, the axial supporting capacity of the sandstone specimen numbered RZ-A2# rapidly decreases towards zero and violent failure occurs in the specimen numbered RZ-A1# after the peak strength is reached, indicating the obvious brittle characteristics of the tested sandstone material. The intact sandstone also shows a typical axial splitting failure mode. Figure 10.4 further shows the effect of specimen variability on the triaxial stress–strain curves of the tested sandstone. In Fig. 10.4, σ_1 and σ_3 represent the major principal stress and the confining pressure, respectively. ε_1 and ε_3 represent the axial strain and the radial strain, respectively. In this study, a positive value of strain suggests compressive deformation and a negative value reflects dilation deformation. From Fig. 10.4, it can be seen that specimen variability has a minimal influence

Fig. 10.3 Axial stress–strain curve of intact sandstone under uniaxial compression

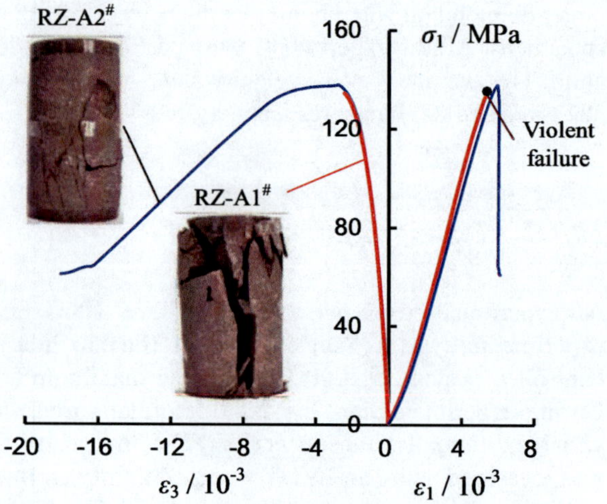

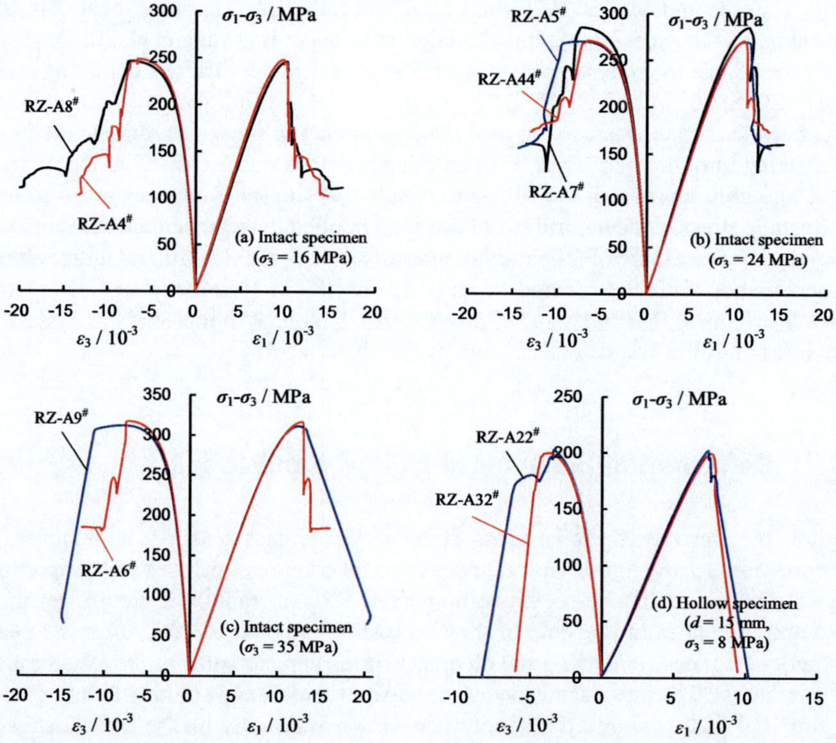

Fig. 10.4 Effect of specimen variability on triaxial stress–strain curves of tested sandstone under different confining pressures

on the deformation behavior of intact and hollow sandstone specimens before the peak strength is reached. Moreover, after the peak strength is reached, sandstone has approximately brittle deformation characteristics, which are not closely related to specimen variability. Therefore, based on a series of conventional triaxial test results of intact and hollow sandstone specimens, the influence of confining pressure and hole diameter on the deformation parameters of the tested sandstone material can be analyzed.

10.2.1 Determination of Deformation Parameters

To determine the deformation parameters of rock materials, a previous study (Yang et al. 2015) used the average slope of the approximately straight line portion (*AB*, i.e., from the elastic initial value to the elastic limit) of the axial deviatoric stress–axial strain curve before the peak strength to calculate the elastic modulus E_{S1}, as shown in Fig. 10.5. However, the secant modulus (E_{50}) is defined as the slope between the original point and the stress at half of peak strength (Fig. 10.5). Poisson's ratio (μ_1) refers to the average value of the ratio of radial strain to axial strain at the stage of elastic deformation.

However in this study, the following method is used to confirm the elastic modulus and Poisson's ratio of the tested sandstone material under conventional triaxial compression. According to the generalized Hooke's law, in conventional triaxial compression ($\sigma_2 = \sigma_3$, $\varepsilon_2 = \varepsilon_3$), the bulk modulus (*K*) and shear modulus (*G*) can be expressed by Eqs. (10.1) and (10.2), respectively.

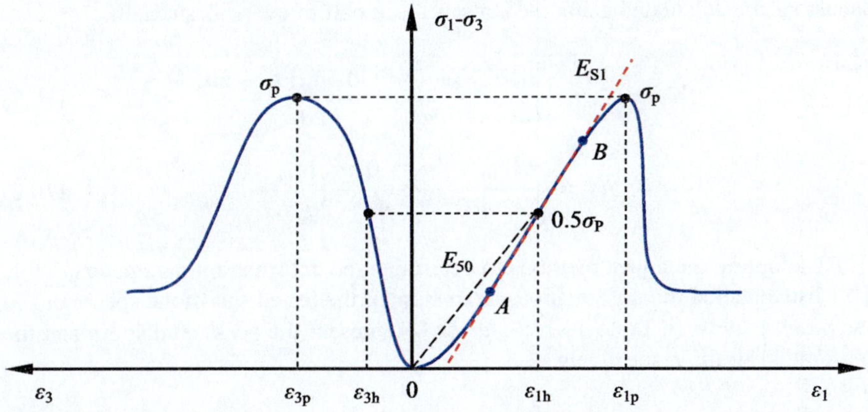

Fig. 10.5 Confirmation of deformation parameters of intact and hollow sandstone specimens in this research

$$K = \frac{\sigma_1 + 2\sigma_3}{3(\varepsilon_1 + 2\varepsilon_3)} \tag{10.1}$$

$$G = \frac{\sigma_1 - \sigma_3}{2(\varepsilon_1 - \varepsilon_3)} \tag{10.2}$$

At the same time, the bulk and shear moduli can be represented in the form of the elastic modulus (E_{S2}) and Poisson's ratio (μ_2), as shown in Eqs. (10.3) and (10.4), respectively.

$$K = \frac{E}{3(1 - 2\mu)} \tag{10.3}$$

$$G = \frac{E}{2(1 + \mu)} \tag{10.4}$$

By substituting Eqs. (10.3) and (10.4) into Eqs. (10.1) and (10.2), the following two equations can be obtained.

$$E = \frac{9KG}{3K + G} = \frac{\sigma_1^2 + \sigma_1\sigma_3 - 2\sigma_3^2}{(\sigma_1 + \sigma_3)\varepsilon_1 - 2\sigma_3\varepsilon_3} \tag{10.5}$$

$$\mu = \frac{3K - 2G}{6K + 2G} = \frac{\sigma_3\varepsilon_1 - \sigma_1\varepsilon_3}{(\sigma_1 + \sigma_3)\varepsilon_1 - 2\sigma_3\varepsilon_3} \tag{10.6}$$

In Eqs. (10.5) and (10.6), the elastic modulus and Poisson's ratio of rock material are not constant and are closely related to the stress level, axial and radial strain. Therefore, in the present study, the elastic modulus (E_{S2}) and Poisson's ratio (μ_2) of sandstone are calculated using the stress state at half of the peak strength.

$$E_{S2} = \frac{(\sigma_3 + 0.5\sigma_p)^2 + (\sigma_3 + 0.5\sigma_p)\sigma_3 - 2\sigma_3^2}{(2\sigma_3 + 0.5\sigma_p)\varepsilon_{1h} - 2\sigma_3\varepsilon_{3h}} \tag{10.7}$$

$$\mu_2 = \frac{\sigma_3\varepsilon_{1h} - (\sigma_3 + 0.5\sigma_p)\varepsilon_{3h}}{(2\sigma_3 + 0.5\sigma_p)\varepsilon_{1h} - 2\sigma_3\varepsilon_{3h}} \tag{10.8}$$

By adopting the above methods to determine the deformation parameters, Table 10.1 lists detailed the deformation parameters of the tested sandstone specimens in the present study. In Table 10.1, ε_{1p} and ε_{3p} represent the peak axial strain and the peak radial strain, respectively.

Table 10.1 Deformation parameters of tested sandstone specimens in this research

Specimen	D / mm	d / mm	σ_3 / MPa	ε_{1p} / 10^{-3}	ε_{3p} / 10^{-3}	E_{S1} / GPa	E_{S2} / GPa	E_{50} / GPa	μ_1	μ_2
RZ-A1#	49.8	0	0	5.408	− 2.424	28.78	23.73	23.73	0.209	0.210
RZ-A2#	49.8	0	0	5.995	− 2.791	27.16	22.29	22.29	0.201	0.200
RZ-A3#	49.8	0	8	8.024	− 5.731	29.95	29.03	28.01	0.230	0.274
RZ-A4#	49.8	0	16	9.814	− 6.546	30.88	31.65	30.08	0.237	0.298
RZ-A5#	49.8	0	24	10.979	− 7.464	30.01	32.32	30.33	0.237	0.316
RZ-A6#	49.8	0	35	12.800	− 7.056	30.29	33.69	31.26	0.231	0.324
RZ-A7#	49.8	0	24	11.239	− 7.102	31.17	33.40	31.42	0.238	0.313
RZ-A8#	49.8	0	16	9.984	− 7.233	29.78	30.45	28.92	0.234	0.297
RZ-A9#	49.8	0	35	12.547	− 8.135	30.83	33.78	31.41	0.241	0.332
RZ-A44#	49.8	0	24	10.651	− 6.530	29.83	32.02	30.00	0.230	0.310
RZ-A12#	49.8	11	0	5.706	− 3.731	26.91	23.16	23.16	0.205	0.204
RZ-A16#	49.8	15	0	5.934	− 3.574	27.90	23.39	23.39	0.188	0.187
RZ-A32#	49.8	15	8	7.635	− 3.657	30.23	28.66	27.57	0.207	0.254
RZ-A22#	49.8	15	8	7.407	− 3.365	30.59	29.89	28.80	0.214	0.259
RZ-A23#	49.8	15	16	8.391	− 3.635	31.43	31.80	30.02	0.220	0.291
RZ-A24#	49.8	15	24	10.030	− 5.153	30.40	32.63	30.52	0.232	0.316
RZ-A25#	49.8	15	35	10.373	− 3.902	30.46	33.39	30.79	0.233	0.336
RZ-A13#	49.8	26	0	6.007	− 3.402	27.24	22.11	22.11	0.164	0.164
RZ-A14#	49.8	26	16	7.293	-2.245	31.74	33.22	31.24	0.214	0.291
RZ-A15#	49.8	26	35	8.513	-2.060	31.59	34.91	31.72	0.213	0.334

Note D: specimen diameter; *d*: hole diameter; σ_3: confining pressure

10.2.2 Influence of Confining Pressure on the Deformation Behavior

Figures 10.6 and 10.7 show the effect of confining pressure on the axial deviatoric stress-axial strain curves of intact and hollow sandstone. From Figs.10.6 and 10.7, it is clear that the confining pressure has a significant effect on the deformation behavior of intact and hollow sandstone. Figures 10.8 and 10.9 show the effect of hole diameter on the axial deviatoric stress-axial strain curves of intact and hollow sandstone, respectively. From Figs. 10.8 and 10.9, it is clear that the deformation

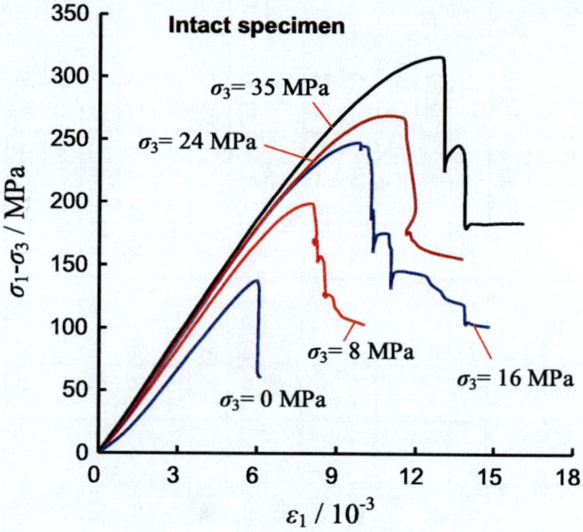

Fig. 10.6 Effect of confining pressure on axial deviatoric stress-axial strain curves of intact sandstone

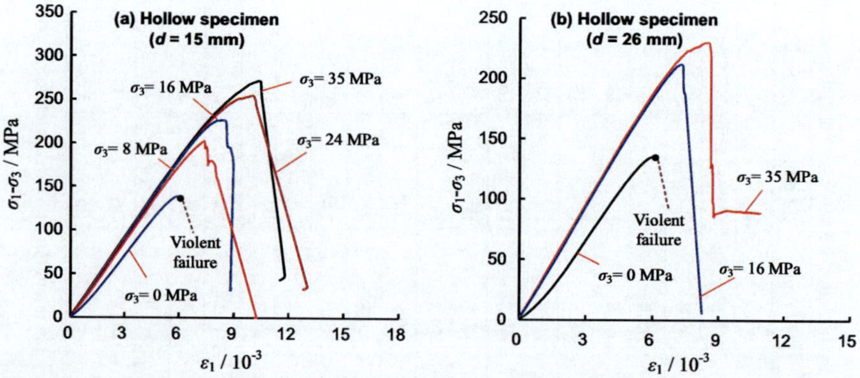

Fig. 10.7 Effect of confining pressure on axial deviatoric stress-axial strain curves of hollow sandstone

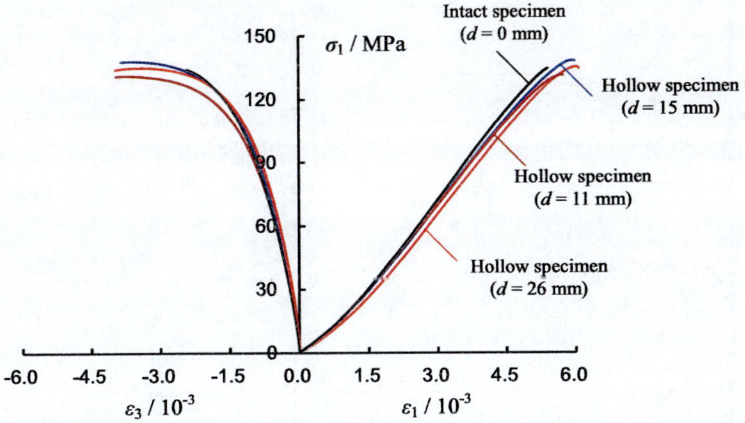

Fig. 10.8 Effect of hole diameter on axial stress–strain curves of hollow sandstone under uniaxial compression

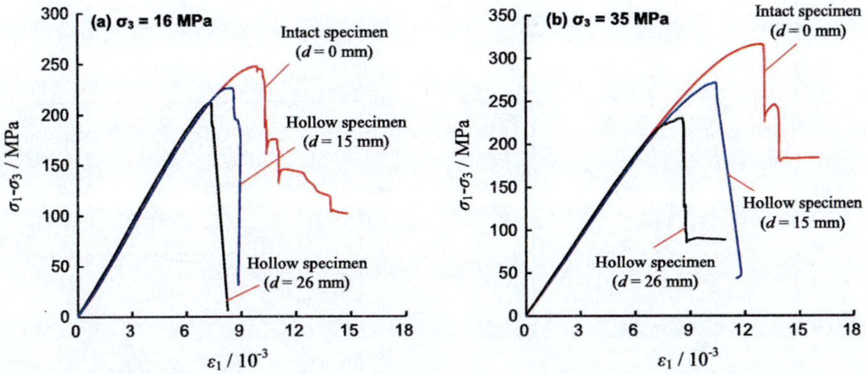

Fig. 10.9 Effect of hole diameter on axial deviatoric stress–strain curves of hollow sandstone under triaxial compression

parameters of hollow sandstone do not depend on the hole diameter, whereas the peak strength of hollow sandstone is closely related to the hole diameter, which will be analyzed in detail in Sect. 10.3.

Based on the data listed in Table 10.1, Fig. 10.10 shows the effect of confining pressure on the Young's modulus values of hollow sandstone specimens with various hole diameters. Figure 10.10a-c show that hole diameter has almost no effect on the Young's modulus value of hollow sandstone under various confining pressures, which is also supported by the axial deviatoric stress-axial strain curves in Figs. 10.8 and 10.9. We take E_{S1} as an example for further analysis. Under uniaxial compression, E_{S1} is respectively 27.97 GPa at $d = 0$ mm, 27.90 GPa at $d = 15$ mm and 27.24 GPa at $d =$ 26 mm. At $\sigma_3 = 16$ MPa, E_{S1} is respectively 30.33 GPa at $d = 0$ mm, 31.43 GPa at d

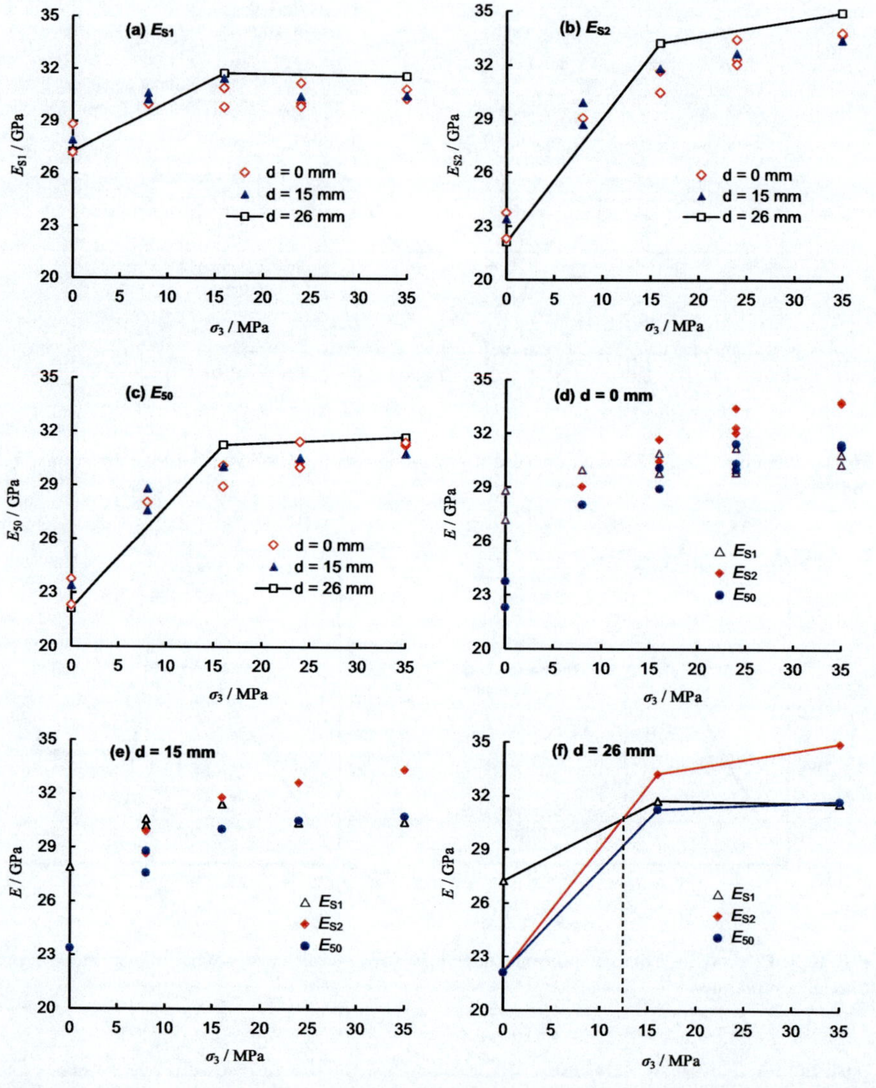

Fig. 10.10 Effect of confining pressure on Young's modulus of hollow sandstone with various hole diameters

= 15 mm and 31.74 GPa at $d = 26$ mm. However, at $\sigma_3 = 35$ MPa, E_{S1} is respectively 30.56 GPa at $d = 0$ mm, 30.46 GPa at $d = 15$ mm and 31.59 GPa at $d = 26$ mm. The above results indicate that the Young's modulus of intact sandstone ($d = 0$ mm) can be directly applied to evaluate the deformation behavior of hollow sandstone with various hole diameters. Figures 10.10d–f show that the Young's modulus of hollow sandstone depends on the confining pressure. With increasing confining pressure, the

Young's modulus of hollow sandstone increases nonlinearly. The elastic modulus E_{S1} and the secant modulus E_{50} of hollow sandstone initially increase as σ_3 increases from 0 to 16 MPa. They remain relatively constant as σ_3 increases from 16 to 35 MPa. However, the elastic modulus E_{S2} of hollow sandstone increases nonlinearly as σ_3 increases from 0 to 35 MPa, and the increasing trend gradually slows, differing from the variations in E_{S1} and E_{50} with confining pressure.

Furthermore, it should be noted that the sensitivity of E_{S1}, E_{S2} and E_{50} of hollow sandstone with the same hole diameter on the confining pressure is different. We take the variance in Young's modulus of hollow sandstone at $d = 15$ mm with confining pressure as an example, as shown in Fig. 10.10e. The elastic modulus E_{S1} of hollow sandstone is obviously higher than the secant modulus E_{50} in the range of σ_3 from 0 to 16 MPa, whereas E_{S1} is approximately equal to E_{50} at confining pressures above 16 MPa, which may result from the more pronounced stage of microcrack closure at lower confining pressures (Yang et al. 2012). However, at higher confining pressures, the stage of pore and fissure closure does not depend on the confining pressure. The elastic modulus E_{S2} of hollow sandstone is obviously lower than E_{S1} at lower confining pressures (approximately < 12.5 MPa), and E_{S2} is approximately equal to E_{S1} at $\sigma_3 = 12.5$ MPa, whereas E_{S2} is significantly larger than E_{S1} at higher confining pressures (above approximately 12.5 MPa).

Based on the data listed in Table 10.1, Fig. 10.11 shows the effect of confining pressure on the Poisson's ratio values of hollow sandstone specimens with various hole diameters. From Fig. 10.11, it is clear the μ_1 and μ_2 values of hollow sandstone under uniaxial compression decrease from 0.205 to 0.164 as d increases from 0 to 26 mm, which indicates that the Poisson's ratio of hollow sandstone under uniaxial compression decreases with increasing hole diameter and does not depend on the

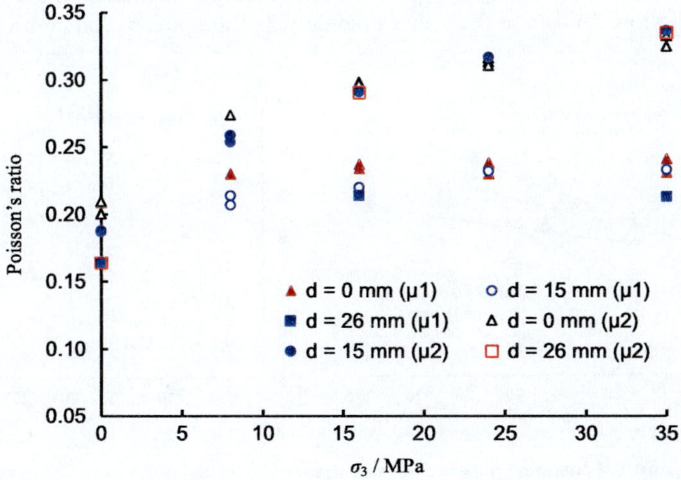

Fig. 10.11 Effect of confining pressure on poisson's ratio of hollow sandstone with various hole diameters

confirmation method. However, in the range of $\sigma_3 = 8 \sim 35$ MPa, the Poisson's ratio of hollow sandstone is independent of the hole diameter at the same confining pressure but dependent on the confirmation method. As shown in Fig. 10.11, the μ_1 value of hollow sandstone under uniaxial compression obtained using the conventional method is slightly lower than that in the range of $\sigma_3 = 8 \sim 35$ MPa. However, the μ_1 value of hollow sandstone ranges from 0.207 to 0.241, which does not depend on the confining pressure in the range of $\sigma_3 = 8 \sim 35$ MPa. However, as σ_3 increases from 0 to 35 MPa, the μ_2 value of hollow sandstone increases from 0.205 to 0.328 at $d =$ 0 mm; from 0.187 to 0.336 at $d = 15$ mm; and from 0.164 to 0.334 at $d = 26$ mm, suggesting that the μ_2 value of hollow sandstone obtained by Eq. (10.8) increases nonlinearly with the confining pressure, and the increasing trend gradually slows.

The above analysis shows that the two methods of confirming the elastic modulus and Poisson's ratio of hollow sandstone will result in substantial engineering mistakes. In shallow rock engineering, it is relatively safe to use E_{S1} and μ_1 obtained by the conventional method; however in deep rock engineering, it is too conservative and more dangerous to use E_{S1} and μ_1. Therefore, it is suggested that E_{S2} and μ_2 should be used to analyze the stability of deep rock engineering in this study.

Based on the data listed in Table 10.1, Fig. 10.12 shows the effect of confining pressure on the peak axial and radial strains of hollow sandstone specimens with various hole diameters. From Fig. 10.12a, one can see that the peak axial strain of hollow sandstone increases linearly with the confining pressure, and the sensitivity of the peak axial strain of hollow sandstone with larger hole diameters to σ_3 is lower than that with smaller hole diameters. From Fig. 10.12b shows that the peak radial strain of hollow sandstone has a nonlinear relationship with increasing confining pressure. Although under uniaxial compression, the peak axial and radial strains of hollow sandstone are independent of hole diameter, as shown in Fig. 10.8. However, under the action of confining pressure, with increasing hole diameter, the peak axial strain of hollow sandstone decreases gradually, but the peak radial strain increases

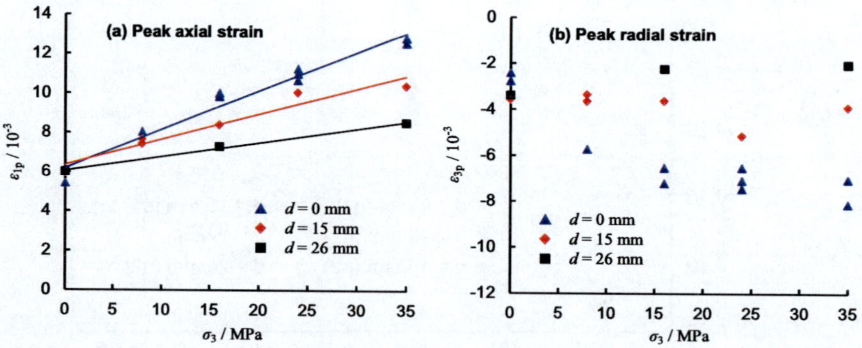

Fig. 10.12 Effect of confining pressure on peak strain of intact and hollow sandstone with various hole diameters

gradually, which results from the confining pressure restricting the radial deformation of hollow sandstone.

10.3 Peak Strength Behavior of Hollow Sandstone

Table 10.2 lists the peak stress (σ_p) values of hollow sandstone from conventional triaxial compression experiments (obtained from the stress–strain curves in Figs. 10.6 and 10.7. Over the entire confining pressure range, σ_S increased from 134.45 to 351.38 MPa at $d = 0$ mm, from 138.41 to 271.76 MPa at $d = 15$ mm, and from 135.44 to 230.32 MPa at $d = 26$ mm. It is clear that with increasing confining pressure, the extent to which the peak strength of hollow sandstone increases is different and depends on the hole diameter. This can be explained as follows. When the confining pressure does not equals to zero and the internal hole pressure is zero, the radial and tangential stresses in the hollow specimen are non-uniform. The tangential stress

Table 10.2 Strength and crack damage threshold parameters of tested sandstone specimens in this research

Specimen	d / mm	σ_3 / MPa	σ_{cd} / MPa	σ_{sd} / MPa	σ_p / MPa	σ_S / MPa	σ_{cd} / σ_p
RZ-A1#	0	0	89.72	89.72	134.45	134.45	0.667
RZ-A2#	0	0	101.91	101.91	137.99	137.99	0.739
RZ-A3#	0	8	137.06	145.06	198.92	206.92	0.689
RZ-A4#	0	16	175.09	191.09	247.71	263.71	0.707
RZ-A5#	0	24	195.33	219.33	269.38	293.38	0.725
RZ-A6#	0	35	254.27	289.27	316.38	351.38	0.804
RZ-A7#	0	24	212.24	236.24	285.22	309.22	0.744
RZ-A8#	0	16	171.13	187.13	246.16	262.16	0.695
RZ-A9#	0	35	238.48	273.48	311.21	346.21	0.766
RZ-A44#	0	24	197.74	221.74	270.16	294.16	0.732
RZ-A12#	11	0	83.36	83.36	131.60	131.60	0.633
RZ-A16#	15	0	95.35	95.35	138.41	138.41	0.689
RZ-A32#	15	8	147.99	155.99	199.40	207.40	0.742
RZ-A22#	15	8	149.96	157.96	202.44	210.44	0.741
RZ-A23#	15	16	184.74	200.74	226.26	242.26	0.816
RZ-A24#	15	24	210.56	234.56	254.59	278.59	0.827
RZ-A25#	15	35	239.64	274.64	271.76	306.76	0.882
RZ-A13#	26	0	96.64	96.64	135.44	135.44	0.714
RZ-A14#	26	16	210.32	226.32	211.89	227.89	0.993
RZ-A15#	26	35	219.22	254.22	230.32	265.32	0.952

Note $\sigma_{sd} = \sigma_{cd} + \sigma_3$; $\sigma_S = \sigma_p + \sigma_3$

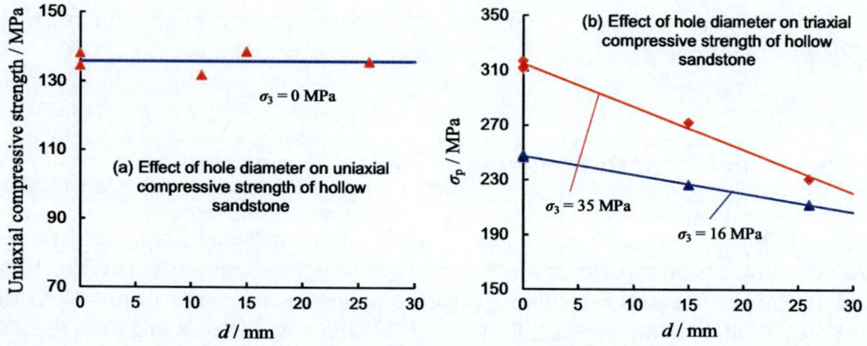

Fig. 10.13 Effect of hole diameter on peak strength of hollow sandstone under different confining pressures

decreases nonlinearly but the radial stress increases gradually with increasing radius (from the internal wall to outer wall). The non-uniform stress distribution results in differences in hollow sandstone specimen with various hole diameters at different confining pressures.

Figure 10.13 presents the effect of hole diameter on the peak strength of hollow sandstone at different confining pressures. From Fig. 10.13a, one can see that the uniaxial compressive strength (UCS) of sandstone (average value) is 136.22 MPa at $d = 0$ mm, 138.41 MPa at $d = 15$ mm, and 135.44 MPa at $d = 26$ mm, indicating that the hole diameter has no effect on the UCS of hollow sandstone. The reason for these trends is as follows. Under uniaxial compression, hollow sandstone specimens with various hole diameters all experience axial splitting tensile failure and the stress distribution is more uniform. Therefore, the UCS of hollow sandstone does not depend on the hole diameter.

However, the triaxial compressive strength of hollow sandstone decreases with increasing hole diameter and the rate of this decrease is dependent on the confining pressure level, as shown in Fig. 10.13b. At $\sigma_3 = 16$ MPa, the peak strength (σ_p) of hollow sandstone decreases from 246.94 to 211.89 MPa as d increases from 0 to 26 mm. However, at $\sigma_3 = 35$ MPa, the σ_p value of hollow sandstone decreases from 313.80 to 230.32 MPa as d increases from 0 to 26 mm. Moreover, the sensitivity of the peak strength of hollow sandstone to hole diameter at $\sigma_3 = 35$ MPa is clearly higher than that at $\sigma_3 = 16$ MPa. Notably, that under confining pressure, hollow sandstone usually fails mainly by a single shear fracture along the direction of the maximum shear stress. At the same hole diameter, the peak strength of hollow sandstone increases with the confining pressure. However, shear fracture area of hollow sandstone is different for various hole diameters. The larger the hole diameter, the smaller the area of the shear fracture. Thus, the corresponding axial supporting capacity is lower for specimen with larger hole diameters. Therefore, under the same confining pressure, with increasing hole diameter, the peak strength will decrease gradually. Moreover, the sensitivity of the peak strength of hollow sandstone to hole diameter at larger confining pressure is obviously higher than that at smaller confining pressure, which is potentially because the non-uniform stress distribution increases with confining pressure at an internal hole pressure equals to zero.

Based on the data listed in Table 10.2, the difference between the peak strengths of two or three different specimens at the same hole diameter is very small at the same confining pressure. To evaluate the discreteness of the tested sandstone specimens, a dispersion coefficient (the ratio of the D-value between the maximum value and the minimum value to the average value) can be used (Yang et al. 2015). The dispersion coefficient of the peak strength of intact sandstone ($d = 0$ mm) is 2.6% for $\sigma_3 = 0$ MPa, 0.63% for $\sigma_3 = 16$ MPa, 5.76% for $\sigma_3 = 24$ MPa and 1.65% for $\sigma_3 = 35$ MPa. In general, the discreteness of the tested sandstone specimens is very small; therefore, one can analyze the relationships between the strength parameters of hollow sandstone and σ_3.

In this section, the Mohr–Coulomb and Hoek–Brown criteria are used to investigate the effect of confining pressure on the strength and crack damage behavior of hollow sandstone. The linear Mohr–Coulomb criterion can be expressed with the following equation:

$$\sigma_S = \sigma_0 + q\sigma_3 = \frac{2C\cos\phi + \sigma_3(1+\sin\phi)}{1-\sin\phi} \tag{10.9}$$

where σ_S is the maximum axial supporting capacity of rock; σ_0 is the uniaxial compressive strength of rock material; q is a coefficient related to the influence of confining pressure on the σ_S of rock. The σ_0 and q are related to the cohesion C and the internal friction angle ϕ of rock material.

The Hoek–Brown criterion describing the peak triaxial compressive strength of a wide range of isotropic rock materials can be written as follows (Hoek and Brown 1980, 1997):

$$\sigma_S = \sigma_3 + (m\sigma_c\sigma_3 + s\sigma_c^2)^{0.5} \tag{10.10}$$

where σ_c is the UCS of the intact rock material and m and s are material constants for a specific rock. When the parameter m is larger, the rock is stronger. The parameter s reflects the fracture extent, ranging from zero to one. When the parameter s is closer to one, the rock is more intact.

By adopting the linear Mohr–Coulomb criterion, the influence of σ_3 on the peak strength of hollow sandstone under triaxial compression can be presented, as shown in Fig. 10.14. Table 10.3 lists the strength parameters of hollow sandstone by the linear Mohr–Coulomb criterion in this study. Figure 10.14 indicates that the peak strength of sandstone under triaxial compression increases with increasing σ_3. To compare the difference between strength criteria with experimental results, the square root of the sum of the least error square is proposed as the discrimination criterion (Yang et al. 2015).

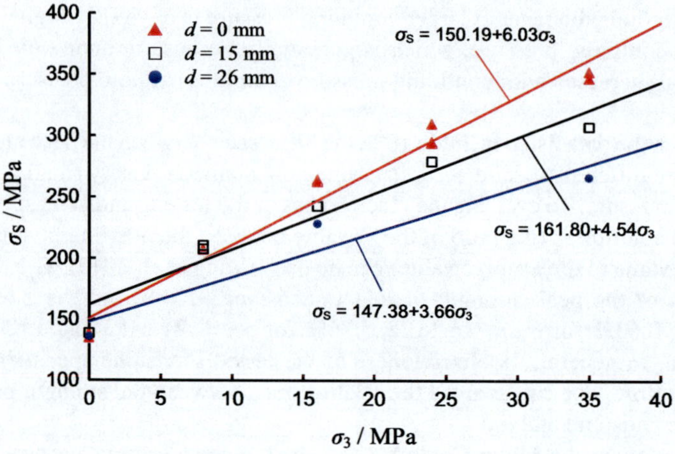

Fig. 10.14 Effect of confining pressure on peak strength of hollow sandstone with various hole diameters by the linear Mohr–Coulomb criterion

Table 10.3 Strength parameters of hollow sandstone by the linear Mohr–Coulomb criterion in this research

Different specimens	σ_0 / MPa	q	C / MPa	ϕ/ (°)	R	δ
Intact specimen ($d = 0$ mm)	150.19	6.03	30.58	45.7	0.985	39.11
Hollow specimen ($d = 15$ mm)	161.80	4.54	37.97	39.7	0.970	33.19
Hollow specimen ($d = 26$ mm)	147.38	3.66	38.52	34.8	0.959	26.97

$$\delta = \sqrt{Q} = \sqrt{\sum_{i=1}^{n}\left(\sigma_{S,i}^{Theory} - \sigma_{S,i}^{Experiment}\right)^2} \tag{10.11}$$

where δ is the square root of the sum of the least error square, Q is the sum of the least error square and i is an experimental point, which usually increases from 1 to n. The δ value is 39.11 MPa at $d = 0$ mm, 33.19 MPa at $d = 15$ mm, but 26.97 MPa at $d = 26$ mm.

From Table 10.3, it can be seen that the σ_0 and q are 150.19 MPa and 6.03 at $d = 0$ mm, 161.80 MPa and 4.54 at $d = 15$ mm, and 147.38 MPa and 3.66 at $d = 26$ mm, respectively. According to Eq. (10.9), the values of C and ϕ of hollow sandstone were calculated using a regression analysis at $d = 0$ mm and found to be 30.58 MPa and 45.7°, respectively. The values of C and ϕ of hollow sandstone at $d = 15$ mm were 37.97 MPa and 39.7°, respectively; whereas the values of C and ϕ of hollow sandstone at $d = 26$ mm were 38.52 MPa and 34.8°, respectively. These results indicate that the peak strength parameters obtained by the linear Mohr–Coulomb criterion depend on the hole diameter, as shown in Fig. 10.15. The cohesion

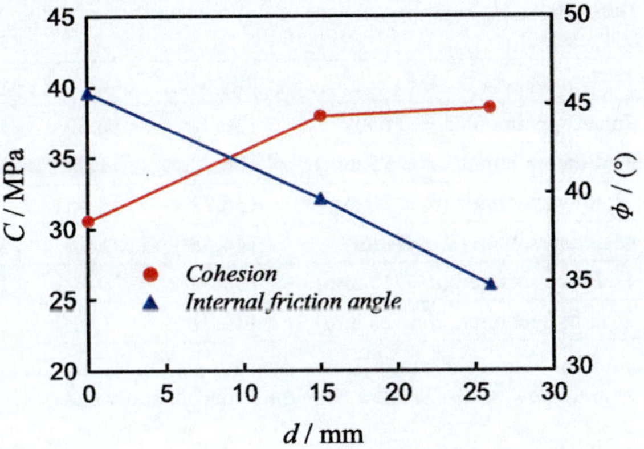

Fig. 10.15 Effect of hole diameter on the cohesion and internal friction angle of sandstone

of sandstone first increases rapidly as the hole diameter increases from 0 to 15 mm. It then remains relatively constant in the range of 15 to 26 mm. However, the internal friction angle of sandstone decreases linearly with increasing hole diameter.

Figure 10.16 shows a comparison between the nonlinear Hoek–Brown criterion and the experimental peak strength of hollow sandstone. When carrying out the regression analysis, parameters m and s are first back-calculated by setting σ_c = 136.22 MPa based on the experimental value of UCS of intact sandstone ($d = 0$ mm). However, $s = 1.125$ at $d = 0$ mm, 1.373 at $d = 15$ mm and 1.197 at $d = 26$ mm. These values are all greater than 1.0, but s must be in the range of 0 ~ 1.0. Therefore, σ_c and m are back-calculated by assuming $s = 1.000$ in this study.

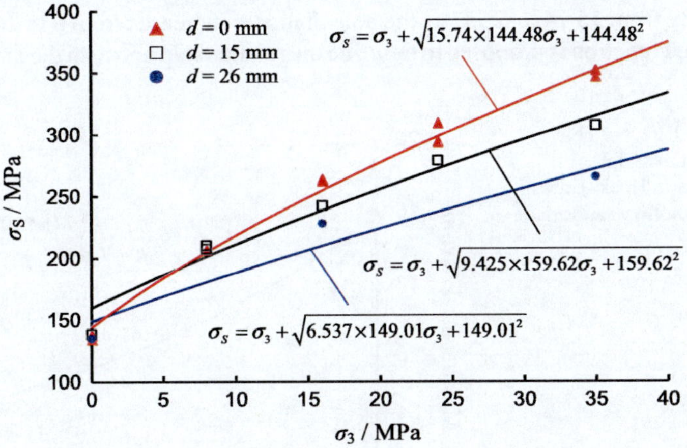

Fig. 10.16 Effect of confining pressure on peak strength of hollow sandstone with various hole diameters by the nonlinear Hoek–Brown criterion

Table 10.4 Strength parameters of hollow sandstone by the nonlinear Hoek–Brown criterion in this research

Different specimens	σ_c / MPa	m	s	a	R	δ
Intact specimen ($d = 0$ mm)*	136.22	16.69	1.125	0.5	0.995	–
Hollow specimen ($d = 15$ mm)*	136.22	11.04	1.373	0.5	0.975	–
Hollow specimen ($d = 26$ mm)*	136.22	7.151	1.197	0.5	0.938	–
Intact specimen ($d = 0$ mm)	144.48	15.74	1.000	0.5	0.995	21.34
Hollow specimen ($d = 15$ mm)	159.62	9.425	1.000	0.5	0.975	25.96
Hollow specimen ($d = 26$ mm)	149.01	6.537	1.000	0.5	0.938	23.21

Note The nonlinear Hoek–Brown strength parameters m and s with * symbol are back-calculated by giving $\sigma_c = 136.22$ MPa from the experimental value of UCS of intact sandstone ($d = 0$ mm)

Table 10.4 shows that the δ value at $d = 0$ mm is 21.34 MPa for the Hoek–Brown criterion, but 39.11 MPa for the Mohr–coulomb criterion, the δ value at $d =$ 15 mm is 25.96 MPa for the Hoek–Brown criterion, but 33.19 MPa for the Mohr–coulomb criterion, and the δ value at $d = 26$ mm is 23.21 MPa for the Hoek–Brown criterion, but 26.97 MPa for the Mohr–coulomb criterion. These results indicate that the non-linear Hoek–Brown criterion reflects the peak strength properties of hollow sandstone better than the linear Mohr–Coulomb criterion does. The above analysis indicates that the linear Mohr–Coulomb criterion predicts the peak strength of hollow sandstone to be higher than its measured value at low and high confining pressures, and lower at intermediate confining pressures (Yang et al. 2012, 2015).

Although the σ_c value in Hoek–Brown criterion has no obvious variance with increasing hole diameter, the m value shows a clear relationship with the hole diameter, as shown in Fig. 10.17. From Fig. 10.17, it is clear that the m value decreases nonlinearly from 15.74 to 6.537 as the hole diameter increases from 0 to 26 mm. An exponential function is proposed to describe the relationship between the m value and

Fig. 10.17 Effect of hole diameter on material parameter m in Hoek–Brown criterion of hollow sandstone

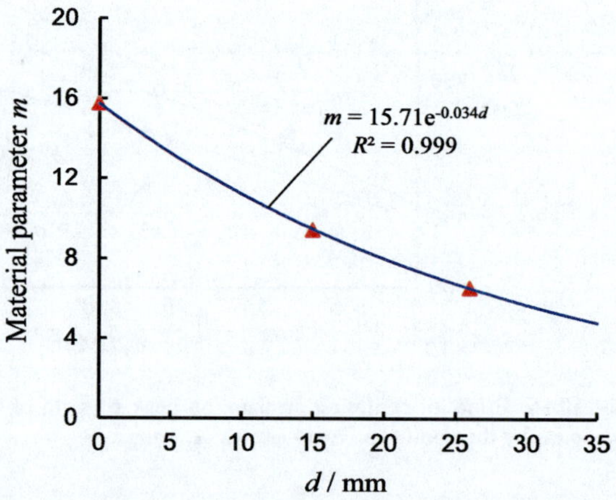

hole diameter, which can predict the effect of hole diameter on material parameter m in nonlinear Hoek–Brown criterion of hollow sandstone.

10.4 Crack Damage Behavior of Hollow Sandstone

10.4.1 Axial Deviatoric Stress-Volumetric Strain Curve

The evolution of the volumetric strain of hollow sandstone under different confining pressures is shown in Fig. 10.18. ε_v refers to the volumetric strain, which is calculated from the sum of the axial strain (ε_1) and twice the radial strain (ε_3) (i.e., $\varepsilon_v = \varepsilon_1 + 2\varepsilon_3$). The volumetric strain was characterized by an initial phase of compaction-dominated behavior followed by a phase of dilatancy-dominated behavior. After point

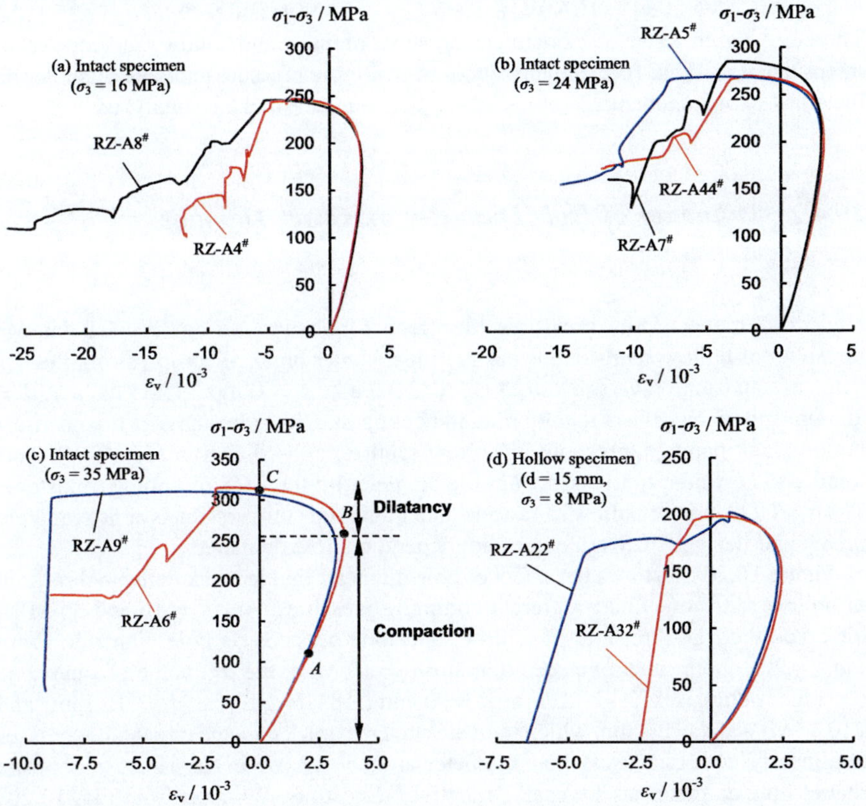

Fig. 10.18 Effect of specimen variability on axial deviatoric stress-volumetric strain curves of tested sandstone under different confining pressures

A, compaction was in direct competition with dilatancy (although the specimen was still in a state of net compaction). Eventually, the rates of compaction and dilatancy became equal at point *B*, resulting in an inflection point in the volumetric strain curve. The crack damage threshold (σ_{cd}) of the specimen (Fairhurst and Hudson 1999; Heap et al. 2009; Yang et al. 2015) is defined as the corresponding axial deviatoric stress at which the volumetric deformation of the specimen switches from compaction-dominated to dilatancy-dominated, as shown in Fig. 10.18c. As the stress increases further, more microcracking occurs and the specimen continues to dilate. The volumetric strain eventually returns to zero strain (Point *C*) and then continues to increase until specimen failure.

Figure 10.18 also shows the effect of specimen variability on the axial deviatoric stress-volumetric strain curves of tested sandstone under different confining pressures. From Fig. 10.18, it is clear that specimen variability has a minor effect on the crack damage threshold of the tested sandstone under different confining pressures. Detailed crack damage threshold for tested sandstone specimens is listed in Table 10.2. The dispersion coefficient of the crack damage threshold of intact sandstone (d = 0 mm) is 2.3% at $\sigma_3 = 16$ MPa, 8.4% for $\sigma_3 = 24$ MPa and 6.4% at $\sigma_3 = 35$ MPa. Therefore, based on the crack damage threshold of intact and hollow sandstone specimens listed in Table 10.2, the influences of confining pressure and hole diameter on the crack damage thresholds of tested sandstone material can be analyzed.

10.4.2 Influence of Hole Diameter on Crack Damage Behavior

Under different confining pressures, the effect of hole diameter on the crack damage threshold of hollow sandstone is also various. Under uniaxial compression, the σ_{cd} value of sandstone (average value) is 95.82 MPa at d = 0 mm, 95.35 MPa at d = 15 mm, and 96.64 MPa at d = 26 mm, indicating that the hole diameter has no effect on the crack damage threshold of hollow sandstone, as shown in Fig. 10.19. This trend is also similar to the effect of hole diameter on the UCS of hollow sandstone. The σ_{cd}/UCS ratio of hollow sandstone under uniaxial compression is approximately 0.689, and the σ_{cd}/UCS value does not depend on hole diameter.

Figure 10.20 illustrates the effect of hole diameter on the crack damage threshold of hollow sandstone under different confining pressures. At σ_3 = 16 and 35 MPa, the crack damage threshold of hollow sandstone depends on hole diameter. From Fig. 10.20a and b, it can be seen that at σ_3 = 16 MPa, the σ_{cd} value of sandstone (average value) is 173.11 MPa at d = 0 mm, 184.74 MPa at d = 15 mm, and 210.32 MPa at d = 26 mm, which indicates that the crack damage threshold increases nonlinearly with increasing hole diameter at σ_3 = 16 MPa. The effect is opposite that of hole diameter on the peak strength of sandstone. Whereas from Fig. 10.20c and d, one can see that at σ_3 = 35 MPa, the σ_{cd} value of sandstone (average value) is 246.38 MPa at d = 0 mm, 239.64 MPa at d = 15 mm, and 219.22 MPa at d

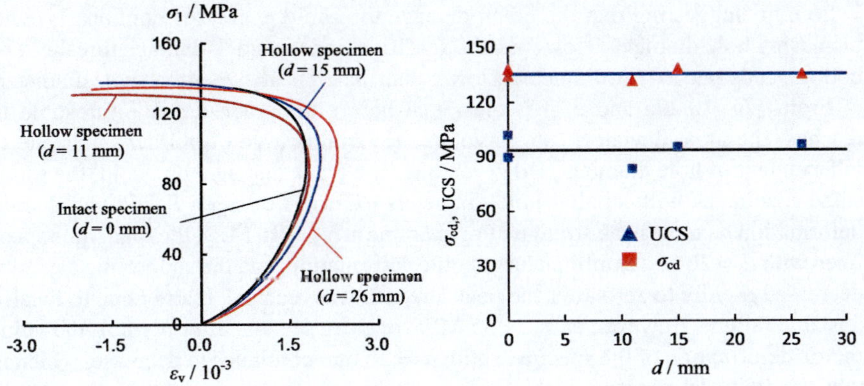

Fig. 10.19 Effect of hole diameter on the crack damage threshold of hollow sandstone under uniaxial compression

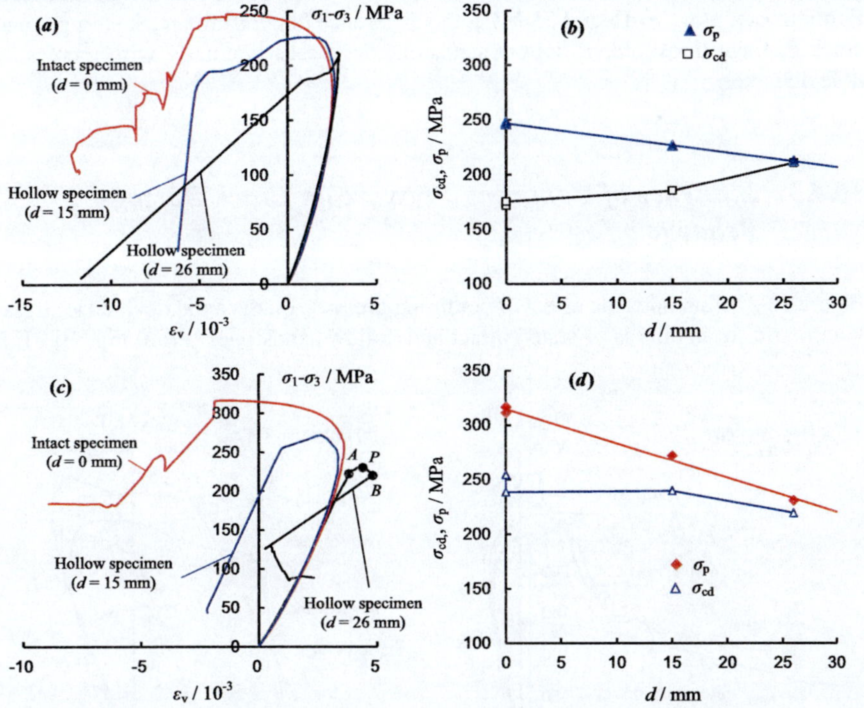

Fig. 10.20 Effect of hole diameter on the crack damage threshold of hollow sandstone under different confining pressures. **a** the axial deviatoric stress-volumetric strain curve at $\sigma_3 = 16$ MPa; **b** the effect of hole diameter on crack damage threshold at $\sigma_3 = 16$ MPa; **c** the axial deviatoric stress-volumetric strain curve at $\sigma_3 = 35$ MPa; **d** the effect of hole diameter on crack damage threshold at $\sigma_3 = 35$ MPa

= 26 mm, suggesting that the crack damage threshold decreases nonlinearly with increasing hole diameter. However, the sensitivity of the crack damage threshold of hollow sandstone to hole diameter is lower than that of peak strength to hole diameter.

From Fig. 10.20a and c, it is clear that before the crack damage threshold is reached, the axial deviatoric stress-volumetric strain curve of hollow sandstone is independent of hole diameter. However, after the crack damage threshold, the sandstone specimens with smaller hole diameters experience larger plastic and ductile deformation near the peak strength. For example at $\sigma_3 = 16$ MPa, the sandstone specimen with $d = 26$ mm exhibited less ductile deformation, and the supporting capacity decreased rapidly to zero after the peak strength was reached, mainly due to locally unstable failure. However, at $\sigma_3 = 35$ MPa, one can see that after point *A*, the volumetric deformation of the specimen with $d = 26$ mm continues to decrease, which is why the radial deformation of this hollow specimen expands toward the internal hole boundary rather than toward the outside boundary. When point *B* (locating post-peak strength) is reached, the volumetric deformation of the hollow specimen reaches a minimum value. After point *B*, the hollow specimen experiences the unstable failure. Furthermore, at $\sigma_3 = 16$ and 35 MPa, the difference between the peak strength and crack damage threshold of hollow sandstone decreases gradually with increasing hole diameter.

10.4.3 Influence of Confining Pressure on Crack Damage Behavior

Figure 10.21 illustrates the effect of confining pressure on the axial deviatoric stress-volumetric strain curves of tested intact and hollow sandstones. From Fig. 10.21, it

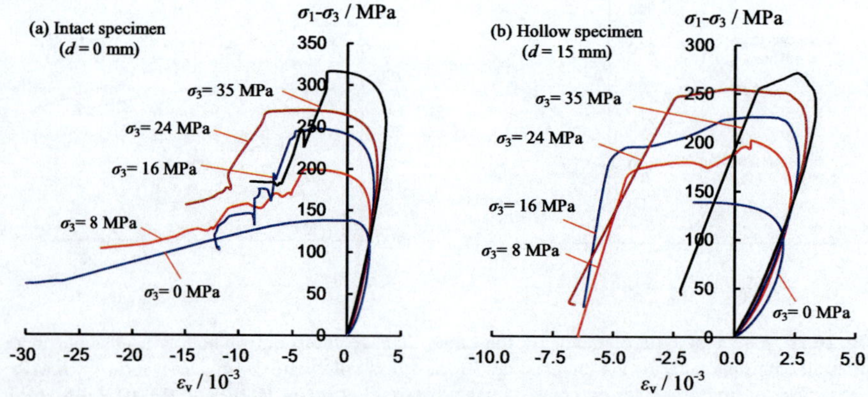

Fig. 10.21 Effect of confining pressure on axial deviatoric stress-volumetric strain curves of tested sandstone

can be seen that, over the whole range of tested confining pressures (0–35 MPa), σ_{cd} varied from 95.82 to 246.38 MPa at $d = 0$ (intact sandstone), from 95.35 to 239.64 MPa at $d = 15$ mm and from 96.64 to 219.22 MPa at $d = 26$ mm. In general, with increasing confining pressure, the crack damage threshold of the hollow sandstone with various hole diameters also increases. Figure 10.16 validated that the non-linear Hoek–Brown criterion reflects the peak strength properties of hollow sandstone better than the linear Mohr–Coulomb criterion. Therefore, the Hoek–Brown criterion was used to analyze the relationship between the crack damage threshold of hollow sandstone and the confining pressure.

By adopting the nonlinear Hoek–Brown criterion, the influence of σ_3 on the crack damage threshold of hollow sandstone under triaxial compression can be determined, as shown in Fig. 10.22. The figure indicates that the crack damage threshold of hollow sandstone exhibits distinctly nonlinear behavior with increasing σ_3, which is approximate to the influence of σ_3 on the peak strength of hollow sandstone under triaxial compression. Table 10.5 lists the nonlinear Hoek–Brown criterion parameters using the crack damage threshold of hollow sandstone.

From Fig. 10.22, it is clear that the effect of σ_3 on σ_{sd} of hollow sandstone depends on the hole diameter. For intact specimen ($d = 0$), the difference between σ_{sd} and

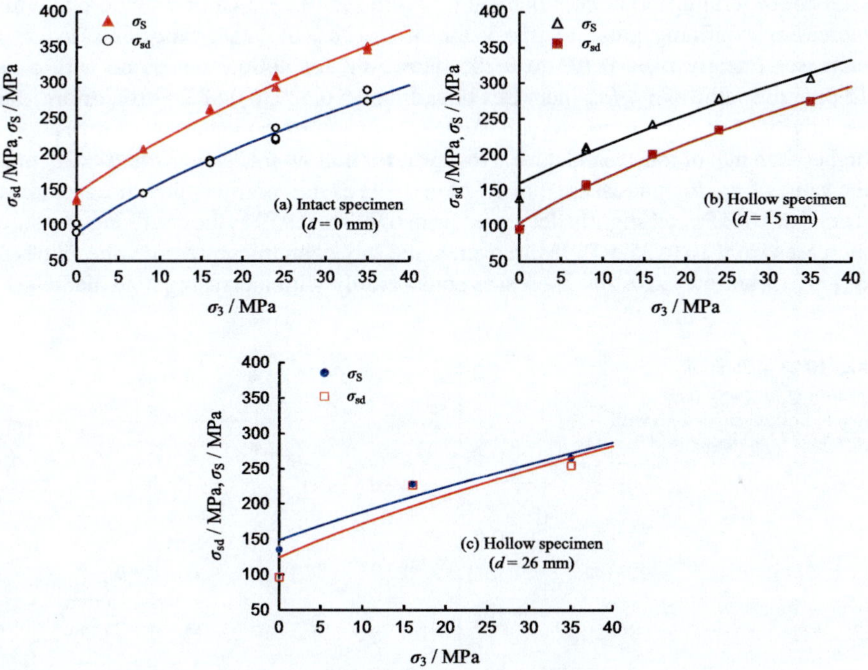

Fig. 10.22 Comparison of crack damage threshold and peak strength of hollow sandstone with various hole diameters. The symbols represent the experimental results, and the lines represent the theoretical values in accordance with the Hoek–Brown criterion

Table 10.5 The nonlinear Hoek–Brown criterion parameters by using the crack damage threshold of hollow sandstone

Different specimens	σ_c / MPa	m	s	a	R
Intact specimen ($d = 0$ mm)	87.68	16.55	1.000	0.5	0.986
Hollow specimen ($d = 15$ mm)	103.81	13.20	1.000	0.5	0.997
Hollow specimen ($d = 26$ mm)	124.69	8.647	1.000	0.5	0.885

σ_S of sandstone slightly increases with increasing σ_3. For hollow sandstone with $d = 15$ mm, the difference between σ_{sd} and σ_S of sandstone does not occur the variance with increasing σ_3. Whereas for hollow sandstone with $d = 26$ mm, the difference between the σ_{sd} and σ_S of sandstone is very small and slightly decreases with increasing σ_3. As d increases from 0 to 26 mm, the nonlinear Hoek–Brown criterion parameter m obtained using σ_{sd} decreases from 16.55 to 8.647, whereas σ_c obtained using σ_{sd} increases from 87.68 to 124.69 MPa. Furthermore, for the same hole diameter, the nonlinear Hoek–Brown criterion parameter m obtained using σ_{sd} is higher than that by using σ_S, but the nonlinear Hoek–Brown criterion parameter σ_c obtained using σ_{sd} is lower than that obtained using σ_S.

Figure 10.23 depicts the effect of confining pressure on the σ_{cd}/σ_p of hollow sandstones with different hole diameters. From Fig10.23, I can conclude that with increasing confining pressure, the value of σ_{cd}/σ_p of intact sandstone ($d = 0$) increases linearly from 0.703 to 0.785. However, for hollow sandstone with $d =$ 15 mm, the value of σ_{cd}/σ_p increases linearly from 0.689 to 0.882. Furthermore, the sensitivity of σ_{cd}/σ_p of hollow sandstone ($d = 15$ mm) to the confining pressure is higher than that of intact sandstone. However, for hollow sandstone with $d = 26$ mm, the value of σ_{cd}/σ_p increases from 0.714 to 0.993 as the confining pressure increases from 0 to 16 MPa and slightly decreases from 0.993 to 0.952 as the confining pressure increases from 16 to 35 MPa. Moreover, at the same confining pressure, the value of σ_{cd}/σ_p of hollow sandstone increases considerably with increasing hole diameter.

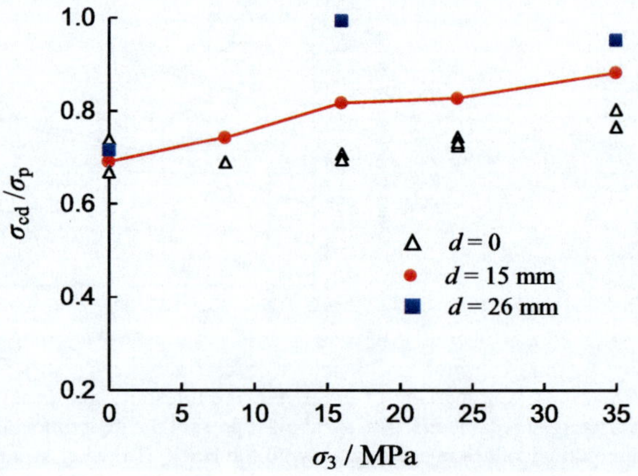

Fig. 10.23 Effect of confining pressure on σ_{cd}/σ_p of hollow sandstone with different hole diameters

10.5 Conclusions

This chapter explores the deformation, peak strength and crack damage behavior of hollow sandstone with various hole diameters using a series of conventional triaxial compression experiments. Based on the experimental results, the following conclusions can be drawn:

(1) Two types of methods confirming the elastic modulus and Poisson's ratio of rock materials are presented to evaluate the deformation characteristics of hollow sandstone under different confining pressures. Two various methods of confirming the elastic modulus and Poisson's ratio of hollow sandstone will result in substantial engineering mistakes. In shallow rock engineering, it is relatively safe to use E_{S1} and μ_1 obtained by the conventional method. However, in deep rock engineering, it is too conservative and dangerous to use E_{S1} and μ_1. Therefore, it is suggested that E_{S2} and μ_2 should be used to analyze the stability in deep rock engineering.

(2) The Young's modulus of hollow sandstone only depends on the confining pressure and is not affected by hole diameter. The elastic modulus E_{S1} of hollow sandstone is obviously higher than the secant modulus E_{50} in the range of σ_3 from 0 to 16 MPa, whereas E_{S1} approximately equals to E_{50} at confining pressures above 16 MPa, which is due to various effect of confining pressure on the more pronounced stage of microcrack closure. However, the elastic modulus E_{S2} of hollow sandstone is obviously lower than E_{S1} at lower confining pressures (approximately < 12.5 MPa), and E_{S2} approximately equals to E_{S1} at $\sigma_3 = 12.5$ MPa, whereas E_{S2} is significantly larger than E_{S1} at higher confining pressures (above approximately 12.5 MPa).

(3) The Poisson's ratio of hollow sandstone under uniaxial compression decreases with increasing hole diameter and does not depend on the confirmation method. However, in the range of σ_3 = 8~ 35 MPa, the Poisson's ratio of hollow sandstone is independent of the hole diameter at the same confining pressure, but it depends on the confirmation method. The μ_1 value of hollow sandstone under uniaxial compression obtained using the conventional method is slightly lower than that obtained in the range of σ_3 = 8 ~35 MPa. However, the μ_1 value of hollow sandstone does not depend on the confining pressure in the range of σ_3 = 8 ~35 MPa. However, as σ_3 increases from 0 to 35 MPa, the μ_2 value of hollow sandstone increases nonlinearly with the confining pressure. Furthermore, the μ_2 value is higher than the μ_1 value at the same confining pressure.

(4) The peak axial strain of hollow sandstone increases linearly with the confining pressure, and the sensitivity of the peak axial strain of hollow sandstone with larger hole diameters to σ_3 is lower than that with smaller hole diameters. However, the peak radial strain of hollow sandstone has a nonlinear relationship with increasing confining pressure. Although under uniaxial compression, the peak axial and radial strains of hollow sandstone are independent of the hole diameter. However, under confining pressure, with increasing hole diameter,

the peak axial strain of hollow sandstone decreases gradually, but the peak radial strain increases gradually due to the confining pressure restricting the radial deformation in the hollow sandstone.

(5) The hole diameter has no effect on the uniaxial compressive strength of hollow sandstone. However, under confining pressure, the triaxial compressive strength of hollow sandstone is significantly dependent on the hole diameter, which decreases linearly with increasing confining pressure. The strength parameters of hollow sandstone depend significantly on the hole diameter. As the hole diameter increases from 0 to 15 mm, the cohesion of sandstone first increases rapidly and then remains relatively constant in the range of 15–26 mm. However, the internal friction angle of sandstone decreases linearly with increasing hole diameter. Generally, the non-linear Hoek–Brown criterion reflects the peak strength properties of hollow sandstone better than the linear Mohr–Coulomb criterion does.

(6) The crack damage threshold of hollow sandstone depends on not only the confining pressure but also the hole diameter. Under uniaxial compression, the ratio of the crack damage threshold (σ_{cd}) to the UCS of hollow sandstone is approximately 0.689, and the σ_{cd}/UCS value does not depend on hole diameter. However, at $\sigma_3 = 16$ MPa, the crack damage threshold increases nonlinearly with increasing hole diameter. At $\sigma_3 = 35$ MPa, the crack damage threshold decreases nonlinearly with increasing hole diameter. Furthermore, the sensitivity of crack damage threshold of hollow sandstone to hole diameter is lower than that of peak strength to hole diameter. The crack damage threshold of hollow sandstone increases nonlinearly with increasing confining pressure, which can be better described by the nonlinear Hoek–Brown criterion.

References

Alsayed MI (2002) Utilising the Hoek triaxial cell for multiaxial testing of hollow rock cylinders. Int J Rock Mech Min Sci 39:355–366

Fairhurst CE, Hudson JA (1999) Draft ISRM suggested method for the complete stress-strain curve for the intact rock in uniaxial compression. Int J Rock Mech Min Sci 36(3):279–289

Haimson B, Kovacich J (2003) Borehole instability in high-porosity Berea sandstone and factors affecting dimensions and shape of fracture-like breakouts. J Eng Geol 69:219–231

Hawkes I, Mellor M (1970) Uniaxial testing in rock mechanics laboratories. Eng Geol 4:177–285

Heap MJ, Baud P, Meredith PG, Bell AF, Main IG (2009) Time-dependent brittle creep in Darley Dale sandstone. J Geophys Res. https://doi.org/10.1029/2008JB006212

Hoek E, Brown ET (1980) Underground excavation in rock. Institution of Mining and Metallurgy, London

Hoek E, Brown ET (1997) Practical estimates of rock mass strength. Int J Rock Mech Min Sci 34(8):1165–1186

Labiouse V, Sauthier C, You S (2014) Hollow cylinder simulation experiments of galleries in Boom clay formation. Rock Mech Rock Eng 47:43–55

Lee DH, Juang CH, Chen JW, Lin HM, Shieh WH (1999) Stress paths and mechanical behavior of a sandstone in hollow cylinder tests. Int J Rock Mech Min Sci 36:857–870

Mogi K (1966) Some precise measurements of fracture strength of rocks under uniform compressive stress. Rock Mech Eng Geol 4:41–55

Monfared M, Sulem J, Delage P, Mohajerani M (2011) A laboratory investigation on thermal properties of the Opalinus claystone. Rock Mech Rock Eng 44(6):735–747

Yang SQ, Jing HW, Wang SY (2012) Experimental investigation on the strength, deformability, failure behavior and acoustic emission locations of red sandstone under triaxial compression. Rock Mech Rock Eng 45:583–606

Yang SQ, Ranjith PG, Gui YL (2015) Experimental study of mechanical behavior and X-ray micro CT observations of sandstone under conventional triaxial compression. Geotech Test J 38(2):179–197

Yang SQ (2016) Experimental study on deformation, peak strength and crack damage behavior of hollow sandstone under conventional triaxial compression. Eng Geol 213:11–24

Chapter 11
Fracture Evolution Mechanism of Hollow Sandstone Under Conventional Triaxial Compression by X-ray Micro-CT Observations and Three-Dimensional Numerical Simulations

Rock is a natural material and usually consists of unequal flaws with different shapes. It is known that the pre-existing flaws significantly affect the mechanical properties of rock (Yang et al. 2008, 2018; Yang and Jing 2011; Tang et al. 2015; Hamidi et al. 2018). Especially in deep rock engineering, rock material is often in a triaxial stress state. A high stress concentration usually leads to the formation of damage zones around flaws. The evaluation of the strength behavior and fracture mechanism of rock material containing flaws is of great significance in all types of rock engineering. However, the internal cracking behavior of rock material sometimes cannot be reliably predicted by exclusively using experimental studies, especially when the rock is in a triaxial loading state (Yu et al. 2018). Two main methods have been developed to visualize the internal fracture process of rock material, that is, the numerical simulation method, and the X-ray micro-computed tomography (CT) scanning and spatial acoustic emission (AE) method (Wang et al. 2019; Skarzynski and Suchorzewski 2018; Lee and Rathnaweera 2016; Yang 2018).

The development of computer technology provides an effective way to properly capture the overall mechanical behavior of rock masses based on meso- or micro mechanics (Shen and Stephansson 1993; Feng et al. 2006; Colombo and Massin 2011; Zhou et al. 2015; Yang et al. 2015a, 2018; Camas et al. 2017). In recent years, the realistic failure process analysis (RFPA) method has been widely used in various rock mechanics behavior simulation studies because it is simple, efficient, and cost-effective, for example, to study: (1) the dynamic disturbance: by using the RFPA^{2D} method, the failure of rock subjected to a complex dynamic loading path was simulated (Niu et al. 2017); (2) the permeability evolution: a series of triaxial compression simulations and permeability measurements was carried out under different confining and pore pressures as well as under different coupling conditions to investigate the mechanical properties and permeability evolution depending on the damage of sandstone (Chen et al. 2017); (3) the creep deformation: Xu et al (2018) proposed a RFPA^{2D} numerical hydro-mechanical model for brittle creep to describe the time-dependent deformation of heterogeneous brittle rock under constant confining and pore pressures; (4) the hydraulic fracture: the effects of the rock brittleness on the

S.-Q. Yang, *Mechanical Behavior and Damage Fracture Mechanism of Deep Rocks*,
https://doi.org/10.1007/978-981-16-7739-7_11

failure mode and stress–strain characteristics of shale specimens were numerically investigated with the RFPA^{2D} method. It was found that hydraulic fractures preferably propagate in brittle minerals (Li et al. 2018); and (5) the cracking behavior (the focus of this chapter): Yang et al. (2015) carried out RFPA^{2D} numerical simulations on a rock specimen containing combined flaws (two square holes and one pre-existing fissure) at seven different fissure angles. Yin et al. (2018) investigated the effect of the pre-existing flaw geometry on mechanical behavior and crack coalescence modes of sandstone specimens containing combined flaws at different fissure angles, ligament and fissure lengths under uniaxial compression using the RFPA^{2D} method.

However, most of the above-mentioned numerical simulations were limited to two-dimensional (2D) conditions. To truly visualize the internal cracking behavior of rock materials, several applications of three-dimensional (3D) simulation models established based on the PFC^{3D} and RFPA^{3D} methods were established. Sarfarazi et al. (2018) evaluated the crack initiation, propagation, and coalescence using a Brazilian model disc containing a single hole or multiple cylindrical holes with PFC^{3D} method. Huang et al. (2019) carried out a series of PFC^{3D} simulations on numerical specimens with two open flaws to investigate the crack initiation, propagation, and coalescence processes under conventional triaxial compression. Wong and Lin (2015) investigated the crack coalescence mechanisms occurring in a heterogeneous solid containing multiple holes under uniaxial compression using the RFPA^{3D} code. Li et al. (2015) studied the failure mechanisms and various fracture morphologies of laboratory-scale rectangular prism rock specimens under uniaxial compression with the RFPA^{3D} method.

As mentioned above, the X-ray micro-CT scanning system which is usually used to observe the micro- or meso-structure of specimens has been gradually in rock mechanics studies with respect to the following two aspects: (1) for the evaluation of complex crack initiation and propagation behaviour, and (2) to establish improved numerical simulation models with structural planes, aggregate voids, and flaw reconstruction. With respect to aspect (1), Huang et al. (2015) investigated the damage evolution and fracture processes of both the surface and interior of the specimen using 3D meso-scale finite element modes of concrete based on in situ X-ray CT images. Using an X-ray micro-CT scanning system, Yang et al. (2015b) conducted a series of uniaxial and triaxial compression experiments to explore the mechanical behavior and internal damage mechanism of sandstone material. With respect to aspect (2), Ren et al. (2015) established 2D meso-scale finite element concrete models with realistic aggregates, cement paste and voids using microscale X-ray CT. By applying the vectorized transformation and volume rendering method, Yu et al. (2018) reconstructed the RFPA^{3D} three-phase structure model for concrete based on processed CT images to simulate the fracture process of concrete under uniaxial compression.

In practical rock engineering, natural or artificial circular holes are ubiquitous in rock masses that are usually in a complex stress state. The high stress concentration causes serious damage around the hole wall, which leads to unpredictable failure behavior. Therefore, it is important to understand the mechanical behavior of hollow rock material with a complex loading path. Lee et al. (1999) investigated the effect of the stress path on hollow cylindrical sandstone under triaxial compression. Alsayed (2002) carried out a series of uniaxial, biaxial, triaxial, and polyaxial compression and indirect tension experiments on Springwell sandstone specimens using a self-developed multiaxial hollow test cell. In our previous experiments, the deformation, peak strength, and crack damage behavior of hollow sandstone specimens under different confining pressures were systemically investigated (Yang 2016). In addition, the internal damage behavior of hollow cylinder sandstone specimens under triaxial compression was evaluated using an X-ray micro-CT scanning system (Yang 2018). However, X-ray micro CT observations and 3D numerical simulations have rarely been used to explore the internal damage failure mechanism of hollow sandstone under conventional triaxial compression.

Therefore, the main aims of this chapter were to conduct a series of X-ray micro-CT observations and 3D numerical simulations on hollow sandstone specimens with various borehole diameters, compare the experimental and numerical results for hollow sandstone specimens under different confining pressures, analyze the effects of borehole diameter and confining pressure on the peak strength and failure behavior of hollow sandstone, and explore the influence of the confining pressure on the fracture evolution mechanism of hollow sandstone with various borehole diameters (Yang et al. 2020).

11.1 Hollow Sandstone and X-ray Micro-CT Scanning Procedure

11.1.1 Hollow Sandstone Specimen and Mechanical Testing Procedure

The material used throughout this research was sandstone collected from Rizhao city in the Shandong Province, China. The sandstone material has a connected porosity of 6.88% and a bulk density of 2410 kg/m^3. The detailed composition of this rock is as follows (Yang 2016): 10.1% wt.% quartz, 11.7% wt.% K-feldspar, 41.6% wt.% plagioclase, 10.7% wt.% calcite, 13.1% wt.% dolomite, 2.4% wt.% hematite and 10.4% wt.% clay mineral. In this experiment, all intact specimens were cored from the same block of material (core diameter of 49.8 mm and length of ~100 mm). The hollow specimens have borehole diameters (*d*) of 15 and 26 mm, as shown in Fig. 10.2 (Chap. 10). All the hollow sandstone specimens are cored according to ASTM (American Society for Testing and Materials) standard D2664 (ASTM 1995). All experiments were performed on dry specimens at room temperature. The average *P*- and *S*-wave velocities of the tested sandstone are 3813 and 2290 m/s, respectively. The measuring method of the axial and radial deformation is the same as that used in a previous study (Yang 2016).

Conventional triaxial experiments on the hollow sandstone specimens were carried out under different confining pressures of 8, 16, 24, and 35 MPa. First, the confining pressure was increased to the desired value at a constant rate of 0.2 MPa/s to ensure that the specimen is under a uniform hydrostatic stress. Deviatoric stress was then applied to the specimen at a constant axial displacement rate of 0.08 mm/min (~1.33 × 10^{-5}/s) until failed, in accordance with the method suggested by the ISRM (Fairhurst and Hudson 1999).

11.1.2 X-ray Micro-CT Scanning Procedure

In the present research, X-ray CT scanning of the sandstone specimen was carried out using a Nanotom 160 high-resolution micro-CT system at a spatial resolution of 30 μm (Yang 2018). The X-ray beam penetrating the specimen was measured with an array of detectors. The X-rays were produced by electrons striking a Mo-W alloy target in an X-ray tube. The electron current was 90 μA, the accelerating voltage was 130 kV, and the exposure time of a single projection was 4 s. The X-ray attenuation degree depends on the density and atomic number of the specimen materials. Material with higher density and higher atomic number generally leads to higher X-ray attenuation. The X-ray projection data were obtained from various directions by 360° rotation of the X-ray source, as shown in Fig. 3.13 (Chap. 3). We collected 2D images at intervals of 0.18°, and thus 2000 slice images were obtained for one specimen. A 2D image representing the linear distribution of the X-ray attenuation was reconstructed using Fourier transformation of the projection data. A 3-D data-set of the specimen was obtained by stacking consecutive 2D images. A cylindrical specimen with a diameter of 50 mm and length of 100 mm was used for the X-ray CT scanning. It should be noted that the capture of the CT images is the same as that presented in Sect. 3.3.1 (Chap. 3). The slice images can be used to explore the internal damage of the sandstone material. Here, the parameters x and z respectively represent the depth (the distance from the front to the back) and the height (the distance from the bottom to the top), as shown in Fig. 3.13d and e, respectively.

11.2 Three-Dimensional Numerical Simulation Methodology

It is well known that rock exhibits discontinuous and inhomogeneous behavior due to natural pre-existing defects and microstructures. Taking this behaviour into consideration is important to establish convincing numerical models. Based on the concepts of physical discontinuity and mathematical continuity, the complex behavior of a discontinuous medium can be simplified in RFPA as behavior of a continuous medium. Hence, the crack initiation, propagation, and coalescence processes of rock can be reasonably and efficiently simulated using RFPA. In the following section, the fundamental principles of RFPA^{3D} will be briefly introduced.

11.2.1 Representative Volume Element Based on the RFPA^{3D} Method

From micro-scale to macro-scale, the mechanical properties of the rock medium are notably scale-dependent (Li and Tang 2015). Exploring the microscopic-behavior accurately reveals the inhomogeneous characteristics of rock. However, it is extremely difficult to implement mathematical descriptions and numerical simulations on inhomogeneous rock with pre-existing micro-defects. The results of a previous study indicated the existence of an intermediate scale (meso-scale), representing a bridge between micro- and macro-scale (Xu et al. 2012). At the mesoscopic level, the rock elements can be considered to be homogenous and exhibit simple mechanical properties. Therefore, macroscopic rock can be discretized into a series of uniform, regularly arranged cubic mesoscopic elements called representative volume elements (RVEs) to statistically analyze of the macroscopic heterogeneity. In this case, the mechanical behavior of quasi-brittle materials considering the effect of inhomogeneity can be described using the Weibull distribution (Basu et al. 2009; Tang et al. 2015):

$$\phi(\sigma)=\frac{m}{\sigma_0}\cdot\left(\frac{\sigma}{\sigma_0}\right)^{m-1}\cdot e^{-\left(\frac{\sigma}{\sigma_0}\right)^m} \tag{11.1}$$

where σ is the elastic modulus or strength property of the RVE; σ_0 is the mean value of σ; and m is the inhomogeneity index of the material, where a greater value indicates a higher degree of homogeneity (Xu et al. 2012, 2014).

11.2.2 Damage Evolution of the RVE

Based on Lemaitre's hypothesis of strain equivalence (Lemaitre 1971), the damage progress can be considered as the degradation of the elastic modulus. Thus, the damage variable $D(0 \leq D \leq 1)$ is defined as:

$$D= 1-\frac{E}{E_0} \tag{11.2}$$

where E and E_0 are the elastic moduli of the damaged and intact RVE, respectively. When $D= 0$, the RVE is intact, whereas the RVE fails when D is close to 1.

The equation can also be rewritten in a slightly different form using the stress–strain relationship as:

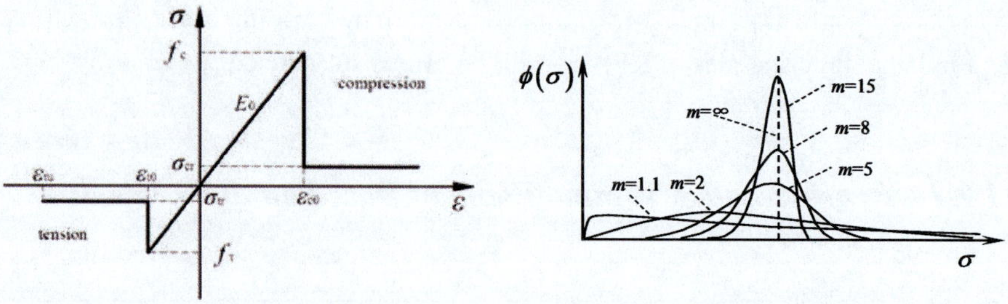

Fig. 11.1 **a** Damage evolution process under axial compression and tension (Tang et al. 2015; Xu et al. 2012); **b** The influence of inhomogeneity index m on the distribution function (Yang et al. 2015c)

$$\sigma = E_0(1 - D)\varepsilon \tag{11.3}$$

Then the damage evolution process for RVE under compression and tension is shown in Fig. 11.1a. Accordingly, the damage models of the RVE within macroscopic material can be classified into the following two categories:

1. When the stress of the RVE reaches the tensile strength, tension-induced damage occurs, which can be expressed by using the maximum tensile stress criterion (Eq. 4). The stress state then enters the residual strength with a brittle stress drop.

$$F_t(\sigma) = -\sigma_3 - f_t = 0 \tag{11.4}$$

where f_t is the tensile strength. Accordingly, the damage variable D can be presented as follows (Tang et al. 2015):

$$D = \begin{cases} 0, \varepsilon < \varepsilon_{t0} \\ 1 - \dfrac{\lambda \varepsilon_{t0}}{\varepsilon}, \ \varepsilon_{t0} \le \varepsilon \le \varepsilon_{tu} \\ 1, \ \varepsilon \ge \varepsilon_{tu} \end{cases} \tag{11.5}$$

where λ is the residual strength coefficient, and ε_{t0} and ε_{tu} are the maximum elastic strain and ultimate tensile strain of the RVE, respectively.

As a general case, if the RVE is in a multi-axial stress state, strain ε in Eq. 11.5 must be modified by using the equivalent strain $\bar{\varepsilon}$ defined as follows (Mazars and Pijaudier 1989; Cerrolaza and Garcia 1997)

$$\bar{\varepsilon} = -\sqrt{\langle -\varepsilon_1 \rangle^2 + \langle -\varepsilon_2 \rangle^2 + \langle -\varepsilon_3 \rangle^2} \tag{11.6}$$

where ε_1, ε_2 and ε_3 are three principal strains. The function $\langle\rangle$ is defined as:

$$\langle x \rangle = \begin{cases} x \; x \leq 0 \\ 0 \; x > 0 \end{cases} \tag{11.7}$$

2. When the stress of the RVE reaches the shear strength, shear-induced damage occurs, which can be described using the Mohr–Coulomb criterion (Eq. 11.8). The stress state then enters the residual strength with a brittle stress drop.

$$F_s(\sigma) = \sigma_1 - \sigma_3 \frac{1 + \sin\phi}{1 - \sin\phi} - f_c = 0 \tag{11.8}$$

where f_c represents the uniaxial compressive strength (UCS) and ϕ is the internal friction angle. Correspondingly, the damage variable D can be defined as follows (Tang et al. 2015):

$$D = \begin{cases} 0, \; \varepsilon < \varepsilon_{c0} \\ 1 - \dfrac{\sigma_{cr}}{\varepsilon E_0}, \; \varepsilon \geq \varepsilon_{c0} \end{cases} \tag{11.9}$$

where σ_{cr} represents the residual uniaxial strength. The compressive threshold strain ε_{c0} can be calculated using $\varepsilon_{c0} = f_c / E_0$.

As a general case, when the RVE is subjected to multi-axial stress, the compressive threshold strain ε_{c0} affected by intermediate principal stresses can be modified as:

$$\varepsilon_{c0} = \frac{1}{E_0}\left[\sigma_{c0} + \frac{1 + \sin\phi}{1 - \sin\phi}\sigma_3 - v(\sigma_1 + \sigma_2)\right] \tag{11.10}$$

where v is the Poisson's ratio.

11.2.3 Three-Dimensional Numerical Model

Based on previous studies (Yang 2016, 2018), numerical simulations on cylindrical sandstone specimens (diameter × height = 50 mm × 100 mm; the same size as the experimental sandstone specimen) were designed in this study using the RFPA^{3D} method. Preformed internal boreholes with different diameters of 0, 15 and 26 mm were created by deleting the related RVEs (Li and Tang 2015). The geometry of the hollow specimens with various borehole diameters generated with RFPA^{3D} are represented in Fig. 11.2. The numerical model for the intact sandstone consists of 857,120 hexahedral elements. The numerical models of hollow sandstone with diameters of 15 and 26 mm have 849,920 and 700,800 hexahedral elements. The length of side of each element in the numerical model is 0.03 mm.

It should be noted that Fig. 11.1b shows the influence of the inhomogeneity index m on the distribution function, which indicates that a smaller m means a more heterogeneous rock material and vice versa. However for the same m, the Weibull's

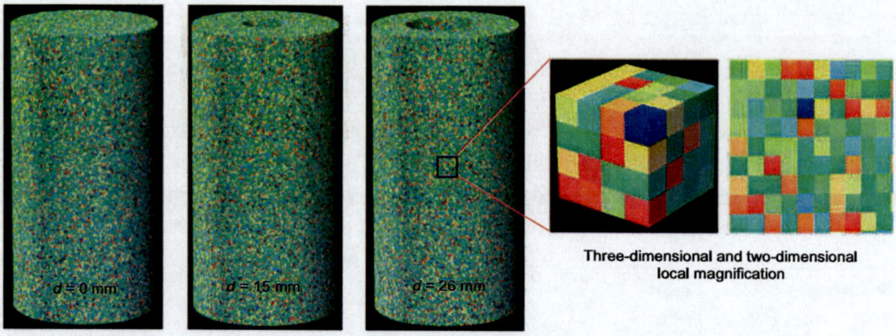

Fig. 11.2 Numerical simulation model designed on RFPA^{3D} method ($m = 3.0$)

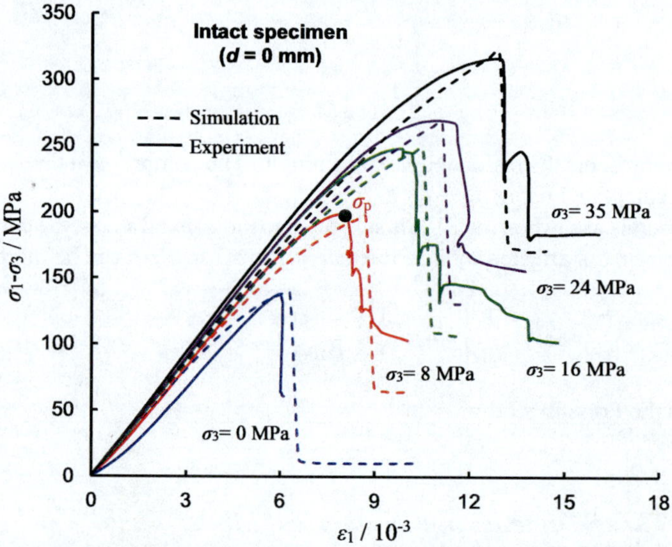

Fig. 11.3 Comparison between experimental and numerical stress–strain curves of intact sandstone specimen under triaxial compression. The experimental curves were obtained from the reference (Yang 2016)

randomness indicated that the numerical simulated results of different rock specimens are very approximate, but not identical (Yang et al. 2015c). Furthermore, the results of a previous study showed that inhomogeneity index *m* is the main factor affecting the crack propagation pattern and macroscopic mechanical behavior (Li and Tang 2015). For models with higher homogeneities, tensile-dominant cracks and brittle characteristics are observed. In contrast, tensile-shear cracks and non-linear stress–strain curve can be observed. Li and Tang (2015) suggested that the inhomogeneity index should be in the range of 1.2–5.0. Therefore, in this study, inhomogeneity index *m* is designed as 3.0, as listed in Table 11.1. In addition, the study also showed

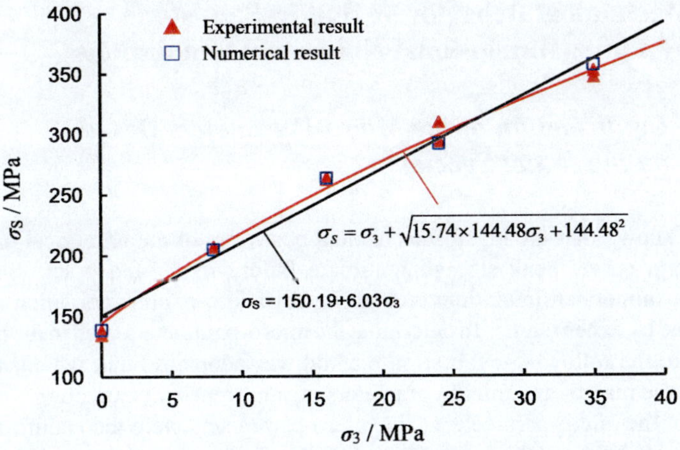

Fig. 11.4 Comparison on peak strength of intact sandstone specimen (d = 0 mm) obtained from triaxial compression experiment and numerical simulation. The linear Mohr–Coulomb and nonlinear Hoek–Brown criterion are also obtained according to the experimental results

Table 11.1 The micro-parameters of RFPA^{3D} simulation for sandstone specimens in this research

Micro-parameters	Value
Inhomogeneity index, m	3
Average value of elastic modulus, E_0, GPa	32
Average value of compressive strength, f_c, MPa	390
Internal friction angle, Φ°	30
Ratio of compressive to tensile strength, C/T	10
Poisson's ratio, μ	0.25
Ultimate tensile strain coefficient, ξ	1.5
Ultimate compressive strain coefficient, δ	200
Residual strength coefficient, λ	0.1
Loading rate in the present study, mm/step	3.0×10^{-2}

that the loading step length within the quasi-static range has no significant effect on the macroscopic strength and crack propagation. Therefore, we used a displacement loading method with a rate of 0.03 mm/step for the numerical simulation, representing a quasi-static loading rate.

11.3 Mechanical Behavior of Hollow Sandstone by Three-Dimensional Numerical Simulations

11.3.1 *Confirmation of the Micro-Parameters Based on the Intact Specimen*

It is well known that the macro-mechanical behaviors of the numerical model (i.e. stress–strain curve, peak strength, ultimate failure mode and crack coalescence process) in numerical simulations is controlled by micro-parameters which can hardly be obtained by experiments. To determine the micro-parameters used in the numerical simulation, the following calibration method was adopted. First, we quantitatively corrected the macro- and micro-parameters of sandstone by conducting a sensitivity analysis of the micro-parameters. The micro-parameters were then confirmed using trial and error. We used the macroscopic behaviour of intact sandstone obtained from the experiment to calibrate the micro-parameters. The macroscopic results obtained by numerical simulation after each trial were used to verify the micro-parameters. This process was repeated until the numerical results of the intact sandstone matched the experimental results. In accordance with the above-mentioned method, we determined the micro-parameters for the sandstone specimens in this study, which are listed in Table 11.1.

Figure 11.3 shows the comparison of axial deviatoric stress-axial strain curves of the intact sandstone specimens under conventional triaxial compression obtained from the experiment and 3D numerical simulation. The parameter ε_1 represents the axial strain; σ_1 represents the maximum principal stress, and σ_1–σ_3 represents the axial deviatoric stress. Figure 11.3 shows that the numerically simulated axial deviatoric stress-axial strain curves (dotted line) are in a good agreement with the experimental curves (solid lines) of the intact sandstone specimens under conventional triaxial compression (Yang 2016). However, the 3D numerical simulation using RFPA^{3D} cannot better reflect the downward concave and initial nonlinear deformation (due to the closure of several primary fissures, pores and voids in the tested sandstone at low stress levels) observed in the results of the uniaxial compression experiments.

Figure 11.4 compares the peak strengths of intact sandstone specimens under different confining pressures obtained by experiments and simulations. The figure shows that the numerical peak strengths of intact sandstone under different confining pressures are quantitatively equal to those obtained by the experiment. The maximum difference in the peak strengths between the experiment and simulation can be observed at the confining pressure of 35 MPa. At $\sigma_3 = 35$ MPa, the peak strength of the intact sandstone obtained in the experiment is 348.80 MPa, whereas the numerical peak strength is 356.67 MPa, representing a difference of 7.87 MPa. It can be seen from Fig. 11.4 that the peak strength of intact sandstone increases nonlinearly with the confining pressure, which can be expressed by the nonlinear Hoek–Brown criterion (Hoek and Brown 1980, 1997), i.e., Eq. (11.11).

$$\sigma_S = \sigma_3 + (m\sigma_c\sigma_3 + s\sigma_c^2)^{0.5} \quad (11.11)$$

where σ_S is the maximum supporting capacity, σ_c is the UCS of the intact rock material and m and s are material constants of a specific rock. When the parameter m is larger, the rock is stronger. The parameter s reflects the fracture extent, which ranges from zero to one. When the parameter s is closer to one, the rock is more intact.

The linear Mohr–Coulomb criterion (Mohr 1914; Yang et al. 2011) is also displayed in Fig. 11.4 for comparison with the nonlinear Hoek–Brown criterion, i.e., Eq. (11.12).

$$\sigma_S = \sigma_c + q\sigma_3 = \frac{2C\cos\phi + \sigma_3(1 + \sin\phi)}{1 - \sin\phi} \quad (11.12)$$

where σ_c and q are related to the cohesion C and the internal friction angle Φ of the rock. Note that σ_S equals the sum of σ_p and σ_3. Here, σ_p means the peak strength obtained by the deviatoric stress-axial strain curve, as shown in Fig. 11.3.

Note that the linear Mohr–Coulomb and nonlinear Hoek–Brown criteria in Fig. 11.4 were obtained from experimental results (Yang 2016). Based on Fig. 11.4, we can conclude that the nonlinear Hoek–Brown criterion better reflects the peak strength of intact sandstone material than the linear Mohr–Coulomb criterion.

Tables 11.2 and 11.3 list the strength parameters of intact sandstone based on the linear Mohr–Coulomb and nonlinear Hoek–Brown Criteria, respectively. In Tables 11.2 and 11.3, the "R" is the correlation coefficient of determination of the regression. It should be noted that σ_c and m are back-calculated by assuming $s =$ 1.000 in this research. In accordance with the linear Mohr–Coulomb criterion, the C and Φ values of intact sandstone material based on the simulation are 30.56 MPa

Table 11.2 Strength parameters of intact sandstone by the linear Mohr–Coulomb criterion in this research

Intact specimen ($d = 0$ mm)	σ_0/MPa	q	C/MPa	ϕ/(°)	R
Experimental result*	150.19	6.03	30.58	45.7	0.985
Numerical result	150.48	6.06	30.56 (−0.07%)	45.8 (+0.22%)	0.991

Note The experimental results in this table are obtained by Yang (2016). The value in () is obtained by the ratio of the difference between the numerical result and experimental result to the experimental result

Table 11.3 Strength parameters of intact sandstone by the nonlinear Hoek–Brown criterion in this research

Intact specimen ($d = 0$ mm)	σ_c/MPa	m	s	R
Experimental result*	144.48	15.74	1.000	0.995
Numerical result	140.94	16.66	1.000	0.995

Note The experimental results in this table are obtained by Yang (2016)

and 45.8°, respectively, and the difference is only 0.07 and 0.22% compared with the experimental results. However, in accordance with the nonlinear Hoek–Brown Coulomb criterion, the σ_c and m values of the sandstone material based on the simulation are 140.94 MPa and 16.66, respectively, and the σ_c and m values of the sandstone material based on the experiment are 144.48 MPa and 15.74, respectively. Therefore, all strength parameters of intact sandstone are very similar, which indicates a good agreement between the experiment and 3D numerical simulation.

11.3.2 Comparison Between the Numerical Simulated and Experimental Results

Figure 11.5 further compares the axial deviatoric stress-axial strain curves of hollow sandstone specimens with various borehole diameters (d = 15 and 26 mm) under conventional triaxial compression obtained by the experiments and 3D numerical simulations. The figure shows that the numerically simulated axial deviatoric stress–strain curves of the hollow sandstone specimens under different confining pressures are in good agreement with the experimental curves (Yang 2016).

Based on the results of the axial deviatoric stress-axial strain curves of the hollow sandstone specimens with various borehole diameters shown in Figs. 11.5 and 11.6 shows the comparison of the peak strength of hollow sandstone specimen obtained from triaxial compression experiments and numerical simulations. The linear Mohr–Coulomb and nonlinear Hoek–Brown criteria were also obtained according to the experimental results (Yang 2016). Figure 11.6 shows that the numerical peak strengths of hollow sandstone under different confining pressures are quantitatively equal to those obtained in the experiment. The maximum difference in the peak strengths between the experiments and simulations was observed at a confining pressure of 35 MPa for d = 15 mm. At σ_3 = 35 MPa, the peak strength of the hollow sandstone with d = 15 mm based on the experiment is 306.76 MPa, whereas the numerical peak strength is 315.16 MPa, representing a difference of 8.4 MPa. Based on Fig. 11.6, we can conclude that the nonlinear Hoek–Brown criterion better reflects the peak strength of hollow sandstone material than the linear Mohr–Coulomb criterion.

Tables 11.4 and 11.5 list the strength parameters of hollow sandstone based on the linear Mohr–Coulomb and the nonlinear Hoek–Brown criteria obtained by the 3D numerical simulations and experiments, respectively. Note that σ_c and m were also back-calculated by assuming s = 1.000. In accordance with the linear Mohr–Coulomb criterion, the C and Φ values of the hollow sandstone material with d = 15 mm based on the simulation are 35.57 MPa and 40.9°, respectively, and the C and Φ values based on the experiment are 37.97 MPa and 39.7°, respectively. Whereas, the C and Φ values of the hollow sandstone material with d = 26 mm based on the simulation are 42.42 MPa and 33.2°, respectively, and the C and Φ values based on the experiment are 38.52 MPa and 34.8°, respectively.

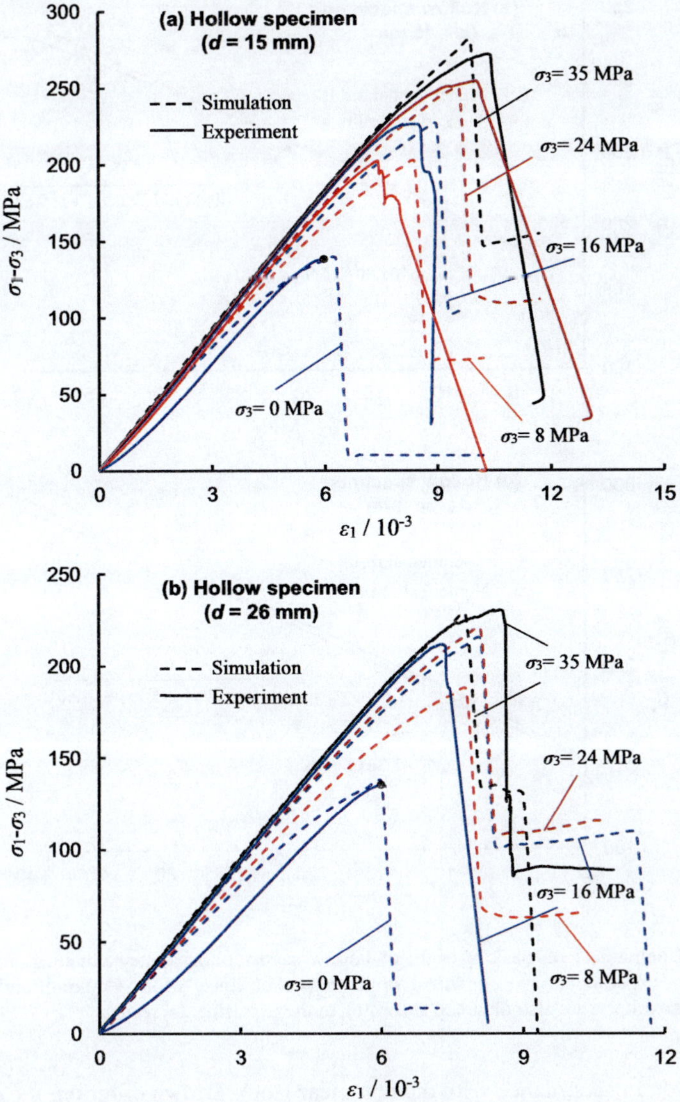

Fig. 11.5 Comparison between experimental and numerical stress–strain curves of hollow sandstone specimen with various borehole diameters under triaxial compression. The experimental curves were obtained from the reference (Yang 2016)

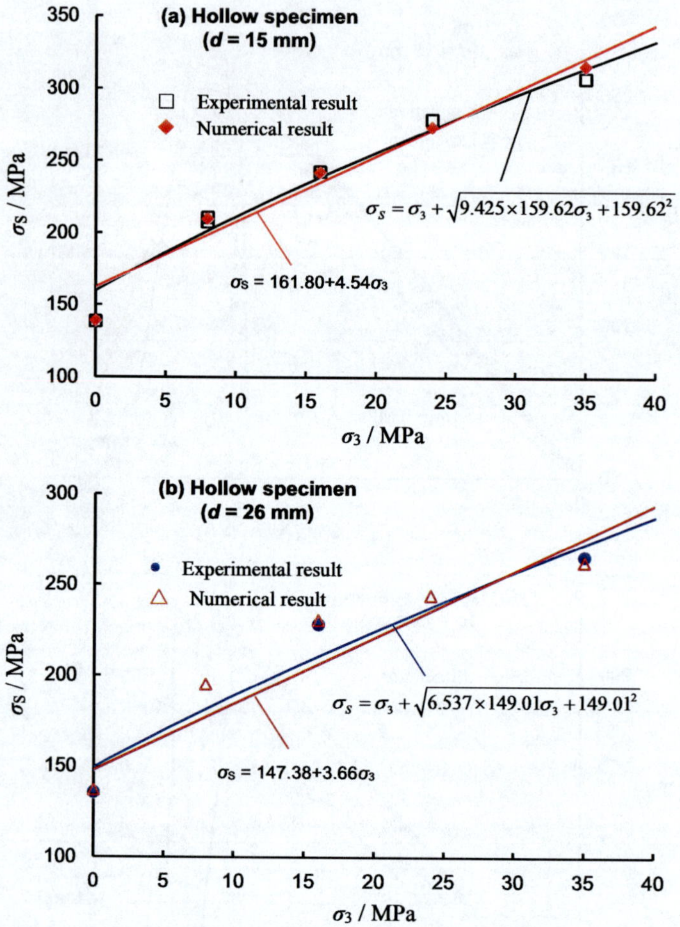

Fig. 11.6 Comparison on peak strength of hollow sandstone specimen obtained from triaxial compression experiments and numerical simulations. The linear Mohr–Coulomb and nonlinear Hoek–Brown criteria are also obtained according to the experimental results

However, in accordance with the nonlinear Hoek–Brown criterion, the σ_c and m values of hollow sandstone material with $d = 15$ mm based on the simulation are 152.87 MPa and 10.59, respectively, and the σ_c and m values based on the experiment are 159.62 MPa and 9.425, respectively. Whereas, the σ_c and m values of hollow sandstone material with $d = 26$ mm based on the simulation are 152.87 MPa and 5.702, respectively, and the σ_c and m values based on the experiment are 149.01 MPa and 6.537, respectively.

Generally, these experimental and numerical strength parameters of hollow sandstone with different borehole diameters obtained by the linear Mohr–Coulomb and

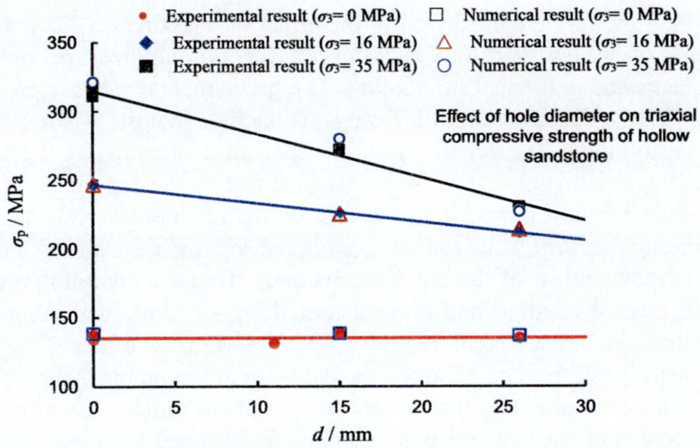

Fig. 11.7 Comparison on the effect of hole diameter on the peak strength of hollow sandstone under different confining pressures obtained by experiment and numerical simulation. The experimental values were obtained from the reference (Yang 2016)

Table 11.4 Strength parameters of hollow sandstone by the linear Mohr–Coulomb criterion in this research

Hollow sandstone	σ_0/MPa	q	C/MPa	ϕ/(°)	R
Experimental result ($d = 15$ mm)*	161.80	4.54	37.97	39.7	0.970
Numerical result ($d = 15$ mm)	155.88	4.80	35.57 (−6.32%)	40.9 (+3.02%)	0.980
Experimental result ($d = 26$ mm)*	147.38	3.66	38.52	34.8	0.959
Numerical result ($d = 26$ mm)	156.88	3.42	42.42 (+10.1%)	33.2 (−4.60%)	0.942

Note The experimental results in this table are obtained by Yang (2016). The value in () is obtained by the ratio of the difference between the numerical result and experimental result to the experimental result

Table 11.5 Strength parameters of hollow sandstone by the nonlinear Hoek–Brown criterion in this research

Hollow sandstone	σ_c/MPa	m	s	a	R
Experimental result ($d = 15$ mm)*	159.62	9.425	1.000	0.5	0.975
Numerical result ($d = 15$ mm)	152.87	10.59	1.000	0.5	0.989
Experimental result ($d = 26$ mm)*	149.01	6.537	1.000	0.5	0.938
Numerical result ($d = 26$ mm)	158.07	5.702	1.000	0.5	0.915

Note The experimental results in this table are obtained by Yang (2016)

nonlinear Hoek–Brown criteria are all very similar, which indicates a good agreement between the experiments and 3D numerical simulations.

Figure 11.7 shows the comparison of the effect of the borehole diameter on the peak strength of hollow sandstone under different confining pressures obtained by the experiments and numerical simulations. The experimental values were obtained from Yang (2016). Based on Fig. 11.7, we can conclude that the effect of the borehole diameter on the peak strength of hollow sandstone obtained by the numerical simulation is approximately the same as that obtained by the experiment. However, the effect of the borehole diameter on the peak strength of hollow sandstone depends on the confining pressure. Under uniaxial compression, the peak strength of hollow sandstone is independent of the borehole diameter. Under a confining pressure of 16 MPa, the peak strength of hollow sandstone decreases with increasing borehole diameter. However, under a confining pressure of 35 MPa, the effect of the borehole diameters on the peak strength of hollow sandstone is most notable in the range of the tested confining pressure. The above-mentioned analysis indicates that the effect of borehole diameter on the peak strength of hollow sandstone becomes more apparent with increasing confining pressure.

11.4 Fracture Evolution Mechanism of Hollow Sandstone Under Different Confining Pressures

Figure 11.8 compares the ultimate failure mode of intact sandstone specimens obtained by the numerical simulations and experiments. Based on Fig. 11.8, the numerical failure modes of the intact sandstone specimens under different confining pressures are very similar to the experimental failure modes. However, there are also some differences, that is, the experimental failure modes involve more lateral tensile cracks. The confining pressure significantly affects the ultimate failure modes of the intact sandstone specimens. Under uniaxial compression, the intact sandstone specimens mainly demonstrate axial splitting fracture as shown by the experimental failure modes. In contrast, under the action of confining pressure, the intact sandstone specimens mainly demonstrate single shear fracture; however locally, several lateral tensile cracks can be observed.

Figures 11.9 and 11.10 show the ultimate failure modes of hollow sandstone specimens with $d = 15$ and 26 mm, respectively, based on the experiments. Based on Fig. 11.9, the ultimate failure modes of the hollow sandstone specimens ($d = 15$ mm) change from axial splitting tensile failure to single shear failure with increasing confining pressure. Furthermore, note that the ultimate failure mode of intact and hollow sandstone specimens ($d = 15$ mm) are very similar under the action of confining pressures (8 –35 MPa). Therefore, we used the CT images of the sandstone specimens at $\sigma_3 = 0$ and 35 MPa to explore the internal damage mechanism of intact and hollow sandstone specimens ($d = 15$ mm) in this research.

However, Fig. 11.10 shows that the ultimate failure modes of hollow sandstone specimens ($d = 26$ mm) differ from those of intact and hollow sandstone specimens ($d = 15$ mm) at the same confining pressure. At $\sigma_3 = 0$, the ultimate failure mode

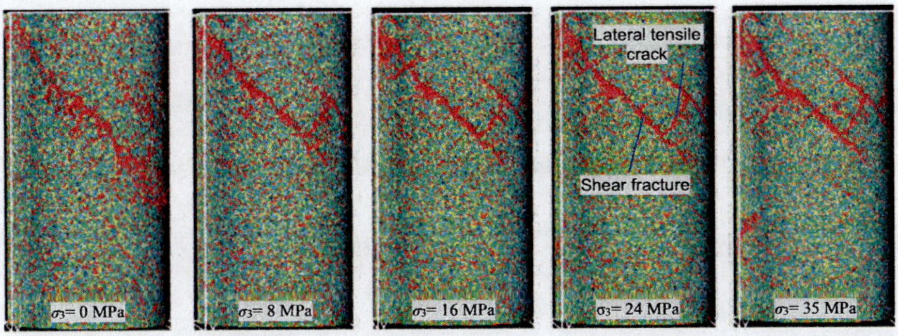

(a) Numerical failure mode

(b) Experimental failure mode

Fig. 11.8 Ultimate failure modes of intact sandstone specimen ($d = 0$) under different confining pressures

(*a*) σ_3= 0 MPa (*b*) σ_3= 8 MPa (*c*) σ_3= 16 MPa (*d*) σ_3= 24 MPa (*e*) σ_3= 35 MPa

Fig. 11.9 Experimental failure modes of hollow sandstone specimen ($d = 15$ mm) under different confining pressures

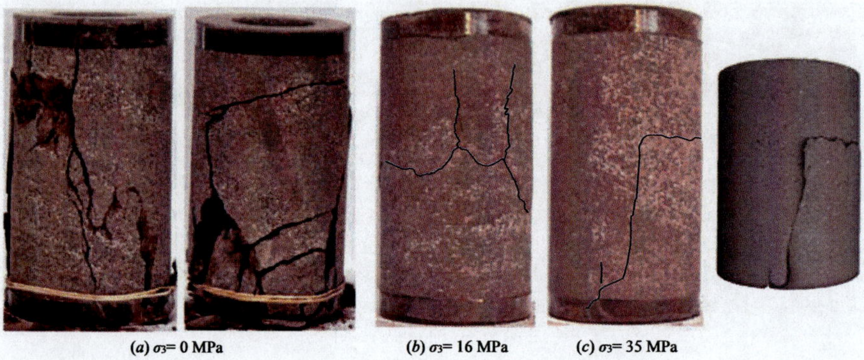

Fig. 11.10 Experimental failure modes of hollow sandstone specimen ($d = 26$ mm) under different confining pressures

of hollow sandstone ($d = 26$ mm) is axial splitting tensile failure, which is similar to the failure mode of the hollow specimen ($d = 15$ mm). However, at $\sigma_3 = 16$ and 35 MPa, the ultimate failure model of hollow sandstone ($d = 26$ mm) is mixed axial and lateral tensile failure. Therefore, for $d = 26$ mm, we also used the CT images of the sandstone specimens at $\sigma_3 = 0$ and 35 MPa to explore the internal damage mechanism of the hollow sandstone specimen ($d = 26$ mm), which will be discussed later.

Figure 11.11 shows the comparison of the X-ray CT scanning images of the surface and actual surface crack photographs of the sandstone specimens after uniaxial and triaxial compression failure ($\sigma_3 = 35$ MPa). Black regions indicate cracks or surface spalling, and other regions reflect areas without surface failure (Fig. 11.11b). The comparison between Fig. 11.11a and b shows that the X-ray CT scanning images of the surface are similar to the actual surface crack photographs, which demonstrates that X-ray micro-CT scanning can be used to explore the internal damage in sandstone material. After the failure of the hollow specimens, the system of cracks inside the specimen stabilizes with increasing axial deformation.

11.4.1 Internal Crack Mechanism of Intact Sandstone

Figure 11.12 shows vertical cross-sections of intact sandstones ($d = 0$) with different depths after uniaxial compression failure. Based on Fig. 11.12, the effect of the depth on the extent of the internal damage of the intact sandstone material after uniaxial compression failure can be analyzed. The fracture modes of the intact specimens are not axially symmetric. At $x = 5$ mm, one large axial splitting tensile crack in the left region and many splitting tensile cracks in the right region of the specimen can be observed. However, at $x = 10$–48 mm, multiple axial splitting tensile cracks can be observed in the right region of the specimen, whereas the left region contains no

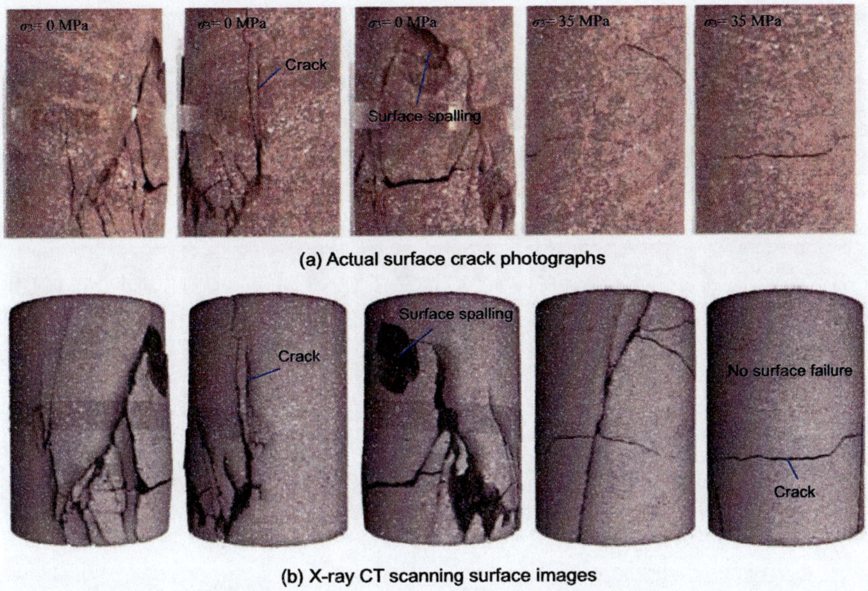

Fig. 11.11 Comparison of X-ray CT scanning surface images and actual surface crack photographs of intact sandstone specimens after uniaxial and triaxial compression failure

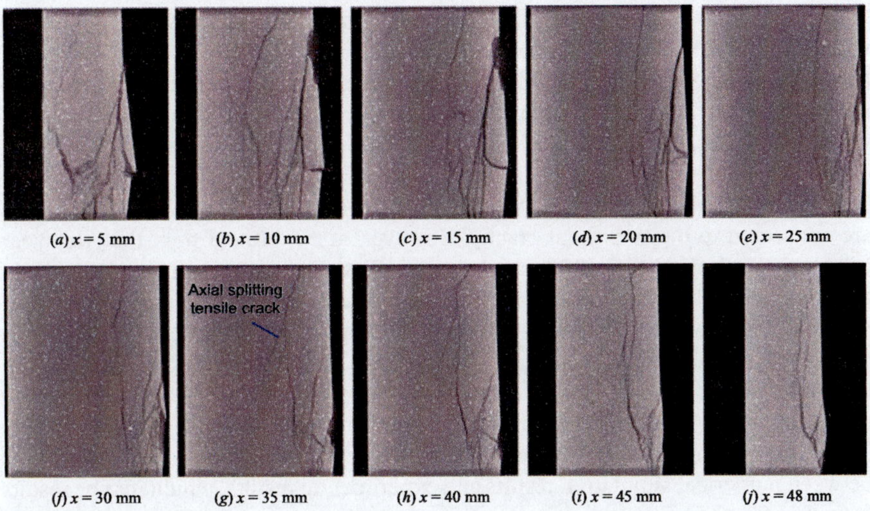

Fig. 11.12 Vertical cross-sections with different depths of intact sandstone ($d = 0$) after uniaxial compression failure

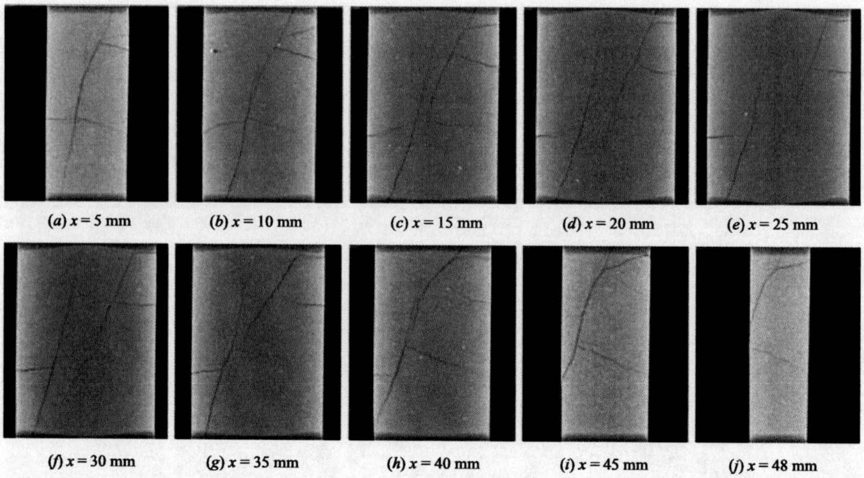

Fig. 11.13 Vertical cross-sections with different depths of intact sandstone ($d = 0$) after triaxial compression failure ($\sigma_3 = 35$ MPa)

any splitting tensile cracks. These splitting tensile cracks are connected and form a macroscopic fault system. Furthermore the crack system becomes simpler with the increase of depth.

Figure 11.13 shows the vertical cross-sections of intact sandstones ($d = 0$) with different depths after triaxial compression failure ($\sigma_3 = 35$ MPa). Based on Fig. 11.13, the fracture modes of the intact specimens are also not axially symmetric. At $x = 5 \sim 10$ mm, one notable shear crack can be observed in the entire regions, in addition to several lateral tensile cracks. At $x = 15 \sim 35$ mm, two notable shear cracks can be observed, but they do not cut through the entire regions; moreover the two shear cracks do not intersect the ligament regions. Several lateral tensile cracks are connected to the two shear cracks. However, at $x =$ 40–48 mm, the two shear cracks observed at $x = 15 \sim 35$ mm start to intersect and finally form one shear crack. Several lateral tensile cracks still exist in the vicinity of the shear crack.

Table 11.6 depicts the horizontal cross-sections of intact sandstone specimens with different heights after uniaxial and triaxial compression failure obtained by the numerical simulations and X-ray micro-CT experiments. Generally, the internal cracks of intact sandstone specimens obtained by the numerical simulations are similar to those obtained by the experiments. Of course, there exist some differences between numerical simulation results and experimental results, which maybe results from the effect of rock heterogeneity on the internal crack modes. Furthermore, Table 11.6 shows that the confining pressure has a great effect on the internal damage of intact sandstone. At the same time, the effect of the height (z) on the extent of the internal damage of the intact sandstone material after uniaxial and triaxial compression failure can be analyzed.

Table 11.6 Horizontal cross-sections with different heights of intact sandstone ($d = 0$) after triaxial compression failure

Intact specimen (d= 0 mm)	Experimental result (σ_3= 0 MPa)	Numerical result (σ_3= 0 MPa)	Experimental result (σ_3= 35 MPa)	Numerical result (σ_3= 35 MPa)
z = 10 mm	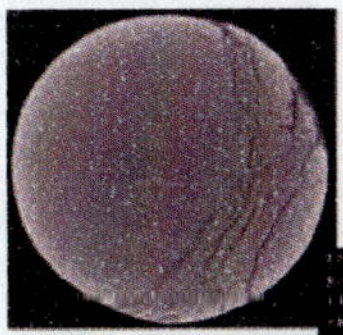	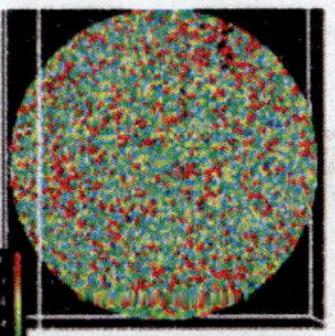	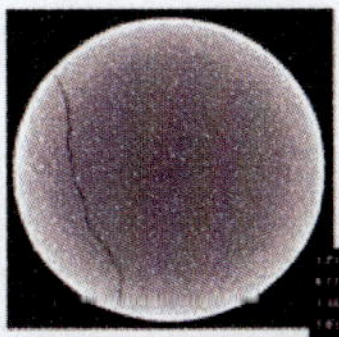	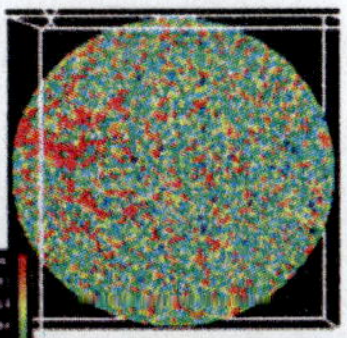
z = 20 mm	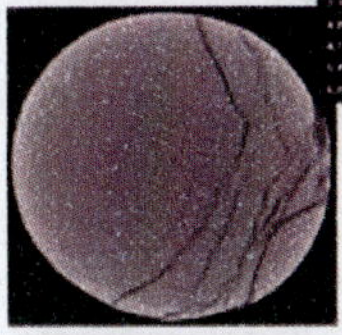	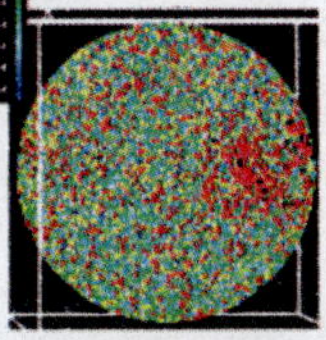	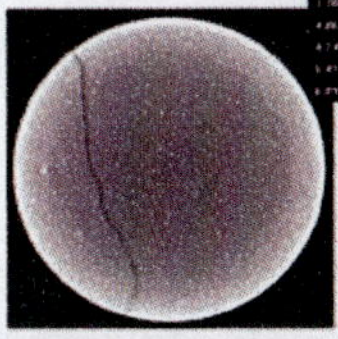	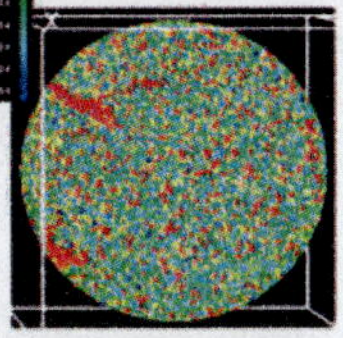
z = 30 mm	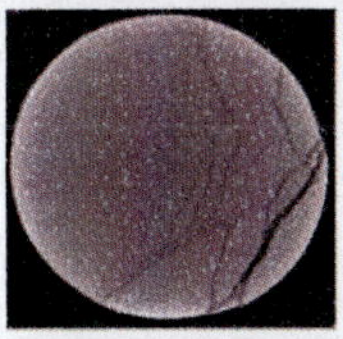	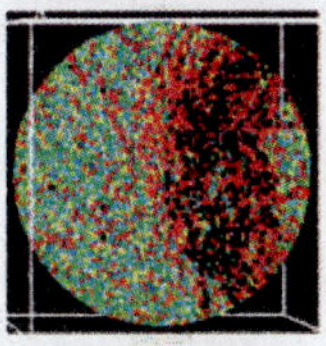	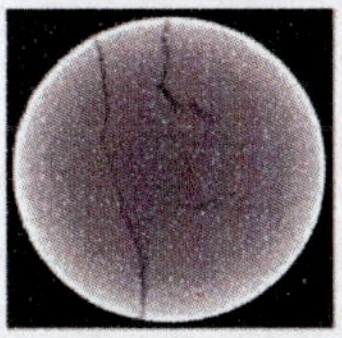	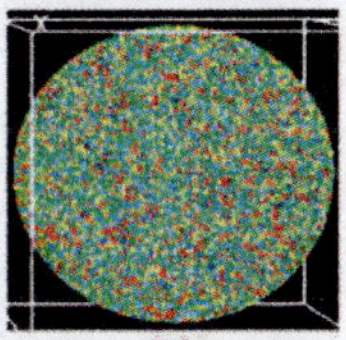
z = 40 mm	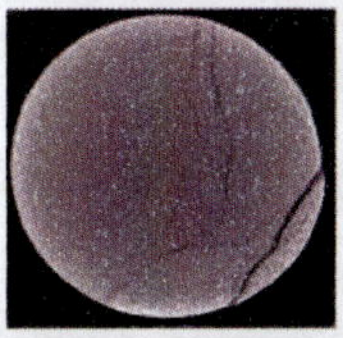	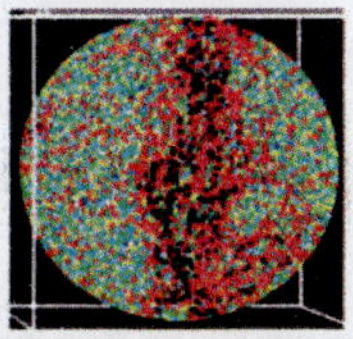	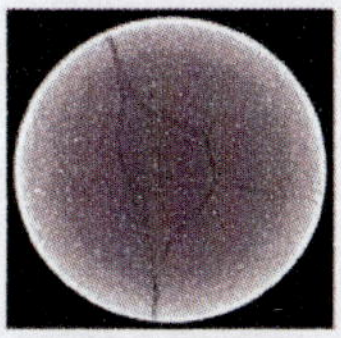	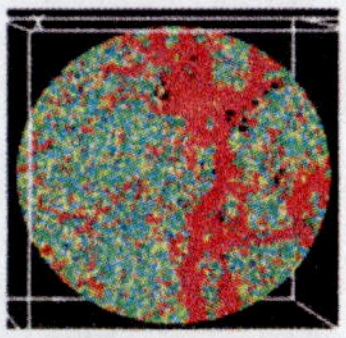
z = 50 mm	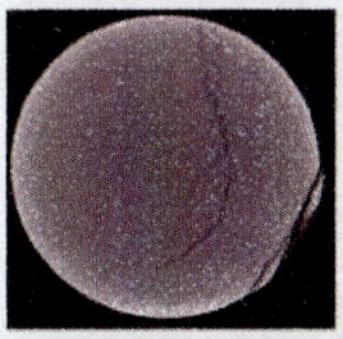	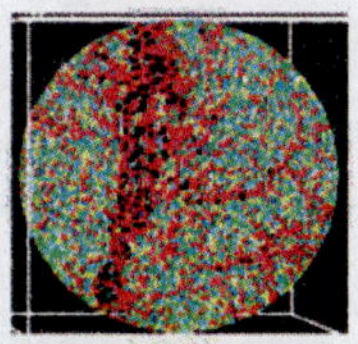	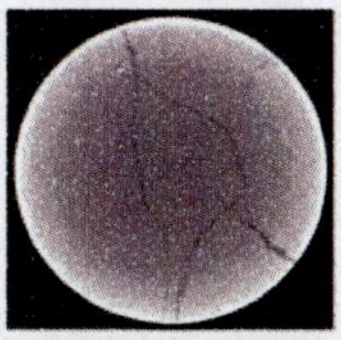	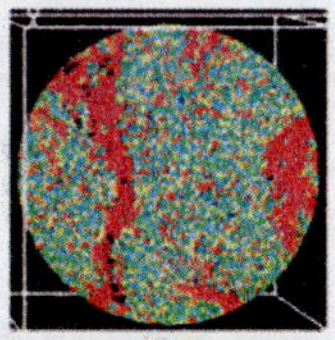
z = 60 mm	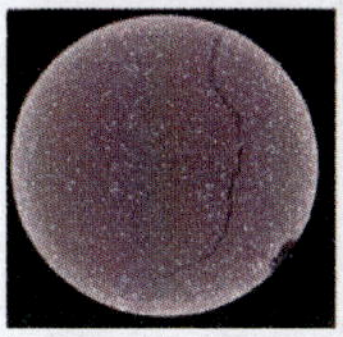	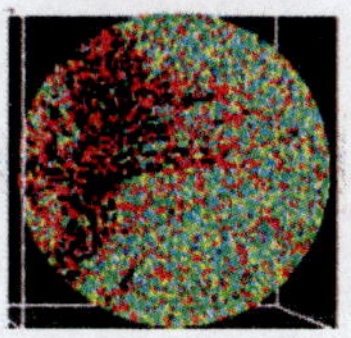	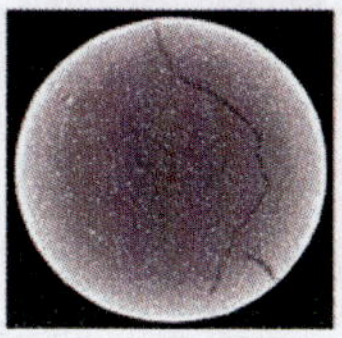	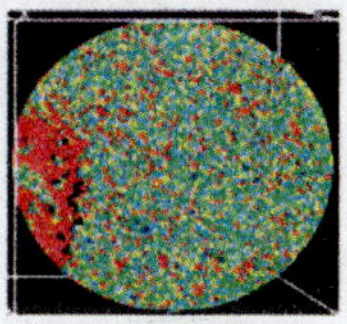

Based on the X-ray micro-CT observations, the horizontal cross-section is in the bottom region of the specimen at $z = 10 \sim 20$ mm under uniaxial compression condition. Several cracks can be observed and the system of cracks is very complex, as shown in the bottom region in Fig. 11.12. Moreover, the widths and lengths of these cracks differ, which indicates various damage extents in the same cross-section. At $z = 30–40$ mm, the horizontal cross-section is in the middle region of the intact specimen and the internal damage of the intact specimen is more serious. With the increase of height, the crack density decreases. But, at $z = 50 \sim 60$ mm, when the horizontal cross-section is in the top region of the specimen, the system of cracks is more simple compared with that at $z = 10–40$ mm, and only one large main crack can be observed. However, under triaxial compression condition ($\sigma_3 = 35$ MPa), at $z = 10–20$ and 60 mm, only one large crack can be observed and the system of cracks is very simple, which can be seen in the bottom and top regions shown in Fig. 11.13. Whereas, at $z = 30$~50 mm, two large cracks can be observed, and several small radial cracks emanate in the disc regions. Although the cracks determined by the numerical simulations are initiated at various heights, the crack systems obtained by the numerical simulations are similar to those observed in the experiments.

11.4.2 Internal Crack Mechanism of Hollow Sandstone with Various Borehole Diameters

Figures 11.14 and 11.15 illustrates the vertical cross-sections of hollow sandstones with $d = 15$ mm and $d = 26$ mm with different depths, respectively, after uniaxial

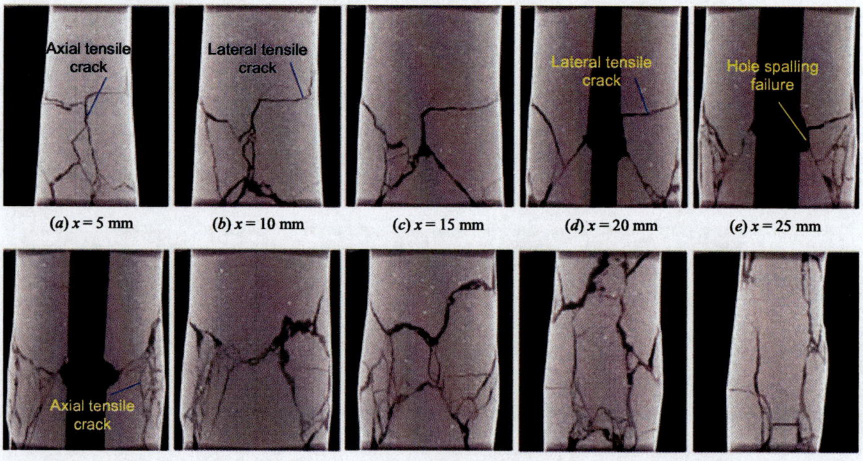

Fig. 11.14 Vertical cross-sections with different depths of hollow sandstone ($d = 15$ mm) after uniaxial compression failure

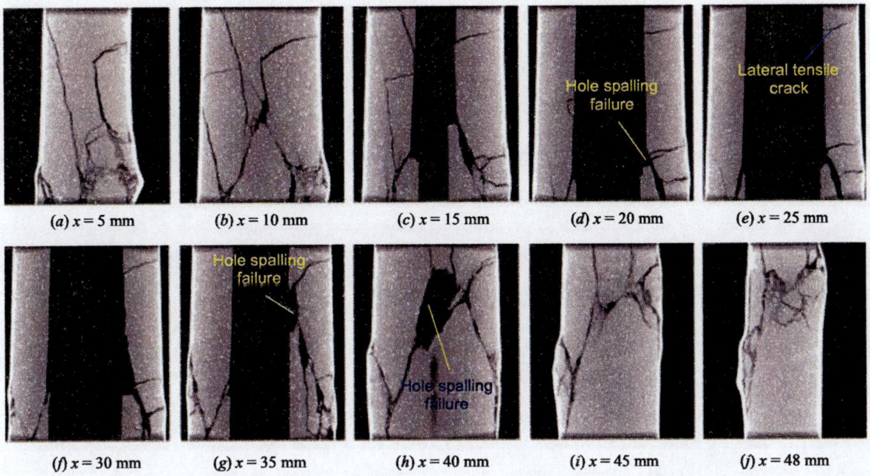

Fig. 11.15 Vertical cross-sections with different depths of hollow sandstone (d = 26 mm) after uniaxial compression failure

compression failure. Based on Figs. 11.14 and 11.15, we can conclude that the crack system mode of hollow sandstone obtained from the vertical cross-sections with different depths differs from that of intact sandstone.

Figure 11.14a–c show several axial and lateral tensile cracks in the central bottom region of the specimen with d = 15 mm at x = 5–15 mm. These cracks do not propagate to the top boundary of the specimen and cracks do not emanate in the central top region of the specimen. At x = 20–30 mm, hole spalling failure and more axial tensile cracks can be observed in this specimen. However, at x = 35–48 mm, complex tensile crack system can be observed in the specimen. The widths of these tensile cracks in the hollow specimens (d = 15 mm) are larger than those in the intact specimen (d = 0). Figures 11.15a–b show a complex tensile crack system in the specimen at x = 5–10 mm. Based on Fig. 11.15, notable axial and lateral tensile cracks can be observed in the specimen with d = 26 mm at x = 5–10 and 45–48 mm, which are very complex, similar to those in the specimen with d = 15 mm shown in Fig. 11.14. However, in addition to notable axial and lateral tensile cracks, many hole spalling failures can be observed at x = 15–40 mm, which are due to the uniaxial compression stress state.

Figures 11.16 and 11.17 illustrates the vertical cross-sections of hollow sandstones with d = 15 and d = 26 mm with different depths, respectively, after triaxial compression failure. Based on Figs. 11.16 and 11.17, we can conclude that the crack system mode of hollow sandstone obtained from the vertical cross-sections with different depths differs from that of intact sandstone due to the effect of the borehole size. Generally, the crack system of hollow sandstone with various borehole diameters after triaxial compression failure is less complex than that after uniaxial compression failure. At σ_3 = 35 MPa, the hollow sandstone mainly displays shear

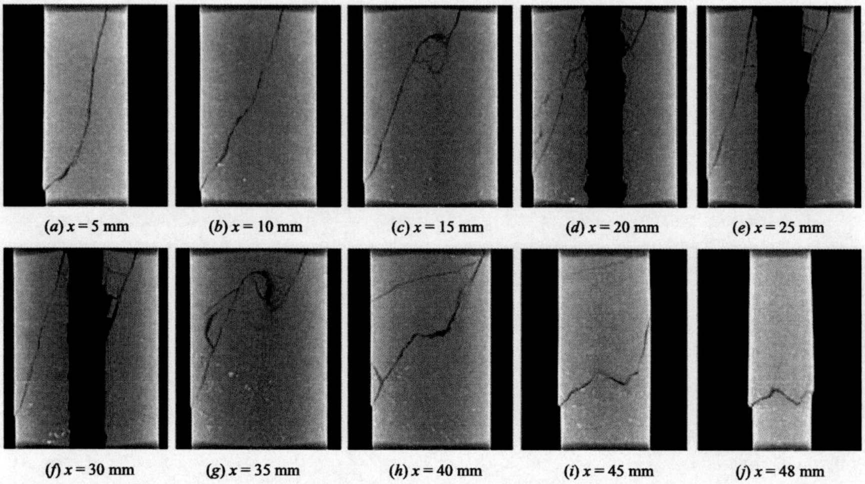

Fig. 11.16 Vertical cross-sections with different depths of hollow sandstone (d = 15 mm) after triaxial compression failure ($\sigma_3 = 35$ MPa)

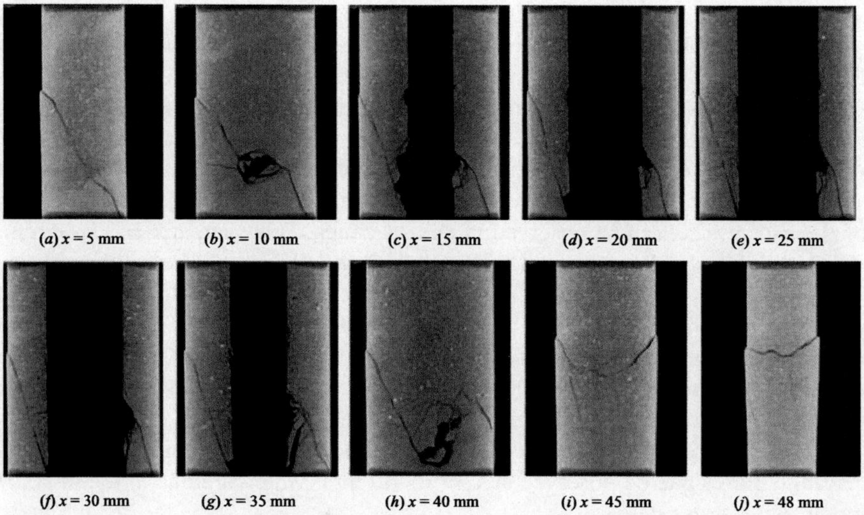

Fig. 11.17 Vertical cross-sections with different depths of hollow sandstone (d = 26 mm) after triaxial compression failure ($\sigma_3 = 35$ MPa)

fracture and almost no tensile cracks can be observed compared with the failure mode under uniaxial compression. Furthermore, compared with uniaxial compression, the hollow specimen at $\sigma_3 = 35$ MPa has more internal wall fractures due to the higher confining pressure. Note that the hollow specimen with $d = 26$ mm has more internal wall fractures than that with $d = 15$ mm at the same confining pressure, which is due to the effect of the borehole size.

Tables 11.7 and 11.8 depict the horizontal cross-sections of hollow sandstone specimens with d = 15 and d = 26 mm with different heights, respectively, after uniaxial and triaxial compression failure obtained by the numerical simulations and

Table 11.7 Horizontal cross-sections with different heights of hollow sandstone (d = 15 mm) after triaxial compression failure

Hollow specimen (d= 15 mm)	Experimental result (σ_3= 0 MPa)	Numerical result (σ_3= 0 MPa)	Experimental result (σ_3= 35 MPa)	Numerical result (σ_3= 35 MPa)
z = 10 mm	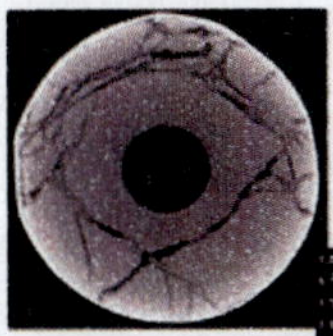	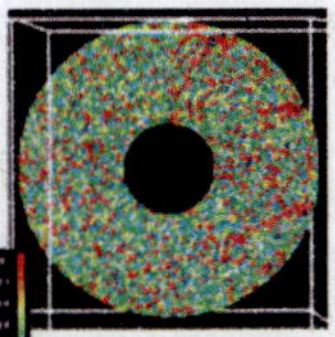	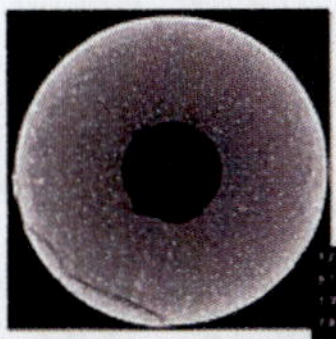	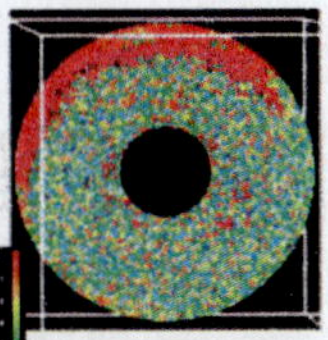
z = 20 mm	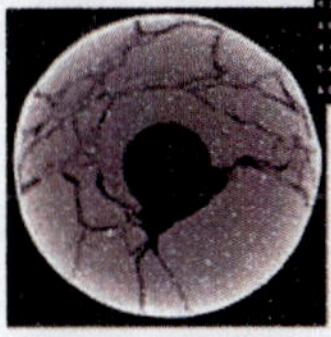	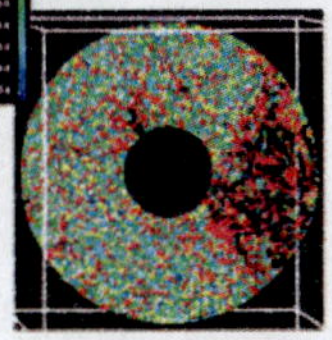	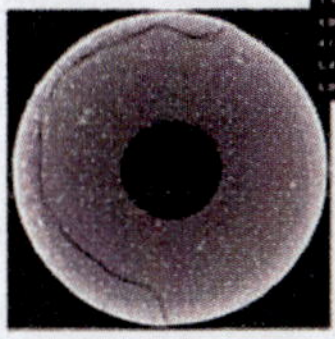	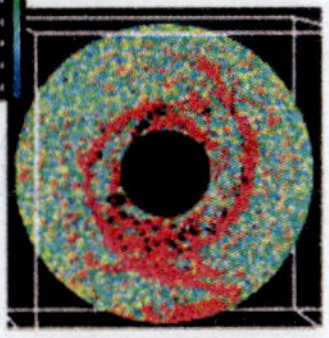
z = 30 mm	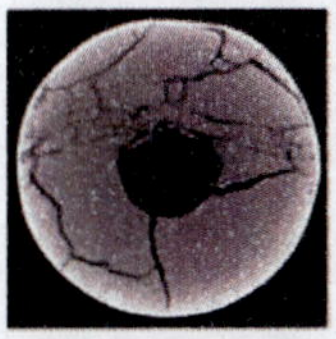	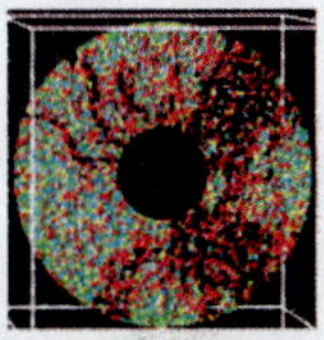	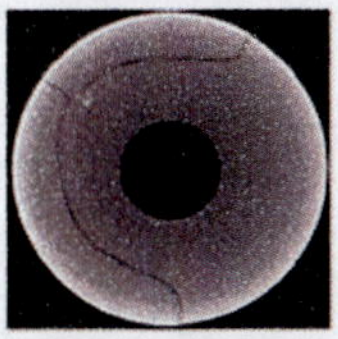	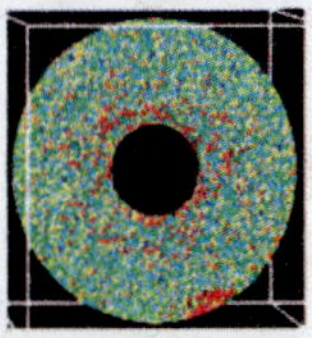
z = 40 mm	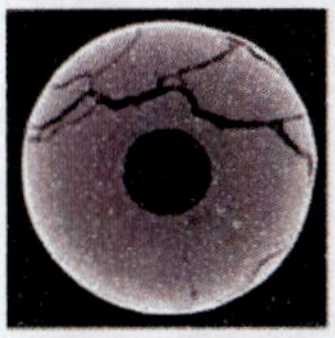	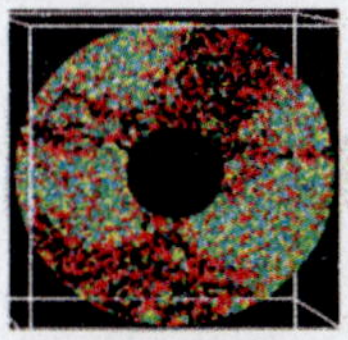	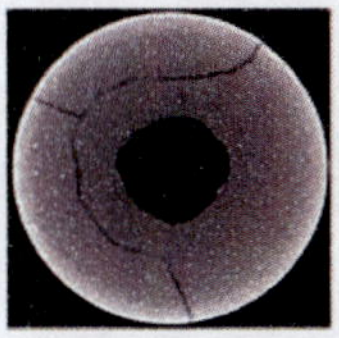	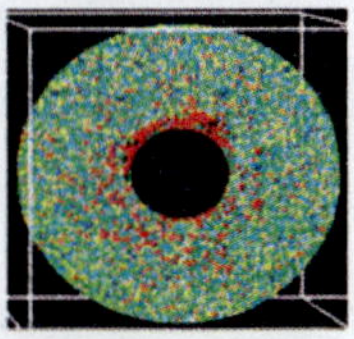
z = 50 mm	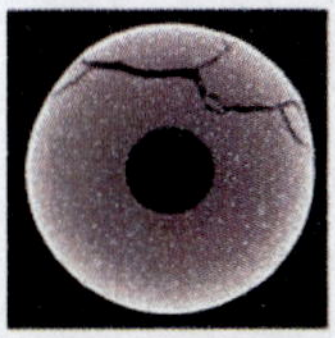	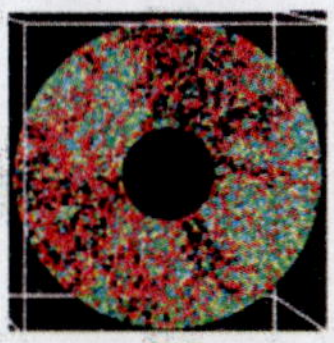	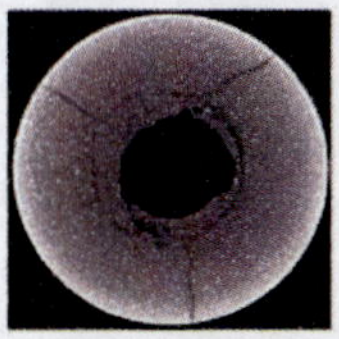	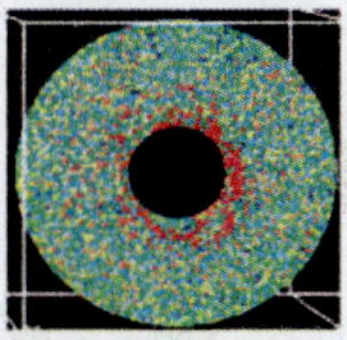
z = 60 mm	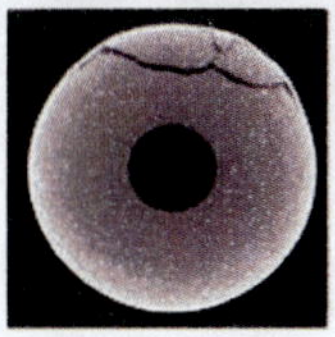	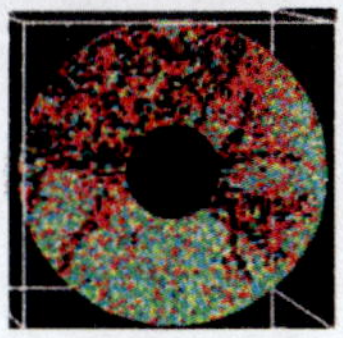	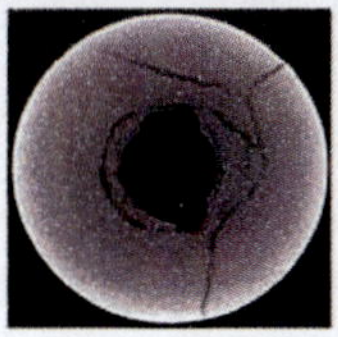	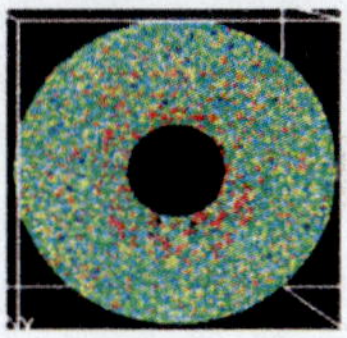

Table 11.8 Horizontal cross-sections with different heights of hollow sandstone ($d = 26$ mm) after triaxial compression failure

Hollow specimen (d= 26 mm)	Experimental result (σ_3= 0 MPa)	Numerical result (σ_3= 0 MPa)	Experimental result (σ_3= 35 MPa)	Numerical result (σ_3= 35 MPa)
z = 10 mm	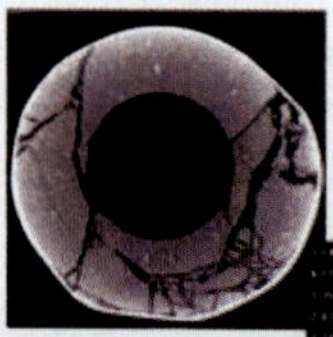	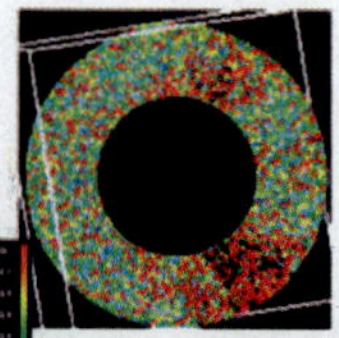	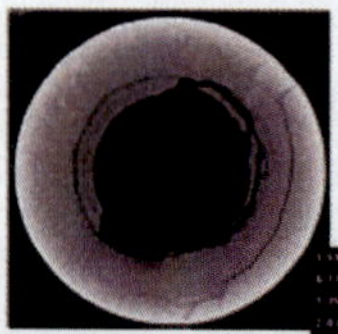	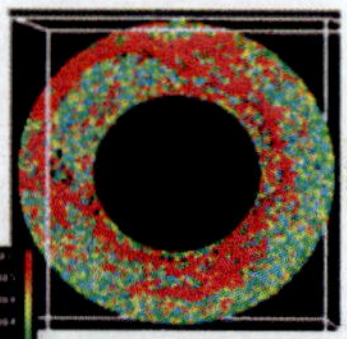
z = 20 mm	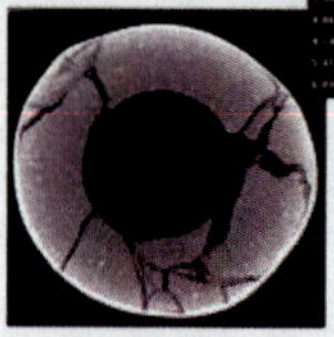	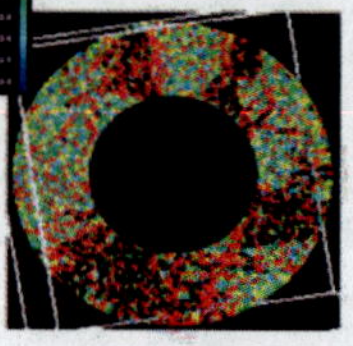	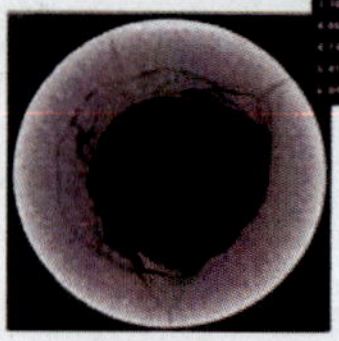	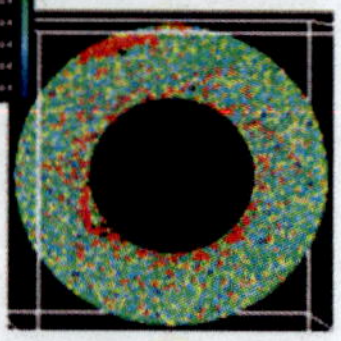
z = 30 mm	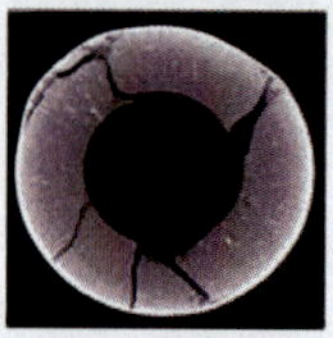	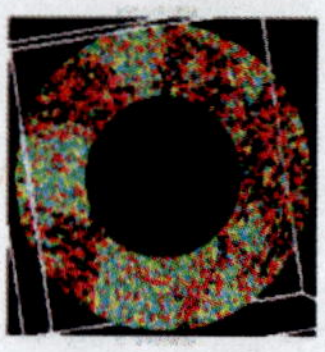	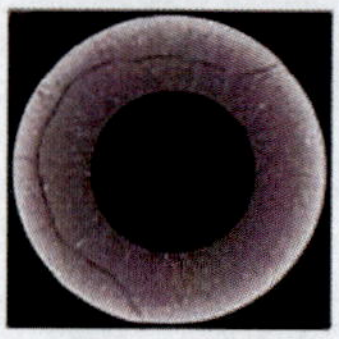	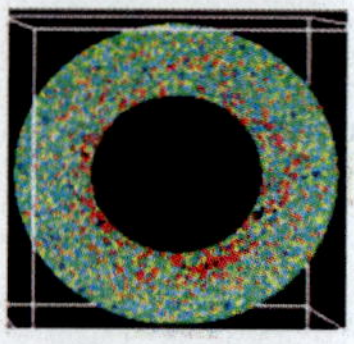
z = 40 mm	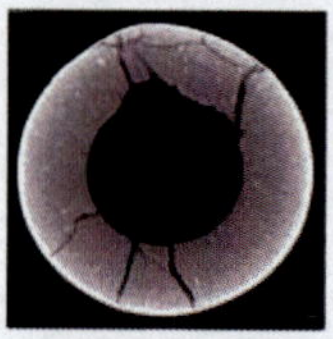	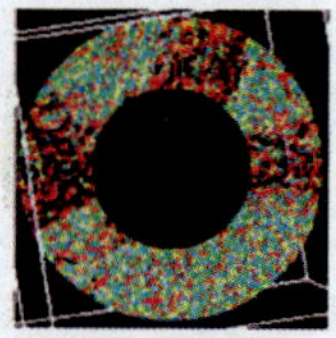	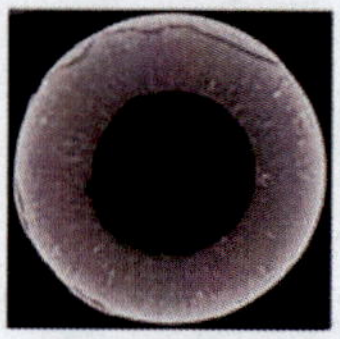	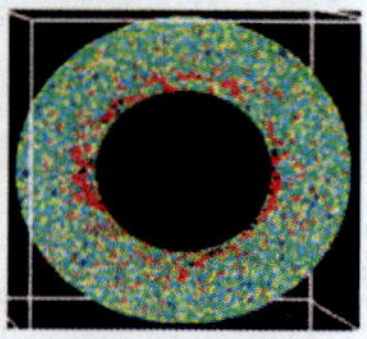
z = 50 mm	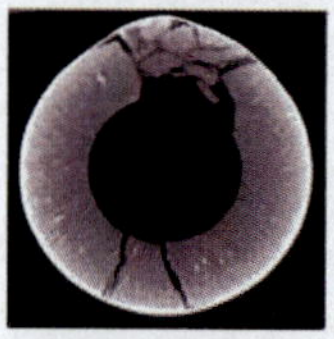	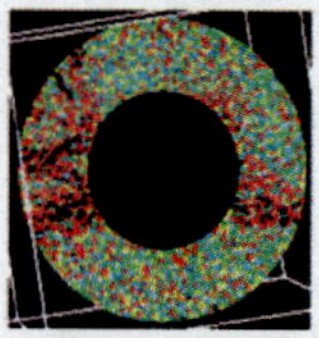	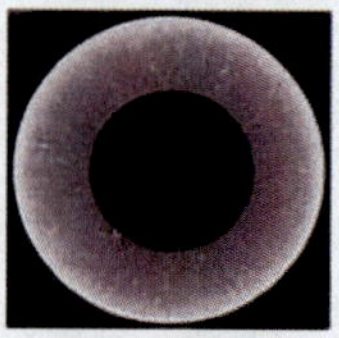	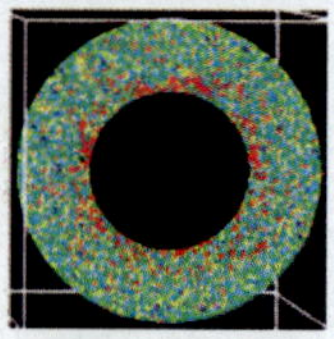
z = 60 mm	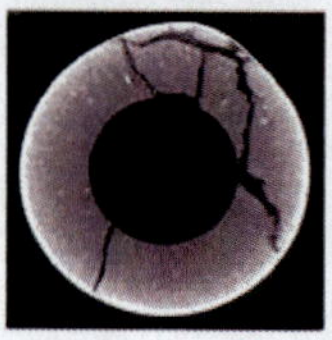	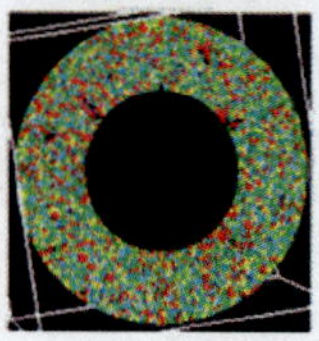	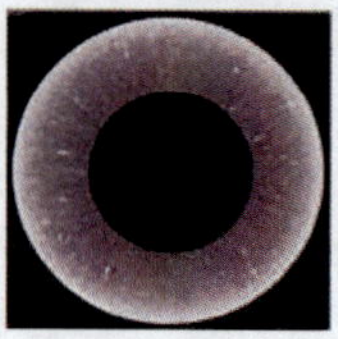	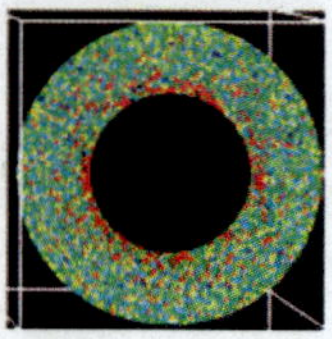

X-ray micro-CT experiments. Generally, the internal cracks of the hollow sandstone specimens obtained by the numerical simulations are very similar to those obtained by the experiments. Tables 11.7 and 11.8 show that the confining pressure and borehole diameter greatly affect the internal damage of hollow sandstone. At the same time, the effect of height (z) on the extent of the internal damage of the hollow sandstone material after uniaxial and triaxial compression failure can be analyzed.

It should be noted that these CT images (Tables 11.7 and 11.8) were obtained on the compressed specimens after they failed. The borehole size has a great effect on the internal damage of hollow sandstone. The results show that the internal damage depends not only on the borehole size but also on the confining pressure. With increasing borehole size, the crack system of the hollow sandstone seems to become more complex at the same confining pressure. However, with increasing confining pressure, the crack system of the hollow sandstone seems to become simpler at the same borehole size.

11.4.3 Effect of Borehole Size on Fracture Evolution Mechanism of Hollow Sandstone

Figure 11.18 further illustrates the evolution of mesoscopic cracks and fracture process of hollow specimen ($d = 15$ mm) under triaxial compression ($\sigma_3 = 16$ MPa). It should be noted that in 3D numerical simulation, each cube volume unit represents one mesoscopic crack once it occurs the failure when exceeding its tensile or compressive strength of RVE. Furthermore, according to the previous study (Zhang and Wong 2012), more than three mesoscopic cracks were interconnected, which may be considered as a macrocrack. In Fig. 11.18, the six points (there points before the peak strength, one point on the peak strength and two points after the peak strength) denoted in Fig. 11.18 correspond to the crack propagation stages. It should be noted that the vertical cross sections of the internal wall and the horizontal cross sections were all obtained in the middle section of the hollow sandstone specimens.

When the hollow specimen is loaded to point 1 (σ_1-$\sigma_3 = 96.15$ MPa $= 43\%\ \sigma_p$), the number of mesoscopic cracks is 4652, which can hardly not observed both on the external and internal wall of hollow specimens. With increasing axial deformation, the axial deviatoric stress linearly increases during the elastic stage. As the axial deviatoric stress reaches point 2 (σ_1–$\sigma_3 = 160.99$ MPa $= 71\%\ \sigma_p$), the mesoscopic internal wall fractures start to evenly form in the axial direction due to the effect of differential stress concentration. However, few mesoscopic cracks can be observed on the external wall. At this time, the number of mesoscopic cracks increases to 24,597. When the axial deviatoric stress increases to point 3 (σ_1-$\sigma_3 = 201.67$ MPa $= 89\%\ \sigma_p$), several minor macro-hole wall fractures consisting of damaged elements start to propagate before the peak strength is reached, which can be clearly observed in the horizontal cross sections. The corresponding number of mesoscopic cracks is 52733. At the peak strength, that is, point 4 (σ_p = 225.70 MPa), the number

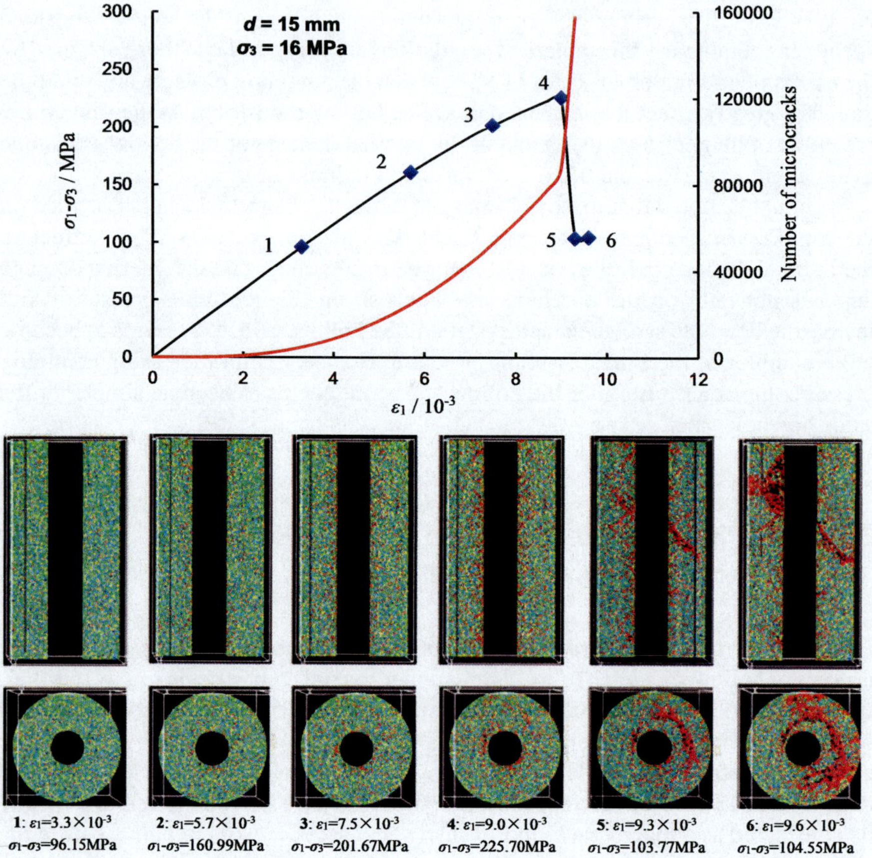

Fig. 11.18 Crack evolution process of hollow sandstone specimen (d=15mm) under triaxial compression (σ_3=16MPa)

of mesoscopic cracks reaches 87,197. In addition, more macro cracks with small angles to the axial stress are initiated at the position at which the mesoscopic internal wall fractures appear. Macro cracks on the external wall cannot be observed. After the peak strength is reached, the mesoscopic cracks start to flourish (the number of mesoscopic cracks is almost twice that of point 4). Two main shear cracks with angles of 40° to the maximum compressive axis propagate towards the external edge of the hollow specimen. Similar to the previous stage, macro cracks are not visible on the external wall. Finally, when the stress–strain curve enters the residual stress, two main shear cracks coalesce between the external and internal walls of the hollow specimens form. In addition, serious damage zones associated with secondary cracks can be observed both on the internal and external walls.

The evolution of the mesoscopic crack number and the fracture process of the hollow specimen ($d = 15$ mm) under triaxial compression ($\sigma_3 = 35$ MPa) are shown in Fig. 11.19. Compared with the hollow specimen at $\sigma_3 = 16$ MPa, the compressive bearing capacity of the hollow specimen improves when the specimen is subjected to a higher confining pressure ($\sigma_3 = 35$ MPa), which leads to a decrease in the number of mesoscopic cracks throughout the loading process. Similar to the hollow specimen at $\sigma_3 = 16$ MPa, only a small number of minor macrohole wall fractures and several internal wall fractures can be observed when the hollow specimen is loaded to points 1, 2 and 3 before reaching the peak strength. As the axial deviatoric stress increases to the peak strength, that is, point 4, more macrocracks with small angles to the axial stress as well as localized failure zones appear along the internal wall compared with the specimen at $\sigma_3 = 16$ MPa due to the tangential stress concentration. At the post-peak residual stress points 5 and 6, two main parallel shear cracks that differ from the connected shear cracks at $\sigma_3 = 16$ MPa are initiated and propagate towards the

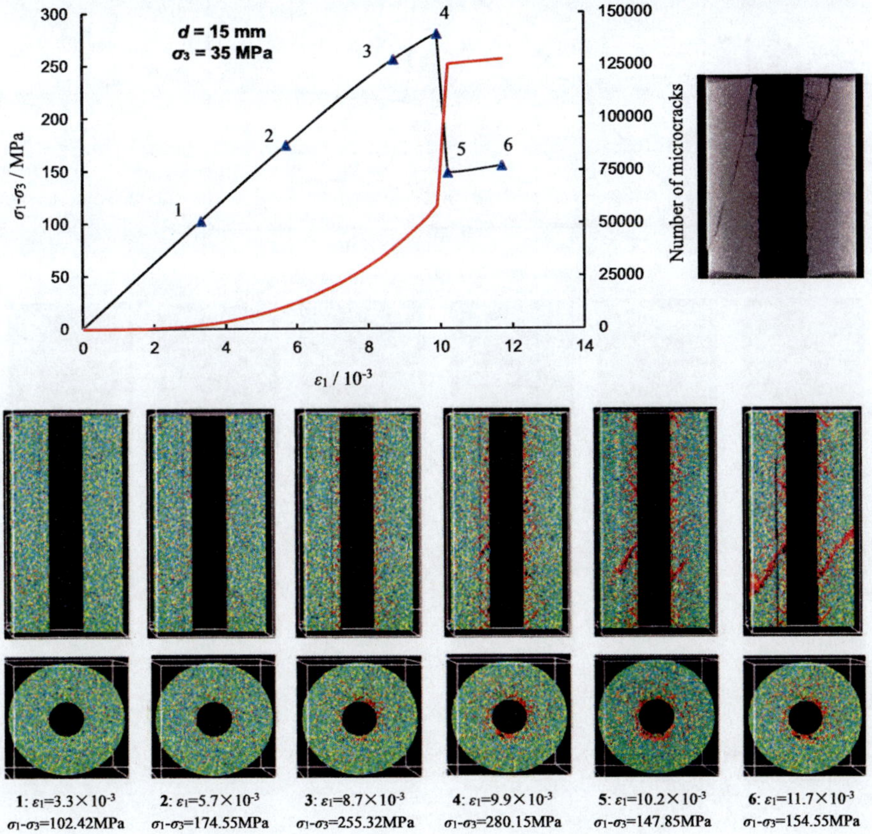

Fig. 11.19 Crack evolution process of hollow sandstone specimen ($d = 15$ mm) under triaxial compression ($\sigma_3 = 35$ MPa)

external wall of the hollow specimen. Note that before the stress–strain curve enters the residual stress, much fewer mesoscopic cracks can be observed on the external wall and forms the macrocracks, which differs from the specimen at $\sigma_3 = 16$ MPa.

The evolution of the number of mesoscopic cracks and the fracture process of the hollow specimen ($d = 26$ mm) under triaxial compression ($\sigma_3 = 16$ MPa) are shown in Fig. 11.20. Similar to the specimen with a borehole diameter of 15 mm, the six points (there points before the peak strength, one point on the peak strength and two points after the peak strength) denoted in Fig. 11.20 correspond to the crack propagation stages. Note that the following comparisons were made between the hollow specimens with borehole diameters of 26 and 15 mm under the same confining pressure of 16 MPa.

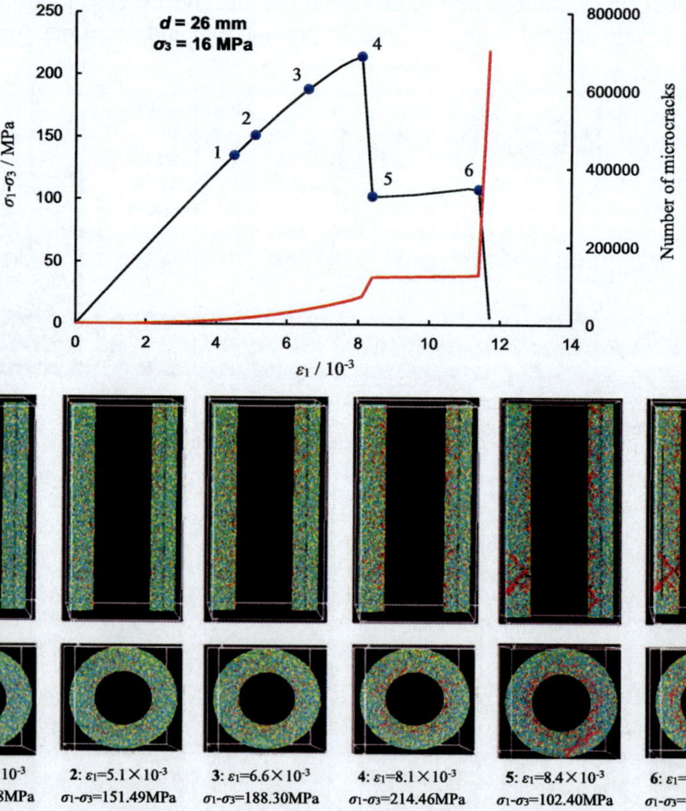

Fig. 11.20 Crack evolution process of hollow sandstone specimen ($d = 26$ mm) under triaxial compression ($\sigma_3 = 16$ MPa)

As shown in Fig. 11.20, the hollow specimen with a larger borehole diameter has a lower compressive bearing capacity and brittleness characteristics. In addition, the number of mesoscopic cracks sharply increases throughout the loading process. When the hollow specimen is loaded to point 1 (σ_1–σ_3 = 135.08 MPa = 63% σ_p), point 2 (σ_1–σ_3 = 151.49 MPa = 71% σ_p) and point 3 (σ_1–σ_3 = 188.30 MPa-89% σ_p), the differential stress is concentrated on the internal wall before the peak strength is reached, which leads to the small localized hole wall failure. As the axial deviatoric stress reaches the peak strength, that is, point 4 (σ_p = 214.46 MPa), more serious damage zones associated with several minor macro-fractures are evenly distributed along the internal wall, which can be observed in the horizontal cross sections. As soon as the stress–strain curve drops to the post-peak stress point 5, two orthogonal shear cracks are rapidly initiated at the lower end and coalesced external wall of the hollow specimen, which cannot be observed in the hollow specimen with d = 15 mm. Finally, when the stress–strain curve enters the residual stress, two additional orthogonal shear cracks connect the internal and external walls of the hollow specimen. In addition, a secondary crack with an angle of 10° to horizontal direction appears on the external wall of the hollow specimen.

The evolution of the number of mesoscopic cracks and the fracture process of the hollow specimen (d = 26 mm) under triaxial compression (σ_3 = 35 MPa) are shown in Fig. 11.21. Generally, the crack evolution process of hollow specimen with d = 26 mm at σ_3 = 35 MPa is similar to that at σ_3 = 16 MPa. When the hollow specimen is loaded to point 1 (σ_1–σ_3 = 86.85 MPa = 38% σ_p), point 2 (σ_1–σ_3 = 142.65 MPa = 63% σ_p) and point 3 (σ_1–σ_3 = 208.76 MPa = 92% σ_p), under higher confining pressure, the tangential stress concentration leads to an increase in localized hole wall failure zones before the peak strength is reached. As the axial deviatoric stress reaches the peak strength, that is, point 4 (σ_p = 227.35 MPa), more minor macro fractures are evenly distributed along the internal wall. Note that two main orthogonal shear cracks are initiated at the lower end and propagate toward external wall of the hollow specimen at the post-peak stress point 5 when the confining pressure is increased to 35 MPa. However, because of the higher confining pressure, the propagation speed is not as fast as that of the hollow specimen (d = 26 mm) at σ_3 = 16 MPa. In addition, crack coalescence between the external and internal walls of the hollow specimen does not occur at this stage. Finally, when the stress–strain curve enters the residual stress, two other orthogonal shear cracks accompanied by many secondary cracks connect the internal and external walls of the hollow specimen.

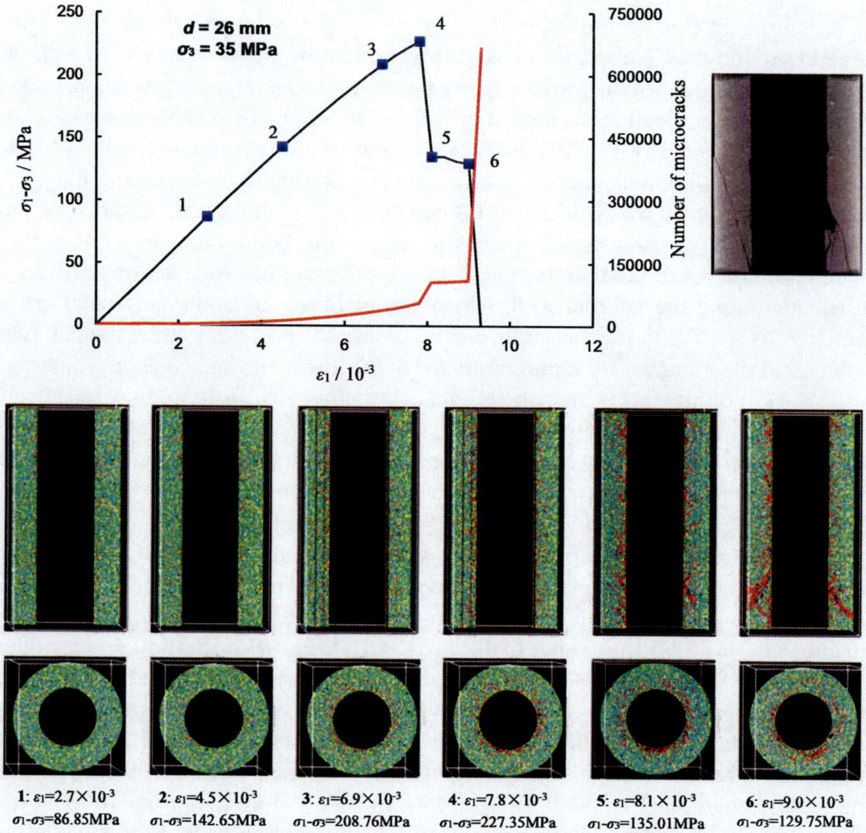

Fig. 11.21 Crack evolution process of hollow sandstone specimen ($d = 26$ mm) under triaxial compression ($\sigma_3 = 35$ MPa)

11.5 Interpretations and Discussions on the Fracture Mechanism of Hollow Sandstone

11.5.1 Interpretations on Effect of Borehole Diameter on the Peak Strength

Figure 11.22a displays a thick-walled hollow cylinder under external pressure and axial load, showing the notation and coordinate system (the internal pressure equals zero). Based on thick-wall cylinder elastic theory, the radial stress (σ_r) and tangential stress (σ_θ) can be expressed by the following two equations, respectively:

$$\sigma_r = \frac{\sigma_3 D^2}{D^2 - d^2} - \frac{\sigma_3 D^2 d^2}{4r^2\left(D^2 - d^2\right)} \tag{11.13}$$

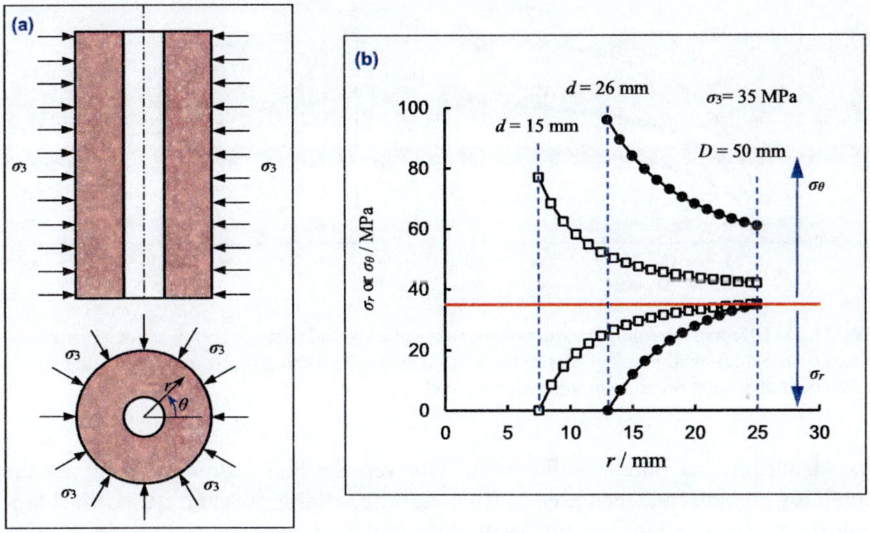

Fig. 11.22 Stress of hollow cylinder based on the thick-wall cylinder elastic theory. **a** An illustration of a thick-walled hollow cylinder specimen under external pressure and axial load, showing notation and coordinate system (internal pressure equals to 0); **b** Radial and tangent stress distribution of hollow cylinder with various hole diameters under the condition of plane strain ($\sigma_3 = 35$ MPa)

$$\sigma_\theta = \frac{\sigma_3 D^2}{D^2 - d^2} + \frac{\sigma_3 D^2 d^2}{4r^2\left(D^2 - d^2\right)} \tag{11.14}$$

Based on Eqs. (11.13) and (11.14), Fig. 11.22b shows the radial and tangent stress distributions of hollow cylinders with various borehole diameters under plane strain ($\sigma_3 = 35$ MPa). Based on Fig. 11.22b, the radial and tangent stresses of the internal thick-wall cylinder specimen are compressive stresses due to the homogenous outer-wall pressure if not considering the principal stress of the vertical plane. At the inter-wall of hollow cylinder, the radial stress equals zero, but the tangent stress is maximal and increases with increasing internal borehole diameter of the hollow cylinder specimen. However, at the outer-wall of the hollow cylinder, the radial stress reaches the maximum value (the value is independent of the borehole diameter), but the tangent stress is minimal. Note that the tangent stress is always greater than the radial stress at any positions of the internal thick-wall cylinder specimen.

Figure 11.23 shows the effect of the confining pressure on the radial and tangent stresses of the hollow cylinder specimen based on the thick-wall cylinder elastic theory. Based on Fig. 11.23a, the radial stress at the internal-wall of the hollow sandstone equals zero and is independent of the confining pressure. Under the same confining pressure, the radial stress gradually increases with increasing internal borehole diameter in the hollow cylinder specimen, but the increasing rate is stable. Furthermore, at a larger confining pressure, the radial stress increases faster with

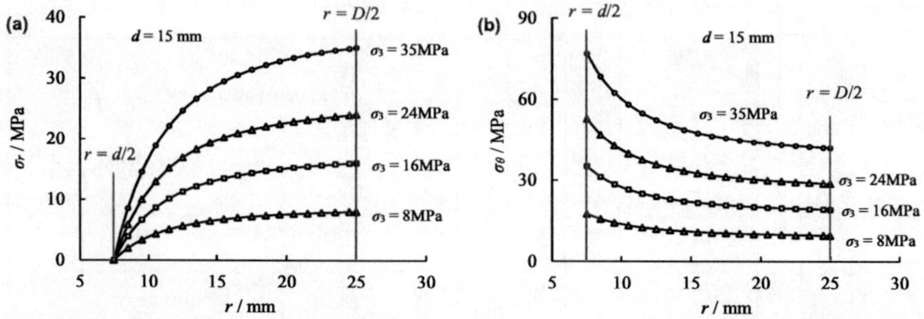

Fig. 11.23 Effect of confining pressure on the radial and tangent stress of hollow cylinder specimen based on the thick-wall cylinder elastic theory. **a** Effect of confining pressure on the radial stress; **b** Effect of confining pressure on the tangent stress

increasing internal borehole diameter. However, the radial stress depends on the confining pressure, and increases with increasing confining pressure at the same borehole diameter except for the inter-wall of the hollow cylinder. Based on Fig. 11.23b, the tangent stress gradually decreases with increasing internal borehole diameter in the hollow cylinder specimen under the same confining pressure, but the decreasing rate is stable. Furthermore, the tangent stress depends on the confining pressure, and increases with increasing confining pressure at the same borehole diameter.

When the confining pressure is not equal to zero and the internal borehole pressure is zero, the radial and tangential stresses of the hollow specimen are non-uniform. The tangential stress nonlinearly decreases but the radial stress gradually increases with increasing radius (from the internal wall to the outer wall). The non-uniform stress distribution results in differences in the hollow sandstone specimens with various borehole sizes at different confining pressures. Under uniaxial compression, hollow sandstone specimens with various borehole sizes experience axial splitting tensile failure and the stress distribution is more uniform. Therefore, the UCS of hollow sandstone is independent of the borehole size (Fig. 11.7). However, under confining pressure, hollow sandstone usually fails mainly via a single shear fracture in the direction of the maximum shear stress. At the same borehole size, the peak strength of the hollow sandstone increases with the confining pressure (Yang 2016). However, the shear fracture areas of hollow sandstones with various borehole sizes differ. The larger the borehole size is, the smaller is the area of the shear fracture. Thus, the corresponding axial supporting capacity of specimens with larger borehole sizes is lower. Therefore, the peak strength gradually decreases with increasing borehole size under the same confining pressure (Fig. 11.7). Moreover, the sensitivity of the peak strength of the hollow sandstone to the borehole size at a larger confining pressure is notably higher than that at a smaller confining pressure, which potentially occurs because the non-uniform stress distribution increases with the confining pressure at an internal hole pressure equal zero, as shown in Fig. 11.23.

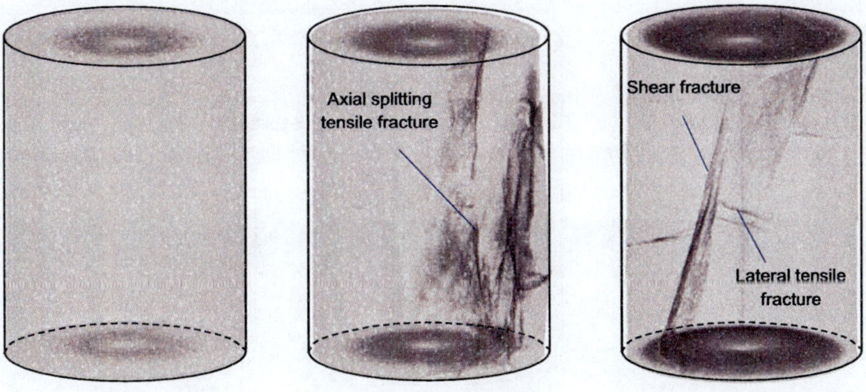

(*a*) *Uncompressed specimen* (*b*) *Failed specimen* ($\sigma_3 = 0$ MPa) (c) *Failed specimen* ($\sigma_3 = 35$ MPa)

Fig. 11.24 3D CT images of uncompressed and failed intact sandstone specimens ($d = 0$)

11.5.2 Discussions on the Effect of Confining Pressure for 3D Fracture Mechanism

Figure 11.24 shows the 3D CT images of uncompressed and failed intact sandstone specimens. In the 3D images, fracture regions with lower CT numbers are black, and other regions with relatively higher CT numbers are transparent. Compared with the 3D image shown in Fig. 11.24a, the sandstone specimen shows a typical axial splitting failure mode under uniaxial compression in Fig. 11.24b, which is identical with the surface fracture shown in Fig. 11.8 and the vertical cross-sections shown in Fig. 11.12. However, the failure mode of the sandstone specimens after triaxial compression failure ($\sigma_3 = 35$ MPa) shown in Fig. Figure 11.24c differ from that after uniaxial compression failure. The specimens show a mixed failure mode, that is, a mixture of shear fracture and lateral tensile failure planes. Furthermore, by rotating the image of the specimen, the trace of the shear fracture plane appears as a curve not a plane, which indicates that the deformation is non-uniform throughout the shear fracture curve. The shear fracture curve and lateral tensile fracture at $\sigma_3 = 35$ MPa can be also observed in the surface fracture shown in Fig. 11.8 and the vertical cross-sections shown in Fig. 11.13.

Figure 11.25 shows the 3D CT images of uncompressed and failed hollow sandstone specimens ($d = 15$ mm). Compared with the 3-D image shown in Fig. 11.25a, the hollow sandstone specimen exhibits a complex and mixed tensile fracture mode (i.e., a mixture of axial splitting and lateral tensile cracks) under uniaxial compression, as shown in Fig. 11.25b, which is identical with the surface fracture shown in Fig. 11.9 and the vertical cross-sections shown in Fig. 11.14. However, the failure mode of the hollow sandstone specimen after triaxial compression failure ($\sigma_3 =$ 35 MPa) differs from that after uniaxial compression failure. The specimens exhibit a shear fracture mode and the trace of the shear fracture is not a plane but a curve,

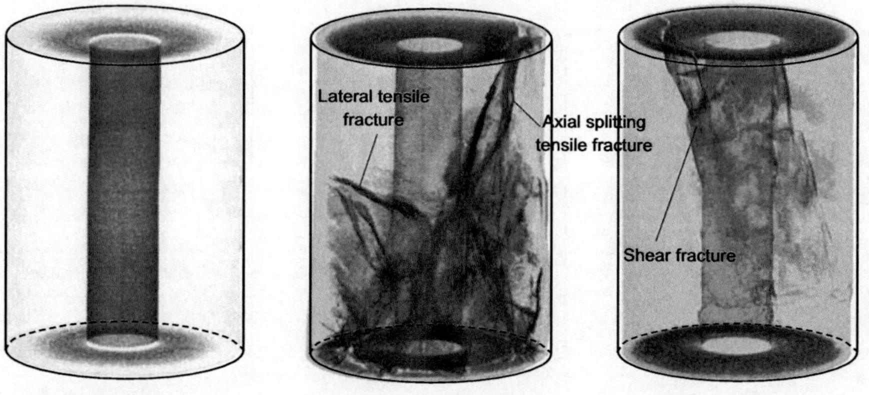

(*a*) *Uncompressed specimen* (*b*) *Failed specimen* ($\sigma_3 = 0$ MPa) (*c*) *Failed specimen* ($\sigma_3 = 35$ MPa)

Fig. 11.25 3D CT images of uncompressed and failed hollow sandstone specimens ($d = 15$ mm)

as shown in Fig. 11.25c. The shear fracture curve of the hollow sandstone at $\sigma_3 =$ 35 MPa is also observable in the surface fracture shown in Fig. 11.9 and the vertical cross-sections shown in Fig. 11.16.

Figure 11.26 shows the 3D CT images of uncompressed and failed hollow sandstone specimens ($d = 26$ mm). Compared with the 3D image shown in Fig. 11.26a, the hollow sandstone specimen exhibits a complex and mixed tensile fracture mode (i.e. a mixture of axial splitting and lateral tensile cracks) under uniaxial compression, as shown in Fig. 11.26b, which is identical with the surface fracture shown in Fig. 11.10 and the vertical cross-sections shown in Fig. 11.15. However, the failure mode of the hollow sandstone specimen after triaxial compression failure ($\sigma_3 =$

(*a*) *Uncompressed specimen* (*b*) *Failed specimen* ($\sigma_3 = 0$ MPa) (*c*) *Failed specimen* ($\sigma_3 = 35$ MPa)

Fig. 11.26 3D sCT images of uncompressed and failed hollow sandstone specimens ($d = 26$ mm)

35 MPa) differs from that after uniaxial compression failure. The hollow specimen shows a local shear fracture and several hole spalling failure modes, which are also observable in the surface fracture shown in Fig. 11.10 and the vertical cross-sections shown in Fig. 11.17.

11.6 Conclusions

To predict the instability around deep wellbores in petroleum engineering and ensure the safety of deep excavation damage zones in tunnel engineering, it is important to explore the fracture evolution mechanism of hollow sandstone material. In this research, a series of X-ray micro-CT observations and 3D numerical simulations was conducted on hollow sandstone specimens with various borehole diameters ($d = 0$, 15 and 26 mm) under different confining pressures (0, 8, 16, 24 and 35 MPa). Based on the mechanical behavior of hollow sandstone under conventional triaxial compression obtained by the experiments and numerical simulations and the analysis of the results of the experiments and numerical simulations, the following conclusion can be drawn.

(1) A RFPA^{3D} was established to simulate the deformation failure behavior of hollow sandstone under conventional triaxial compression. Based on the results of the triaxial compression experiments on intact sandstone specimens, we calibrated several micro-parameters in RFPA^{3D}, which better reflect the macro-mechanical response of the tested sandstone material. The axial deviatoric stress–strain curves of intact sandstone under different confining pressures obtained by the simulations are in good agreement with those obtained from the experiments. Based on the calibrated micro-parameters, a series of 3D numerical simulations was conducted for hollow sandstone specimens with various borehole diameters under different confining pressures. The quantitative comparison of the axial deviatoric stress–strain curve and peak strength of the hollow sandstone obtained by the experiments and simulations indicates that the 3D numerical results agree very well with the experimental results. The nonlinear Hoek–Brown criterion better reflects the peak strength of hollow sandstone material than the linear Mohr–Coulomb criterion.

(2) The effect of the borehole diameter on the peak strength of hollow sandstone depends on the confining pressure, which was interpreted by radial and tangent stress distribution around the borehole. The tangential stress nonlinearly decreases but the radial stress gradually increases with increasing radius (from the internal wall to the outer wall). The non-uniform stress distribution results in differences in the hollow sandstone specimens with various borehole sizes at different confining pressures. Under uniaxial compression, all hollow sandstones with various borehole sizes experience axial splitting tensile failure and the stress distribution is more uniform. Therefore, the UCS of hollow sandstone is independent of the borehole size. However, hollow sandstone usually fails under confining pressure, mainly via a single shear fracture in the direction of the maximum shear stress. The shear fracture areas of hollow sandstones with various borehole sizes differ. The larger the borehole

size is, the smaller is the area of the shear fracture. Thus, the corresponding axial supporting capacity is lower for specimens with larger borehole sizes. Therefore, the peak strength gradually decreases with increasing borehole size under the same confining pressure.

(3) The surface and internal fracture characteristics of hollow sandstone obtained by the X-ray micro-CT observations and simulations were compared. The internal cracks of intact and hollow sandstone specimens obtained by the numerical simulations are very similar to those obtained by the X-ray micro-CT observations. The influence of the confining pressure on the fracture evolution mechanism of hollow sandstones with various borehole diameters was analyzed in detail based on the numerical simulation results. Cracks are first initiated at the hole wall and gradually propagate toward the external wall of the hollow specimen. The internal damage of the hollow sandstone depends not only on the borehole size but also on the confining pressure. With increasing borehole size, the crack system of the hollow sandstone seems to become more complex at the same confining pressure. However, with increasing confining pressure, the crack system of the hollow sandstone seems to become simpler at the same borehole size. These investigated conclusions are important for the prediction of the instability around deep wellbores in petroleum engineering and to ensure the safety of deep excavation damage zones in tunnel engineering.

References

ASTM (1995) Standard test method for triaxial compressive strength of undrained rock core specimens without pore pressure measurements. Test Designation D2664–95 Annual book of ASYM standards, American Society for Testing and Materials, Philadelphia (14)02

Alsayed MI (2002) Utilising the Hoek triaxial cell for multiaxial testing of hollow rock cylinders. Int J Rock Mech Min Sci 39:355–366

Basu B, Tiwari D, Kundu D, Prasad R (2009) Is Weibull distribution the most appropriate statistical strength distribution for brittle materials? Ceram Int 35:237–246

Cerrolaza M, Garcia R (1997) Boundary elements and damage mechanics to analyze excavations in rock mass. Eng Anal Bound Elem 20:1–16

Chen X, Yu J, Li H, Wang S (2017) Experimental and numerical investigation of permeability evolution with damage of sandstone under triaxial compression. Rock Mech Rock Eng 50:1529–1549

Camas D, Moreno B (2017) Numerical and experimental study of the plastic zone in cracked specimens. Eng Fract Mech 185:20–32

Colombo D, Massin P (2011) Fast and robust level set update for 3D non-pla-nar X-FEM crack propagation modelling. Comput Methods Appl Mech Eng 200(25–28):2160–2180

Feng XT, Pan PZ, Zhou H. Simulation of the rock microfracturing process under uniaxial compression using an elasto-plastic cellular automaton. Int J Rock Mech Min Sci 43:1091–1108

Fairhurst CE, Hudson JA (1999) Draft ISRM suggested method for the complete stress-strain curve for the intact rock in uniaxial compression. Int J Rock Mech Min Sci 36:279–289

Hamidi F, Scheuermann A, Galindo-Torres SA, Pedroso D, Li L (2018) Synthetic low-strength materials as rock substitutes for physical model studies. Geomech Geophys Geo-Energy and Geo-Res 4(3):273–297

Huang YH, Yang SQ, Tian WL (2019) Crack coalescence behavior of sandstone specimen containing two pre-existing flaws under different confining pressures. Theoret Appl Fract Mech 99:118–130

Hoek E, Brown ET (1980) Underground Excavation in Rock. Institution of Mining and Metallurgy, London

Hoek E, Brown ET (1997) Practical estimates of rock mass strength. Int J Rock Mech Min Sci 34(8):1165–1186

Lemaitre J (1971) Evaluation of dissipation and damage in metals subjected to dynamic loading. Proc Int Conf Mech 1:15–20

Lee DH, Juang CH, Chen JW, Lin HM, Shieh WH (1999) Stress paths and mechanical behavior of a sandstone in hollow cylinder tests. Int J Rock Mech Min Sci 36:857–870

Lee B, Rathnaweera TD (2016) Stress threshold identification of progressive fracturing in Bukit Timah granite under uniaxial and triaxial stress conditions. Geomech Geophy Geo-Energy Geo-Resources 2(4):301–330

Li G, Tang C (2015) A statistical meso-damage mechanical method for modeling trans-scale progressive failure process of rock. Int J Rock Mech Min Sci 74:133–150

Li G, Liang ZZ Tang CA (2015) Morphologic interpretation of rock failure mechanisms under uniaxial compression based on 3D multiscale high-resolution numerical modeling. Rock Mech Rock Eng 48:2235–2262

Li Z, Li L, Li M, Zhang L, Zhang Z, Huang B (2018) A numerical investigation on the effects of rock brittleness on the hydraulic fractures in the shale reservoir. J Nat Gas Sci Eng 50:22–32

Mazars J, Pijaudier CG (1989) Continuum damage theory-application to concrete. J Eng Mech-ASCE 115:345–365

Mohr O (1914) Abhandlungen aus dem Gebiete der Technischen Mechanik, 2nd edn. Ernst, Berlin

Niu L, Zhu W, Cheng Z, Guan K, Qin T (2017) Numerical simulation on excavation-induced damage of rock under quasi-static unloading and dynamic disturbance. Environ Earth Sci 76:1–14

Ren WY, Yang ZJ, Sharma R, Zhang C, Withers PJ (2015) Two-dimensional X-ray CT image based meso-scale fracture modelling of concrete. Eng Fract Mech 133:24–39

Wong RHC, Lin P (2015) Numerical study of stress distribution and crack coalescence mechanisms of a solid containing multiple holes. Int J Rock Mech Min Sci 79:41–54

Shen B, Stephansson O (1993) Numerical analysis of mixed mode I and mode II fracture propagation. International Journal of Rock Mechanics and Mining Sciences and Geomechanics Abstracts, 30:861–867

Sarfarazi V, Haeri H, Shemirani AB (2018) Simulation of fracture mechanism of pre-holed concrete model under Brazilian test using PFC^{3D}. Smart Struct Syst 22:675–687

Skarzynski L, Suchorzewski J (2018) Mechanical and fracture properties of concrete reinforced with recycled and industrial steel fibers using Digital Image Correlation technique and X-ray micro-computed tomography. Constr Build Mater 83:283–299. https://doi.org/10.1016/j.conbuildmat.2018.06.182

Tang CA, Tang SB, Gong B, Bai HM (2015) Discontinuous deformation and displacement analysis: from continuous to discontinuous. Sci China Tech Sci 58:1567–1574

Wang J-P, Andò E, Charrier P, Salager S, Lambert P, François B (2019) Micro investigation of unsaturated sand in mini-triaxial shearing using X-ray CT. Géotechnique Lett. https://doi.org/10.1680/jgele.18.00214

Xu T, Tang CA, Zhao J, Li LC, Heap MJ (2012) Modelling the time-dependent rheological behaviour of heterogeneous brittle rocks. Geophys J Int 189(3):1781–1796

Xu T, Xu Q, Deng M, Ma T, Yang T, Tang C-A (2014) A numerical analysis of rock creep-induced slide: a case study from Jiweishan Mountain China. Environ Earth Sci 72(6):2111–2128

Xu T, Zhou G, Heap MJ, Yang S, Konietzky H, Baud P (2018) The modeling of time-dependent deformation and fracturing of brittle rocks under varying confining and pore pressures. Rock Mech Rock Eng

Yang SQ, Jiang YZ, Xu WY, Chen XQ (2008) Experimental investigation on strength and failure behavior of pre-cracked marble under conventional triaxial compression. Int J Solids Struct 45:4796–4819

Yang YM, Ju Y, Sun YH, Zhang DX (2015a) Numerical study of the stress field during crack growth in porous rocks. Geomech Geophys Geo-Energ Geo-Resour 1:91–101

Yang SQ, Ranjith PG, Gui YL (2015b) Experimental study of mechanical behavior and X-ray micro CT observations of sandstone under conventional triaxial compression. Geotech Test J 38:179–197

Yang SQ, Xu T, He L, Jing HW, Wen S, Yu QL (2015c) Numerical study on failure behavior of brittle rock specimen containing pre-existing combined flaws under different confining pressure. Arch Civ Mech Eng 15:1085–1097

Yang SQ, Jing HW (2011) Strength failure and crack coalescence behavior of brittle sandstone samples containing a single fissure under uniaxial compression. Int J Fract 168:227–250

Yang SQ, Jing HW, Li YS, Han LJ (2011) Experimental investigation on mechanical behavior of coarse marble under six different loading paths. Exp Mech 51(3):315–334

Yang SQ (2016) Experimental study on deformation, peak strength and crack damage behavior of hollow sandstone under conventional triaxial compression. Eng Geol 11–24

Yang SQ (2018) Fracturing Mechanism of Compressed Hollow Sandstone Evaluated by X-ray Micro-CT Scanning. Rock Mech Rock Eng 51:2033–2053

Yang SQ, Tian WL, Huang YH, Ma ZG, Fan LF, Wu ZJ (2018) Experimental and discrete element modeling on cracking behavior of sandstone containing a single oval flaw under uniaxial compression. Eng Fract Mech 194:154–174

Yang SQ, Yang Z, Jing HW, Xu T (2020) Fracture evolution mechanism of hollow sandstone under conventional triaxial compression by X-ray micro-CT observations and three-dimensional numerical simulations. Int J Solids Struct 190:156–180

Yin Q, Jing HW, Su HJ (2018) Investigation on mechanical behavior and crack coalescence of sandstone specimens containing fissure-hole combined flaws under uniaxial compression. Geosci J 22:825–842

Yu QL, Liu HY, Yang TH, Liu HL (2018) 3D numerical study on fracture process of concrete with different ITZ properties using X-ray computerized tomography. Int J Solids Struct 147:204–222

Zhang XP, Wong LNY (2012) Cracking processes in rock-like material containing a single flaw under uniaxial compression: a numerical study based on parallel bonded-particle model approach. Rock Mech Rock Eng 45:711–737

Zhou XP, Bi J, Qian QH (2015) Numerical simulation sof crack growth and coalescence in rock-like materials containing multiple pre-existing flaws. Rock Mech Rock Eng 48:1097–1114

Chapter 12
Fracturing Mechanism of Compressed Hollow-Cylinder Sandstone Evaluated by X-ray Micro CT Scanning

The testing of hollow cylinder specimens in rock engineering is very common. However, revealing the internal damage failure mechanism of hollow rock material subjected to triaxial compression is very significant for evaluating the stability occurring around deep well bores in petroleum engineering or the safety of deep excavation damage zones in tunnel engineering. Lee et al. (1999) carried out a series of hollow cylindrical triaxial tests to investigate the potential effect of the stress path on the mechanical behaviour of sandstone. Alsayed (2002) developed a new multiaxial hollow cylinder test cell and demonstrated its versatility by performing uniaxial, biaxial, and triaxial compression tests on hollow-cylinder sandstone specimens. Based on a series of triaxial compression experiments using hollow-cylinder sandstone specimens with various hole diameters, Yang (2016) investigated the deformation, peak strength and crack damage behaviour of hollow sandstone specimens under different confining pressures.

However, in the previous studies, few X-ray micro CT techniques have been used to explore the internal damage failure mechanism of hollow-cylinder sandstone compressed to various deformation positions before or after the peak strength. Therefore, in this chapter, we report the triaxial compression experimental results of compressed hollow-cylinder sandstone under two different confining pressures of 16 and 35 MPa. Based on the experimental results, the internal damage failure characteristics of the compressed hollow-cylinder sandstone are analysed in detail using an X-ray micro CT scanning system. Then, the three-dimensional reconstructions of X-ray CT images of compressed hollow-cylinder sandstone are also performed. Finally, to better explain the internal crack evolution mechanism of hollow-cylinder sandstone during the entire loading, a conceptual model of hollow-cylinder sandstone material under the action of confining pressure is proposed.

S.-Q. Yang, *Mechanical Behavior and Damage Fracture Mechanism of Deep Rocks*,
https://doi.org/10.1007/978-981-16-7739-7_12

Table 12.1 Tested sandstone specimens in this research

Specimen	d/mm	σ_3/MPa	σ_{cd}/MPa	σ_p/MPa	Note
RZ-A18#	15	16	186.59	–	Compressed specimen until point *a* before the peak strength
RZ-A19#	15	16	183.24	229.18	Compressed specimen until point *b* after the peak strength
RZ-A30#	15	35	252.08	–	Compressed specimen until point *c* before the peak strength
RZ-A31#	15	35	243.36	280.13	Compressed specimen until point *d* after the peak strength

12.1 Sandstone Material and Testing Procedure

12.1.1 Experimental Material

The material used throughout this research was sandstone collected from Rizhao city in Shandong province, China (Yang 2016). The sandstone material has a connected porosity of 6.88% and a bulk density of 2410 kg/m^3. The detailed composition of this rock is as follows: the content (weight %) of quartz is 10.1%, K-feldspar 11.7%, plagioclase 41.6%, dolomite 13.1%, calcite 10.7%, clay mineral 10.4% and hematite 2.4%. It should be noted that in the previous experimental study (Yang 2016), the cohesion and internal friction angle of the tested intact sandstone material were 30.58 MPa and 45.7°, respectively. For this research, all specimens were cored from the same block of material to an actual diameter (D) of 49.8 mm and approximate 100 mm in length (L). Into the cylindrical specimen a through-going hole of 15 mm diameter (d) was drilled. All experiments were performed on dry specimens at ambient temperature. The average P-wave velocity and S-wave velocity of the tested sandstone were 3813 m/s and 2290 m/s, respectively. Table 12.1 lists the sandstone specimens tested in this research.

12.1.2 Mechanical Testing Equipment and Procedure

All experiments were performed in a TAW-1000 servo-controlled rock mechanics experimental system (Yang et al. 2015). The maximum loading capacity of the servo-controlled system is 1000 kN, and the maximum confining pressure is 60 MPa. During experimentation, we measured the axial force and the axial deformation using an axial linear variable differential transducer (LVDT) with a range of 8 mm. The radial deformation was measured using an LVDT wrapped tightly around the specimen, for which the maximum displacement capacity was 4 mm. To eliminate end friction effects (Mogi 1996; Hawkes and Mellor 1970), the axial and radial LVDTs were located in the central part of the specimen.

The conventional triaxial experiments were carried out under different confining pressures of 16 and 35 MPa and consisted of the following two steps: First, the confining pressure was increased to the desired value at a constant rate of 0.5 MPa/s to ensure that the specimen was under uniform hydrostatic stresses. The deviatoric stress was then applied to the specimen at a constant axial displacement rate of 0.08 mm/min (approximately 1.33×10^{-5}/s) until various deformation positions were achieved, as shown in Table 12.1, in accordance with ISRM suggested method (Fairhurst and Hudson 1999). It should be noted that in this research, the internal pressure of the hollow-cylinder sandstone specimens is kept at atmospheric pressure.

12.1.3 X-ray Micro CT Scanning Procedure

In the present research, X-ray CT scanning of the sandstone specimen was carried out using a Nanotom 160 high-resolution micro-CT at a spatial resolution of 30 μm. The X-ray beam penetrating the specimen was measured by an array of detectors. The X-rays were produced by electrons striking a Mo-W alloy target in an X-ray tube. The electron current was 90 μA, the accelerating voltage was 130 kV, and the scanning time was 4 s. The degree of X-ray attenuation depends on the density and atomic number of the materials composing the specimens. Materials with higher density and higher atomic number generally cause higher attenuation of X-rays. A two-dimensional image representing the linear distribution of X-ray attenuation was reconstructed using Fourier transformation of the projection data. We collected two-dimensional images at intervals of 0.18°; thus, 2000 slice images could be obtained for one specimen by performing a 360° rotation of the specimen as shown in Fig. 3.13a (Chap. 3). A 3-D data set of the specimen was obtained by stacking consecutive two-dimensional images.

For the 100 mm long specimens only the centre portion (50 mm in diameter and 68 mm long) of the specimen was scanned, as shown in Fig. 3.13b and c. For the vertical cross-section shown in Fig. 3.13d, the slice images were captured at 0.5 mm intervals and the scanning sequence was from front to back. For the horizontal cross-section shown in Fig. 3.13e, the slice images were captured at 0.03 mm intervals, and each image was 0.03 mm thick. Therefore, a total of 2266 slice images were obtained in the X-ray micro CT testing of each sandstone specimen. These slice images can be used to explore internal damage in the sandstone material.

12.2 Mechanical Behaviour of Compressed Hollow-Cylinder Sandstone

In the previous experimental study (Yang 2016), the size effect of the borehole of sandstone specimens under different confining pressures was investigated in detail

(Fig. 12.1). In this research, σ_1 and σ_3 represent the major principal stress and the confining pressure respectively. ε_1 and ε_3 represent the axial strain and the radial strain, respectively. The experimental results show that the uniaxial compressive strength (UCS) of sandstone (average value) is 136.22 MPa at $d = 0$ mm, 138.41 MPa at $d = 15$ mm, and 135.44 MPa at $d = 26$ mm, indicating that the hole diameter has no effect on the UCS of hollow sandstone, as shown in Fig. 12.1a. However, under confining pressure, the triaxial compressive strength of hollow sandstone is significantly dependent on the hole diameter, which decreases linearly with increasing confining pressure (Fig. 12.1b).

Figure 12.2 shows the experimental results of hollow-cylinder sandstone specimens compressed to various deformation positions under two different confining pressures. In Fig. 12.2, a positive value for the strain indicates compressive deformation and a negative value indicates dilative deformation. ε_v refers to the volumetric strain, which is calculated from the sum of the axial strain (ε_1) and twice the radial strain (ε_3) (i.e., $\varepsilon_v = \varepsilon_1 + 2\varepsilon_3$). The crack damage threshold (σ_{cd}) of the specimen (Wong et al. 1997; Fairhurst and Hudson 1999; Yang et al. 2015) is defined as the

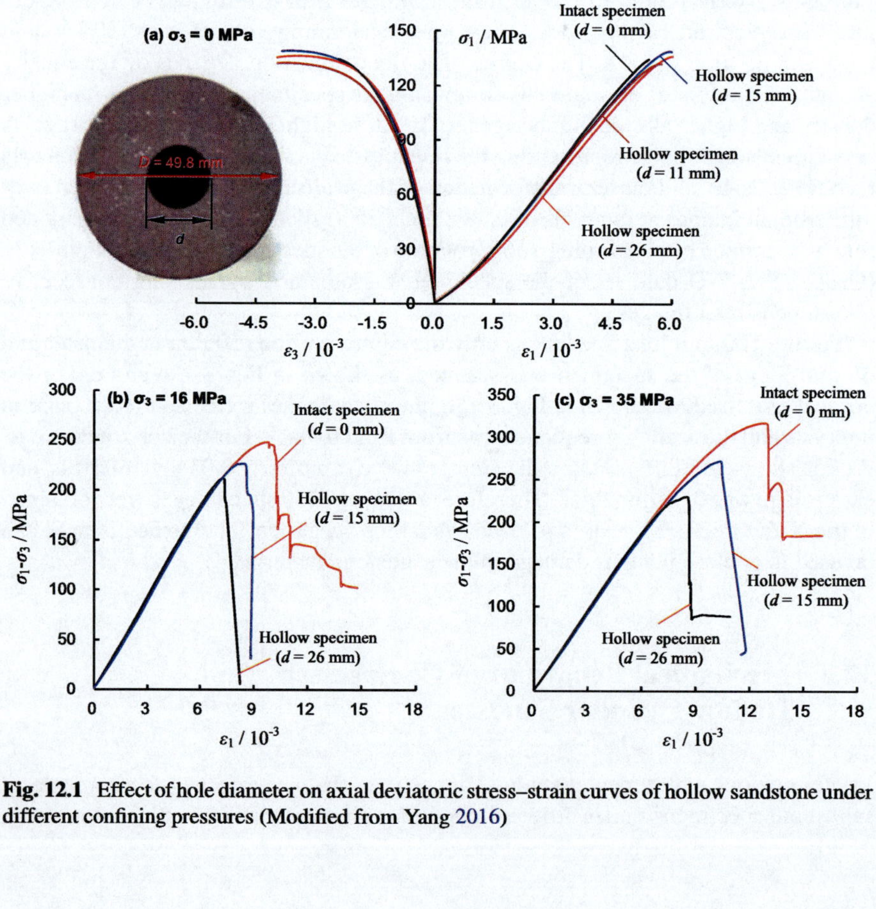

Fig. 12.1 Effect of hole diameter on axial deviatoric stress–strain curves of hollow sandstone under different confining pressures (Modified from Yang 2016)

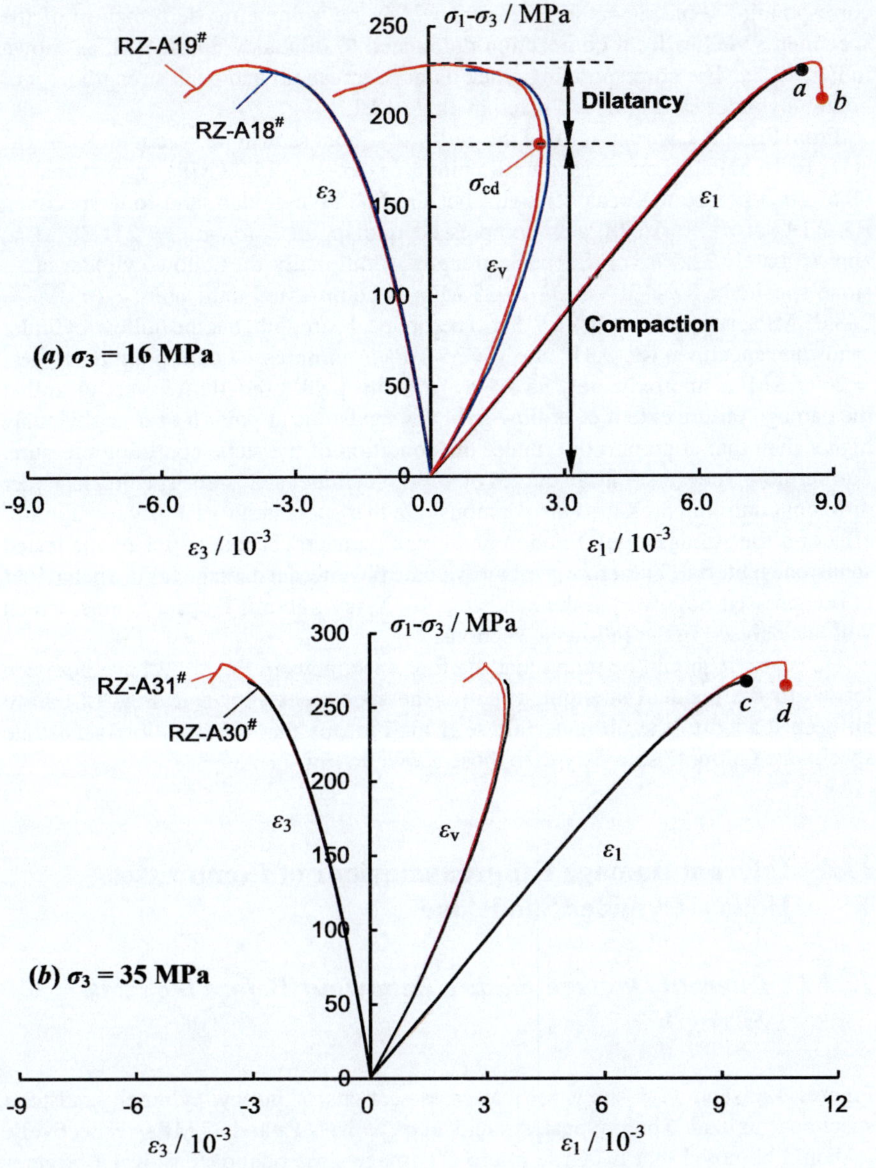

Fig. 12.2 Axial deviatoric stress–strain curves of hollow-cylinder sandstone specimens (d = 15 mm). **a** $\sigma_3 = 16$ MPa; **b** $\sigma_3 = 35$ MPa

corresponding axial deviatoric stress at which the volumetric deformation of the specimen switches from compaction-dominated to dilatancy-dominated, as shown in Fig. 12.2a. The corresponding crack damage threshold and peak strength (σ_p) of hollow-cylinder sandstone are listed in Table 12.1.

From Fig. 12.2, we can see that the hollow-cylinder sandstone specimen RZ-A18# at $\sigma_3 = 16$ MPa is compressed until point *a* (σ_1-$\sigma_3 = 225.99$ MPa, approximately 98.6% σ_p) before the peak strength, but the hollow-cylinder sandstone specimen RZ-A19# at $\sigma_3 = 16$ MPa is compressed until point *b* (σ_1-$\sigma_3 = 211.88$ MPa, approximately 92.5% σ_p) after peak strength. Additionally, the hollow-cylinder sandstone specimen RZ-A30# at $\sigma_3 = 35$ MPa is compressed until point *c* (σ_1-$\sigma_3 =$ 268.01 MPa, approximately 95.7% σ_p) before peak strength, but the hollow-cylinder sandstone specimen RZ-A31# at $\sigma_3 = 35$ MPa is compressed until point *d* ($\sigma_1-\sigma_3$ $= 267.61$ MPa, approximately 95.5% σ_p) after the peak strength. It is very clear that the damage failure extent of hollow-cylinder sandstone at point *b* or *d* is obviously higher than that at point *a* or *c* under the condition of the same confining pressure. Furthermore, the stress–strain curves of hollow-cylinder sandstone specimens under different confining pressures also demonstrate that specimen variability has a minor effect on the strength, deformation and crack damage characteristics of the tested sandstone material. Therefore, we can evaluate the internal damage failure behaviour of compressed hollow-cylinder sandstone by X-ray micro CT observations, which will be analysed in the following sections.

However, it should be noted that the four specimens in Fig. 12.2 have not been loaded to the residual strength, whereas the specimens shown in Fig. 12.1 have all been loaded to the ultimate failure, which means that these hollow sandstone specimens fail more seriously than those shown in Fig. 12.2.

12.3 Internal Damage Failure Behaviour of Compressed Hollow-Cylinder Sandstone

12.3.1 Internal Damage Failure Behaviour Before the Peak Strength

Figures 12.3 and 12.4 show vertical cross-sections of hollow-cylinder sandstone specimens unloaded before peak strength at $\sigma_3 = 16$ MPa and 35 MPa, respectively. It should be noted that in X-ray micro CT images, the boundary shows a brighter color, which may not refer to higher density and is due to the image processing method adopted to filter out the ring noise in this research. In our previous studies on damage mechanical behavior of sandstone (Yang et al. 2015), marble (Yang et al. 2016) and granite (Yang et al. 2017), the brighter colors in the boundary of X-ray micro CT images are also found. However, this does not affect the analysis of internal damage and crack system by using X-ray micro CT images. From Figs. 12.3 and

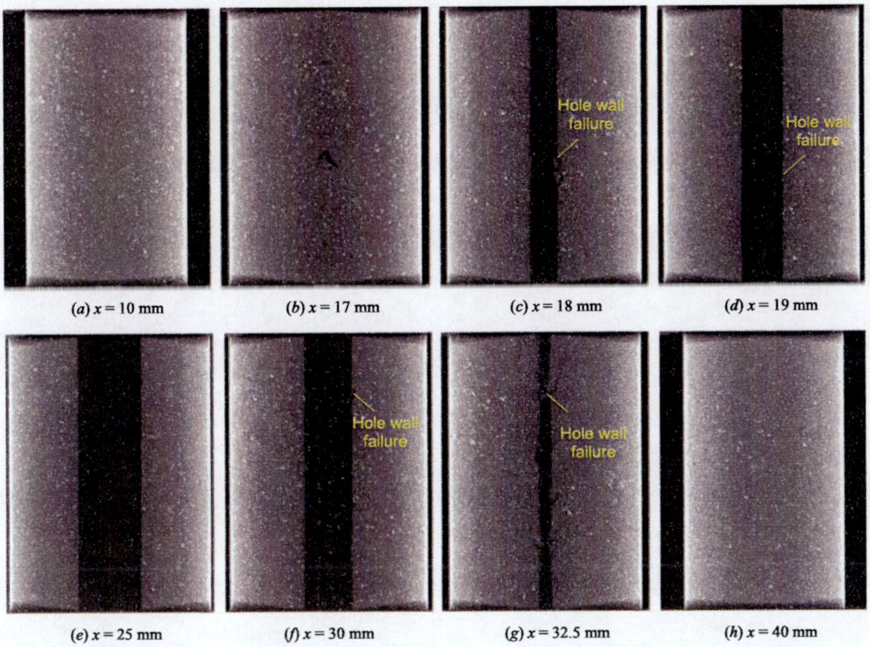

Fig. 12.3 Vertical cross-sections with different depths of hollow-cylinder sandstone specimen RZ-A18# ($\sigma_3 = 16$ MPa)

12.4, we can analyse the effect of depth (x) on the extent of internal damage to compressed hollow-cylinder sandstone specimens before the peak strength.

For hollow specimen RZ-A18#, at $x = 17$–19 mm and 30–32.5 mm, some minor hole wall fractures were observed. However at other depths, the hollow-cylinder sandstone specimen has no cracks. Compared with hollow specimen RZ-A18#, the hollow specimen RZ-A30# has more hole wall fractures and a more serious damage due to the action of higher confining pressure even though with approximate stress states before the peak strength, which can be validated by vertical cross-sections at $x = 17$–32.5 mm in Fig. 12.4. Figure 12.5 further demonstrates that for the same axial stress level, hollow-cylinder sandstone under a higher confining pressure has a more complicated crack system than that under a lower confining pressure, which results from the higher tangential stresses of hole walls for a greater confining pressure. Furthermore, it should be noted that in Fig. 12.5, these cracks are all horizontal surface cracks with no larger depths along the radial direction, as shown in Figs. 12.3 and 12.4, and are actually the result of shear slippage.

Figures 12.6 and 12.7 illustrate horizontal cross-sections of hollow-cylinder sandstone specimens unloaded before the peak strength at $\sigma_3 = 16$ MPa and 35 MPa, respectively. From Figs. 12.6 and 12.7, we can analyse the effect of height (z) on the extent of internal damage to the compressed hollow-cylinder sandstone specimen before the peak strength.

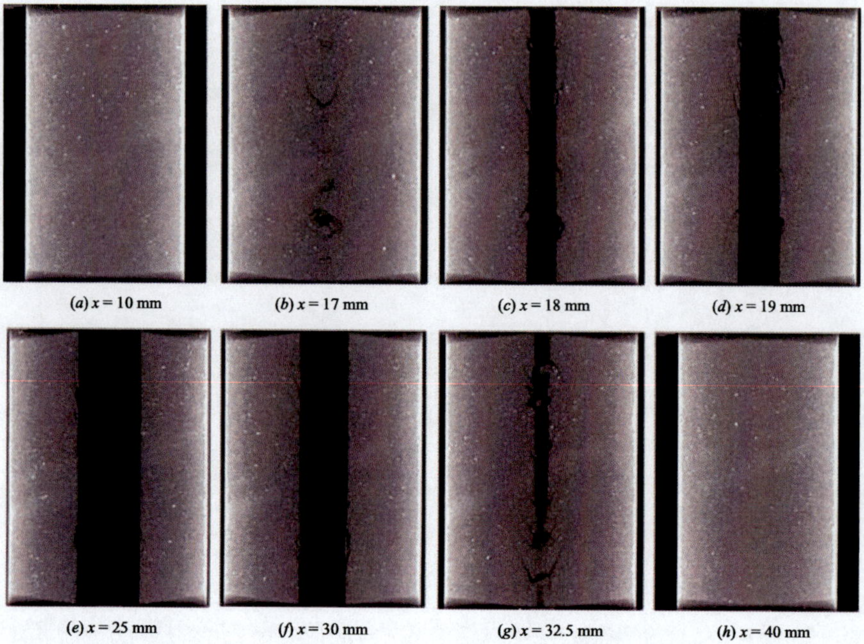

Fig. 12.4 Vertical cross-sections with different depths of hollow-cylinder sandstone specimen RZ-A30# ($\sigma_3 = 35$ MPa)

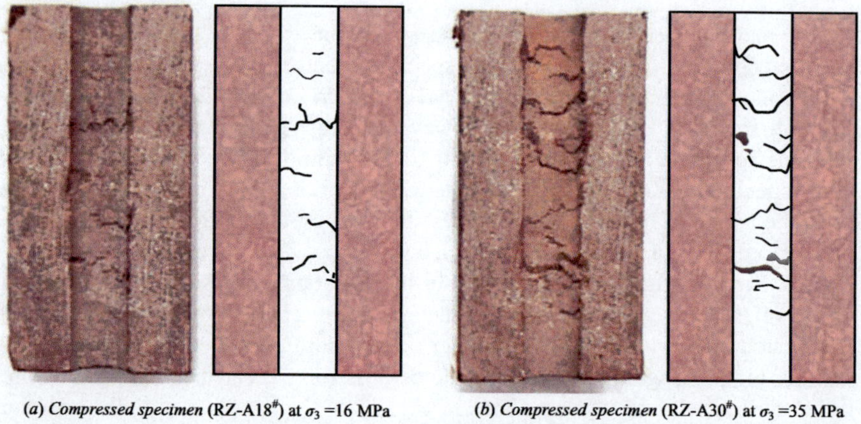

Fig. 12.5 Internal crack mode of hollow-cylinder sandstone specimens RZ-A18# ($\sigma_3 = 16$ MPa) and RZ-A30# ($\sigma_3 = 35$ MPa)

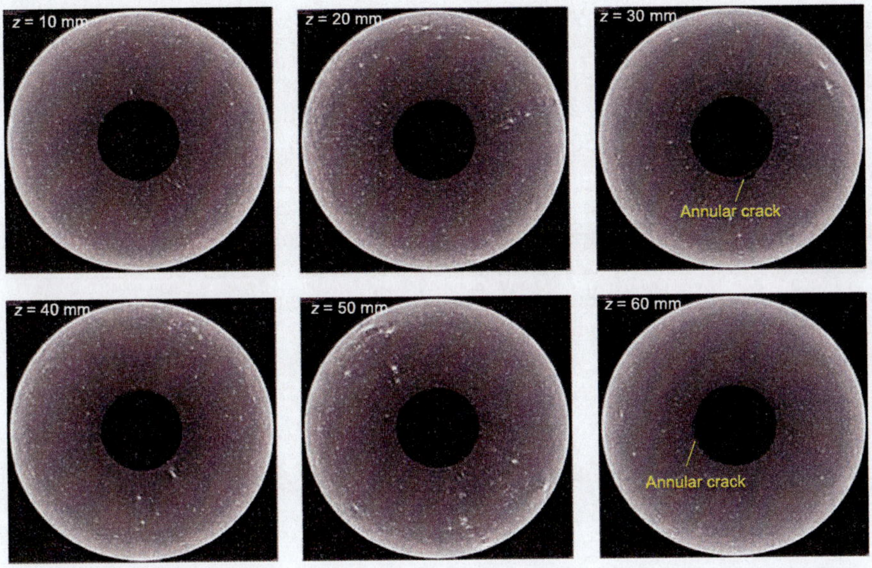

Fig. 12.6 Horizontal cross-sections different heights of hollow-cylinder sandstone specimen RZ-A18# ($\sigma_3 = 16$ MPa).

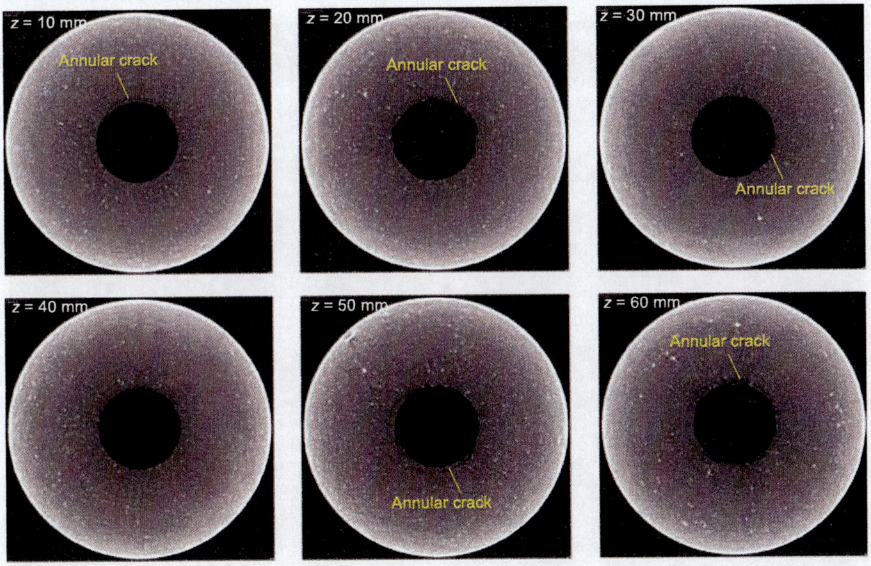

Fig. 12.7 Horizontal cross-sections with different heights of hollow-cylinder sandstone specimen RZ-A30# ($\sigma_3 = 35$ MPa)

For hollow specimen RZ-A18#, the crack system is very simple. At $z = 30$ and 60 mm, one annular crack is observed and a minor damage to the specimen occurs. However, at other heights, no hole wall fractures are observed. However, for hollow specimen RZ-A30#, at $z = 40$ mm, no internal wall fractures are observed, but at $z = 10 \sim 30$ mm and $z = 50 \sim 60$ mm, there are several annular cracks emanating close to internal walls. Compared with hollow specimen RZ-A18#, hollow specimen RZ-A30# has more annular cracks and more serious damage due to the action of the higher confining pressure even though with approximate stress states before the peak strength. Moreover, these annular cracks result mainly from the shear slippage along the direction of the hole wall due to the higher tangent stress concentration. In addition, hole wall fractures in hollow specimens RZ-A18# and RZ-A30# are not axisymmetric.

12.3.2 Internal Damage Failure Behaviour After the Peak Strength

Figures 12.8 and 12.9 show vertical cross-sections of hollow-cylinder sandstone specimens unloaded after the peak strength at $\sigma_3 = 16$ MPa and 35 MPa, respectively.

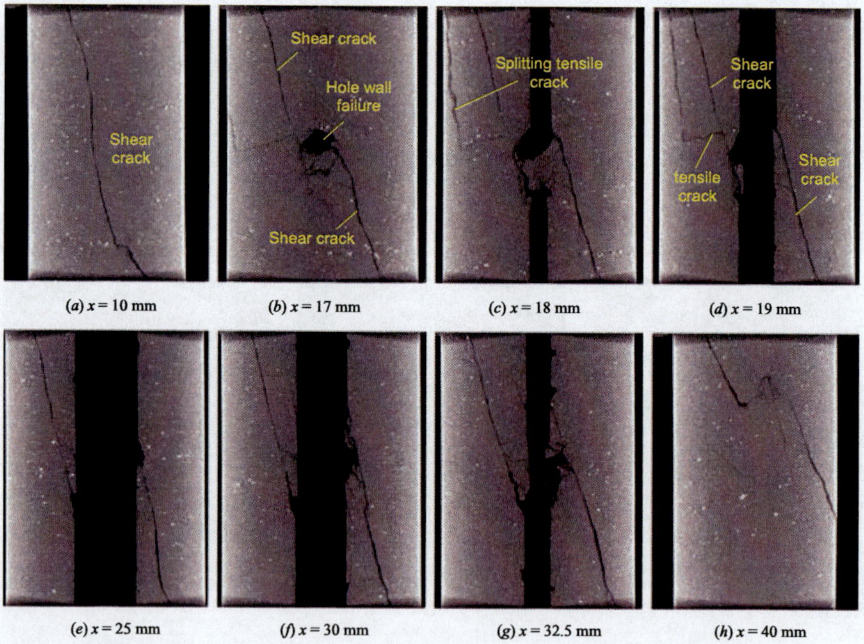

Fig. 12.8 Vertical cross-sections with different depths of hollow-cylinder sandstone specimen RZ-A19# ($\sigma_3 = 16$ MPa)

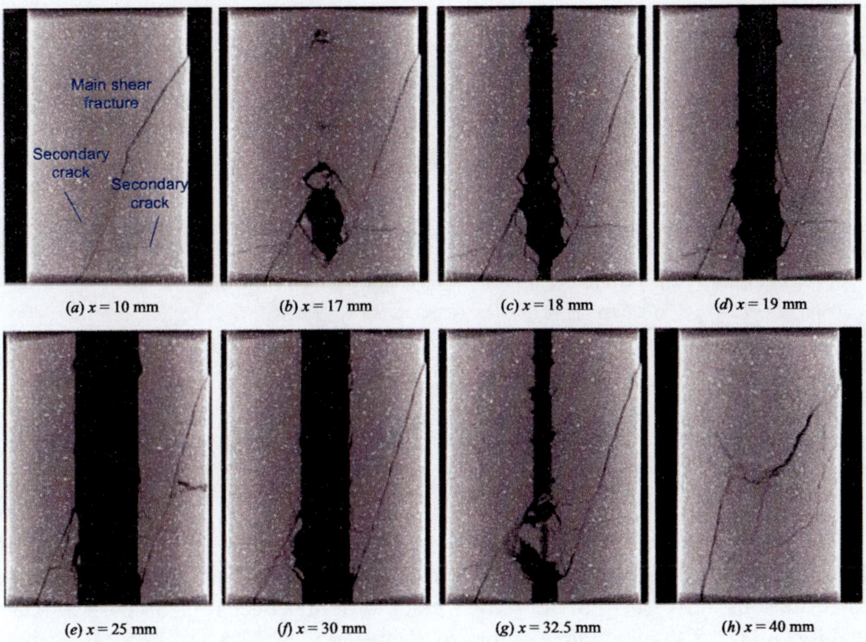

Fig. 12.9 Vertical cross-sections with different depths of hollow-cylinder sandstone specimen RZ-A31# ($\sigma_3 = 35$ MPa)

From Figs. 12.8 and 12.9, we can analyse the effect of depth (x) on the extent of internal damage to compressed hollow-cylinder sandstone specimens after the peak strength.

For hollow specimen RZ-A19#, the crack system is very complex compared with hollow specimen RZ-A18#, which means that hollow-cylinder sandstone specimen with higher compressed deformation after the peak strength have a more complicated vertical crack system than those with lower compressed deformation before the peak strength. From Fig. 12.8, it can be seen that at $x = 10$ mm, the cross-section is divided into two halves by one large crack. It should be noted that this large crack at $x =$ 10 mm is actually a shear crack even though the crack direction is approximately parallel to the major principal stress, as demonstrated in Fig. 12.10a. At $x = 17$ mm, larger hole fractures are observed. Some crushing fractures also emanate at $x =$ 18 mm, as shown in Fig. 12.8c. One larger axial splitting tensile crack is observed at $x = 18$ and 19 mm. At $x = 25 \sim 30$ mm, the hollow-cylinder sandstone specimen is mainly dominated by two shear cracks, as shown in Fig. 1212.8and f. Furthermore, in the middle region, the hole wall fractures are very serious. The shear crack has an angle of approximately 75° with horizontal direction (i.e., confining pressure). When the depth reaches 32.5 mm, in addition to the two main shear cracks, the hole wall fractures are also very obvious. Moreover, from Fig. 12.8e–h, we can conclude that the fracture modes of the hollows specimens are not axially symmetric.

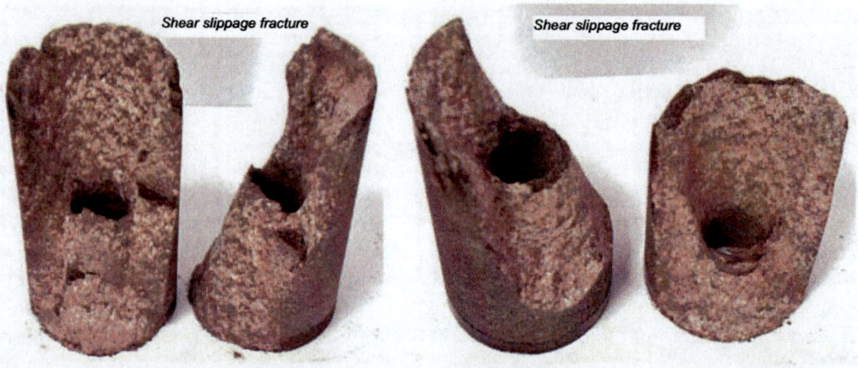

(*a*) *Compressed specimen* (RZ-A19#) at $\sigma_3 = 16$ MPa (*b*) *Compressed specimen* (RZ-A31#) at $\sigma_3 = 35$ MPa

Fig. 12.10 Shear fracture mode of hollow-cylinder sandstone specimens RZ-A19# ($\sigma_3 = 16$ MPa) and RZ-A31# ($\sigma_3 = 35$ MPa)

For hollow specimen RZ-A31#, the crack system is approximately the same as that of hollow specimen RZ-A19#. However, due to the difference in confining pressures, there are some differences in the vertical crack system between specimens RZ-A19# and RZ-A31#, as shown in the comparison between Figs. 12.8 and 12.9.

At $x = 10$ mm, the angle of the large shear crack in the horizontal direction in specimen RZ-A31# at $\sigma_3 = 35$ MPa is lower than that in specimen RZ-A19# at $\sigma_3 = 16$ MPa. Even though the specimen failed mainly by the shear fracture, a lot of secondary tiny cracks are also observed at $\sigma_3 = 35$ MPa, as shown in Fig. 12.9 (a). At $x = 17$–32.5 mm, the hollow-cylinder sandstone specimen at $\sigma_3 = 35$ MPa is mainly dominated by two shear cracks, as shown in Fig. 12.10b, which is similar to the result at $\sigma_3 = 16$ MPa. The right shear crack has an angle of approximately 71° with the horizontal direction, 4° lower than that at $\sigma_3 = 16$ MPa, whereas the left shear crack has an angle of approximately 64° with the horizontal direction, 11° lower than that at $\sigma_3 = 16$ MPa. Furthermore, the extents of the hole wall fractures at $\sigma_3 = 35$ MPa are more serious than those at $\sigma_3 = 16$ MPa, which results from the larger tangent stresses under higher confining pressures. In general, the crack system is more complicated at $\sigma_3 = 35$ MPa than at $\sigma_3 = 16$ MPa.

From Fig. 12.10, it can be seen that the main fracture mode of compressed hollow specimen is shear fracture, but there are also a lot of tiny secondary cracks that may be initiated in shear or tensile mode. Figures 12.11 and 12.12 illustrate horizontal cross-sections of hollow-cylinder sandstone specimens unloaded after the peak strength at $\sigma_3 = 16$ MPa and 35 MPa, respectively. In Figs. 12.11 and 12.12, these arrows represent mainly the moving directions of main shear crack in the horizontal CT cross-sections of hollow-cylinder sandstone specimen with the increase of height. From Figs. 12.11 and 12.12, we can analyse the effect of height (z) on the extent of internal damage to compressed hollow-cylinder sandstone specimens after the peak strength.

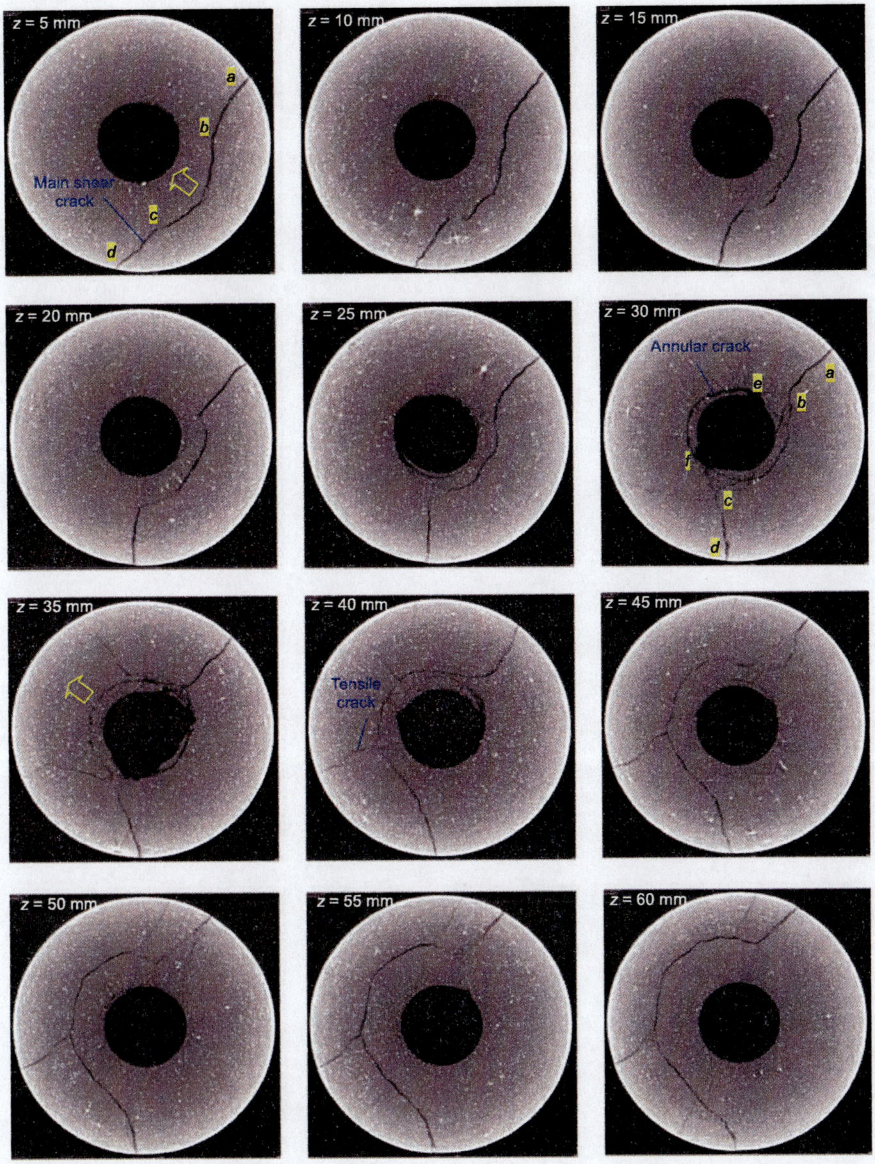

Fig. 12.11 Horizontal cross-sections with different heights of hollow-cylinder sandstone specimen RZ-A19# ($\sigma_3 = 16$ MPa). These arrows represent mainly the moving directions of main shear crack in the horizontal CT cross-sections of hollow-cylinder sandstone specimen with the increase of height

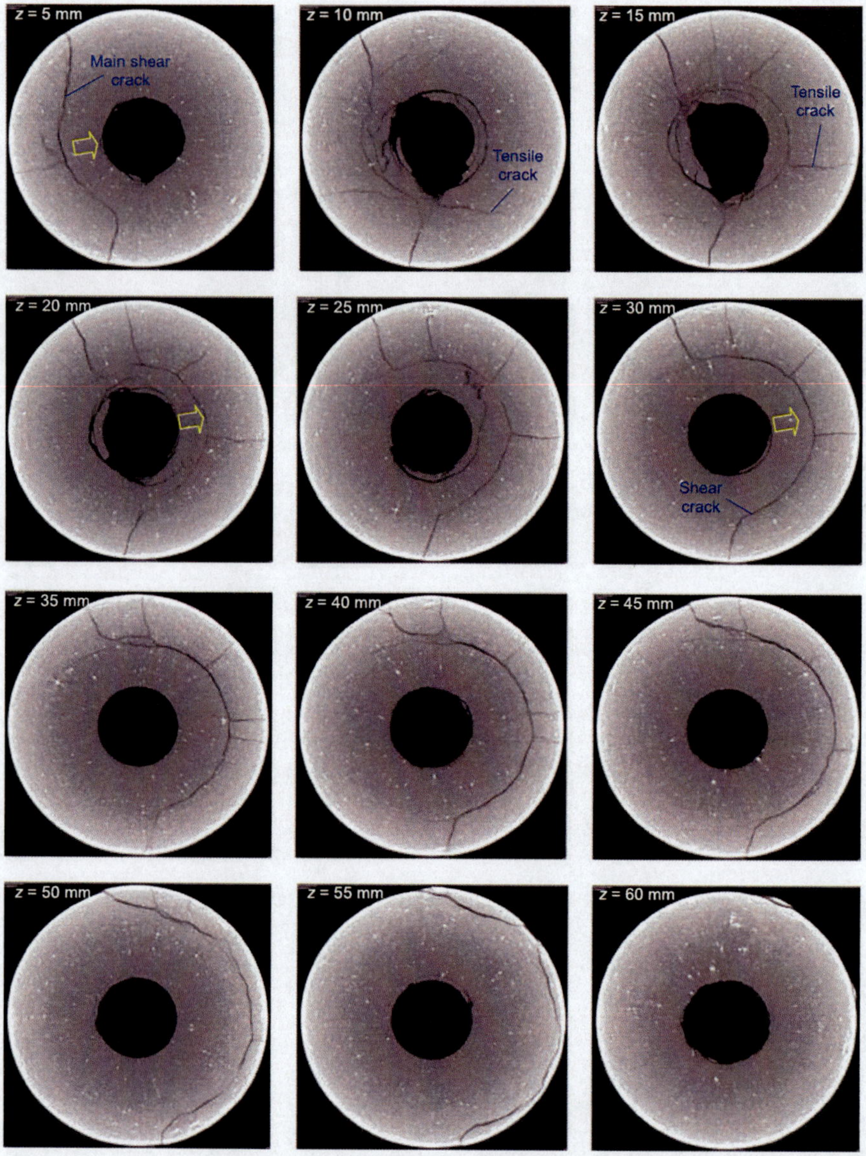

Fig. 12.12 Horizontal cross-sections with different heights of hollow-cylinder sandstone specimen RZ-A31# ($\sigma_3 = 35$ MPa). These arrows represent mainly the moving directions of main shear crack in the horizontal CT cross-sections of hollow-cylinder sandstone specimen with the increase of height

For hollow specimen RZ-A19#, at $z = 5$–25 mm, the horizontal cross-section is located in the bottom region of the specimen, no hole wall fractures are observed but one large inclined crack is observed. Moreover the trace of the crack (curve a-b-c-d) is not linear. With the increase in height, the crack (a-b-c) moves towards the internal wall gradually, but the crack c-d occurs the deviation from 30° to 90° at an angle with the direction of the confining pressure. When the height reaches 30 mm, crack b–c approaches the internal wall, and three tensile cracks coalesce between crack b–c and the internal wall. At the same time, close to the right region of the internal wall, one large annular crack e–f is observed, which indicates serious damage fracture. It should be noted at $z = 30$–35 mm, the horizontal cross-section is located in the middle region of the specimen, and the internal damage to the specimen is more serious. At $z = 35$ mm, all around the internal wall, the specimen all emanates the damage fracture zone (i.e., annular cracks), which result from the shear fracture. Some serious localized crushing zones are also observed. At the same time, we can see that at $z = 35$–60 mm, some tensile cracks are observed between the main crack a-b-e-f-c-d and the outer wall, which are different from those at $z = 5$–30 mm. Moreover, with the increase of height from 35 to 60 mm, the main crack a-b-e-f-c-d moves towards the outer wall along the direction of the denoted arrow. Furthermore at $z = 45$–60 mm, almost no internal wall fractures are observed. In addition, from the horizontal cross-section at $z = 30$ and 35 mm, it is clear that the internal damage failure zones are not axisymmetric, which is difficult to explain by using thick-wall cylinder theory.

For hollow specimen RZ-A31#, the crack system is similar to that of hollow specimen RZ-A19#. However, due to the difference in the confining pressures, there is some difference in the horizontal crack systems between specimens RZ-A19# and RZ-A31#, as shown in the comparison between Figs. 12.11 and 12.12. At $z =$ 5 mm, there is one large shear crack in specimen RZ-A31# at $\sigma_3 = 35$ MPa, which is similar to that at $\sigma_3 = 16$ MPa. With the increase of height, at z = 10–25 mm, many annular cracks around the hole wall are observed, which is the result of multiple shear slippage fractures along the direction of axial stress. Moreover, for the specimen at $\sigma_3 = 35$ MPa more tensile cracks emanate along the radial direction than for that at $\sigma_3 = 16$ MPa, and the damage fracture zones at $\sigma_3 = 35$ MPa are more serious than those at $\sigma_3 = 16$ MPa, which results from the increase of confining pressure.

Afterwards, with the increase in height, the main shear crack moves towards the right edge of the specimen and less annular cracks around the hole wall can be found, as at $z = 30$–60 mm shown in Fig. 12.12. Furthermore, the hole sandstone specimen at $\sigma_3 = 35$ MPa exhibits on more non-axis symmetry than that at $\sigma_3 =$ 16 MPa. Generally, the crack system of hollow-cylinder sandstone specimens with higher compressed deformation after the peak strength are also more complicated than those with lower compressed deformation before peak strength. The hollow-cylinder sandstone specimens with lower compressed deformation are only dominated by some internal wall fractures due to the action of shear slippage; whereas the hollow-cylinder sandstone specimens with higher compressed deformation are mainly dominated by two shear fractures, also including some internal wall fractures.

12.3.3 *Effect of Borehole Size on Internal Damage Behaviour After the Peak Strength*

Figures 12.13, 12.14 depict the effect of borehole size on the internal damage behaviour of compressed specimens via horizontal cross-section CT images obtained after uniaxial compression failure. It should be noted that these CT images are obtained for the compressed specimens after they were loaded to the ultimate failure, as shown in Fig. 12.1b and c. According to Figs. 12.13, 12.14, the borehole size has a great effect on the internal damage of hollow sandstone. Comparing Figs. 12.13 and 12.14, it can be seen that the internal damage depends not only on the borehole size but also on the confining pressure. With the increase in borehole size, the crack system of hollow sandstone seems to become more complex for the same confining pressure. However, with the increase of confining pressure, the crack system of hollow sandstone seems to become simpler for the same borehole size. The reason can be explained as follows.

When the confining pressure does not equal to zero and the internal hole pressure is zero, the radial and tangential stresses in the hollow specimen are non-uniform according to thick-wall cylinder elastic theory. The tangential stress decreases nonlinearly but the radial stress increases gradually with increasing radius (from the internal wall to outer wall). The non-uniform stress distribution results in differences in hollow sandstone specimens with various borehole sizes at different confining pressures. Under uniaxial compression, hollow sandstone specimens with various borehole sizes all experience axial splitting tensile failure and the stress distribution is more uniform (Fig. 12.13). Therefore, the UCS of hollow sandstone does not depend on the borehole size (Fig. 12.1a).

However, under confining pressure, hollow sandstone usually fails mainly via a single shear fracture along the direction of the maximum shear stress. At the same borehole size, the peak strength of hollow sandstone increases with the confining pressure (Yang 2016). However, the shear fracture area of hollow sandstone is different for various borehole sizes. The larger the borehole size is, the smaller the area of the shear fracture. Thus, the corresponding axial supporting capacity is lower for specimens with larger borehole sizes. Therefore, under the same confining pressure, with increasing borehole size, the peak strength will decrease gradually (Fig. 12.1b, c). Moreover, the sensitivity of the peak strength of hollow sandstone to borehole size at a larger confining pressure is obviously higher than that at a smaller confining pressure, which potentially occurs because the non-uniform stress distribution increases with confining pressure at an internal hole pressure equal to zero.

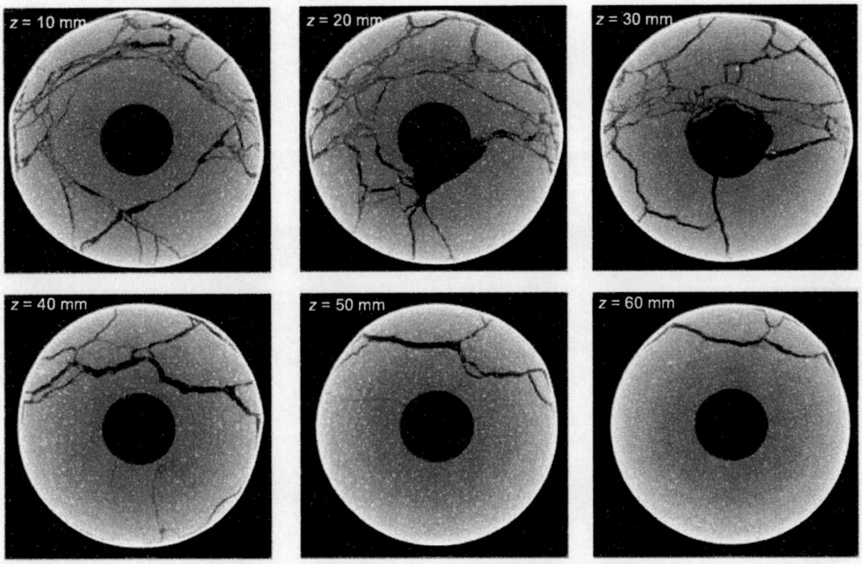

(a) Hollow sandstone ($d = 15$ mm) after uniaxial compression failure

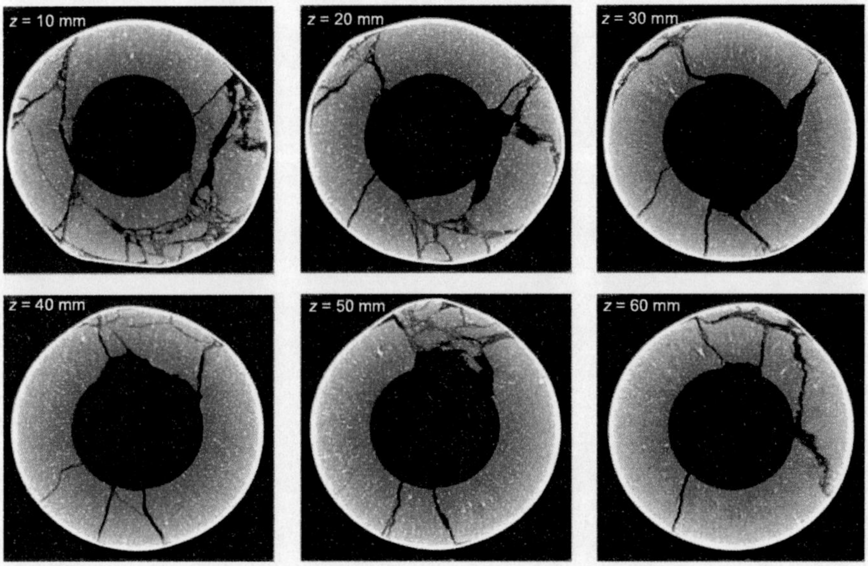

(b) Hollow sandstone ($d = 26$ mm) after uniaxial compression failure

Fig. 12.13 Effect of borehole size on horizontal cross-sections with different heights of hollow sandstone after uniaxial compression failure

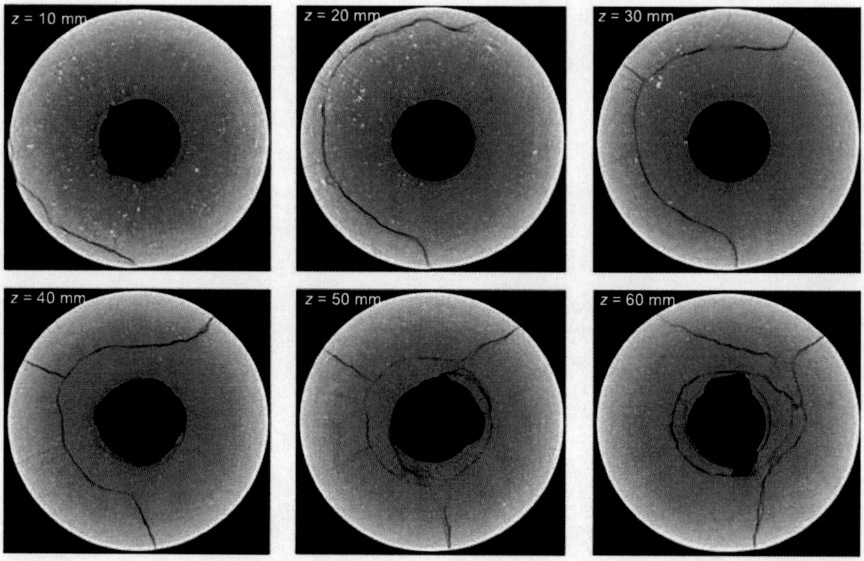

(a) Hollow sandstone ($d = 15$ mm) after triaxial compression failure ($\sigma_3 = 35$ MPa)

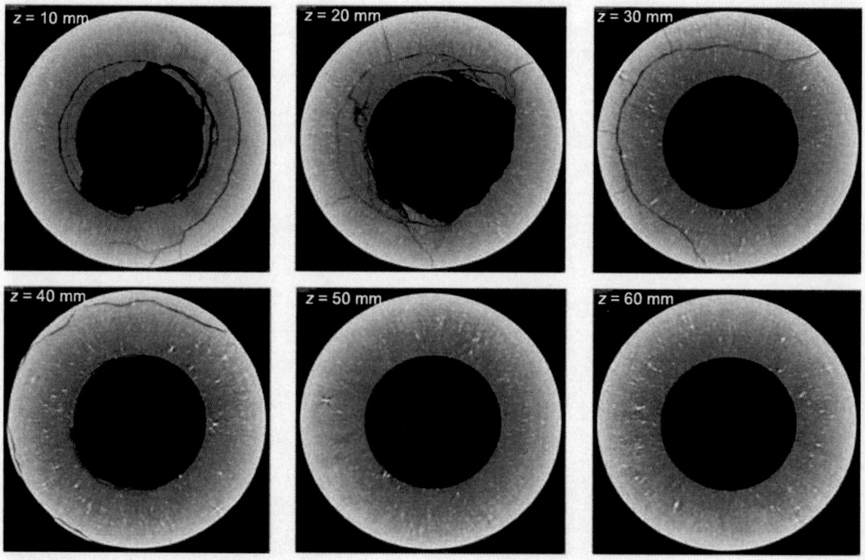

(b) Hollow sandstone ($d = 26$ mm) after triaxial compression failure ($\sigma_3 = 35$ MPa)

Fig. 12.14 Effect of borehole size on horizontal cross-sections with different heights of hollow sandstone after triaxial compression failure ($\sigma_3 = 35$ MPa)

12.4 Discussion on the Fracturing Mechanism of Hollow-Cylinder Sandstone

12.4.1 Three-Dimensional Reconstruction of X-ray CT Images

Figure 12.15 shows 3-D CT images of uncompressed and compressed hollow-cylinder sandstone specimens (d = 15 mm) after triaxial compression failure (σ_3 = 16 MPa). In the 3-D images, fracture regions with lower CT numbers are black, and other regions with relatively higher CT numbers are transparent. Compared with the 3-D image of the uncompressed specimen shown in Fig. 12.15a, it can be seen that when the hollow-cylinder sandstone specimen RZ-A18# is loaded to point *a* before the peak strength, the hollow-cylinder sandstone specimen shows some local fracture zones at the internal hole wall as in Fig. 12.15b. However, when the hollow-cylinder sandstone specimen RZ-A19# is loaded to point *b* after the peak strength, the failure mode is obviously different from that at point *a* before the peak strength. The hollow specimen shows two main shear fracture modes as shown in Fig. 12.15c. Furthermore, by rotating the image of the specimen, the trace of the shear fracture plane appears as a curve, not a plane as demonstrated by Fig. 12.8a, which indicates that the deformation is non-uniform throughout the shear fracture curve, as demonstrated by Fig. 12.10a.

Figure 12.16 shows 3-D CT images of uncompressed and compressed hollow-cylinder sandstone specimens (d = 15 mm) after triaxial compression failure (σ_3 = 35 MPa). Compared with Figs. 12.16a and 12.15b, even though the two hollow-cylinder sandstone specimens reach the stress level of the triaxial peak strength (Fig. 12.2), the extent of internal wall fracture in the hollow-cylinder sandstone

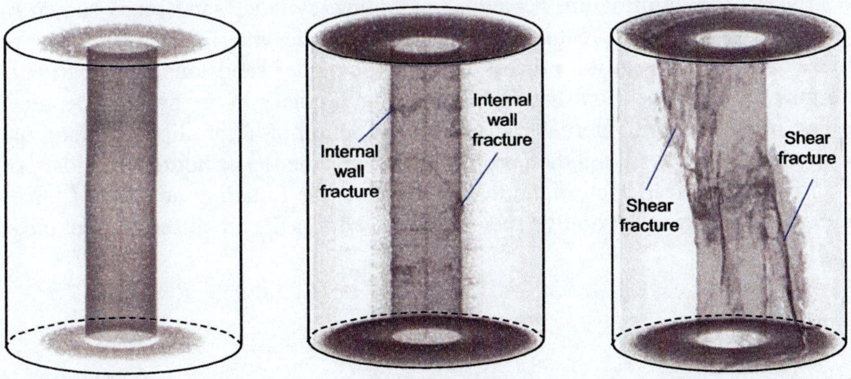

(*a*) *Uncompressed specimen* (*b*) *Compressed specimen* (RZ-A18#) (*c*) *Compressed specimen* (RZ-A19#)

Fig. 12.15 Three-dimensional reconstruction of X-ray CT images of uncompressed and compressed hollow-cylinder sandstone specimens (d = 15 mm, σ_3 = 16 MPa)

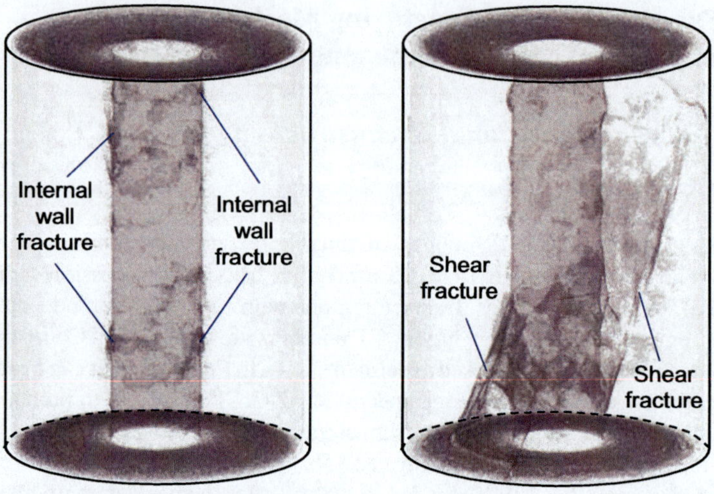

(*a*) *Compressed specimen* (RZ-A30#) (*b*) *Compressed specimen* (RZ-A31#)

Fig. 12.16 Three-dimensional reconstruction of X-ray CT images of uncompressed and compressed hollow-cylinder sandstone specimens ($d = 15$ mm, $\sigma_3 = 35$ MPa)

specimen after triaxial compression failure ($\sigma_3 = 35$ MPa) is obviously higher than that at $\sigma_3 = 16$ MPa. However, when the hollow-cylinder sandstone specimen RZ-A31# is loaded to point *d* after the peak strength, the failure mode is obviously different from that at point *c* before the peak strength. The hollow specimen shows two main shear fracture modes, as demonstrated by Fig. 12.10b.

Even though two different hollow-cylinder sandstone specimens shown in Fig. 12.15b and c or in Fig. 12.16a and b, due to the very good consistency of the strength and deformation behaviour of hollow-cylinder sandstones shown in Fig. 12.2, from Figs. 12.15 and 12.16, we can predict the approximately three-dimensional crack evolution process of hollow-cylinder sandstone during triaxial deformation. It can be seen that three-dimensional cracks in the hollow specimen are first initiated at the internal wall due to the action of shear slippage along the direction of axial stress, and then propagate towards the top or bottom boundary of the specimen in shear fracture mode. Furthermore, the initiating position of three-dimensional cracks in the hollow specimen at $\sigma_3 = 35$ MPa are different from those at $\sigma_3 = 16$ MPa.

12.4.2 Discussion on Fracturing Mechanism of Hollow-Cylinder Sandstone

To analyse the internal crack evolution mechanism of hollow-cylinder sandstone specimens during the entire loading, conceptual failure models (see Figs. 12.17 and 12.18) of hollow-cylinder sandstone under different confining pressures were proposed in this research. In Figs. 12.17 and 12.18, σ_a, σ_p, σ_b and σ_{cr} indicate the yielding stress, peak strength, post-peak drop stress and residual strength, respectively. The arrows represent the coalescing direction of shear fractures in the rock specimens. The sandstone is very homogeneous at the macroscopic scale, but strongly heterogeneous at the mesoscopic scale. The macroscopic fracture phenomenon may be regarded as the average effect of mesoscopic failure and heterogeneous mesoscopic failure can be described by probability and statistics theory such as the Weibull distribution (Yang et al. 2008). Thus, we can regard the mesoscopic element material as a random variable and assume that it distributes randomly in the rock specimen. The strength and stiffness of the mesoscopic element material in the specimen vary strongly from one element to another.

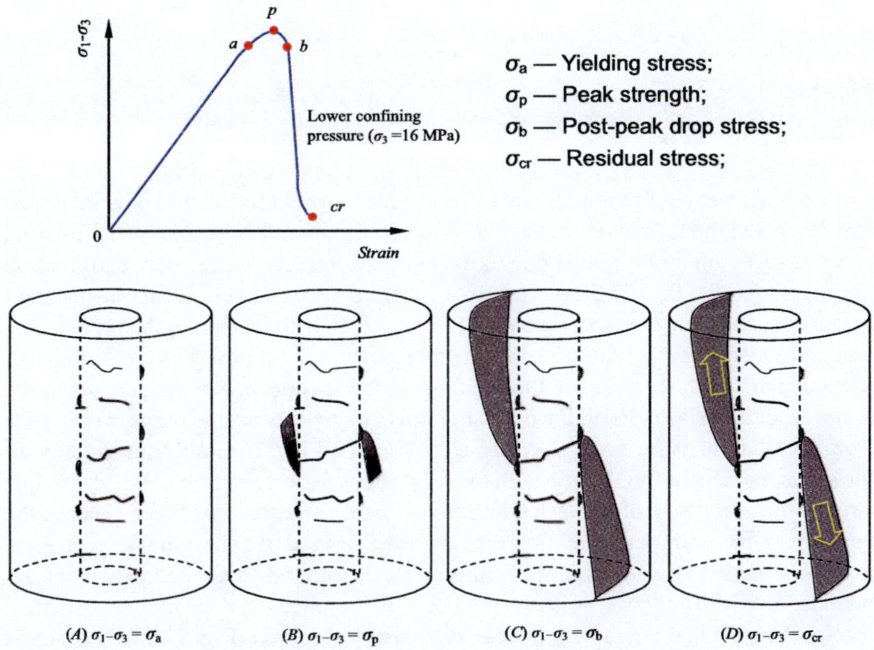

Fig. 12.17 A conceptual model of hollow-cylinder sandstone material under a lower confining pressure (σ_3 = 16 MPa). The arrows represent the coalescing direction of shear fractures in the rock specimens

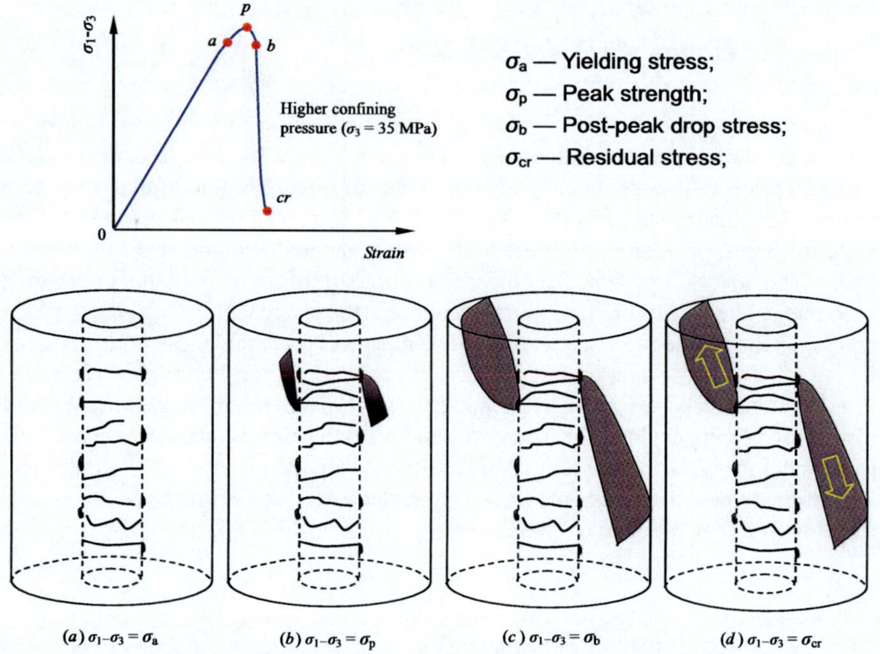

Fig. 12.18 Three-dimensional reconstruction of X-ray CT images of uncompressed and compressed hollow-cylinder sandstone specimens ($d = 15$ mm, $\sigma_3 = 35$ MPa)

Under a lower confining pressure ($\sigma_3 = 16$ MPa), with the increase of axial deformation, when the hollow-cylinder sandstone specimen is loaded to σ_a, the internal wall fractures are first initiated due to the higher differential stress concentration on the internal wall (Fig. 12.17a). However, internal wall fracture occurs at different heights because these mesoscopic material elements with low strength and stiffness gradually reach their maximum supporting capacity, exhibit post-yield softening and produce damage failure during the yielding of the specimen. Afterwards, when the hollow specimen is loaded to the peak strength (σ_p), two shear cracks begin to initiate from the position of the internal wall fracture (Fig. 12.17b). The initiation direction of the shear crack has a small angle with the axial stress. However, after the macroscopic stress–strain response of the tested hollow-cylinder sandstone specimen reaches the peak stress, larger internal wall fractures are observed and two shear cracks propagate at a lower angle to the maximum compressive axis until reaching the top and bottom boundaries of the hollow specimen (Fig. 12.17c). Finally, when the hollow specimen reaches the residual stress (σ_{cr}), shear slippage is observed (Fig. 12.17d), which is further validated by the fracture surface of the hollow specimen (Fig. 12.10a).

Under a higher confining pressure ($\sigma_3 = 35$ MPa), with the increase of axial deformation, when the hollow-cylinder sandstone specimen is loaded to σ_a, the internal wall fracture is first initiated due to the strength concentration with maximum tangential and radial stress on the internal wall (Fig. 12.18a). However, compared with the hollow specimen at $\sigma_3 = 16$ MPa, the hollow specimen at $\sigma_3 = 35$ MPa has more internal wall fractures due to the action of the higher confining pressure, which is also validated by Fig. 12.9. Afterwards, when the hollow specimen is loaded to the peak strength (σ_p), two shear cracks begin to initiate from the position of the internal wall fracture (Fig. 12.18b). It should be noted that the initiation position of the shear cracks in Fig. 20B is different from that at $\sigma_3 = 16$ MPa, as shown in Fig. 12.18b. However, after the macroscopic stress–strain response of the tested hollow-cylinder sandstone specimen reaches the peak stress, larger internal wall fractures are observed and two shear cracks propagate at a lower angle to the maximum compressive axis until reaching the top boundary and right bottom region of the hollow specimen (Fig. 12.18c). But compared with that at $\sigma_3 = 16$ MPa (Fig. 12.17b), the propagation directions of the shear crack in hollow-cylinder sandstone specimen at $\sigma_3 = 35$ MPa have a larger angle with the axial stress, which is also supported by the comparison of the vertical cross-section CT image between Figs. 12.8 and 12.9, or the comparison of the 3D CT image between Figs. 12.15c and 12.16b. Finally, when the hollow specimen reaches the residual stress (σ_{cr}), shear slippage is also observed (Fig. 12.18d), which is further validated by the fracture surface of the hollow specimen (Fig. 12.10b).

12.5 Conclusions

(1) Conventional triaxial compression experiments under two different confining pressures of 16 and 35 MPa were carried out on hollow-cylinder sandstone specimens (internal hole diameter of 15 mm and internal pressure equal to zero) under dry conditions at ambient temperature to investigate the internal damage failure mechanism of hollow rock in detail. The fracture process of hollow-cylinder sandstone was controlled using the axial displacement loading method, and four specimens were unloaded rapidly and compressed to various deformation positions, in which two specimens were unloaded before the peak strength and two specimens after the peak strength. Specimen variability has a minor effect on the strength, deformation and crack damage characteristics of the tested sandstone material under different confining pressures, which can be used to evaluate the internal damage failure behaviour of compressed hollow-cylinder sandstone.

(2) After uniaxial and triaxial compression failure, the hollow-cylinder sandstone specimens were analysed using an X-ray micro CT scanning system. Many X-ray CT images of the hollow specimens were generated, and three-dimensional images of hollow specimens were reconstructed. Based on the horizontal and vertical cross-section CT images, the internal damage behaviour of hollow-cylinder sandstone specimens was evaluated. It can be seen that the crack

system of hollow-cylinder sandstone specimens with higher compressed deformation after the peak strength are more complicated than those with lower compressed deformation close to the peak strength. In the present study, the hollow-cylinder sandstone specimens with lower compressed deformation are only dominated by some internal wall fractures; whereas the hollow-cylinder sandstone specimens with higher compressed deformation are mainly dominated by two shear fractures, as well as some internal wall fractures.

(3) The horizontal and vertical cross-section CT images and three-dimensional reconstructed images of hollow-cylinder sandstone all demonstrated that the three-dimensional cracks in the hollow specimen under triaxial compression are first initiated from the internal wall, and then propagate towards the top or bottom boundary of the specimen in a shear fracture mode. A conceptual model of hollow-cylinder sandstone material under the action of confining pressure are proposed, which can better explain the internal crack evolution mechanism of hollow-cylinder sandstone during the entire loading. Furthermore, for the same axial stress level, hollow-cylinder sandstone under a higher confining pressure has a more complicated crack system than that under a lower confining pressure. The investigated conclusions are significant for predicting the instability occurring around deep well bores in petroleum engineering and for ensuring the safety of deep excavation damage zones in tunnel engineering.

References

Alsayed MI (2002) Utilising the Hoek triaxial cell for multiaxial testing of hollow rock cylinders. Int J Rock Mech Min Sci 39:355–366

Fairhurst CE, Hudson JA (1999) Draft ISRM suggested method for the complete stress-strain curve for the intact rock in uniaxial compression. Int J Rock Mech Min Sci 36(3):279–289

Hawkes I, Mellor M (1970) Uniaxial testing in rock mechanics laboratories. Eng Geol 4:177–285

Lee DH, Juang CH, Chen JW, Lin HM, Shieh WH (1999) Stress paths and mechanical behavior of a sandstone in hollow cylinder tests. Int J Rock Mech Min Sci 36:857–870

Mogi K (1966) Some precise measurements of fracture strength of rocks under uniform compressive stress. Rock Mech Eng Geol 4:41–55

Wong TF, David C, Zhu W (1997) The transition from brittle faulting to cataclastic flow in porous sandstones: mechanical deformation. J Geophys Res 102(B2):3009–3025

Yang SQ (2016) Experimental study on deformation, peak strength and crack damage behavior of hollow sandstone under conventional triaxial compression. Eng Geol 213:11–24

Yang SQ, Jiang YZ, Xu WY, Chen XQ (2008) Experimental investigation on strength and failure behavior of pre-cracked marble under conventional triaxial compression. Int J Solids Struct 45:4796–4819

Yang SQ, Ranjith PG, Gui YL (2015) Experimental study of mechanical behavior and X-ray micro CT observations of sandstone under conventional triaxial compression. Geotech Test J 38(2):179–197

Yang SQ, Ju Y, Gao F, Gui YL (2016) Strength, deformability and X-ray micro-CT observations of deeply buried marble under different confining pressures. Rock Mech Rock Eng 49(11):4227–4244

Yang SQ, Ranjith PG, Jing HW, Tian WL, Ju Y (2017) An experimental investigation on thermal damage and failure mechanics behavior of granite after exposure to different high temperature treatments. Geothermics 65:180–197

Chapter 13 Thermal Damage and Failure Mechanical Behavior of Granite After Exposure to Different High Temperature Treatments Under Uniaxial Compression

In the past, uniaxial compression experiments have provided no satisfactory answers on the thermal damage and failure mechanism of granite. In addition, X-ray computer tomography (CT) observations have rarely considered granites after various high temperature treatments. Therefore, in this chapter, we report the results of a series of uniaxial compression tests on granite subjected to various high temperature treatments ranging from 25 to 800 °C (Yang et al. 2017). Based on the experimental results, we first investigate the influence of high temperature on the crack damage, strength and deformation parameters of granite material. Then, we analyze the effect of high temperature on the thermal damage characteristics of granite by using optical microscopic observation. Finally, using an X-ray micro CT scanning system and acoustic emission (AE) monitoring, the internal crack and failure process of deformed granite specimens with different high temperatures are analyzed in detail.

13.1 Experimental Material and Testing Procedure

13.1.1 Granite Material

The material used throughout this research was granite collected from Rizhao city in Shandong province, China, which was used to investigate the mechanical and thermal damage behavior of rock material subjected to high temperature heating. In this experiment, all specimens were cored from the same block of material to an actual diameter of 50 mm and approximately 100 mm in length, which resulted in a length: diameter ratio of 2: 1 for mechanical tests in accordance with the ISRM (International Society for Rock Mechanics) suggested method (Fairhurst and Hudson 1999).

In the present research, the high temperature treatments were administered to the specimens using a high-temperature furnace in the following procedure (Yang et al. 2014). Firstly, the specimens were heated to their target temperature (200, 300, 400,

S.-Q. Yang, *Mechanical Behavior and Damage Fracture Mechanism of Deep Rocks*,
https://doi.org/10.1007/978-981-16-7739-7_13

Table 13.1 Granite specimens after different high temperature treatments

Specimen	D_b/mm	L_b/mm	D_a/mm	L_a/mm	ρ_b/(kg/m^3)	ρ_a/(kg/m^3)	V_P/(m/s)	V_S (m/s)	T/°C
GA-1#	49.80	99.82	49.80	99.82	2643	2643	4517	1258	25
GA-2#	49.10	99.83	49.15	99.83	2645	2633	3781	1013	200
GA-8#	51.29	100.45	51.36	100.6	2650	2632	3517	941	300
GA-3#	49.70	98.13	49.76	98.16	2644	2628	3020	878	400
GA-9#	51.26	100.95	51.39	101.34	2640	2608	2717	787	500
GA-4#	49.50	99.08	49.96	100.07	2650	2565	1651	501	600
GA-6#	49.80	98.47	50.94	101.18	2647	2449	670	177	700
GA-5#	49.80	98.79	51.68	102.03	2638	2358	441	131	800

Note *D*-Diameter; *L*-Length; *ρ*-Density; V-Wave velocity; T-Temperature

500, 600, 700 and 800 °C) at a rate of 5 °C/min, and the heating rate is also same as the previous experiments for Strathbogie granite (Shao et al. 2015) and for sandstone (Ding et al. 2016; Zhu et al. 2016). They were then left at their target temperature for two hours. Finally, the specimens were cooled naturally to room temperature (25 °C) in a confined space. Table 13.1 lists the tested granite specimens and conditions in this research. In Table 13.1, V_P and V_S were measured using digital sonic equipment. The subscripts "*a*" and "*b*" in the diameter and length indicate the values obtained "after" and "before" high temperature treatment.

13.1.2 Mineral Composition of Granite

According to the results of X-ray diffraction (XRD), the minerals in the granite material at room temperature are quartz (Q), feldspar (F), biotite (B), amphibole, chlorite and dolomite. The detailed composition of this rock is as follows: 11.12% quartz, 59.85% feldspar, 21.56% biotite, 6% amphibole, 1.01% chlorite and 0.46% dolomite. Three main mineral grains, i.e. Q, F and B, are marked in Fig. 13.1. Some thermal cracks in granite specimen under higher temperatures can be observed, which will be analyzed in detail in Sect. 13.2. Figure 13.2 and Table 13.2 show the effect of high temperature treatment on the mineral composition of granite. From Fig. 13.2 and Table 13.2, it is clear that the mineral composition of granite does not change with increasing temperature, which is similar to the previous findings for tuff (Heap et al. 2012). For each grain, some differences of mineral composition result from the heterogeneity of the granite material.

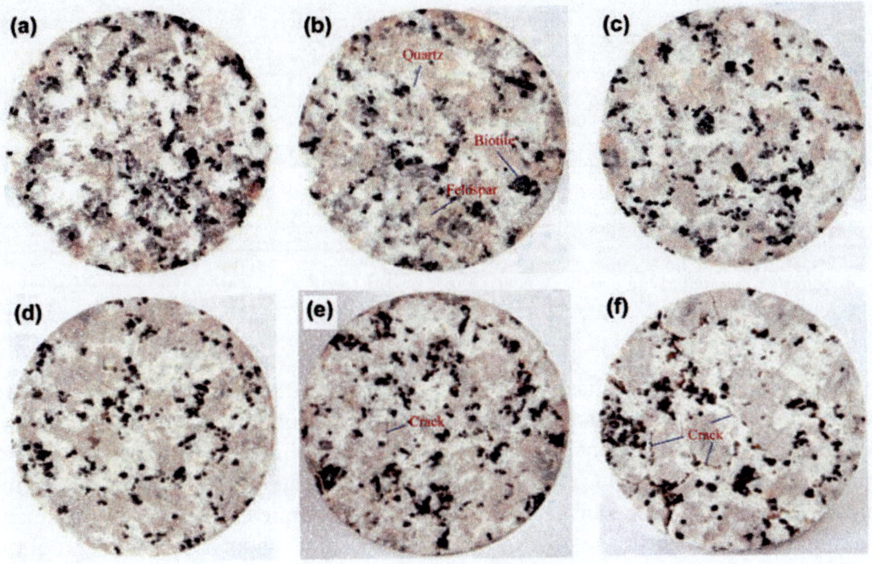

Fig. 13.1 The end surface of granite after different high temperature treatment. **a** T = 25 °C. **b** T = 200 °C. **c** T = 400 °C. **d** T = 600 °C. **e** T = 700 °C. **f** T = 800 °C

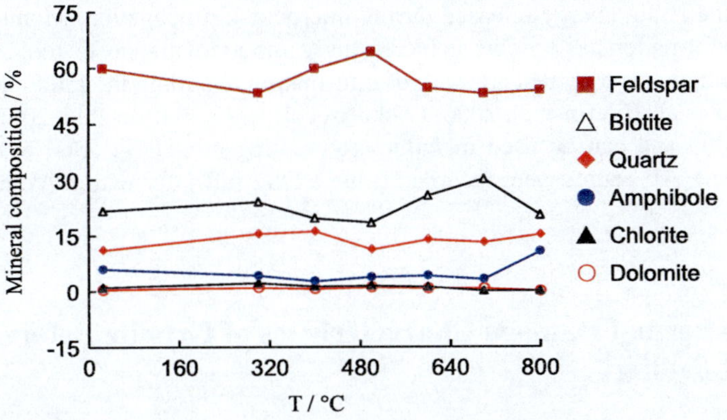

Fig. 13.2 Effect of temperature on the mineral composition of granite

13.1.3 Mechanical Testing and AE Procedure

All experiments were performed in a TAW-1000 servo-controlled rock mechanics experimental system. The maximum loading capacity of the servo-controlled system is 1000 kN, and the maximum confining pressure is 60 MPa. During experimentation, we measured the axial force and the axial deformation with an axial linear

Table 13.2 Mineral compositions of granite specimens after different high temperature treatments

T/°C	Feldspar/%	Biotite/%	Quartz/%	Amphibole/%	Chlorite/%	Dolomite/%
25	59.85	21.56	11.12	6.00	1.01	0.46
300	53.18	23.96	15.45	4.22	2.21	0.98
400	59.70	19.53	16.09	2.75	1.28	0.65
500	64.12	18.33	11.21	3.84	1.47	1.03
600	54.45	25.56	13.88	4.21	1.14	0.76
700	52.94	29.96	13.12	3.22	0.00	0.76
800	53.80	20.35	15.21	10.64	0.00	0.00

variable differential transducer (LVDT) with a range of 8 mm. The radial deformation was measured using an LVDT wrapped tightly around the specimen, for which the maximum displacement capacity was 4 mm. To eliminate end friction effects (Mogi 1996; Hawkes and Mellor 1970), the axial and radial LVDTs were located in the central part of the specimen. The axial stress was applied to the granite specimen at a constant axial displacement rate of 0.08 mm/min (about 1.33×10^{-5}/s) until failure.

During experimentation, in order to study the fracture evolution characteristics of the deforming specimens, the output of AEs was continuously monitored. AEs are transient elastic waves generated by the rapid release of energy within a material, such as the strain energy released during microcrack propagation. Monitoring AE during deformation has become an increasingly important diagnostic tool in material science which has provided a wealth of information regarding the failure process in brittle materials (Fortin et al. 2006; Lockner et al. 1991; Benson et al. 2007; Brantut et al. 2011) and can be used in failure forecasting modelling (Bell et al. 2011a and 2011b). AE counts were recorded using a DS2 full-information AE measuring system.

13.2 Thermal Damage Characteristics of Granite Before Loading

13.2.1 X-ray Micro CT Scanning Procedure

In this research, X-ray CT scanning of the granite specimens was carried out using a Nanotom 160 high-resolution micro-CT at a spatial resolution of 30 μm. The X-ray micro CT scanning procedure is the same as that in Chap. 3. The X-ray micro CT scanning parameter in this chapter is as follows: the electron current was 90 μA, the accelerating voltage was 130 kV, and the scanning time was 4 s.

Figure 13.3 depicts the comparison of X-ray CT scanning surface images and actual surface crack photographs of granite specimens before loading and after

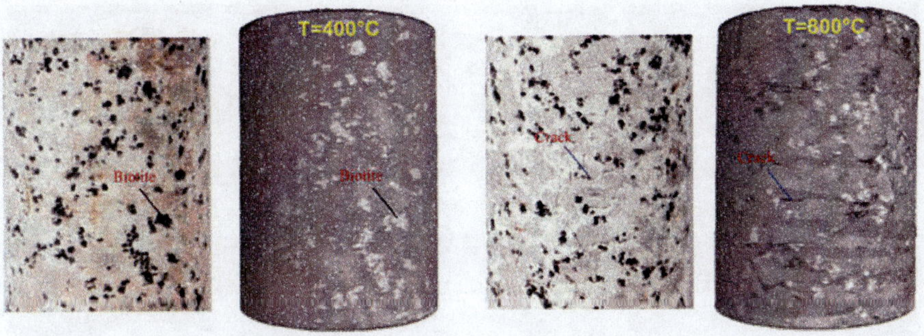

Fig. 13.3 Comparison of X-ray CT scanning surface images and actual surface crack photographs of granite specimens before uniaxial compression loading

uniaxial compression failure. The CT number is a function of the density and chemical composition of the material in a voxel (Hirono et al. 2008). If the CT number of air is defined as 0, the CT number of water is 1000. Moreover, the density of grain is larger; the color of CT image is brighter. In the present research, the granite contained mainly three mineral grains: quartz, feldspar and biotite. The density of biotite is about 3120 kg/m^3, the density of quartz is about 2650 kg/m^3, and the density of feldspar is about 2570 kg/m^3. Therefore, the biotite in the CT image is brighter than quartz and feldspar, as denoted in Fig. 13.4. However, it is very difficult to distinguish the feldspar and quartz due to their approximate density in the CT image. In this research, it should be noticed that black regions mean cracks or surface spalling, and other regions indicate no surface failure. From Fig. 13.3, it can be seen that the X-ray CT scanning surface images approximate the actual surface crack photographs, which demonstrates that X-ray micro-CT scanning can be used to explore internal damage in granite material.

13.2.2 Thermal Damage Characteristics by CT and Optical Microscopic Observation

Figure 13.4 presents the horizontal cross-sections with different heights of uncompressed granite specimens after different high temperature treatments (before uniaxial compression loading). From Fig. 13.4, we can see that the biotite grains are distributed randomly in the granite specimen. Due to the effect of rock heterogeneity, the distribution of biotite grains varies with the difference of heights. Moreover, it seems that the high temperature treatment has almost no effect on the distribution of biotite grains. However, we could not observe the cracks due to thermal damage failure in the range of T = 25–600 °C using our X-ray micro CT equipment. Despite this, the crack density of granite increases substantially with increasing temperature from 600 to 800 °C, as evidenced by Fig. 13.4.

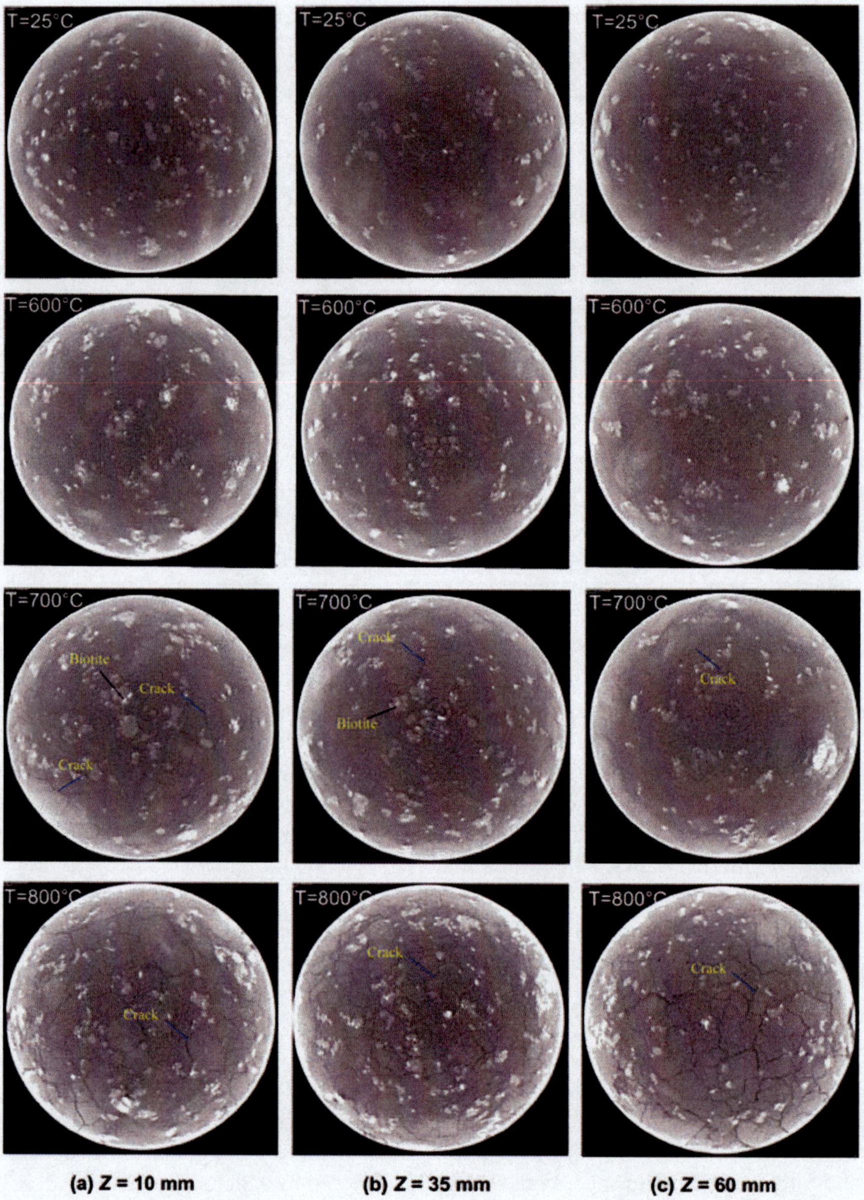

Fig. 13.4 Horizontal cross-sections with different heights of uncompressed granite specimens after different high temperature treatments (before uniaxial compression loading)

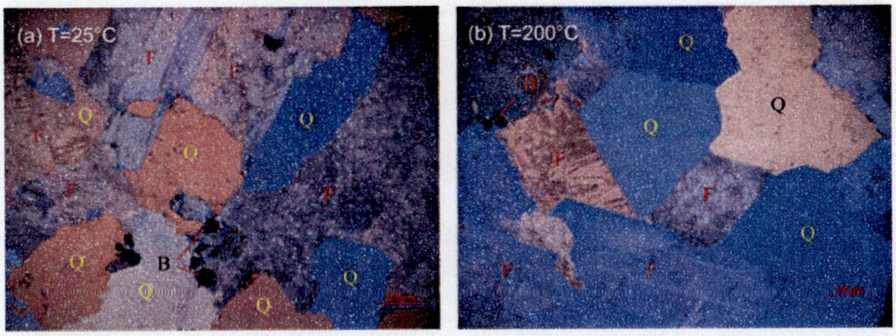

Fig. 13.5 Optical microscopic observation of granite specimens at **a** T = 25 °C and **b** T = 200 °C

In order to analyze the influence of high temperature on thermal cracks in uncompressed granite specimens, an advanced optical microscopic observation was adopted in the present study. Figures 13.5, 13.6, 13.7 and 13.8 illustrate the optical micro-

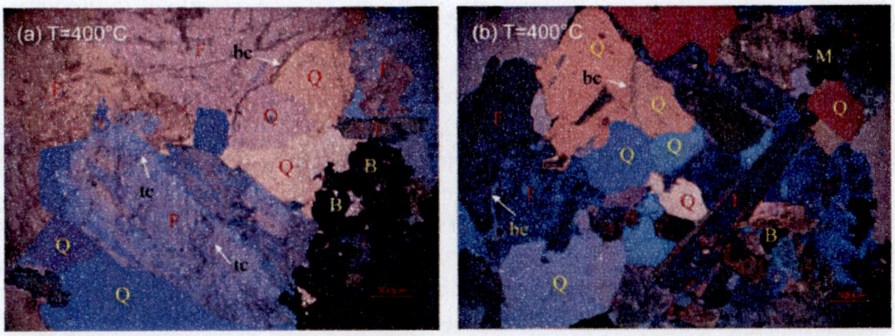

Fig. 13.6 Optical microscopic observation of granite specimens at T = 400 °C. "bc" means "boundary crack", and "tc" means "transgranular crack"

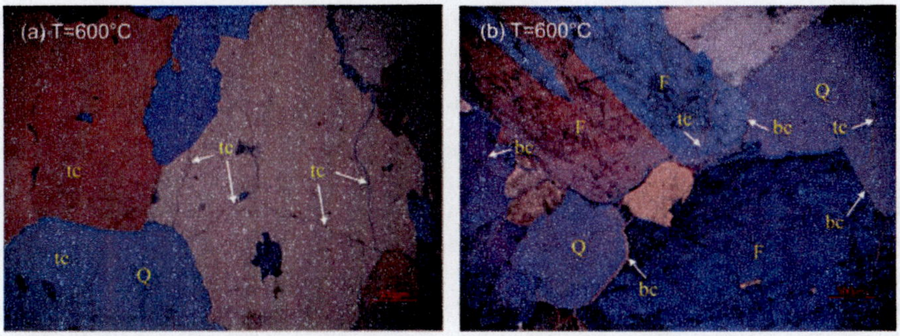

Fig. 13.7 Optical microscopic observation of granite specimens at T = 600 °C

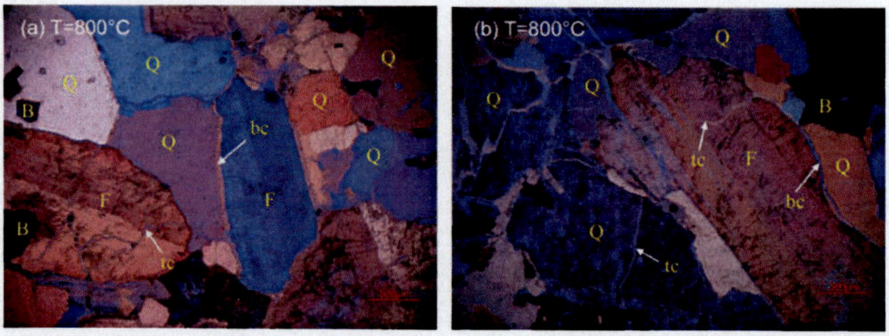

Fig. 13.8 Optical microscopic observation of granite specimens at T = 800 °C

scopic observation results in granite specimens at different temperatures. From Figs. 13.5, 13.6, 13.7 and 13.8, we can clearly see the influence of high temperature treatment on the mineral grains and the cracks.

At T = 25 °C, three mineral grains are closely arrayed and almost no initial pores and fissures can be observed, as shown in Fig. 13.5a, which is obviously different to sandstone material (Yang et al. 2015). When the temperature is increased to 200 °C, compared with the result at T = 25 °C, we cannot observe any micro-structural alterations from the distribution of three main mineral grains as shown in Fig. 13.5b, which indicates that 200 °C has not caused any significant thermally-induced microcracks. When the temperature is increased to 400 °C, the differences in the thermal-expansion behaviors of various mineral grains can induce great thermal stresses within the granite material matrix. Once the maximum thermal stress exceeds the tensile or shear strength limits of the rock, new cracks will emanate. In addition, differences in thermal expansion along different crystallographic axes of the same mineral can also cause structural damage upon heating (Somerton 1992). In Fig. 13.6, we can observe grain boundary cracks and transgranular cracks. The boundary cracks mainly occur between two quartz grains, two feldspar grains, or between quartz and feldspar grains; whereas it is difficult to occur between quartz and biotite, or feldspar and biotite grains (Li et al. 2002). However, the transgranular cracks are mainly observed in feldspar grains, and these results are mainly due to the lower strength of feldspar than the quartz.

When the temperature is increased to 600 °C, the optical microscopic observations shown in Fig. 13.7 are different to those in Fig. 13.6. The internal stresses built up as a result of the thermal expansion mismatch between the minerals within granite are sufficient to generate significant thermal cracking. From Fig. 13.7, we can observe many boundary cracks and transgranular cracks. These transgranular cracks are induced not only within the feldspar grains but also the quartz grains. This reason is that at T = 600 °C, the quartz will have passed the α/β transition ~573 °C (Somerton 1992), which results in a large increase in the volume of quartz, and therefore a substantial amount of thermal microcracking occurs (Glover et al. 1995). Furthermore, thermal stresses can also lead to the propagation of the original cracks, which may result in the structural damage of rocks. Therefore, the crack density of granite at T = 600 °C is obviously higher than that at T = 400 °C.

It should be noticed that at T = 400 and 600 °C, the cracks observed using optical microscopic equipment cannot be captured by X-ray micro CT technique with a minimum resolution of 30 μm, which indicates that the widths of the cracks observed in granite at 400 and 600 °C are all smaller than 30 μm. However, when the temperature is increased to 700 °C, because of the influence of greater thermally-induced damages, some cracks larger than 30 μm can be observed in horizontal cross-sections with various heights and these cracks are very isolated from each other, as shown in Fig. 13.4. Furthermore, due to the influence of more induced thermal damages, a considerable number of isolated cracks begins to propagate and coalesce at triple junctions.

Figure 13.8 presents further optical microscopic observations of granite specimens at T = 800 °C. From Fig. 13.8, we can see that more and more thermally-damaged cracks have formed, again due to the stresses caused by mineral thermal expansion. Many thermal microcracks in the 800 °C specimens have also encouraged crack interaction and coalescence, which leads to the further increase of crack density and aperture thickness compared with that at T = 600 °C. The above analysis means that with the increase of temperature from 400 to 800 °C, more and more thermal cracks are observed in granite specimens. Furthermore, the crack density and crack aperture thickness of granite all increase with increasing temperature from 400 to 800 °C.

Figure 13.9 illustrates the effect of high temperature on the porosity of granite. From Fig. 13.9, we can see that the porosity of granite at room temperature is 0.828% (a low value), which also leads to the original pores and fissures in the granite being very difficult to observe before loading. As the temperature increases from 25 to 300 °C, the porosity of granite decreases from 0.828% to 0.685%, which also verifies that the thermal expansion of mineral grains acts as the main effect on the increase of strength and stiffness. However, as the temperature increases from 300 to 800 °C, the porosity of granite increases from 0.685 to 3.460%, which also verifies that

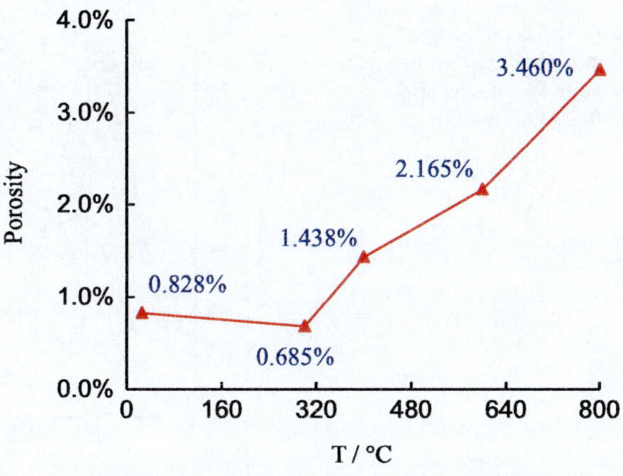

Fig. 13.9 Effect of high temperature on the porosity of granite

more and more thermal microcracks occur in the granite, as the variance of optical microscopic observations shown in Figs. 13.6, 13.7 and 13.8 indicates. The difference in the thermal microcracks and the porosity of granite also results in the variation of strength and deformation behavior with increasing temperature, which is analyzed in the next section.

13.3 Strength and Deformation Behavior of Granite Under Uniaxial Compression

13.3.1 Axial Stress–strain Behavior of Granite

Figure 13.10 shows the axial stress-axial strain curves for granite specimens after different thermal treatments. It is clear from Fig. 13.10 that the strength and deformation behavior of granite depend greatly on the temperature. In Fig. 13.10, σ_1 and ε_1 represent the axial stress and the axial strain, respectively. According to Fig. 13.10, it can be seen that the axial stress-axial strain curves of granite at lower temperatures (T = 25–500 °C) show obvious brittleness after the peak strength, and the post-peak behavior of granite has more ductility at higher temperatures (T = 600–800 °C), which are associated with the density of thermal cracks in the specimen.

The axial stress-axial strain curves of granite at higher temperatures show initially a non-linear deformation, which usually results from the closure of open microcracks (see Fig. 13.1 e, f). It is known that a minor load may result in a large deformation due to the closure of open microcracks in the rock specimen, and thus the macroscopic stress–strain curve is non-linear. Moreover, this stage of microcrack closure depends on various high temperature treatments. With the increase of temperature, the stage of initially nonlinear deformation becomes more and more distinct. This is likely to be a result of more thermal cracks induced after higher temperature treatments. After

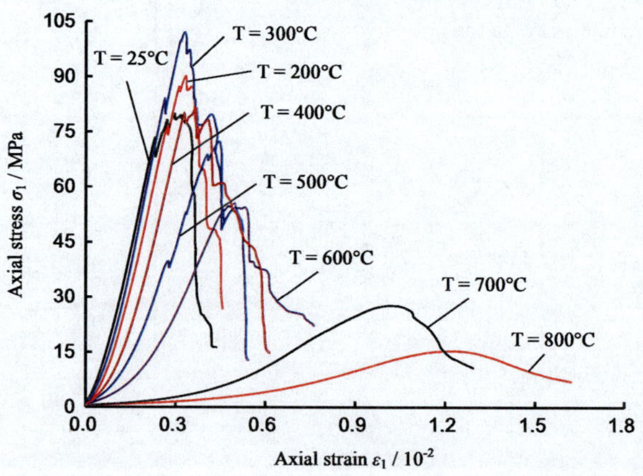

Fig. 13.10 Axial stress–strain curves for granite specimens after different thermal treatments

the stage of microcrack closure, the stage of elastic deformation dominates the linear portions of the stress–strain curves. With the continuous increase of axial deformation, the stress–strain curves of granite begin to depart from the linear behavior, which marks the yielding of the specimens. After that, the specimen will reach the peak strength and then enter the post-peak deformation stage. It should be noticed that the higher the temperature, the specimen fails more slowly after the peak strength with the increase of axial deformation.

Table 13.2 lists the corresponding mechanical parameters of granite specimens after thermal treatments. In Table 13.2, ε_{1c} and σ_c are respectively defined as the peak axial strain and uniaxial compressive strength (UCS) of rock, and E_S and μ_S respectively represent the static elastic modulus and static Poisson's ratio of rock. Note that, E_S is calculated by the average slope in the range of elastic deformation (Fig. 13.10), but μ_S is obtained by the average of the absolute value of the ratio of radial strain and axial strain in the range of elastic deformation. The crack damage threshold (σ_{cd}) of the specimen (Wong et al. 1997; Fairhurst and Hudson 1999; Heap et al. 2009; Yang et al. 2015) is defined as the corresponding axial deviatoric stress where the volumetric deformation of the specimen switches from compaction-dominated to dilatancy-dominated. The axial strain at reaching σ_{cd} is defined as ε_{1cd}. E_d and μ_d are the dynamic elastic modulus and dynamic Poisson's ratio of rock, respectively, which can be expressed by the following two equations (Table 13.3).

$$\mu_d = \frac{(V_P/V_S)^2 - 2}{2\left[(V_P/V_S)^2 - 1\right]} \tag{13.1}$$

$$E_d = \frac{\rho V_S^2 (3V_P^2 - 4V_S^2)}{V_P^2 - V_S^2} \tag{13.2}$$

Table 13.3 Strength and deformation parameters of granite specimens after thermal treatments

Specimen	T/°C	$\varepsilon_{1cd}/10^{-2}$	$\varepsilon_{1c}/10^{-2}$	σ_{cd}/MPa	σ_c/MPa	E_S/GPa	μ_S	E_d/GPa	μ_d
GA-1#	25	0.215	0.297	68.30	80.06	37.35	0.127	12.20	0.458
GA-2#	200	0.259	0.336	72.11	90.19	36.26	0.098	7.90	0.461
GA-8#	300	0.248	0.331	77.99	102.04	40.73	0.109	6.81	0.461
GA-3#	400	0.284	0.369	64.49	81.73	32.45	0.095	5.89	0.454
GA-9#	500	0.264	0.446	38.37	72.28	25.18	0.041	4.70	0.454
GA-4#	600	0.372	0.500	38.28	55.59	16.88	0.038	1.87	0.449
GA-6#	700	0.542	1.016	8.36	27.70	4.842	0.244	0.22	0.462
GA-5#	800	0.584	1.210	3.86	15.35	2.068	0.367	0.12	0.452

13.3.2 *Effect of Temperature on Strength and Deformation Parameters of Granite*

Based on the experimental data for granite specimens after thermal treatments listed in Table 13.2, Fig. 13.11a shows the evolution of crack damage threshold and UCS with increasing heat treatment temperature for the granite specimens. At a room temperature of 25 °C, the σ_c of granite was 80.06 MPa, and increased with the temperature in the range of T = 25–300 °C, and at a heat treatment temperature of 300 °C, the specimen was strongest (the σ_c was increased to 102.04 MPa). However, the σ_c decreased to 81.73 MPa (approximately equal to that for room temperature) at 400 °C and to 15.35 MPa (marginally lower than that for room temperature) at 800 °C. The trend of the crack damage threshold of granite specimens with increasing temperature is similar to that of UCS. With the increase of temperature from 25 to 300 °C, the σ_{cd} value increased from 68.30 to 77.99 MPa. The σ_{cd} value then decreased from 77.99 to 3.86 MPa as the temperature increased from 300 to 800 °C. Furthermore, the σ_{cd} values of granite specimens were all lower than the σ_c values, and the reductions depended on the temperature. However, the σ_{cd}/σ_c ratio value of granite decreased nonlinearly from 0.85 to 0.25 as the temperature increased from 25 to 800 °C, as shown in Fig. 13.11b.

Figure 13.12 shows the influence of temperature on the elastic modulus and Poisson's ratio of granite specimens. According to Fig. 13.12a, we can see that the alteration of dynamic elastic modulus is different to that of static elastic modulus. The E_d value of granite decreases gradually with the increase of heat treatment temperature, which decreased from 12.20 to 0.12 GPa as the temperature increased from 25 to 800 °C. In contrast, the E_S value of granite first increases and then decreases with increasing heat treatment temperature, having a maximum value at T = 300 °C. At a room temperature of 25 °C, the E_S of granite was 37.35 GPa, and increased with the temperature in the range of T = 25–300 °C, and at a heat treatment temperature of

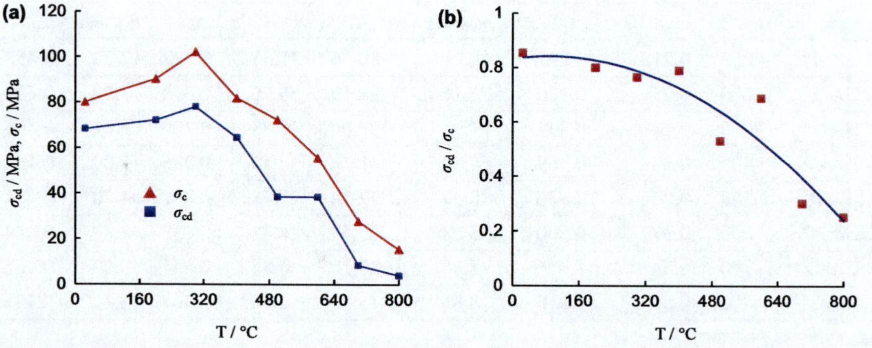

Fig. 13.11 Influence of temperature on crack damage threshold and peak strength of granite specimens. **a** Uniaxial compressive strength and crack damage threshold; **b** Ratio of crack damage threshold to uniaxial compressive strength

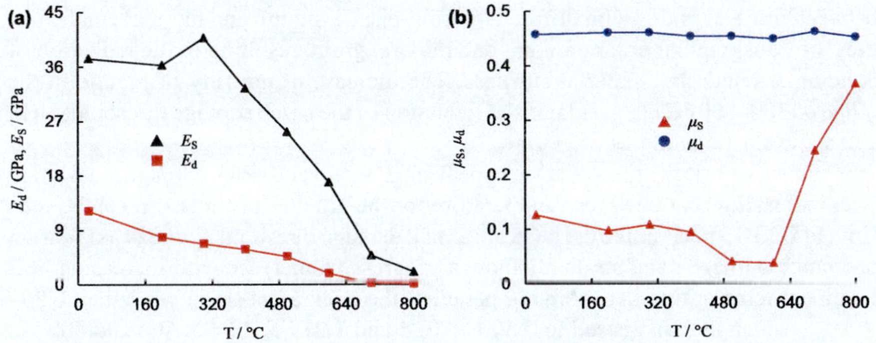

Fig. 13.12 Influence of temperature on elastic modulus and Poisson's ratio of granite specimens. **a** Dynamic and static elastic modulus; **b** Dynamic and static Poisson's ratio

300 °C, the specimen was stiffest (the E_S was increased to 40.73 GPa). However, the E_S decreased to 32.45 GPa (a little lower than that for room temperature) at 400 °C and to 2.068 GPa (marginally lower than that for room temperature) at 800 °C. Furthermore, the static elastic moduli of granite specimens are all higher than the dynamic elastic moduli.

According to Fig. 13.12b, we can conclude that the dynamic Poisson's ratio of granite ranges from 0.454 to 0.462, and the average μ_d value is about 0.456, which indicates that the μ_d of granite does not depend on the temperature. However, the static Poisson's ratio of granite obviously depends on the temperature. With the increase of temperature, the μ_S value first decreases and then increases, having a lowest value at T = 600 °C. The μ_S value of granite decreased from 0.127 to 0.038 as the temperature increased from 25 to 600 °C, and then increased from 0.038 to 0.367 as the temperature increased from 600 to 800 °C. Furthermore, the static Poisson's ratios of granite specimens are all lower than the dynamic Poisson's ratios.

Therefore, the previous experimental results demonstrated that, in all cases, the crack damage threshold, UCS and static elastic modulus of granite were increased at 300 °C, before decreasing up to the maximum temperature of 800 °C. This can be explained as follows. Even though at T = 300 °C, it is not large enough to initiate any boundary cracks or transgranular cracks with the thermally induced strain, which made the specimens heat-treated at 300 °C stronger without producing any micro-cracks and this also agrees very well with the decrease of porosity (see Fig. 13.9). Moreover, in the range of 300 °C, it is believed that the thermal expansion which occurs with increasing temperature causes the reduction of the distance between the interfaces of individual minerals, and this increases their mutual attraction and bonding strength. However, above 300 °C, the crack damage threshold, UCS and static elastic modulus of granite decrease sharply with increasing temperature, for the following reason. The increase of temperature from 300 to 800 °C causes the crystalline particles of the rock mass to fracture through pre-existing grain boundaries between mineral grains with different thermo-elastic moduli and thermal conductivities, or transgranular cracks among one mineral grain, resulting in the reduction of bonding strength and material stiffness. The increase of porosity of granite in the range of 300–800 °C also explains the reduction of the crack damage threshold, UCS and static elastic modulus of granite.

Figure 13.13 shows the variation in peak axial strain and crack damage axial strain with increasing heat treatment temperature for the granite specimens. It can be seen from Fig. 13.13 that, unlike the UCS and crack damage threshold, the peak axial strain and crack damage axial strain all show a progressive increase with increasing heat treatment temperature. At room temperature, the peak axial strain was about 0.297×10^{-2}, which then increased to 0.500×10^{-2} and 1.210×10^{-2} at 600 and 800 °C, respectively. Changes in peak axial strain were modest to 600 °C, but increased dramatically from 600 to 800 °C. The crack damage axial strain shows a similar relationship with the peak axial strain. The crack damage axial strain had a slight increase from 0.215×10^{-2} to 0.372×10^{-2} as the temperature increased from 25 to 600 °C, and then increased greatly from 0.372×10^{-2} to 0.584×10^{-2} as the temperature increased from 600 to 800 °C. Regardless of these changes in crack damage threshold and peak strength in the specimens, the crack damage axial strain and peak axial strain always increased as the temperature was increased, indicating that the specimens could support a larger axial deformation before catastrophic failure. This is likely to be the result of a large increase in thermal microcracks, and their potential interaction and coalescence, from 600 to 800 °C.

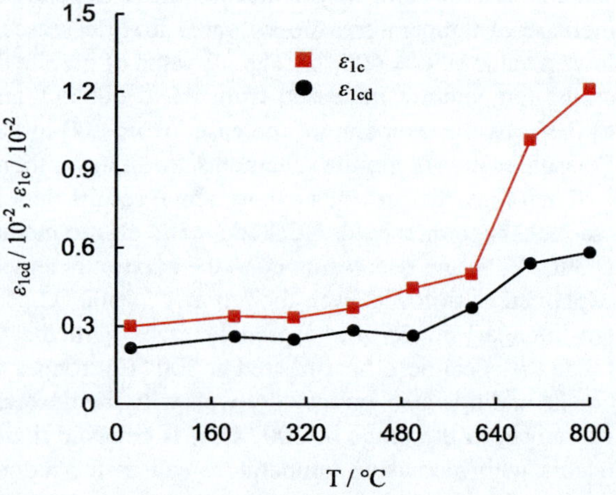

Fig. 13.13 Influence of temperature on peak axial strain and crack damage axial strain of granite specimens

13.4 AE Behavior of Granite During Uniaxial Compression

The AE monitoring technique is very useful for exploring the failure process and crack evolution mechanism of granite (Yang et al. 2014). Therefore, during the uniaxial compression, the AE technique was adopted for the granite specimens subjected to various high temperature treatments in this research. Figure 13.14 presents the axial stress-time curves, the output of AE counts and accumulated AE counts of granite specimens for different temperatures. In accordance with the AE results shown in Fig. 13.14, a detailed investigation of the AE characteristics of granite specimens with different temperatures was carried out. It is clear that the AE characteristics of granite specimens are obviously dependent to the temperature.

At T = 25–400 °C, the AE characteristics of granite specimens at various temperatures are very approximate, and can be divided into two typical periods, i.e. a quiet period and an active period. In the quiet period, the uniaxial compression stress-time curves of granite specimens with respect to various temperatures all show downward concave and initial nonlinear deformation at low stress levels. Note that since, the quiet period corresponds to the stage of fissure closure and elastic deformation, the AE events are not very active and the AE counts are also very rare compared with the active period, which results only from the occurrence of crack closure at this stage. Moreover, we can see that for hard and brittle granite material, the AE behavior in the quiet period is not dependent to the temperature. In the active period, the stress–strain curves of granite specimens undergo two different stages, i.e. the stage of stable crack propagation, and the stage of unstable crack propagation. The AE counts of granite specimens are decentralized and have several larger peak values, which correspond to crack initiation and propagation in the specimens. Furthermore, the AE characteristics have a close relationship with the crack initiation, propagation and coalescence processes in the rock specimen. Next, we take specimen at T = 25 °C as an example to analyze the AE behavior.

In accordance with Fig. 13.14a, we can analyze the real-time crack evolution process of a granite specimen at room temperature. Before point *a* ($\sigma_1 = 73.44$ MPa $= 91.7\% \sigma_c$) as shown in Fig. 13.14, the stress concentration at the tips of fissures does not reach the material strength, which does not result in crack initiation. However, when the axial stress is loaded to point *a*, the specimen begins to initiate the first tensile crack. Notice the initiated stress of the first tensile crack is identified by AE counts, which leads to the axial supporting capacity dropping to 68.02 MPa in a smaller range of axial strain. Afterwards, with the increase of axial deformation, the axial stress begins to increase, while due to some minor damage to the supporting structure, the reloading average modulus ~ E_a (about 30.08 GPa) is a little lower than the elastic modulus (−37.35 GPa) before point *a*. When the axial stress is increased to point *b*, a larger AE event is observed, which indicates the initiation of one new tensile crack or the propagation of one previous tensile crack. After point *b*, the reloading E_a (about 23.27 GPa) is obviously lower than the previous average modulus (30.08 GPa), which means obvious damage to the supporting structure. When the specimen is loaded to point *c*, the axial stress is about 78.39 MPa and the corresponding axial

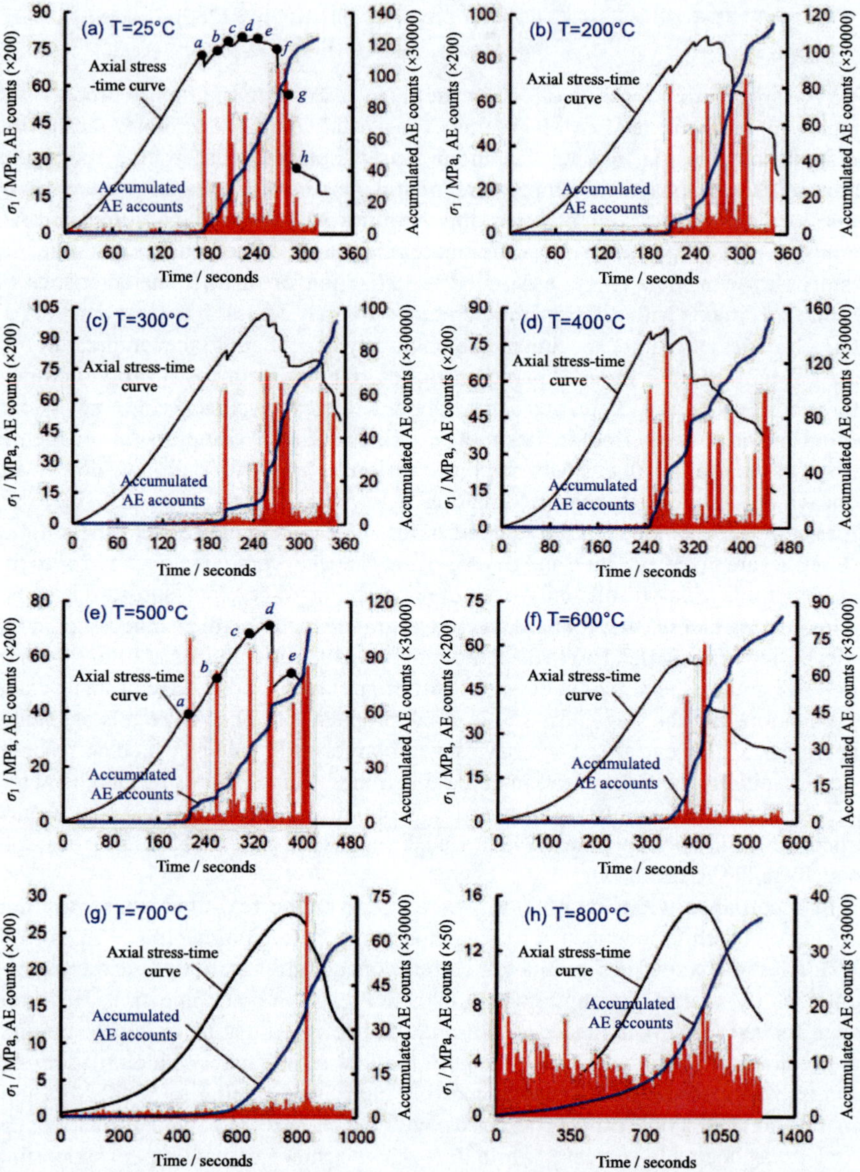

Fig. 13.14 AE counts and accumulated AE counts of granite specimens with respect to various temperatures during the whole deformation failure. The number in double bracket in *vertical axis* labels is the multiplier of coordinates

strain is – 2.72×10^{-3}. For the moment, a bigger AE event is also observed, which corresponds to a minor stress drop from 78.39 to 76.45 MPa.

Afterwards, with the increase of axial deformation, the axial stress begins to increase slowly and the reloading average modulus is about 20.48 GPa, which is a little lower than the previous average modulus (23.27 GPa). The continuous increase of deformation leads to the specimen being loaded to 80.06 MPa (point *d*, 100%σ_c). After the peak strength, the axial stress drops from 80.06 to 78.37 MPa with the increase of axial deformation. Then, due to obvious damage to the supporting structure, the reloading average modulus ~ E_a (about 6.64 GPa) is distinctly lower than the average modulus before peak strength. At reaching point *e*, the axial stress of this specimen is about 79.60 MPa and the corresponding axial strain is – 3.19×10^{-3}. At this time, the axial supporting capacity begins to decrease slowly with increasing axial deformation. At point *f*, a larger stress drop occurs from 74.24 to 61.51 MPa in an almost constant axial strain, which also results in a larger AE event. Similarly, at point *g*, there is also a larger stress drop from 57.10 to 38.71 MPa, which also accompanies with a larger AE event. After point *h*, due to no initiation of new cracks or the propagation of old cracks, there are not any larger AE events.

It should be noted that the AE behavior of granite at T = 200–400 °C is very similar to that at T = 25 °C, and some differences result mainly from the variance of crack initiation and propagation modes in the specimen. Therefore, here we do not analyze in detail the AE behavior of granite at T = 200–400 °C any further. In addition, we found that at a certain temperature, it was not true that the greater the axial stress dropped, the larger the corresponding AE event. For example, at T = 25 °C, although there was a smaller stress drop at point *c*, the AE event was higher than at point *a*, which can be analyzed as follows. The energy dissipated in fracturing the rock can be closely related with the cracking mode. At point *a*, the elastic energy in the specimen has released a part, but more elastic energy was stored in the specimen. Therefore, with the increase of axial deformation, more gathered energy in the specimen is dissipated to emanate cracks at point *c*, and the AE event was greater, although with a smaller stress drop.

However, at T = 500 °C, the AE characteristics of granite specimen are very similar to those at T = 25–400 °C, and they also have a quiet period and an active period. But the quiet period only corresponds to the stage of fissure closure and the front region at the stage of elastic deformation. Before point *a* ($\sigma_1 = 39.88$ MPa = 55.2% σ_c) as shown in Fig. 13.14e, the stress concentration at the tips of fissure does not reach the material strength, which does not result in a crack initiation. However, when the axial stress is loaded to point *a*, the specimen begins to initiate the first crack. Note that the initiated stress of the first crack is identified by AE counts, which leads to the axial supporting capacity dropping to 37.88 MPa in a smaller range of axial strain. Afterwards, with the increase of axial deformation, the axial stress begins to increase, but the reloading average modulus ~ E_a (about 27.38 GPa) is a little higher than the elastic modulus (–25.18 GPa) before point *a*, which may be a result of the closure of original cracks. When the axial stress is increased to point *b*, a bigger AE event is observed and the specimen also shows a minor stress drop from 53.10 to 52.39 MPa, which indicates the initiation of one new crack or the propagation of one previous crack. After point *b*, the reloading E_a (about 23.85 GPa) is obviously lower than the previous average modulus (27.38 GPa), which means obvious damage to the supporting structure. When the specimen is loaded to point *c*, the axial stress of this specimen is about 68.76 MPa and the corresponding axial strain is -4.08×10^{-3}. For the moment, a bigger AE event is also observed, which corresponds to a minor stress drop from 68.76 to 67.04 MPa. Afterwards, with the increase of axial deformation, due to obvious damage to the supporting structure, the axial stress begins to increase slowly and the reloading average modulus is about 17.36 GPa, which is obviously lower than the previous average modulus (23.85 GPa). The continuous increase of deformation leads to the specimen being loaded to 72.28 MPa (point *d*, 100%σ_c). After the peak strength, the axial stress drops from 72.28 to 48.19 MPa with the increase of axial deformation, which also leads to a larger AE event. At this time, the axial supporting capacity begins to increase slowly with increasing axial deformation by a lower average modulus of about 17.20 GPa. At point *e*, the specimen drops to a smaller value, which is also accompanied with some larger AE events.

At T = 600 °C, the AE characteristics of granite specimens are different to those at T = 25–500 °C. Although the AE behavior has a quiet period and an active period, only three larger AE events are observed, which are located after the peak strength. From the comparison of the accumulated AE counts curves of granite at T = 600–800 °C, it can be seen that the accumulated AE counts curves become more and more smooth with the increase of temperature, as demonstrated by Fig. 13.14f–h. The results indicate that less and less elastic energy is released with the increase of temperature.

At T = 700 °C, one largest AE event is also located after the peak strength, which indicates that one big macroscopic crack is formed. At the stage of initial loading, many AE events are observed in the granite specimen at T = 700 °C, which is obviously higher than those observed at T = 25–600°, and this results from the thermal cracking of granite before loading. However, at T = 800 °C, the AE events of granite specimens are very active during the entire deformation and the values of AE events are higher than those at T = 700 °C, whereas no larger AE events are observed, which indicates that at this temperature, the ductile failure mechanism dominates the granite specimens.

Figure 13.15 illustrates the evolution of accumulated AE counts for granite specimens after different thermal treatments. From Fig. 13.15, we can see that with the increase of temperature, the evolution of accumulated AE counts for granite specimens becomes slower and slower, which indicates that at higher temperatures, due to serious thermal damage, the granite specimens undergo more ductile failure and released less elastic energy. However, for the same temperature, the accumulated AE counts of granite increases slowly with the increase of time, followed by a sharp increase, which agrees very well with an exponential increase.

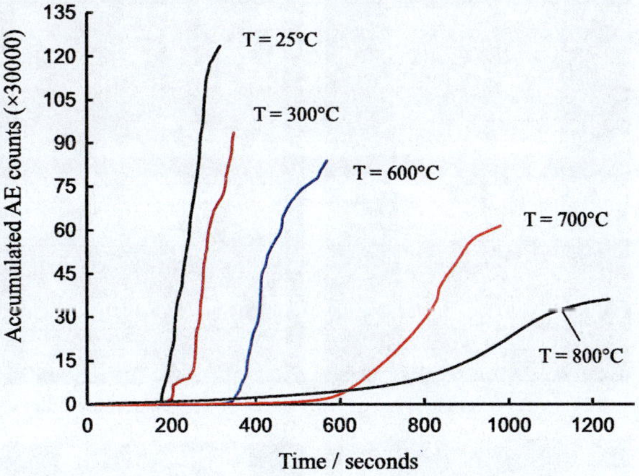

Fig. 13.15 Evolution of accumulated AE counts for granite specimens after different thermal treatments

13.5 Internal Crack Mechanism of Granite Under Uniaxial Compression

Figures 13.16 and 13.17 show vertical and horizontal cross-sections of a compressed granite specimen at T = 25 and 300 °C after uniaxial compression failure, respectively. From Figs. 13.16 and 13.17, we can see that the main failure and observed tensile cracks all occur at local regions. The propagation and coalescence of tensile cracks cause the ultimate failure mode, but in the other regions of the specimens, no macrocracks are observed, even though the granite specimen has undergone unstable failure. This reason can be explained as follows. Rock is a kind of natural heterogeneous material. In a rock specimen, the strength of local material is unequal for various regions. Once the applied stress concentration exceeds the strength of the local material, cracking will begin to initiate. After the occurrence of local cracking, the stress concentration transfers to the tip of the crack (Yang et al. 2014) and promotes the failure to propagate continuously in the region. It should be noted that the tensile cracks all propagate along the direction of axial stress. Furthermore, due to the boundary effect of mineral grains, the propagating paths of tensile cracks are not linear.

From Fig. 13.16, we can see that the granite at room temperature is a kind of typically brittle rock material, showing axial splitting tensile failure mode, and many axial tensile cracks are observed, both at the surface and internally, which also results in a typical brittle deformation after peak stress, as shown in Fig. 13.10. It is clear that the cross-section with various depths and heights has different crack system modes. However, due to the effect of three mineral grains on the crack propagation path, the splitting tensile cracks are not linear. Furthermore, one tensile crack has different apertures at different regions. At $x = 5$ mm, two obvious axial tensile cracks are observed, and the cracks pass through the entire specimen height. At the same time, between two larger axial tensile cracks, there are some minor tensile cracks in the central region of this specimen. Furthermore, the crack system mode of granite

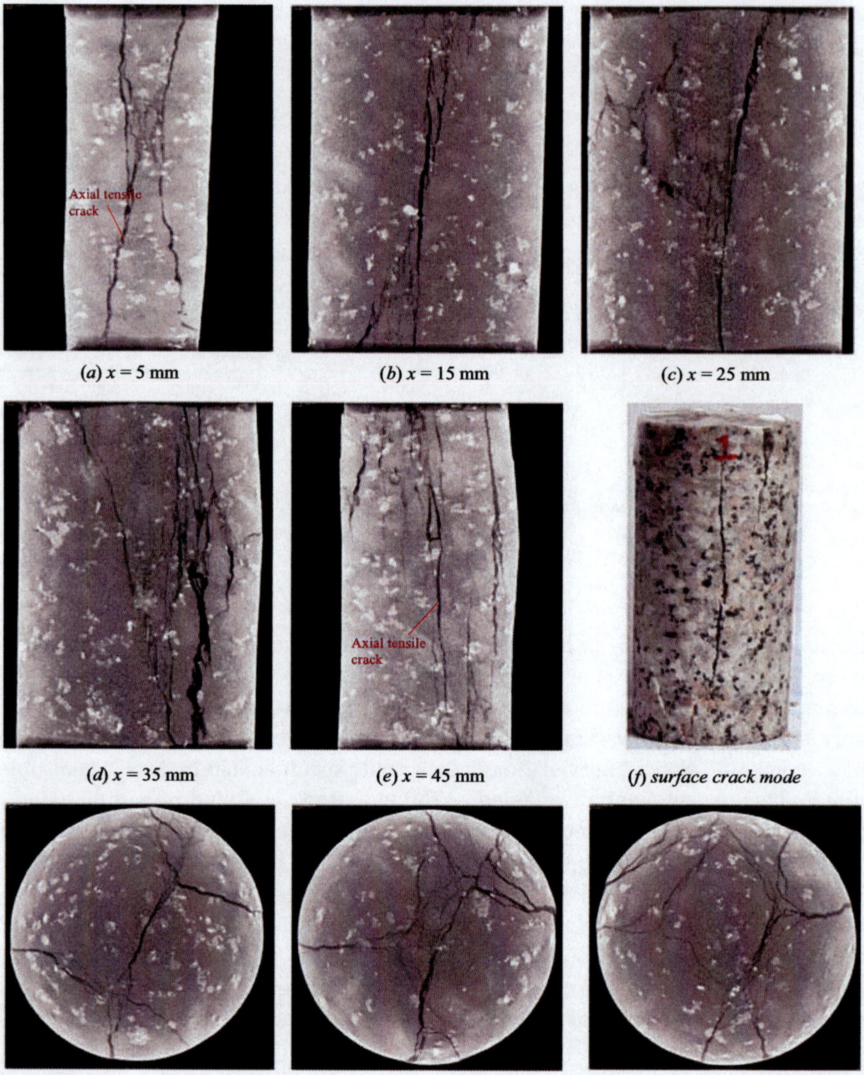

(*a*) x = 5 mm (*b*) x = 15 mm (*c*) x = 25 mm
(*d*) x = 35 mm (*e*) x = 45 mm (*f*) *surface crack mode*
(*g*) z = 10 mm (*h*) z = 35 mm (*i*) z = 50 mm

Fig. 13.16 Vertical and horizontal cross-sections of compressed granite specimen at T = 25 °C after uniaxial compression failure

at $x = 5$ mm approximates the surface crack mode shown in Fig. 13.16f. At other depths, the crack system mode of granite is obviously different to the surface crack mode. At $x = 15$ mm, only one big axial tensile crack is observed, but on the top and bottom regions, near this large axial crack, there are clusters of many minor tensile cracks along the direction of axial stress. At $x = 25$ mm, with the exception of one

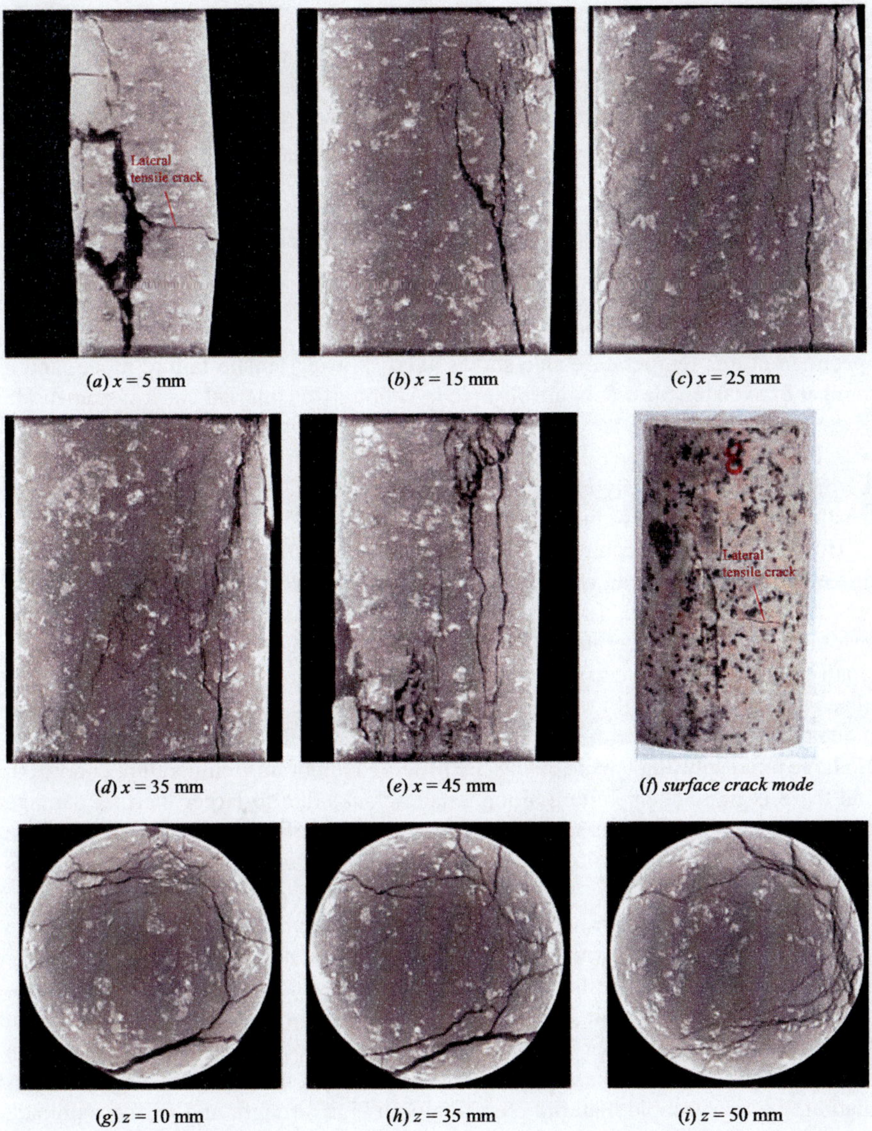

Fig. 13.17 Vertical and horizontal cross-sections of compressed granite specimen at T = 300 °C after uniaxial compression failure

large axial tensile crack, there are many short tensile cracks on the left side of the large axial tensile crack. At $x = 35$ mm, the crack system mode of granite is very complicated, consisting of one inclined tensile crack and multiple axial tensile cracks passing through the entire specimen. Furthermore, the crack aperture of each tensile crack is different at various regions. However, at $x = 45$ mm, multiple vertical tensile cracks are observed and propagate along the direction of axial stress. Some tensile cracks pass through the entire specimen and some tensile cracks only propagate a short distance, which can also be verified by horizontal cross-sections with different heights as shown in Fig. 13.16g–i.

Figure 13.17 illustrates vertical and horizontal cross-sections of a compressed granite specimen at T = 300 °C after uniaxial compression failure. The granite specimen at this temperature also shows axial splitting tensile failure mode, and a number of axial tensile cracks are observed. Although the internal crack system mode of granite at T = 300 °C is very similar to that at T = 25 °C by comparing Fig. 13.17 with Fig. 13.16, the crack aperture thicknesses of granite at some local regions at T = 300 °C are obviously larger than that at T = 25 °C. Furthermore, some lateral tensile cracks are observed in the granite specimen at T = 300 °C.

However, once the granite specimen contains thermally induced cracks, the fracture evolution process can be obviously affected by these cracks, especially at 700–800 °C, as shown in Figs. 13.18, 13.9 and 13.20, which is different to the process for specimens at lower temperatures. At T = 700 °C, except for axial tensile cracks, some small cracks initiated due to smaller thermal damage before the compression are also observed, as testified by Fig. 13.1e and Fig. 13.4. At T = 800 °C, the ultimate failure mode of the granite specimen is obviously different to those at other temperatures. No large axial splitting tensile cracks are observed under this temperature condition, and there is more shear failure along tensile cracks due to larger thermal damage before the compression, as testified by Fig. 13.1f and Fig. 13.4. The deformation failure mechanism of granite at various high temperatures can be summarized as follows.

The strength and stiffness of the mesoscopic element material in the granite specimen varies greatly from one element to another. At lower temperatures (e.g. 25–300 °C), due to insignificant thermal cracks, a small proportion of the mesoscopic material elements with low strength and stiffness first reach their maximum supporting capacities under uniaxial compression, producing plastic deformation during the yielding of the specimen. Afterwards, with the increase of axial deformation, the mesoscopic material elements with high strength and stiffness unload, because they do not reach their maximum supporting limit, and plastic deformation in the specimen is not uniform. Therefore, following a decrease in the macroscopic axial deviatoric stress, the plastic deformation that is initiated further concentrates on the failed mesoscopic material elements with low strength and stiffness, which results in localized deformation and some axial splitting tensile fractures are observed.

However, it should be noted that the crack system modes of a granite specimen at T = 700 and 800 °C are more complicated than those at T = 25 and 300 °C. Due to a considerable number of thermal cracks in the granite specimen at higher temperatures before loading (see Fig. 13.19), under uniaxial compression, only a small axial stress

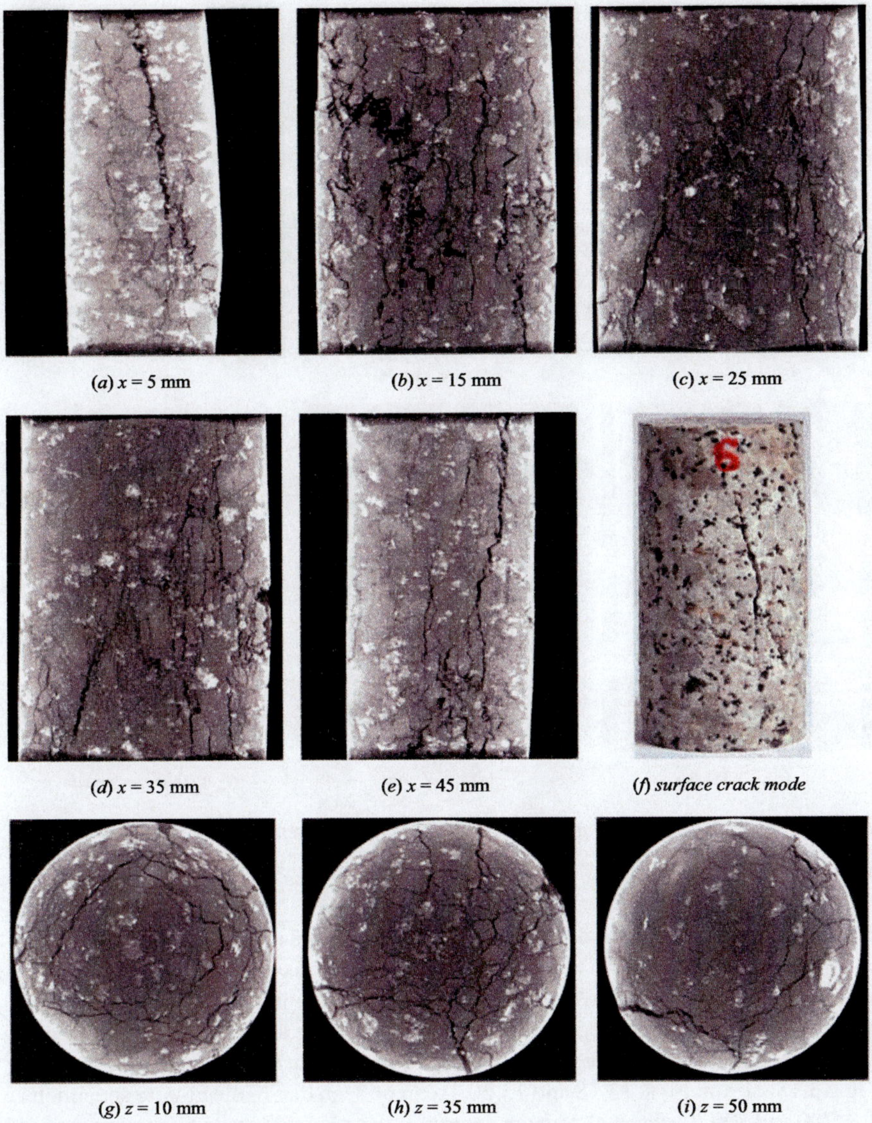

Fig. 13.18 Vertical and horizontal cross-sections of compressed granite specimen at T = 700 °C after uniaxial compression failure

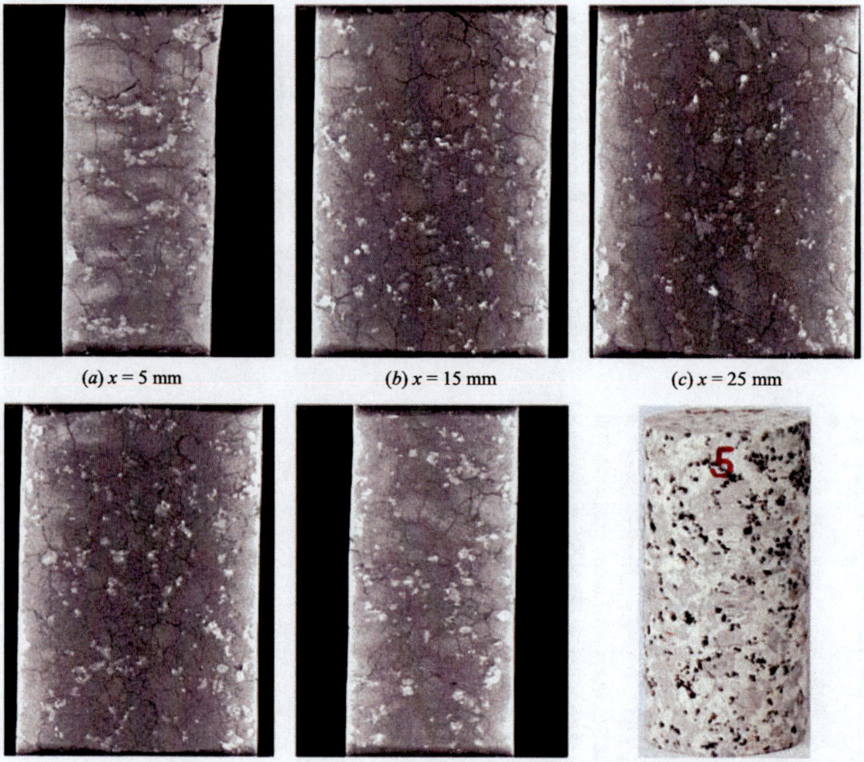

(*a*) x = 5 mm (*b*) x = 15 mm (*c*) x = 25 mm

(*d*) x = 35 mm (*e*) x = 45 mm (*f*) *surface crack mode*

Fig. 13.19 Vertical cross-sections with different depths of uncompressed granite specimen at T = 800 °C before loading

may result in the propagation and coalescence of thermal cracks, which also induces a larger axial deformation. Therefore, the elastic modulus of granite at above 300 °C also becomes lower and lower with the increase of temperature. At the same time, due to the effect of the direction of uniaxial compressive stress on crack evolution, a granite specimen can produce more axial splitting tensile cracks, as demonstrated in Fig. 13.18b. From Figs. 13.18 and 13.20, it can be seen that in the granite specimen at T = 700 and 800 °C, thermal damage determined the larger ductile fracture process of granite at 700–800 °C, which can be explained as follows. Under higher temperatures, the positions where induced thermal cracks have lost their cohesion, so a lower local shear stress concentration at a grain contact is sufficiently high to rupture the weakening positions formed thermal cracks, and frictional slips and grain rotations may occur. At the same time, it should be noted that 600 °C is a critical temperature for granite moving from brittle fracture to ductile fracture in this research.

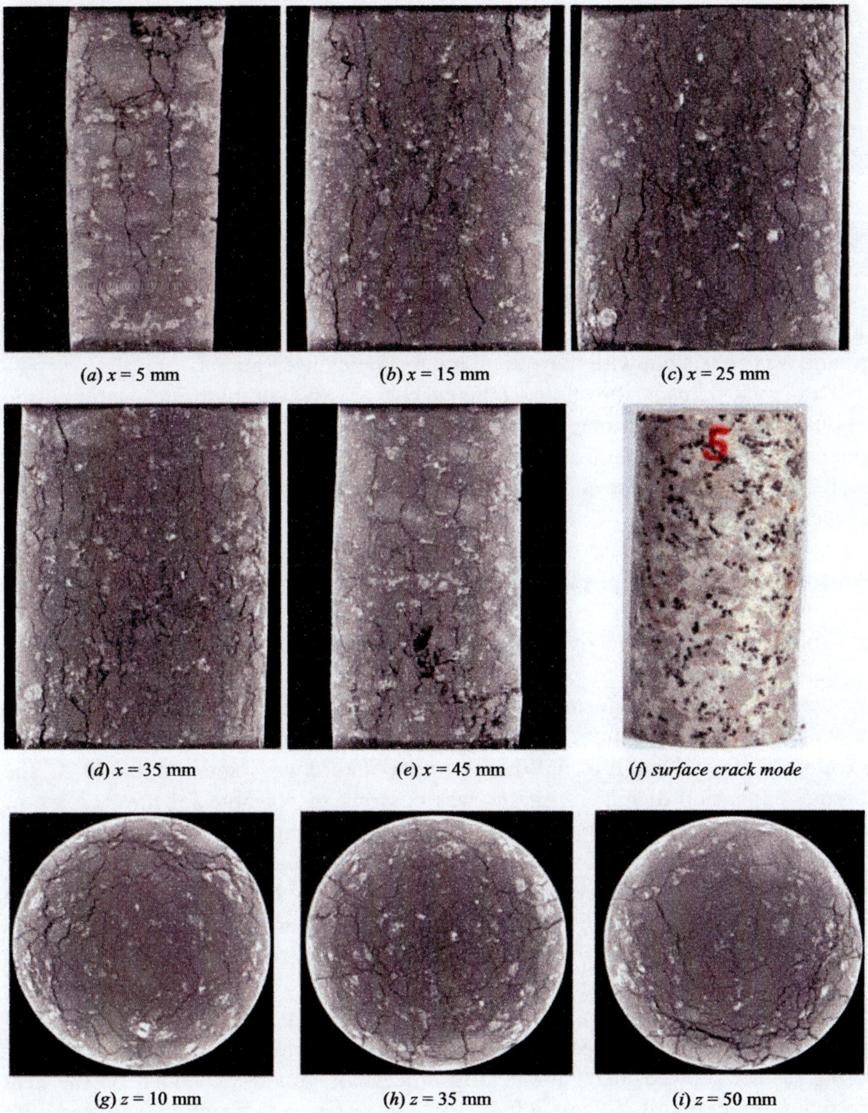

Fig. 13.20 Vertical and horizontal cross-sections of compressed granite specimen at T = 800 °C after uniaxial compression failure

13.6 Conclusions

To investigate the thermal damage and failure mechanical behavior of granite, uniaxial compression experiments, X-ray micro CT observation and AE monitoring were all adopted on granite specimens after exposure to different high temperature treatments (200, 300, 400, 500, 600, 700 and 800 °C). Based on our study, the following conclusions can be summarized.

The mineral composition of the tested granite material at room temperature is described as follows: 11.12% quartz, 59.85% feldspar, 21.56% biotite, 6% amphibole, 1.01% chlorite and 0.46% dolomite, which is not related to high temperature and some differences of mineral composition result mainly from the effect of the heterogeneity of rock material. The porosity of granite first at room temperature was 0.828%, but the porosity of granite decreased to 0.685% as the temperature was increased to 300 °C. However, as the temperature increased from 300 to 800 °C, the porosity of granite increased from 0.685 to 3.460%. The variance of porosity of granite was consistent with the optical microscopic observations.

The crack damage threshold, uniaxial compressive strength and static elastic modulus of granite specimens increased from 25 to 300 °C, before both decreased up to the maximum temperature of 800 °C. However, the crack damage axial strain and peak axial strain at failure always showed an increase with increasing heat treatment temperature before 600 °C, following a sharp increase after 600 °C. The dynamic elastic modulus of granite decreased with the temperature whereas the dynamic Poisson's ratio was independent of the temperature. At room temperature, a granite specimen shows typically brittle failure characteristics, whereas the gradual increase of temperature results in a more ductile failure of granite after the peak stress.

Based on optical microscopic observations, the mechanism by which the high temperature treatment affected the mechanical parameters of the granite was interpreted. At room temperature, three main mineral grains (quartz, feldspar and biotite) are closely arrayed and linked. When the temperature was increased to 300 °C, the thermal expansion of mineral grains was possibly responsible for the increase of strength and stiffness but no microcracks within the granite material were observed. At 400–600 °C, the mechanisms were contributed by boundary cracks and transgranular cracks in feldspar and quartz grains, which results in the decrease of strength and stiffness. However, at 700–800 °C, the mechanisms are associated with the interactions and coalescences of boundary cracks and transgranular cracks, which acted to reduce the strength and stiffness of the granite.

Using acoustic emission (AE) monitoring, the internal crack evolution process of granite at various high temperatures was monitored during deformation. The monitoring results showed that the cracking process of granite depended on the heat treatment temperature. A large number of AE events were usually associated with crack initiation, propagation and coalescence in the granite specimens. Once the granite specimen contained thermally induced cracks, the fracture evolution process could be obviously affected by these cracks, especially at 700–800 °C, which was different to those for specimens at lower temperatures.

The deformation mechanism of failed granite at various temperatures was analyzed using X-ray micro CT. During the loading, the uniaxial compression stress direction dominated the more brittle fracture process of granite at 25–600 °C, which led to splitting tensile main cracks induced along the axial stress, and thermal damage determined the larger ductile fracture process of granite at 700–800 °C, which resulted in a more ductile deformation after the peak strength. It should be noted that 600 °C is a critical temperature of granite from brittle fracture to ductile fracture in this research.

The findings of this research demonstrate the clear dependence of deformation, strength, crack damage, acoustic emission behavior and failure mechanism of granite on high temperature treatment. Understanding of the variation in thermal damage and failure mechanical behavior of granite will assist greatly in the process of characterization of the temperature response in nuclear waste deposition engineering, underground coal gasification, and heat mining in enhanced geothermal energy around the world.

References

Benson PM, Thompson AB, Meredith PG, Vinciguerra S, Young RP (2007) Imaging slow failure in triaxially deformed Etna basalt using 3D acoustic-emission location and X-ray computed tomography. Geophys Res Lett 34:L03303. https://doi.org/10.1029/2006GL028721

Bell AF, Greenough J, Heap MJ, Main IG (2011) Challenges for forecasting based on accelerating rates of earthquakes at volcanoes and laboratory analogues. Geophys J Int 185:718–723

Bell AF, Naylor M, Heap MJ, Main IG (2011b) Forecasting volcanic eruptions and other material failure phenomena: an evaluation of the failure forecast method. Geophys Res Lett 38:L15304. https://doi.org/10.1029/2011GL048155

Brantut N, Schubnel A, Guéguen Y (2011) Damage and rupture dynamics at the brittle-ductile transition: the case of gypsum. J Geophys Rese 116:B01404. https://doi.org/10.1029/2010JB007675

Ding QL, Ju F, Mao XB, Ma D, Yu BY, Song SB (2016) Experimental investigation of the mechanical behavior in unloading conditions of sandstone after high-Temperature treatment. Rock Mech Rock Eng 49(7):2641–2653

Fairhurst CE, Hudson JA (1999) Draft ISRM suggested method for the complete stress-strain curve for the intact rock in uniaxial compression. Int J Rock Mech Min Sci 36(3):279–289

Fortin J, Stanchits S, Dresen G, Gueguen Y (2006) Acoustic emission and velocities associated with the formation of compaction bands in sandstone. J Geophys Res 111:B10203. https://doi.org/10.1029/2005JB003854

Glover PWJ, Baud P, Darot M, Meredith PG, Boon SA, LeRevelec M, Zoussi S, Reuschle T (1995) Phase transition in quartz monitored using acoustic emissions. Geophys J Int 120:775–782

Hawkes I, Mellor M (1970) Uniaxial testing in rock mechanics laboratories. Eng Geol 4:177–285

Heap MJ, Lavallée Y, Laumann A, Hess KU, Meredith PG, Dingwell DB (2012) How tough is tuff in the event of fire? Geology 40:311–314

Heap MJ, Baud P, Meredith PG (2009) The influence of temperature on brittle creep in sandstones. Geophys Res Lett 36:L19305. https://doi.org/10.1029/2009GL039373

Hirono T, Sakaguchi M, Otsuki K, Sone H, Fujimoto K, Mishima T, Lin W, Tanikawa W, Tanimizu M, Soh W, Yeh EC, Song SR (2008) Characterization of slip zone associated with the 1999 Taiwan Chi-Chi Earthquake: X-Ray CT image analyses and microstructural observations of the Taiwan chelungpu fault. Tectonophysics 449(1–4):63–84

Li L, Tsui Y, Lee PKK, Tham LG, Li TJ, Ge XR (2002) Progressive cracking of granite plate under uniaxial compression. Chin J Rock Mech Eng 21(7):940–947 (In Chinese)

Lockner D, Byerlee J, Kuksenko V, Ponomarev A, Sidorin A (1991) Quasi-static fault growth and shear fracture energy in granite. Nature 350:39–42

Mogi K (1966) Some precise measurements of fracture strength of rocks under uniform compressive stress. Rock Mech Eng Geol 4:41–55

Shao SS, Wasantha PLP, Ranjith PG, Chen BK (2014) Effect of cooling rate on the mechanical behavior of heated Strathbogie granite with different grain sizes. Int J Rock Mech Min Sci 70:381–387

Somerton WH (1992) Thermal properties and temperature-related behavior of rock/fluid systems. Elsevier, Amsterdam, pp 22–29

Wong TF, David C, Zhu W (1997) The transition from brittle faulting to cataclastic flow in porous sandstones: Mechanical deformation. J Geophys Res 102:3009–3025

Yang SQ, Jing HW, Huang YH, Ranjith PG, Jiao YY (2014) Fracture mechanical behavior of red sandstone containing a single fissure and two parallel fissures after exposure to different high temperature treatments. J Struct Geol 69(Part A):245–264

Yang SQ, Ranjith PG, Huang YH, Yin PF, Jing HW, Gui YL, Yu QL (2015) Experimental investigation on mechanical damage characteristics of sandstone under triaxial cyclic loading. Geophys J Int 201:662–682

Yang SQ, Ranjith PG, Jing HW, Tian WL, Ju Y (2017) An experimental investigation on thermal damage and failure mechanical behavior of granite after exposure to different high temperature treatments. Geothermics 65:180–197

Zhu TT, Jing HW, Su HJ, Yin Q, Du MR, Han GS (2016) Physical and mechanical properties of sandstone containing a single fissure after exposure to high temperatures. Int J Min Sci Technol 26(2):319–325

Chapter 14
Triaxial Mechanical and Permeability Behavior of Sandstone After Exposure to Different High Temperature Treatments

Although the effects of heat treatment on the mechanical properties of rock have been studied systematically in the previous studies, limited experiments on conventional triaxial compression test results are available for heated sandstone with respect to different high temperature treatments. Moreover, for permeability of rock under both stress changes and high-temperature effects, little information has been reported in the literature (Ding et al. 2016a, 2016b). Consequently, the aim of this chapter is to investigate the physical, mechanical and permeability properties of sandstone exposed to high temperatures.

In the present study, sandstone material was used to study the physical, mechanical and permeability properties of rock exposed to high temperatures through a suite of laboratory tests involving thermal treatment at temperatures up to 800 °C and triaxial compressive tests under confining pressures of 5, 10, 20 and 30 MPa (Yang et al. 2017). First, the physical properties (weight, volume, density and dynamic parameters) of the sandstone specimens were analyzed after high temperature treatment. Second, the influences of thermal treatment temperature on the strength and failure behaviors were analyzed in detail based on the triaxial compression results. Furthermore, the permeability evolution behavior of sandstone during the complete stress–strain process after exposure to different high temperature treatments was also investigated using the transient pulse technique. Finally, the microscopic mechanism of high temperature on the physical, mechanical and permeability behaviors of the tested sandstone were revealed using X-ray diffraction (XRD) and scanning electron microscopy (SEM) analysis. Furthermore, a series of empirical relations between the temperatures and the physical and mechanical properties of sandstone were derived and are expected to aid in geothermal energy extraction from super-critical temperature resources.

S.-Q. Yang, *Mechanical Behavior and Damage Fracture Mechanism of Deep Rocks*,
https://doi.org/10.1007/978-981-16-7739-7_14

14.1 Experimental Material and Testing Procedure

14.1.1 Sandstone Material and Heating Procedure

The material used throughout this research was sandstone collected from Rizhao city in China, which is located near the Yellow Sea in the Circum-Pacific geothermal belt. In this area, the primary surface formations are the Paleoproterozoic Fenzishan group and Jingshan group and the Cretaceous period Wangshi group consisting mainly of dolomite marble, purple sandstone and silty mudstone that can be easily identified. Rizhao sandstone is a high-level hard stone with good homogeneity and high porosity (Yang et al. 2014).

In the present research, high temperature treatments were administered to the specimens with a high-temperature furnace using the following procedure. First, the specimens were heated to their target temperature (100, 200, 300, 400, 500, 600, 700 and 800 °C) at a rate of 5 °C /min, and this heating rate was also the same as that used in the previous experiments for sandstone (Zhu et al. 2016; Ding et al. 2016a, 2016b). The specimens were then held at their target temperature for two hours. Finally, the specimens were cooled naturally to room temperature (25 °C) in a confined space.

14.1.2 Measurement of Physical Properties

The length, diameter, weight and wave velocity (both P-wave and S-wave) measurements were conducted on the tested sandstone before and after thermal treatment with Vernier calipers, an electronic scale and an ultrasonic concrete tester with a reported accuracy of 0.01 g and 0.02 mm and 1 m/s respectively. The volume can be derived using the measured length and diameter for each specimen, and the bulk density can be calculated using the measured weight and calculated volume.

Different acoustic transducers were used to measure the velocities of different sound waves. According to elastic mechanics theory, the dynamic elastic modulus (E_d) and dynamic Poisson's ratio (μ_d) can be derived from the measured P-wave velocity and S-wave velocity of rock using the following equations:

$$\mu_d = \frac{(V_P/V_S)^2 - 2}{2\left[(V_P/V_S)^2 - 1\right]} \tag{14.1}$$

$$E_d = \frac{\rho V_S^2\left(3V_P^2 - 4V_S^2\right)}{V_P^2 - V_S^2} \tag{14.2}$$

To analyze the change of sandstone in physical properties, the rates of change for physical parameters were derived and calculated as

$$\omega = \frac{\delta(T) - \delta_0}{\delta_0} \tag{14.3}$$

where ω is the rate of change of the physical variable, and δ_0 and $\delta(T)$ are the physical variables of a specimen before and after thermal treatment, respectively.

14.1.3 Conventional Triaxial Compression and Seepage Testing Procedure

For sandstone specimens, conventional triaxial compression and permeability experiments with four different confining pressures (5, 10, 20 and 30 MPa) were all performed on servo-controlled triaxial equipment. In the triaxial compression experiment, all specimens were cored from the same block of material to an actual diameter of 50 mm and approximately 100 mm in length, which resulted in a length-diameter ratio of 2: 1 for mechanical tests in accordance with the ISRM suggested method (Fairhurst and Hudson 1999). To investigate the seepage evolution of sandstone under triaxial compression after thermal treatment, the transient pulse technique was applied to measure the permeability of rock during the process of the triaxial conventional compression test.

The loading procedure for the short-term conventional compression-seepage test can be described as follows. First, the specimens are subjected to the confining pressure at the desired value at a rate of 20 MPa/min, and at this stage, the axial stress is proportionally increased to the value of the confining pressure, bringing the specimens to an initial isotropic stress or zero deviatoric stress state. Subsequently, the specimens are saturated with gas fluid until the pore pressure is increased and held at a constant value (p_0), which means that the upstream pressure (p_u) and downstream pressure (p_d) are balanced. This step is necessary to ensure that the fluid exists in a single-phase flow state during the seepage tests. The deviatoric stress is increased stepwise in the axial load control condition at a rate of 5 MPa/min up to each designed deviatoric stress. The axial load is held constant, and the transient flow procedure is subsequently invoked for measurement of the permeability (see Fig. 14.1).

For the transient-pulse permeability test with constant volume reservoirs, the pressure on the upstream side, p_u, is suddenly increased by a small increment Δp_0, a pressure gradient is generated between the two ends of the specimens, and the permeability of the specimens is calculated according to the following equations (Brace et al. 1968):

$$k = -\frac{\mu L}{At}\frac{V_u V_d}{V_u + V_d} n \tag{14.4}$$

$$n = \ln\left(1 - \frac{p_d(t) - p_0}{\Delta p_0}\frac{V_u + V_d}{V_u}\right) \tag{14.5}$$

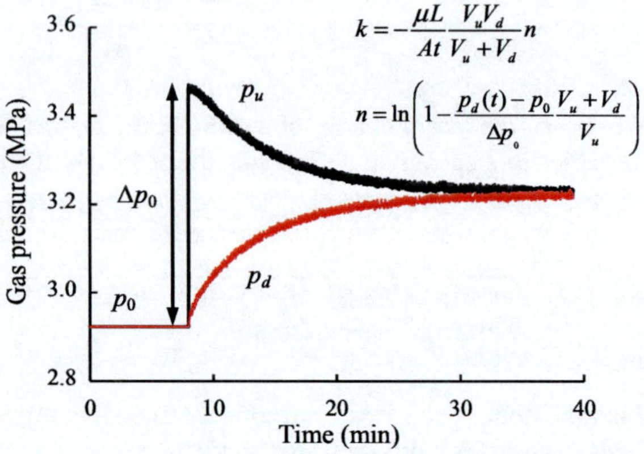

Fig. 14.1 Example of a pressure pulse decay test and calculation of permeability

where p_0 is the initial saturated pressure, Δp_0 is the pressure increment on the upstream side, p_u is the pressure on the upstream side, p_d is the pressure on the downstream side, V_u and V_d are the volumes of the upstream and downstream reservoirs, respectively ($V_u = 17.6$ ml and $V_d = 20.5$ ml in this setup), μ is the dynamic viscosity of the fluid ($\mu \approx 1.8 \times 10^{-5}$ Pa s for pure nitrogen at room temperature) (Brown 1981), and A and L are the cross-sectional area and the length of the specimens, respectively. Equations (14.4) and (14.5) can be applied only when the compressive storage in the specimen is negligible compared with the compressive storage in the upstream reservoir according to previous studies (Hsieh et al. 1981; Lin 1982). The storage of the upstream reservoir of the equipment used in this research is considerably larger than the compressive storage in the sandstone sample, which means that this condition approximately holds for the hard Rizhao sandstone.

14.1.4 Brazilian Testing Procedure

To determinate the indirect tensile strength (σ_t), a series of Brazilian tests was conducted. The Brazilian test is a simple testing method that produces a good estimate of the tensile strength of brittle material. When a thin circular disk is diametrically compressed between two curved loading jaws, the resulting stress field is biaxial, and a tensile stress component can be observed in the central region of the specimen, perpendicular to the loading direction (Brown 1981). A series of 27 tests were performed on disk specimens with diameter $D = 50$ mm and approximate thickness $c = 25$ mm, according to the ISRM suggested methods (Brown 1981). The indirect tensile strength is calculated based on the assumption that failure occurs at the point of maximum tensile stress (i.e., at the center of the disk) as follows:

$$\sigma_t = \frac{2P_{\max}}{\pi Dc} \tag{14.6}$$

where P_{max} is the maximum force applied to the specimen.

14.1.5 Equipment for XRD and SEM Analysis

The composition and internal structure of a material directly affect the mechanical properties. To support and analyze the influence of temperature on the mechanical properties of sandstone, the material composition and micro-structural morphology characteristics of sandstone after various thermal treatments were investigated to create a qualitative understanding of the associated mechanism.

XRD analysis was used in identification of the rock composition at ambient temperature. The XRD analysis was performed using an X-Ray PANalytical Diffractometer (BRUKER) instrument from a start position (2θ) of 3° to the end position (2θ) of 80° with a step size of 0.02°. The mineralogical analysis was performed using the X-Pert system software package and the existence of minerals was detected from the position (2θ) and intensity of peaks in the XRD spectra.

The micro-structural characteristics of sandstone after thermal treatment can be revealed via SEM analysis. The SEM analysis was performed with a FEI Quanta TM 250 SEM instrument and observations were collected under different magnifications (from 500 to 8000) to analyze the changes in microstructure induced by thermal treatment.

14.2 Results of Basic Physical Tests of Sandstone Specimen After Thermal Treatment

The basic physical properties, i.e. weight, volume, density, and P-wave and S-wave velocities of sandstone specimens are investigated before and after different heat treatment temperatures, and the results are shown in Table 14.1.

14.2.1 Effect of Temperature on Weight, Volume and Bulk Density of Sandstone

According to the test results for the physical properties of tested sandstone before and after thermal treatment, the rates of change in weight (ω_m), volume (ω_V) and bulk density (ω_ρ) for each thermal treatment temperature are derived from Eq. (14.3) and plotted in Figs. 14.2, 14.3 and 14.4.

As shown in Fig. 14.2, the rate of change in weight for sandstone specimens is less than zero, i.e., the weight of specimens decrease with increasing thermal treatment temperature. In the range of 400 °C and below, the weight of the specimen decreases

Table 14.1 Tested sandstone specimens used for conventional triaxial compression

Specimen	T/°C	Before thermal treatment				After thermal treatment			
		V/cm^3	m/g	V_P/(m/s)	V_S/(m/s)	V/cm^3	m/g	V_P/(m/s)	V_S/(m/s)
A-B-1	25	192.85	471.85	4030	1502	–	–	–	–
A-B-2	25	189.97	463.46	3995	1511	–	–	–	–
A-B-3	25	192.29	466.7	4013	1496	–	–	–	–
A-B-4	25	193.19	471.6	4033	1501	–	–	–	–
A-B-5	100	187.53	459.8	3800	1526	189.10	457.92	3952	1536
A-B-6	100	194.46	475.17	3800	1506	195.34	473.41	3956	1530
A-B-7	100	190.46	464.64	3836	1517	191.69	462.88	3962	1568
A-B-8	100	194.06	471.95	3683	1532	194.68	469.88	3866	1539
A-B-9	200	193.69	475.88	3720	1502	194.30	473.17	3731	1518
A-B-10	200	189.15	461.09	3809	1508	189.48	458.87	3838	1549
A-B-11	200	191.67	471.45	3865	1506	192.13	469.14	3853	1556
A-B-12	200	192.52	470.72	3748	1528	192.76	468.21	3780	1542
A-B-13	300	191.75	469.41	3685	1492	194.51	466.18	3401	1853
A-B-14	300	192.75	472.77	3766	1503	194.90	469.44	3478	1838
A-B-15	300	194.44	470.57	3802	1508	194.73	468.10	3539	1338
A-B-16	300	190.59	466.51	3818	1500	192.37	463.08	3565	1874
A-B-17	400	192.86	468.06	3806	1872	193.28	464.36	3236	1187
A-B-18	400	192.18	467.36	3797	1506	192.56	463.75	3184	1221
A-B-19	400	191.85	468.11	3840	1519	192.25	464.75	3344	1331
A-B-20	400	191.15	469.24	3695	1481	192.13	465.65	3241	1197
A-B-21	500	191.48	473.31	3876	1538	194.16	469.35	2541	1223
A-B-22	500	189.19	466.84	3873	1519	191.83	463.1	2541	1302
A-B-23	500	191.33	469.16	3773	1511	193.11	465.37	2706	1261
A-B-24	500	192.85	471.79	3738	1510	194.96	468.29	2604	1306
A-B-25	600	192.76	472.31	3790	1523	195.13	467.13	2107	1163
A-B-26	600	190.92	467.36	3790	1517	192.91	462.13	2072	1179
A-B-27	600	191.67	470.02	3731	1503	193.91	464.76	2109	1134
A-B-28	600	194.06	471.67	3791	1522	194.82	466.73	2087	1206
A-B-29	700	189.35	462.69	3798	1519	192.53	456.18	1657	1147
A-B-30	700	193.13	472.51	3786	1522	195.74	465.10	1784	1261
A-B-31	700	192.70	471.12	3770	1503	195.19	465.01	1850	1177
A-B-32	700	188.68	463.63	3876	1525	190.89	457.60	1836	1291
A-B-33	800	192.58	469.3	3731	1498	196.25	461.83	1149	1087
A-B-34	800	188.83	465.42	3646	1500	193.07	457.05	1351	1201
A-B-35	800	182.57	453.34	3758	1546	186.85	445.03	1298	1181
A-B-36	800	190.94	465.12	3834	1518	194.46	457.05	1162	1094

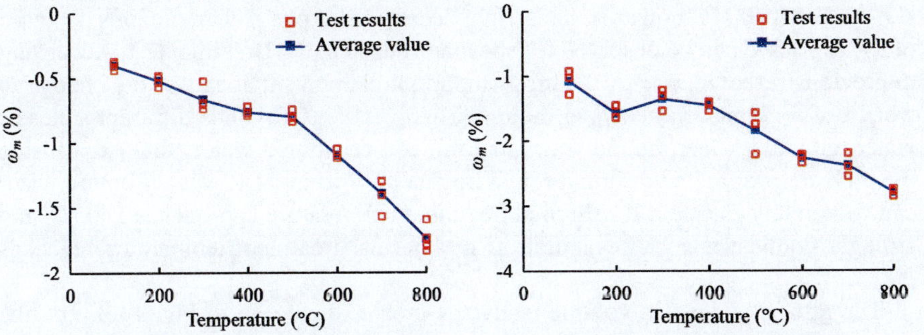

(a) Specimens for triaxial compression tests (b) Specimens for Brazilian tests

Fig. 14.2 Rate of change in weight (ω_m) of tested sandstone specimens versus temperature

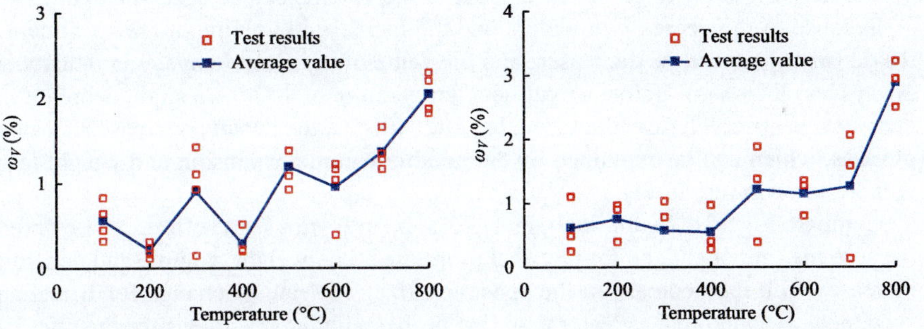

(a) Specimens for triaxial compression tests (b) Specimens for Brazilian tests

Fig. 14.3 Rate of change in volume (ω_V) of tested sandstone specimens versus temperature

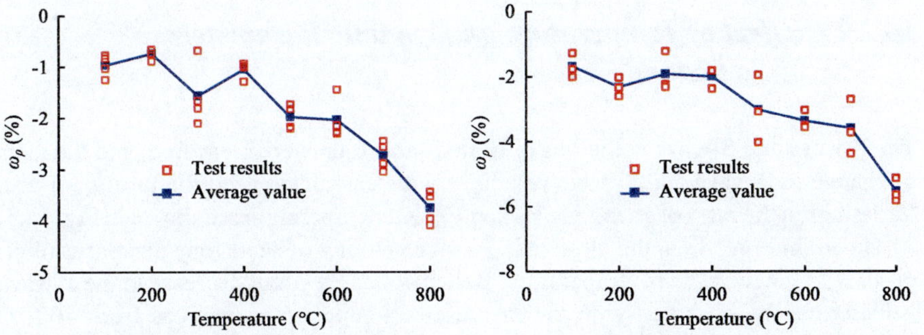

(a) Specimens for triaxial compression tests (b) Specimens for Brazilian tests

Fig. 14.4 Rate of change in density (ω_ρ) of tested sandstone specimens versus temperature slowly. From 400 °C upwards, the weight decrease sharply, and an obvious decrease of 3% at 800 °C can be observed for specimens used in the Brazilian test. According to previous research, one of the most important elements that causes the change in weight of rock material is water escape (Zhang et al. 2016). The different states of water (attached water, bound water and mineral-combined water) that exist inside the rock are evaporated during thermal treatment. Specifically, weakly bound water can completely escape if the thermal treatment temperature approaches 150 °C, and strongly bound water can evaporate if the thermal treatment temperature exceeds 200 °C.

The rate of change in volume is always greater than zero in Fig. 14.3, i.e., the volumes of the thermally treated specimens are always larger than those of the virgin specimens. The slope of the rate of change in volume varies slightly with temperature (Fig. 14.3) in the range of 400 °C and below. The volume increases obviously from 400 °C, and a notably large increase of 3% in the volume can be observed at 800 °C.

Because of the decrease in weight and the increase in volume of the specimens, the density of sandstone decreases, and the value of the rate of change in density is always less than zero. Below a treatment temperature of 400 °C, a slight decrease in density was observed (less than 1%), but after 400 °C, the density decreases became obvious, which can be explained by the notable volume expansion and weight loss at those temperature levels.

Compare Figs. 14.2b, 14.3a, b, and 14.4a, b, an obvious size effect can be noted because the rate of change for physical properties (i.e., weight, volume and density) increases with the decrease in the specimen size. The rate of change for Brazilian specimens is found to be greater than that of the cylinder specimens for the rates of change of weight, volume and density.

14.2.2 Effect of Temperature on Dynamic Parameters of Sandstone

The P-wave and S-wave velocities of heated sandstone were measured, and the rates of change in P-wave and S-wave velocities were calculated according to Eq. (14.3). Variations in the rates of change in V_P and V_S with temperature are shown in Fig. 14.5. As shown in Fig. 14.6a, the ultrasonic P-wave velocity of sandstone decreases obviously with the increase in temperature. Below 400 °C, a slight decrease in the P-wave velocity can be observed with the increase of temperature, whereas from 400 °C onwards, the P-wave velocity decreases sharply.

As shown in Fig. 14.5b, the rate of change in V_S is greater than zero in the range of 300 °C and below, i.e., the S-wave velocities of the specimens slightly increase after thermal treatment. When the temperature exceeds 300 °C, the rate of change in V_S is less than zero, and an obvious reduction of the S-wave velocities of sandstone can be observed with the increase of thermal treatment temperature. Furthermore, the S-wave velocities of sandstone change minimally with the increase of temperature

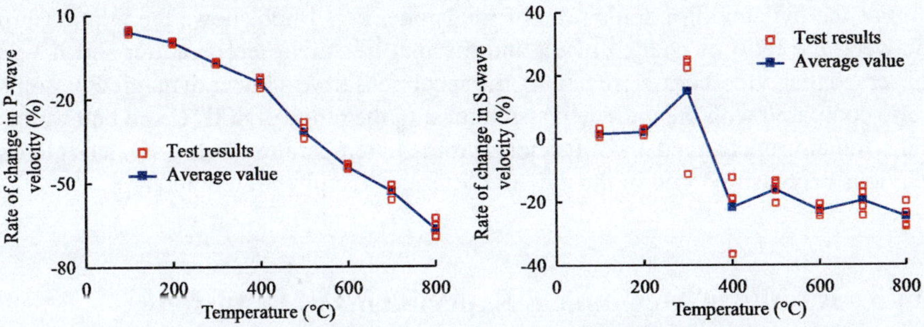

(a) Rate of change in P-wave velocity (b) Rate of change in S-wave velocity

Fig. 14.5 Wave velocity-temperature curves of sandstone after thermal treatment

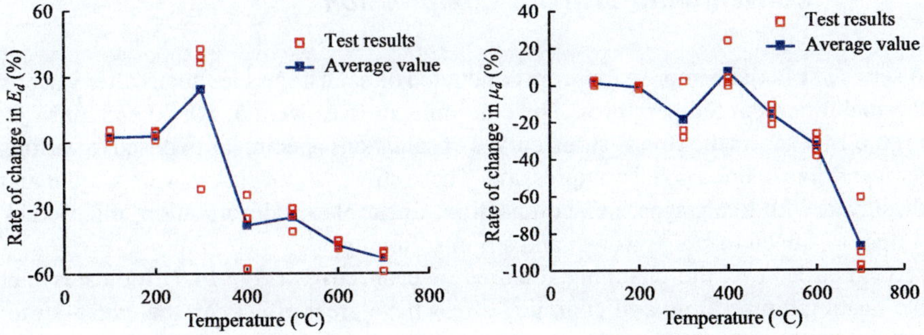

(a) Dynamic elastic modulus (b) Dynamic Poisson's ratio

Fig. 14.6 Dynamic parameters-temperature curves of sandstone after thermal treatment

above 400 °C, i.e., the S-wave velocities of sandstone do not change with the thermal temperature after 400 °C.

To further analyze the dynamic elastic modulus and Poisson's ratio of sandstone after high temperature treatment, the ratios of change in E_d and μ_d were calculated according to Eqs. (14.1) and (14.2) and the results are shown in Fig. 14.6. It should be noted that E_d and μ_d can be derived under the condition of $V_P \geq 1.41 V_S$ according to Eqs. (14.1) and (14.2), whereas in Table 14.1, the measured V_P and V_S of sandstone after 800 °C treatment are similar that E_d and μ_d can not be determined in this manner. Therefore, E_d and μ_d of sandstone after 800 °C treatment are not discussed in this work. As shown in Fig. 14.6 (a), below 300 °C, a slight increase in the dynamic elastic modulus can be observed with the increase of temperature, whereas from 400 °C onwards, the dynamic elastic modulus decreases sharply. Especially for the 700 °C thermal treatment, the dynamic elastic modulus decreases by 53% after treatment at 700 °C. It can be concluded from Fig. 14.6a that greater deformation can be derived with a higher thermal treatment temperature.

For the dynamic Poisson's ratio of sandstone, Fig. 14.6b shows the relationship between the ratio of change in μ_d and thermal treatment temperature, and it was observed that after thermal treatment, the specimens have a lower dynamic Poisson's ratio compared with the untreated specimens. In the range of 400 °C and below, this ratio remains unchanged and a further increase in temperature (>400 °C) leads to an obvious decreasing trend in the dynamic Poisson's ratio with temperature.

14.3 Results of Mechanical Experiments of Sandstone Specimens After Thermal Treatment

14.3.1 Stress–strain Curves of Sandstone Under Conventional Triaxial Compression

A series of triaxial compression were conducted on sandstone specimens after various thermal treatment temperatures. The confining stresses were 5, 10, 20 and 30 MPa. Figure 14.7 shows the stress–strain curves of sandstone specimens exposed to various temperatures under triaxial compression. These stress–strain curves were similar and displayed with five stages, i.e., compaction, linear elastic deformation, microcrack initiation, unstable crack growth and strain softening.

Comparison of the changing features of each curve (Fig. 14.7) indicates that the heating temperature and confining stress have great effects on the stress–strain characteristic curves. The compaction stage is more obvious with a lower confining stress and a higher temperature. As the temperature increases from 25 °C to 800 °C, the percentage of compaction increases in the pre-peak section of the stress–strain curve with the same confining stress, but it clearly decreases with the increase of confining stress from 5 to 30 MPa and the same thermal treatment temperature.

An obvious strain softening occurs in the post-peak stress–strain curve when specimens are exposed to various temperatures after strain softening. The specimen maintains the residual strength at each confining stress after different thermal treatment.

14.3.2 Effect of Temperature on the Strength and Deformation Parameters of Sandstone

Table 14.2 lists the strength and deformation parameters of sandstone specimens after exposure to various temperatures. In Table 14.2, the static elastic modulus (E_S) of rock is defined as the average slope of the approximately straight line portion of the axial deviatoric stress–axial strain curve before the peak strength. The crack damage threshold (σ_{cd}) of rock (Fairhurst and Hudson 1999; Yang et al. 2015) is defined as the corresponding axial deviatoric stress at which the volumetric deformation of the

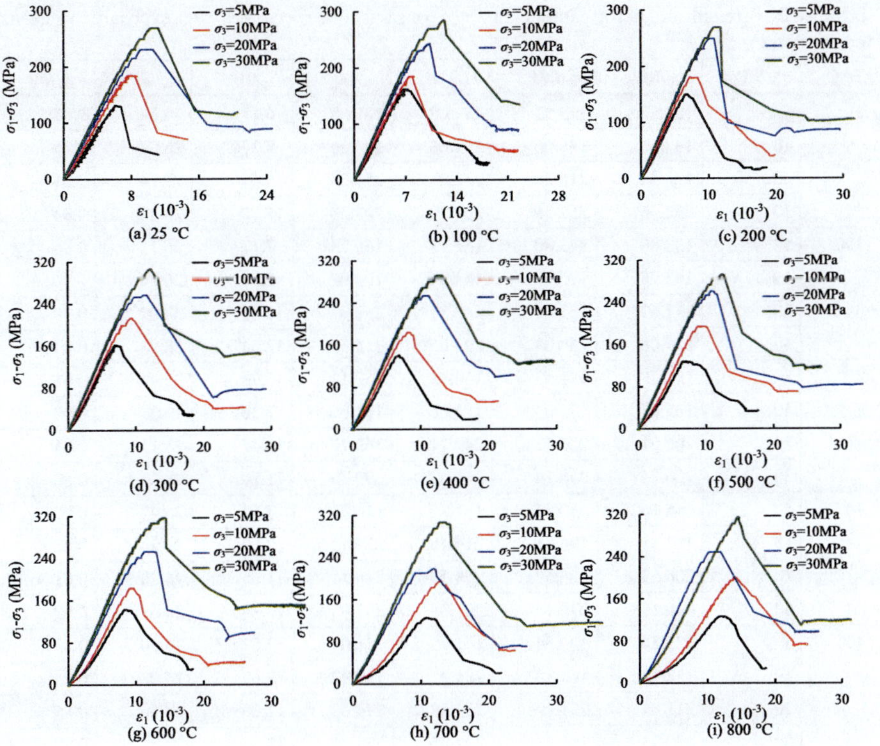

Fig. 14.7 Stress–strain curves of sandstone under triaxial compression after various thermal treatments

specimen switches from compaction-dominated to dilatancy-dominated. The peak stress (σ_p) values of sandstone are obtained from the maximum deviatoric stress in the deviatoric stress-axial strain curves shown in Fig. 14.7. The peak axial strain (ε_{1p}) refers to the axial strain corresponding to the peak stress (σ_p).

Based on the experimental data listed in Table 14.2, Fig. 14.8 shows the influence of high temperature on the peak strength and crack damage threshold of sandstone specimens under triaxial compression. It is clear that the peak strength and crack damage threshold of sandstone depend on both the confining pressure and the temperature. From Fig. 14.8a, at $\sigma_3 = 5$ MPa, σ_S increased from 136.08 to 166.81 MPa as the temperature increased from 25 to 300 °C, but subsequently decreased from 166.81 to 131.82 MPa as the temperature increased from 300 to 800 °C. However, at $\sigma_3 = 10$, 20 and 30 MPa, σ_S increased nonlinearly with the increase of temperature. The higher the confining pressure, the more obvious the increasing trend of σ_S will be. However, from Fig. 14.8b, we observe that at $\sigma_3 = 5$ MPa, the crack damage threshold (σ_{sd}) of sandstone increases from 113.53 MPa at T = 25 °C to 150.16 MPa at T = 300 °C and subsequently decreases linearly to 94.75 MPa at T = 800 °C. With the increase of temperature, the evolutionary trend of σ_{sd} at $\sigma_3 = 10$ MPa is

Table 14.2 Strength and deformation parameters of sandstone specimens after exposure to various temperatures

T/°C	σ_3/MPa	σ_{cd}/MPa	σ_{sd}/MPa	σ_p/MPa	σ_S/MPa	$\varepsilon_{1p}/10^{-3}$	E_S/GPa	σ_{sd}/σ_S
25	5	108.53	113.53	131.08	136.08	6.19	24.34	0.83
	10	138.66	148.66	184.66	194.66	8.37	26.07	0.76
	20	196.44	216.44	231.08	251.08	9.65	27.24	0.86
	30	222.36	252.36	269.28	299.28	11.22	28.34	0.84
100	5	138.80	143.80	161.73	166.73	7.17	26.12	0.86
	10	165.78	175.78	184.94	194.94	7.94	26.50	0.90
	20	217.62	237.62	242.87	262.87	10.40	28.02	0.90
	30	234.08	264.08	284.65	314.65	12.34	28.88	0.84
200	5	136.59	141.59	150.13	155.13	6.73	25.44	0.91
	10	161.03	171.03	181.06	191.06	7.30	27.39	0.90
	20	206.61	226.61	249.71	269.71	10.47	28.11	0.84
	30	226.74	256.74	269.43	299.43	11.70	28.21	0.86
300	5	145.16	150.16	161.81	166.81	7.52	26.47	0.90
	10	184.81	194.81	214.40	224.40	9.60	27.76	0.87
	20	206.59	226.59	258.15	278.15	11.65	28.29	0.81
	30	246.16	276.16	307.05	337.05	12.35	28.61	0.82
400	5	125.54	130.54	141.05	146.05	7.07	24.50	0.89
	10	152.18	162.18	184.92	194.92	8.69	26.28	0.83
	20	211.82	231.82	254.08	274.08	11.30	27.66	0.85
	30	250.23	280.23	292.36	322.36	13.31	27.79	0.87
500	5	115.76	120.76	127.37	132.37	7.90	23.30	0.91
	10	163.54	173.54	195.12	205.12	10.13	26.07	0.85
	20	218.26	238.26	261.64	281.64	11.26	28.15	0.85
	30	245.28	275.28	292.29	322.29	12.67	28.41	0.85
600	5	111.87	116.87	142.69	147.69	9.25	21.75	0.79
	10	148.25	158.25	185.03	195.03	9.90	24.68	0.81
	20	210.50	230.50	253.94	273.94	12.90	25.69	0.84
	30	250.34	280.34	318.02	348.02	14.38	28.33	0.81
700	5	102.51	107.51	125.66	130.66	11.13	17.09	0.82
	10	138.78	148.78	196.40	206.40	14.17	20.23	0.72
	20	191.85	211.85	219.34	239.34	12.18	24.69	0.89
	30	217.48	247.48	307.56	337.56	13.53	27.12	0.73
800	5	89.752	94.752	126.82	131.82	12.60	14.11	0.72
	10	129.98	139.98	201.85	211.85	16.78	19.46	0.66
	20	224.27	244.27	249.57	269.57	11.34	26.59	0.91
	30	238.73	268.73	315.15	345.15	14.64	25.09	0.78

Note $\sigma_{sd} = \sigma_{cd} + \sigma_3$; $\sigma_S = \sigma_p + \sigma_3$

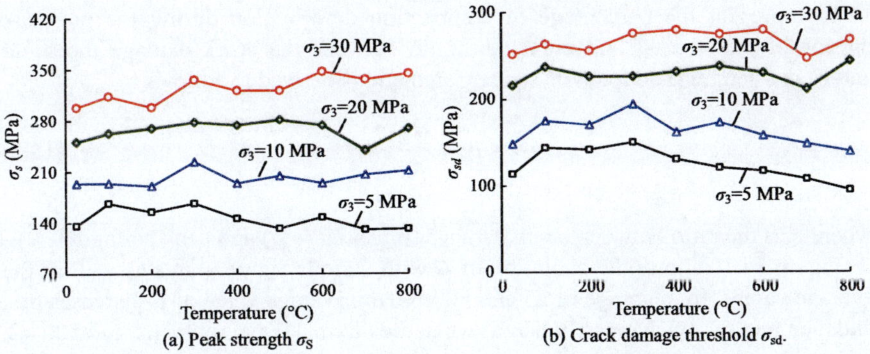

Fig. 14.8 Characteristic stress of sandstone under triaxial compression after various thermal treatments

similar to that at $\sigma_3 = 5$ MPa. However under higher confining pressures ($\sigma_3 = 20$ and 30 MPa), the σ_{sd} of sandstone does not depend on the temperature.

According to the linear Mohr–Coulomb criterion and Fig. 14.8, we analyzed the relationship between peak strength and crack damage parameters of sandstone and temperature, as shown in Fig. 14.9. The evolution law of cohesion and internal friction angle separately obtained from σ_S and σ_{sd} are approximate. The cohesions of σ_S and σ_{sd} first increases and subsequently decreases as the temperature continues to exceed 300 °C, whereas the internal fiction angle first decreases in the range of 300 °C and below and subsequently increases with temperature. The evolution of cohesion and internal friction angle can explain the change laws of σ_S and σ_{sd} with different confining pressures. Because of the decreasing cohesion and increasing internal friction angle after 800 °C compared with 25 °C, the peak strength and crack damage threshold after 800 °C treatment are lower at $\sigma_3 = 5$ MPa but greater at $\sigma_3 = 30$ MPa than those of the non-thermal damage specimen.

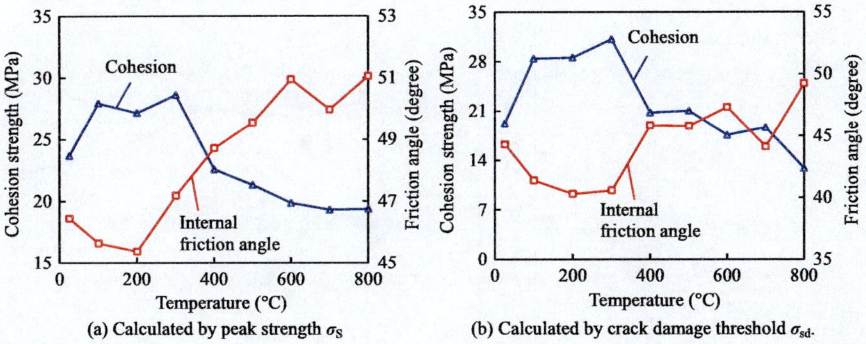

Fig. 14.9 Evolution of cohesion and internal friction angle of sandstone with various thermal treatment temperatures

To represent the percentage of compaction deformation during the pre-peak portion of the stress–strain curve, the ratio between the crack damage threshold and peak strength is adopted in this research and is derived as follows

$$\beta = \frac{\sigma_{\rm sd}}{\sigma_S} \tag{14.7}$$

where β is the ratio between crack damage threshold ($\sigma_{\rm sd}$) and peak strength ($\sigma_{\rm S}$).

Figure 14.10 shows the evolution of β with various temperatures as well as the average values. In the range of 25 and 500 °C, the average value of β increases first and then remains unchanged, whereas when the thermal temperature exceed 500 °C, an obvious decreases of the average value of β can be observed with increasing temperature. Corresponding with the evolution of β, no observable change occurs in the compaction phenomenon in the stress–strain curves (Fig. 14.8) when the temperature reaches 500 °C, whereas the compaction phenomenon gradually appears as the temperature exceeds 500 °C.

The uniaxial compressive strength ($\sigma_{\rm UCS}$) of rock is a highly important design parameters in geotechnical engineering design and the $\sigma_{\rm UCS}$ of sandstone can be obtained according to the triaxial compressive strengths of sandstone with different confining stresses. Figure 14.11 shows the tendency of $\sigma_{\rm UCS}$ with the increasing temperature in which $\sigma_{\rm UCS}$ is observed to first increase and then decrease, and reach a maximum at 200–300 °C. From regression analysis (see Fig. 14.11), the uniaxial compressive strength ($\sigma_{\rm UCS}$) of sandstone and temperature are expressed in the following equation:

$$\frac{\sigma_{ucs}}{\sigma_{ucs,25}} = 0.89 + 0.71 \times \left(\frac{T/100 + 0.69}{3.68}\right)^{1.25} \exp\left(\left(\frac{T/100 + 0.69}{3.68}\right)^{2.25}\right) \tag{14.8}$$

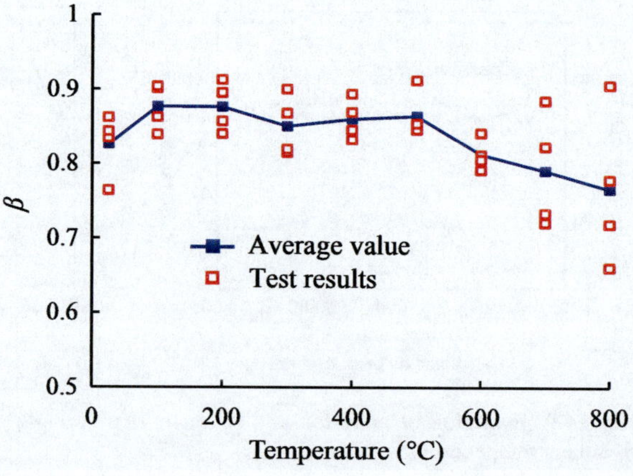

Fig. 14.10 Ratio of $\sigma_{\rm s}$ and $\sigma_{\rm sd}$ after various thermal treatments

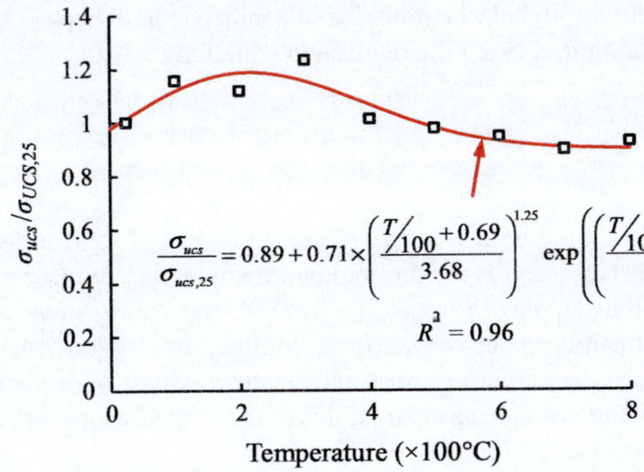

Fig. 14.11 The relationship between the average σ_{UCS} and different thermal-treatment temperatures of sandstone

where σ_{UCS} is the uniaxial compressive strength, $\sigma_{UCS,25}$ is the uniaxial compressive strength of sandstone without thermal treatment (25 °C), and T is the thermal treatment temperature.

The elastic modulus (E_S) offers a measure of the anti-deformability of sandstone under triaxial compression. Figure 14.12 depicts the sandstone elasticity modulus E_S curve relative to the thermal treatment temperature. Overall, the elastic modulus of sandstone changes minimally, and subsequently decreases obviously as temperature continues to exceed 500 °C. The average elastic modulus of sandstone with different confining stresses changes slightly from 26.50 to 26.48 GPa as the temperature increases from 25 to 500 °C, and subsequently decreases linearly from 26.48 to 21.31 GPa as the temperature increases from 500 to 800 °C.

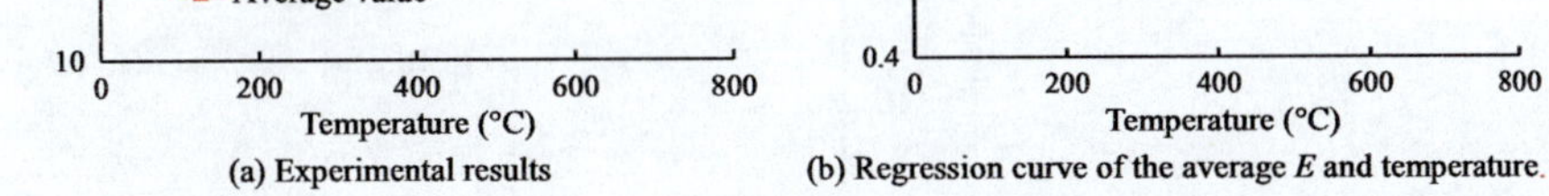

Fig. 14.12 Changing curve of static elastic modulus of sandstone with various thermal treatment temperatures

According to Fig. 14.12, the elastic modulus of sandstone and thermal treatment temperature can be configured using the regression equations:

$$\frac{E}{E_{25}} = -8 \times 10^{-7}T^2 + 4 \times 10^{-4}T + 1 \qquad (14.9)$$

where E is the elastic modulus, E_{25} is the elastic modulus of sandstone without thermal treatment (25 °C), and T is the thermal treatment temperature.

According to previous studies (Yang et al. 2014), the confining stress affect the elastic modulus significantly under conventional compression. Figure 14.13 shows the relationship between the confining stress and elastic modulus of sandstone after thermal treatment, and the average elastic modulus of those specimens with the same confining stress is calculated and plotted in Fig. 14.13.

It can be observed from Fig. 14.13 that the average elastic modulus increases with increasing confining stress. The elastic modulus of sandstone with different temperature treatments varies significantly when $\sigma_3 = 5$ MPa, and with the increase of σ_3, the discreteness of the elastic modulus decreases obviously. When the confining stress reaches 30 MPa, little difference can be found in the elastic moduli of sandstone with different temperature treatments. This result means that the influence of temperature on the elastic modulus of sandstone weakens with a high confining stress. Via regression analysis, the elastic modulus and confining stress (σ_3) conform to the following equation:

$$E = 28.25 - 9.80\exp(-0.11\sigma_3) \qquad (14.10)$$

In addition to its influence on the deformation resistance and ultimate bearing capacity of rock, thermal treatment also affects the deformation characteristics of sandstone under triaxial compression. Figure 14.14 shows the changes in the peak

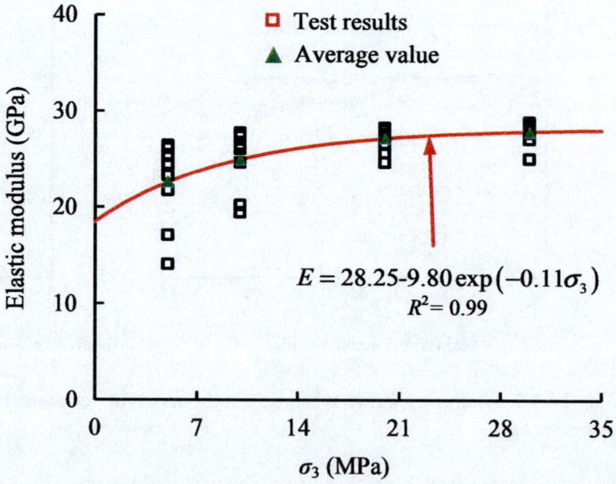

Fig. 14.13 Elastic modulus of sandstone versus confining stress

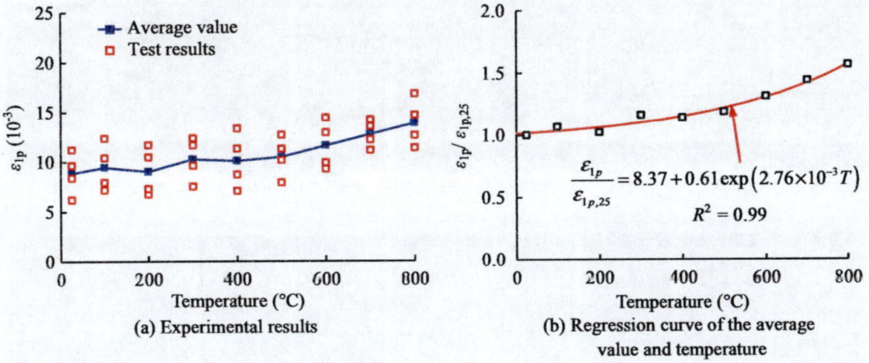

Fig. 14.14 Changing curve of peak strain (ε_{1p}) of sandstone with various thermal treatment temperature

strain ε_{1p} with thermal treatment temperature. In contrast to the variation of the elastic modulus and characteristic stresses, the average peak strain increase linearly with increasing temperature. As the temperature increases from 25 to 800 °C, the dynamic peak strain increases linearly from 8.86 × 10^{-3} to 13.84 × 10^{-3}.

The following equation describes the relationship between peak strain (ε_{1p}) and thermal treatment temperature:

$$\frac{\varepsilon_{1p}}{\varepsilon_{1p,25}} = 8.37 + 0.61\exp\left(2.76 \times 10^{-3}T\right) \quad (14.11)$$

where ε_{1p} is the peak strain under compression, $\varepsilon_{1p,25}$ is the peak strain of sandstone without thermal treatment (25 °C), and T is the thermal treatment temperature.

14.3.3 Effect of Temperature on the Failure Behavior of Sandstone

Rock failure types mostly depend on rock lithology, anisotropy, brittleness, rock microstructures and applied stresses. Figure 14.15 shows the macroscopic failure modes of sandstone under different confining stresses after thermal treatment.

As shown in Fig. 14.15, under a confining stress of 5 MPa at room temperature, rock failure is primarily accompanied by somewhat irregular longitudinal splitting. With increasing thermal temperature up to 800 °C, more vertical and horizontal tensile cracks can be observed, and connected cracks instead of a clear shear fracture are observed on the specimen after 800 °C treatment. With a confining stress of up to 20 MPa, the sandstone specimens without thermal treatment demonstrated a typical shear fracture mode, whereas a mixed tension and shear fracture mode was observed

Fig. 14.15 Macro-fractures of sandstone specimens after triaxial compression with various thermal treatment temperatures

with lateral tensile cracks and shear cracks in the specimen treated at 800 °C. Under a confining pressures of up to 30 MPa, longitudinal fracturing is suppressed and failure occurs along one major through-going shear (brittle) fracture can be observed with thermal treatment up to 800 °C.

Therefore, thermal treatment below 800 °C has little influence on the failure type of sandstone if the confining stress is up to 30 MPa. With a confining stress below to 30 MPa, more vertical and horizontal tensile cracks form with the increasing thermal temperature and the failure mode changes correspondently.

14.3.4 *Effect of Temperature on the Tensile Strength of Sandstone*

The tensile strength of sandstone after different temperature thermal treatment is shown in Fig. 14.16. When Fig. 14.16 was analyzed, the following results were obtained. For temperature-exposed sandstone, tensile strength decreased with increasing temperature. The tensile strength of sandstone changes minimally with treatments below 400 °C, but an obvious downward trend can be observed from 400 °C onwards. The average tensile strength of sandstone at room temperature is 6.00 MPa, and changed to 5.91 MPa at 400 °C, however, it decreased to 3.37 MPa at 800 °C, which indicates an obvious decrease of 43.89% for the tensile strength of sandstone as the thermal temperature increases to 800 °C.

The following equation describes the relationship between tension strength (σ_t) and thermal treatment temperature:

$$\frac{\sigma_t}{\sigma_{t,25}} = -8 \times 10^{-7}T^2 + 1 \times 10^{-4}T + 0.98 \tag{14.12}$$

where σ_t is the tension strength, $\sigma_{t,25}$ is the tension strength of sandstone measured without thermal treatment (25 °C), and T is the thermal treatment temperature.

Representative examples after different thermal treatments are presented in Fig. 14.17 to illustrate crack propagation. It can be observed from Fig. 14.17 that tensile failure occurs along the loading direction, an observation consistent with previous studies at room temperature. The cracks initiate from the center and the specimen is separated into two pieces along the impact direction. This situation occurs because crack initiation is more likely to occur at the tips of an internal defect such as a rock granule boundary, soft medium, micro-voids and cracks, etc. The initial cracks extend along the loading direction to both sides in a short time, long

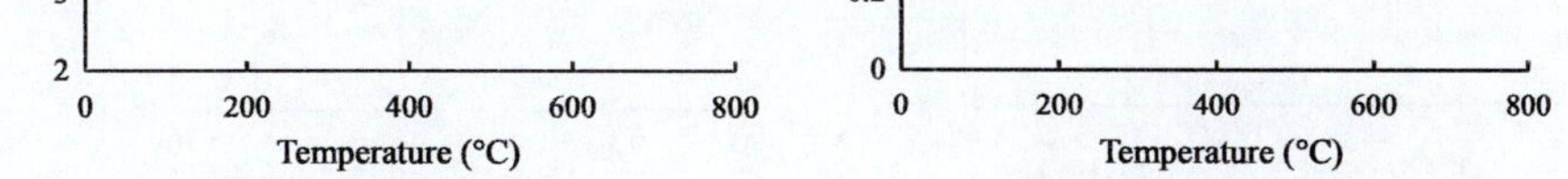

(a) Experimental results (b) Regression curve of the average value and temperature

Fig. 14.16 Tensile strength versus temperature for sandstone

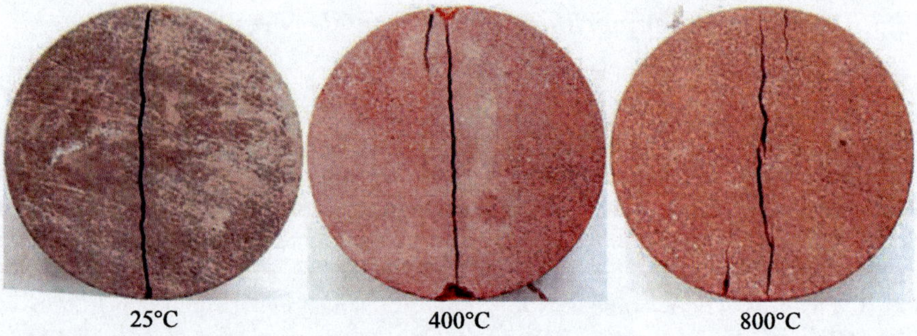

Fig. 14.17 Failure modes of sandstone specimen after Brazilian test

before other micro-cracks are initiated. During the process, the cracks along the radial direction connect, merge, and finally form a macro-crack, which causes the specimen to split into two roughly identical halves.

Different failure modes can be observed and are shown in Fig. 14.18 with different temperatures. A single crack along the line of loading direction can be observed on the specimen after failure at room temperature, whereas for the specimen after thermal treatment at 800 °C, a primary crack occurs near the center of the specimen along the loading direction, and subsequently another crack parallel to the line of loading direction is produced on both ends of the specimen. This situation occurs because when the temperature reaches 800 °C, the rock experiences a series of complex changes, including decomposition of mineral constituents, evaporation of constituent water and the quartz phase transition. Compared with those at room temperature, the deformation resistance and toughness of the granite both decrease to a certain extent, which leads to serious degradation of the mechanical properties. Before the loading stress reaches the peak value, it exceeds the granite strength, and the specimen begins to yield before the primary crack is connected.

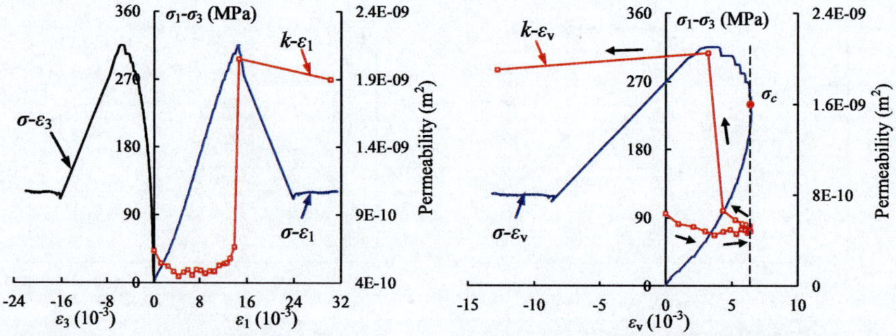

(a) Relation between permeability and axial strain (b) Relation between permeability and volumetric strain

Fig. 14.18 Permeability evolution of sandstone during triaxial compression with a confining stress of 30 MPa and temperature 800 °C

14.4 Results of Permeability Experiments on Sandstone Specimens Exposed to Heat Treatment

To investigate the permeability behavior of sandstone under triaxial compression, a series conventional triaxial compression tests coupled with transient-pulse permeability tests are conducted on sandstone specimens treated with various temperatures from 25 to 800 °C. The influence of thermal treatment and confining stress on the permeability property of sandstone has been analyzed and discussed.

14.4.1 Permeability Evolution of Sandstone Under Conventional Triaxial Compression

The typical evolution curve of the permeability of sandstone during conventional triaxial compression and the relationship between the permeability and volumetric strain are presented in Fig. 14.18, which shows the test results of the sandstone specimen treated at 800 °C with a confining pressure of 30 MPa. It should be noted that a series of small platforms in the stress–strain curves are induced by holding the stress constant to measure the permeability of sandstone specimen.

The permeability evolution trend of sandstone in each phase varies significantly and can be easily calculated and divided into five phases using the volumetric strain-permeability curve in Fig. 14.18b: the micro-defect closure phase, the linear elastic deformation phase, the nonlinear deformation phase, the post-peak stress softening phase and the residual deformation phase. In the micro-defect closure phase, the permeability decreases with the increase in compressive volumetric strain, which can be explained with the closure of the pre-existing micro-defects under compression. In the linear elastic deformation phase, the compressive volumetric strain increases linearly with increasing axial deviatoric stress, and the pre-existing micro-defects and newly formed micro-cracks change only slightly. Therefore, the permeability of sandstone changes slightly in this phase. In the nonlinear deformation phase, the volumetric strain changes from compression to dilatation, and the microcracks start to propagate and coalesce, resulting in the obvious increase of permeability with the increasing axial deviatoric stress. In the post-peak stress softening phase, the dilatancy volumetric strain increases, and the macro-cracks are induced because of the coalescence of the micro-cracks and the new propagation cracks. As a result, the permeability increases rapidly in this phase. In the residual deformation phase, a macro-failure mode is formed in the specimen, the seepage channel is along the fracture plane and changes little in this phase. Therefore, the permeability of sandstone remains constant.

14.4.2 Effect of Confining Pressure on Permeability Behavior of Sandstone

The variations in permeability with increasing axial strain are plotted in Fig. 14.19. Corresponding to the five distinct regions in the stress–strain relations, the permeability-axial strain curves of the sandstone in the triaxial loading process show a clear decrease in the micro-defects closure region and a dramatic increase in the crack growth region. This variation pattern of permeability is completely the same as that of the conceptual model given in (Souley et al. 2001) and is similar to that of other types of rocks, e.g. granite (Chen et al. 2014).

The influence of confining stress on the permeability variation is also illustrated in Fig. 14.20. At a given thermal treatment temperature, the permeability decreases with increasing confining pressure, as a result of more significant closure of preexisting

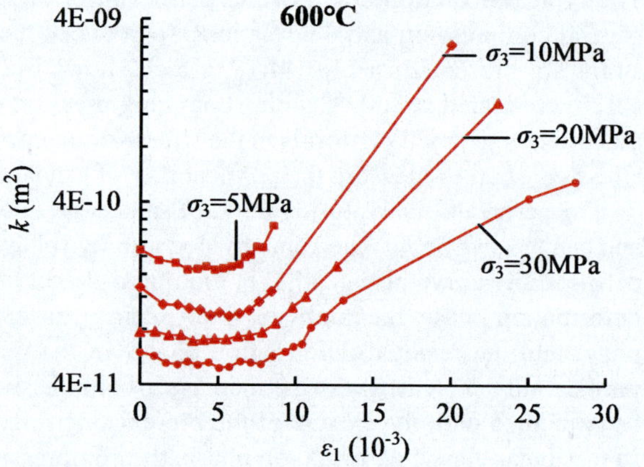

Fig. 14.19 Permeability evolution under triaxial compression with different confining stresses ($T = 600$ °C)

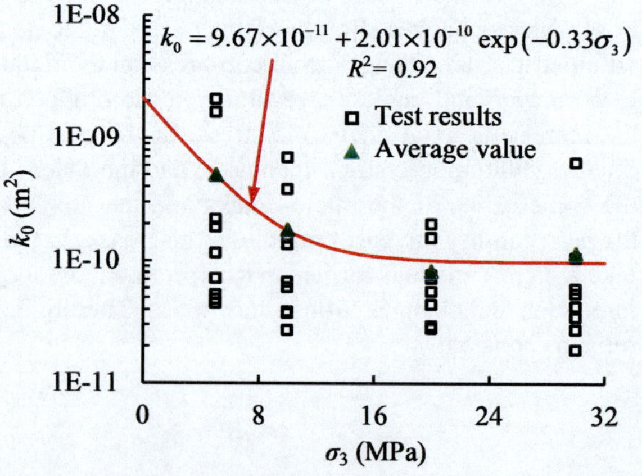

Fig. 14.20 Initial permeability of sandstone versus confining stress after thermal treatment

microcracks and a larger threshold for damage initiation. As the confining pressure increases to 10 MPa, the permeability decreases by approximately one order of magnitude. Figure 14.20 shows the relationship between the initial permeability and confining stress. An obvious decreasing trend can be observed in Fig. 14.20 with the average value, especially for the confining stresses of 5 and 10 MPa. When the confining stress increases from 20 to 30 MPa, the average initial permeability changes slightly.

According to regression analysis, the compressive strength and temperature can be described by the following equation:

$$k_0 = 9.67 \times 10^{-11} + 2.01 \times 10^{-10} \exp(-0.33\sigma_3) \quad (14.13)$$

where k_0 is the initial permeability of sandstone prior to triaxial compression.

14.4.3 *Effect of Temperature on the Permeability Behavior of Sandstone*

The thermal treatment has a great influence on the porosity and cracks in sandstone, and the deterioration of rock properties, such as the permeability, changes accordingly. Figure 14.21 shows the relationship between initial permeability and temperature, and the average values are calculated and shown in Fig. 14.21 for analysis. Comparing Fig. 14.12 and Fig. 14.21, it can be found that the initial permeability and elastic modulus have the opposite evolution law with temperature. Overall, the initial permeability of sandstone first increases slightly, and subsequently increases obviously as the temperature continues to exceed 500 °C. At this point, it should be

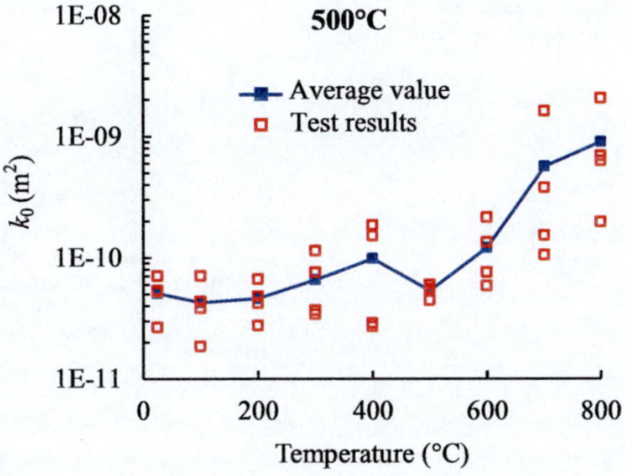

Fig. 14.21 Relationship between thermal treatment temperature and initial permeability of sandstone specimen

noted that the initial permeability is the permeability of sandstone under hydrostatic compression after the confining stress is applied without the axial deviatoric stress.

The increase of permeability is closely related to variation in the parameters of the crystal structure from the previous study. With the increase in temperature, internal moisture continuously escapes from the rock, leading to changes in the internal mineral composition of rock and the increasing micro-cracks. When the thermal temperature exceeds 500 °C, the variation of micro-cracks and the internal porosity of rock increase, the original small network structure is destroyed, the connectivity of the penetration system is greatly enhanced, and the ability of channel flow is improved accordingly. These results are attributed to the rapid increase in the rock permeability.

The influence of thermal treatment temperature on the permeability variation is also illustrated in Fig. 14.22. At a given confining stress, the permeability-axial strain curves of the sandstone are similar for different thermal treatment temperatures. As the thermal treatment temperature increases from 25 to 800 °C, the initial permeability decreases in the range of 25 to 400 °C and subsequently increases by approximately one order of magnitude.

According to regression analysis (see Fig. 14.23), the initial permeability of sandstone and temperature conforms to the following equation:

$$\frac{k_0}{k_{0.25}} = 0.6 + 0.04\exp\left(7.7 \times 10^{-3}T\right) \tag{14.14}$$

where k_0 is the initial permeability of sandstone, and $k_{0,25}$ is the initial permeability of sandstone without thermal treatment (25 °C).

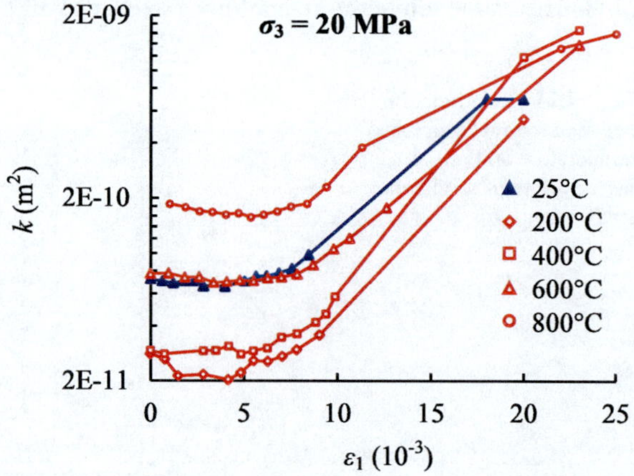

Fig. 14.22 Permeability evolution under triaxial compression with different thermal-treatment temperatures ($\sigma_3 = 20$ MPa)

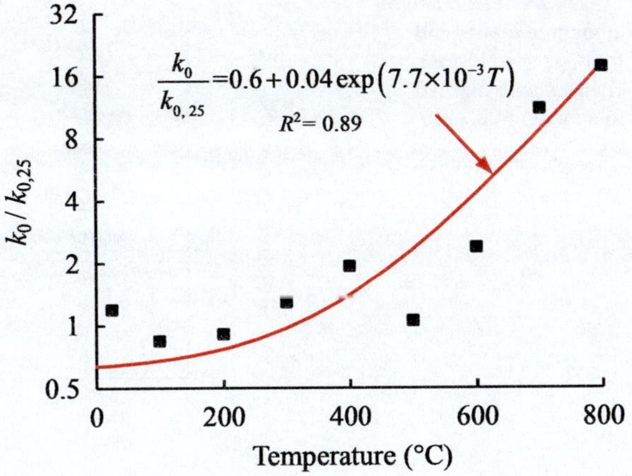

Fig. 14.23 Relationship between thermal treatment temperature and initial permeability of sandstone specimen

14.5 Results of SEM and XRD Analysis

14.5.1 Effect of Temperature on the Material Composition of Sandstone

It is well known that calcite decomposes into calcium oxide and carbon dioxide when the temperature exceeds 700 °C (Heap et al. 2013). Additionally, kaolinite (a water aluminum silicate) is subject to physical dehydration and chemical changes when exposed to high temperatures (Li et al. 2016). To prove the occurrence of the chemical reaction in sandstone above 700 °C, X-ray diffraction (XRD) data were obtained from fine sandstone powders at 9 different heat-treatment temperatures. Figure 14.24 shows the test results under 5 different temperatures (25, 200, 400, 600, 800 °C) with a position (2θ) of 20° to 30°.

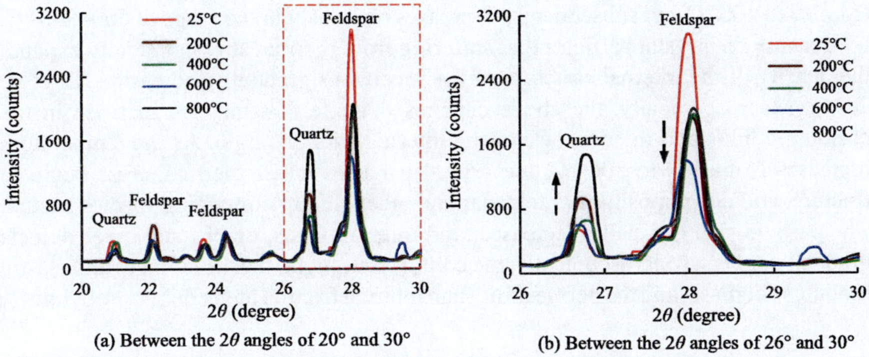

(a) Between the 2θ angles of 20° and 30°

(b) Between the 2θ angles of 26° and 30°

Fig. 14.24 X-ray diffraction (XRD) spectra for the powder of sandstone under various heat-treatment temperatures

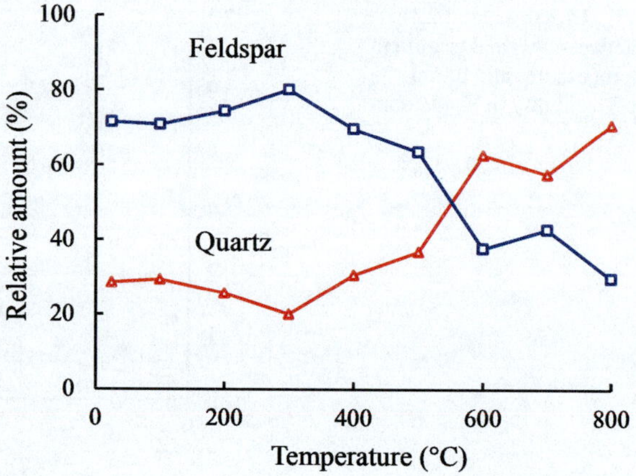

Fig. 14.25 The relative amount of quartz and feldspar of sandstone changes with thermal treatment temperature

According to the results of X-ray diffraction (XRD), the main minerals in the sandstone material are recognized as quartz, feldspar, calcite, dolomite, hematite and clay minerals. In addition, as shown in Fig. 14.24, the XRD spectra clearly indicate that the peak for quartz increases obviously in the sandstone powder at the 800 °C heat-treatment temperature. To illustrate the change of content in the sandstone after thermal treatment, relative amounts of quartz and feldspar after different high temperature treatments are calculated and plotted in Fig. 14.25. It can be observed that the relative amounts of quartz and feldspar increase obviously after 500 °C, which illustrates that the decomposition of feldspar and other silicon-containing species occurs when the sandstone specimen is treated at 800 °C and is a non-reversible chemical reaction.

The average particle size of sandstone d (particle diameter) after different thermal treatment temperatures can be obtained from X-ray diffraction according to the Scherrer function, and the average particle sizes of sandstone after high temperature treatment were calculated and plotted in Fig. 14.26. It can be observed from Fig. 14.26 that the average particle size increases with the increase in temperature from 25 to 400 °C and subsequently decreases obviously in the range of 500–800 °C.

Because the crystal particles that suffering from thermal stress gradually expand, the density of the internal structure of the specimens gradually increasing (Li et al. 2016). Macroscopically, the above changes indicate that that the increase in the elastic modulus and the decrease of the internal friction angle. As the temperature increases from 500 to 800 °C, the crystal particles under thermal stress begin to fracture, and decomposition of feldspar and other silicon-containing species occurs. The particle size gradually decreases, and internal voids, cracks and other defects gradually appear. Macroscopically, the cohesion strength, tensile strength and elastic modulus decrease, and the permeability and internal friction angle increase obviously.

Fig. 14.26 The relationship between the average particle size of sandstone and thermal treatment temperature

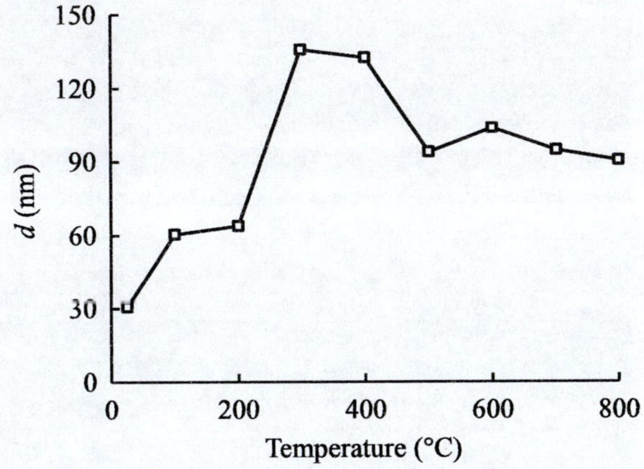

14.5.2 *Effect of Temperature on the Micro-Structure of Sandstone*

SEM images of the topography and morphology of the grain structures and micro-fractures are shown in Fig. 14.27. According to Fig. 14.27, the internal structure of sandstone changes dramatically as the heat-treatment temperature increases, identifying temperature as one of the primary factors that affects the mechanical properties of sandstone.

The sandstone used in the tests was a typical sedimentary rock with a certain amount of original pores and cracks because many pores and cracks can be observed in Fig. 14.27a of the sandstone specimen after air-drying treatment. When the thermal treatment temperature increases from 25 to 400 °C, the number and average size of the original pores decreases, as well as the initial cracks, and the average size of the particles of sandstone increase obviously. At a temperature of 25 °C, certain original pores can be observed at a magnification of 500, whereas at greater magnification, few cracks can be observed. At a magnification of 2000, a large number of small and isolated cracks and pores can be observed in the specimens, and the average size of the isolated grains is small. When the temperature increases to 300 °C, fewer pores and cracks can be observed, and the grain size increases obviously, as shown by comparing Fig. 14.27a, d.

When the thermal treatment temperature increases from 400 to 800 °C, the number of pores and cracks continues to increase, as well as the grain size, and the boundaries of different grains become more obvious. At a temperature of 400 °C, the specimen appears to be intact and dense, different grains are closely connected, and few pores and cracks can be observed in Fig. 14.27e. When the thermal temperature increases to 800 °C, plentiful interconnected defects (pores and cracks) between different grains obviously appear on the specimen profile, and with a magnification of 8000, numerous glassy structures can be observed.

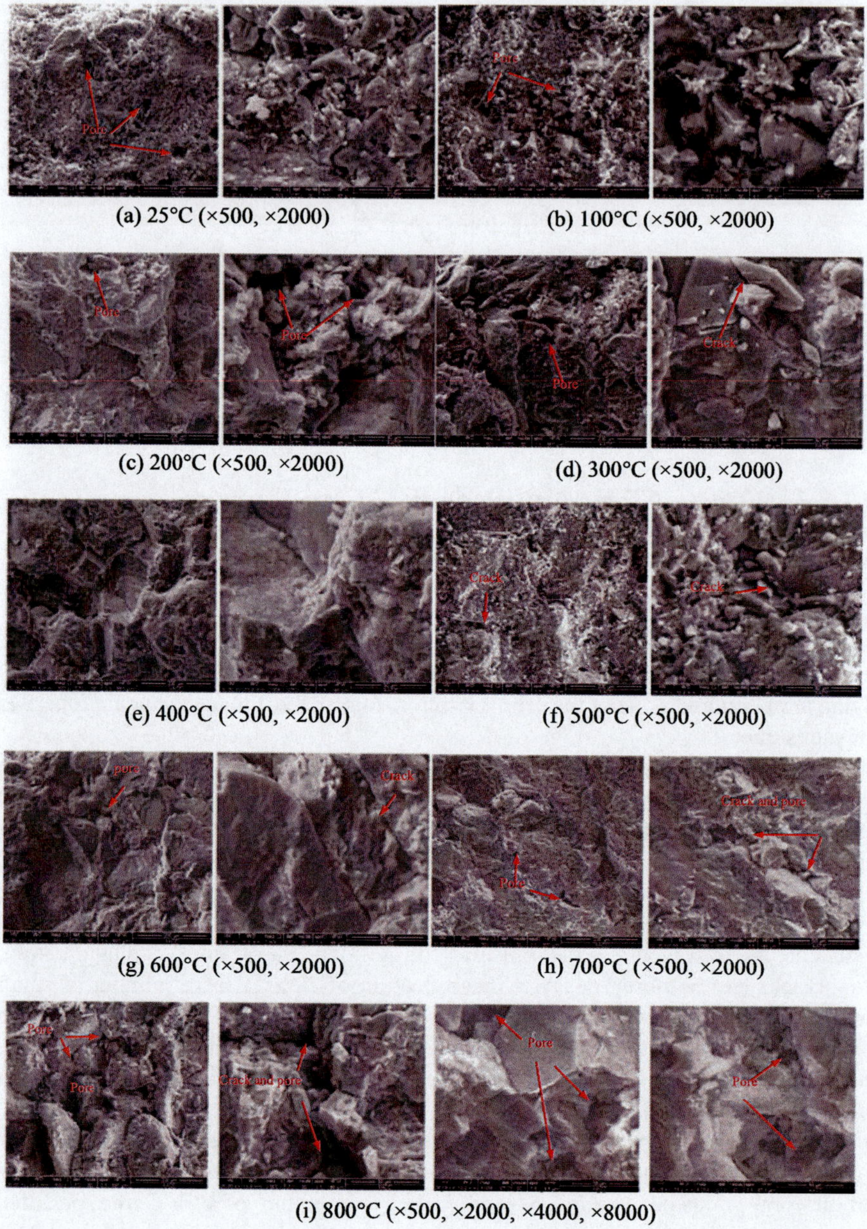

Fig. 14.27 SEM observations of sandstone after different high-temperature treatments

As the temperature increases, the thermal expansion of the sand particles inside the crystal increases the volume of sandstone specimen (Fig. 14.3), and the internal free water and organic matter in the sandstone gradually evaporate, which leads to a decrease in the mass (Fig. 14.2) and density (Fig. 14.4) of the sandstone specimen.

When the temperature increases from 25 to 400 °C, mineral components are expanded, and thermal stress prevents the extension of internal cracks and causes the closure of cracks, thus leading to increased compactness (Ding et al. 2016a, 2016b). Clay minerals, which are susceptible to heat, are often altered after high temperature treatment (Mccabe et al. 2010). As a result, the 400 °C treatment improved the physical and mechanical properties of the sandstone, including a decrease in permeability (Fig. 14.21) and increases in the dynamic elastic modulus (Fig. 14.6a), static elastic modulus (Fig. 14.12) and cohesion strength (Fig. 14.9). Alternatively, the internal friction angle decreases (Fig. 14.9) because of the loss of clay mineral and the coalescence of original cracks and pores.

When the temperature increased from 400 to 800 °C, the thermal stress was further increased because of the increasing constraints induced by the different expansion rates of the components. Additionally, the loss of hydroxyl structural water (Zhang and Zhao 2013) and the α/β transition of quartz (Glover et al. 1995) also contributed to fracture propagation. New cracks were generated when the thermal stress reached the tensile strength or shear strength of the sandstone materials, thus decreasing the carrying capacity. As a result, more numerous new cracks formed and extended along the boundary of different grains, and as the original cracks become widened and connected, the mechanical characteristics were further negatively affected. The 600 and 800 °C treatments cause a rapid decrease in the cohesion strength (Fig. 14.9), static and dynamic elastic modulus (Figs. 14.6a, Fig. 14.12), whereas the permeability (Fig. 14.21), axial peak strain (Fig. 14.14) and the internal friction angle (Fig. 14.9) increase obviously because of the increase of evolution of cracks, and a higher temperature led to a more significant rate of decrease.

14.6 Conclusions

The physical and mechanical characteristics of sandstone after exposure to high temperatures were separately observed and analyzed in this paper. Conventional triaxial compression experiments were accompanied by transient pulse permeability tests, and Brazilian tests were performed. Based on the experimental results, the following conclusions can be drawn.

(1) A thermal treatment of 400 °C was identified as the transition temperature that induces changes in the physical properties of sandstone. In the range of 400 °C and below, the mass and density of the sandstone specimen decreases slowly, and a slight decrease in P-wave velocity can be observed with the increase of temperature, from 400 °C onwards. Additionally, the mass, density and P-wave velocity decrease sharply, and the volumes of the thermally treated specimens

are always larger than those of the virgin specimens, especially at 600 °C and above. An obvious size effect shows that the rate of change for physical properties (i.e. mass, volume and density) increases with the decrease in the size of the sandstone specimen.

(2) For the mechanical properties of sandstone, a thermal treatment of 500 °C was identified as the transition temperature that induces obvious changes. The variance rates of elastic modulus, ratio of crack damage strength and peak strength, peak strain and initial permeability of sandstone increased obviously after 500 °C thermal treatment. The different stress–strain curves of sandstone specimens exposed to various temperatures under triaxial compression are similar and can be divided into five stages, i.e. compaction, linear elastic deformation, microcrack initiation, unstable crack growth and unloading stage. The compaction stage is more obvious with a lower confining stress and a higher temperature. Because of the different influence of thermal treatment temperature on the cohesion and fiction angle, the evolution curves of the triaxial compressive strength with thermal temperature are varied under different confining stresses. The cohesion is observed to first increase and subsequently decrease as the temperature exceeds 300 °C, whereas the fiction angle decreases in the range of 300 °C and below, and then increases obviously.

(3) The evolution of permeability during conventional triaxial compression can be divided into five stages according to the deformation phase of sandstone under compression. Generally, the permeability decreases before the crack damage strength, and then greatly increases. The evolution curves of permeability under triaxial compression with different confining stresses and thermal treatment temperatures are similar, and the values vary to a large extent. A high confining stress can weaken the influence of thermal temperature on the elastic modulus of sandstone. Under a certain thermal treatment temperature, the initial permeability of sandstone decreases sharply with increasing of confining stress in the range of 0 to 10 MPa. With a certain confining stress, the initial permeability of sandstone increases with the increase in temperature, especially in the range of 400–800 °C.

(4) With increasing thermal treatment temperature, the material composition and the state of sandstone change dramatically according to X-ray diffraction analysis. After 500 °C, the relative amount of quartz and feldspar increases and the average particle size decreases obviously, which leads to changes in the physical and mechanical characteristics of sandstone after thermal treatment. Electron microscopy revealed that the internal cracks of sandstone gradually decreased and the internal structure became more compact as the thermal treatment temperature increased from 25 to 400 °C. Additionally, in the range of 400–800 °C, porosity, micro-cracking and other defects gradually increased. Certain empirical models reflect the relationships between the thermal treatment temperature and the compressive strength, tension strength, static elastic modulus and permeability of sandstone. Equations for these relationships are derived in this research, and can be used in engineering practice for protection and safety design in sandstone caverns.

References

Brace WF, Walsh JB, Frangos WT (1968) Permeability of granite under high pressure. J Geophys Res 73(6):2225–2236

Brown ET (1981) Rock characterization, testing and monitoring: ISRM suggested methods. International Society for Rock Mechanics by Pergamon Press

Chen YF, Hu SH, Wei K, Hu R, Zhou CB, Jing LR (2014) Experimental characterization and micromechanical modeling of a damage-induced permeability variation in Beishan granite. Int J Rock Mech Min Sci 71:64–76

Ding QL, Ju F, Mao XB, Ma D, Yu BY, Song SB (2016) Experimental investigation of the mechanical behavior in unloading conditions of sandstone after high-temperature treatment. Rock Mech Rock Eng 49:2641–2653

Ding QL, Ju F, Song SB, Yu BY, Ma D (2016) An experimental study of fractured sandstone permeability after high-temperature treatment under different confining pressures. J Nat Gas Sci Eng 34:55–63

Fairhurst CE, Hudson JA (1999) Draft ISRM suggested method for the complete stress-strain curve for intact rock in uniaxial compression. Int J Rock Mech Min Sci 36:279–289

Glover PWJ, Baud P, Darot M, Meredith PG, Boon SA, Leravalec M (1995) α/β phase transition in quartz monitored using acoustic emissions. Geophys J Int 120(3):775–782

Heap MJ, Mollo S, Vinciguerra S (2013) Thermal weakening of the carbonate basement under Mt. Etna volcano (Italy): implications for volcano instability. J Volcanol Geoth Res 250:42–60

Hsieh PA, Tracy JV, Neuzil CE, Bredehoeft JD, Silliman SE (1981) A transient laboratory method for determining the hydraulic properties of 'tight' rocks-I. Theory. Int J Rock Mech Min Sci Geomech Abstr. 18:245–252

Li M, Mao XB, Cao LL, Pu H, Mao RR, Lu AH (2016) Effects of thermal treatment on the dynamic mechanical properties of coal measures sandstone. Rock Mech Rock Eng 49:3525–3539

Lin W (1982) Parametric analyses of the transient method of measuring permeability. J Geophys Res 87(B2):1055–1060

Mccabe S, Smith BJ, Warke PA (2010) Exploitation of inherited weakness in fire damaged building sandstone: the 'fatiguing' of 'shocked' stone. Eng Geol 115:217–225

Souley M, Homand F, Pepa S, Hoxha D (2001) Damage-induced permeability changes in granite: a case example at the URL in Canada. Int J Rock Mech Min Sci 38:97–310

Yang SQ, Jing HW, Cheng L (2014) Influences of pore pressure on short-term and creep mechanical behavior of red sandstone. Eng Geol 179(10):10–23

Yang SQ, Huang YH, Jiao YY, Zeng W, Yu QL (2015) An experimental study on seepage behavior of sandstone material with different gas pressures. Acta Mech Sin 31(6):837–844

Yang SQ, Xu P, Li YB, Huang YH (2017) Experimental investigation on triaxial mechanical and permeability behavior of sandstone after exposure to different high temperature treatments. Geothermics 69:93–109

Zhang WQ, Sun Q, Hao SQ (2016) Experimental study on the variation of physical and mechanical of rock after high temperature treatment. Appl Therm Eng 98:1297–1304

Zhu TT, Jing HW, Su HJ, Yin Q, Du MR, Han GS (2016) Physical and mechanical properties of sandstone containing a single fissure after exposure to high temperatures. Int J Min Sci Technol 26:319–325

Zhang Q, Zhao J (2013) Determination of mechanical properties and full-field strain measurements of rock material under dynamic loads. Int J Rock Mech Min Sci 60:423–439

Chapter 15
Effect of High Temperature on the Permeability Evolution and Failure Response of Granite Under Triaxial Compression

Granite is potentially an excellent medium for the disposal of high-level nuclear waste (HLW) because it exhibits a low native permeability and high integrity (Yang et al. 2017, 2018; Zhao et al. 2015). Beishan granite has been selected as a candidate for HLW disposal in China because of its low native permeability. However, the safe isolation of nuclear waste from human underground water is very important, thus, it is of great significance to study the permeability evolution of granite subjected to high temperature. During the decay of radionuclides, a large amount of heat is released, heating the rock mass around the emplaced place and causing damage (Chen et al. 2017a; b), and thus, it is valuable to find out the failure response of granite after high-temperature treatment under triaxial compression. Therefore, the permeability evolution and failure response of granite are important and must be known to assure the stable and safe operation of the HLW repository. At the same time, because an HLW repository is underground, the deformation and failure of the surrounding rock mass are notably affected by in situ stresses. Thus in this chapter, it is crucially important to incorporate the roles of stress on failure and the resulting failure upon permeability evolution (Yang et al. 2020). First, the effect of high temperature on the mechanical characteristics and permeability evolution behaviour of granite was analysed in detail. Subsequently, the effect of high temperature on triaxial deformation and AE behaviour of granite was investigated and the peak strength and crack damage threshold was analysed. Finally, mechanisms responsible for these impacts of high temperature influences the triaxial strength parameters of granite are discussed.

15.1 Experimental Methodology

15.1.1 Granite Material

As shown in Fig. 15.1, the granite used in this study was collected from Rizhao, Shandong Province, China. Based on X-ray diffraction (XRD), the granite comprises

S.-Q. Yang, *Mechanical Behavior and Damage Fracture Mechanism of Deep Rocks*,
https://doi.org/10.1007/978-981-16-7739-7_15

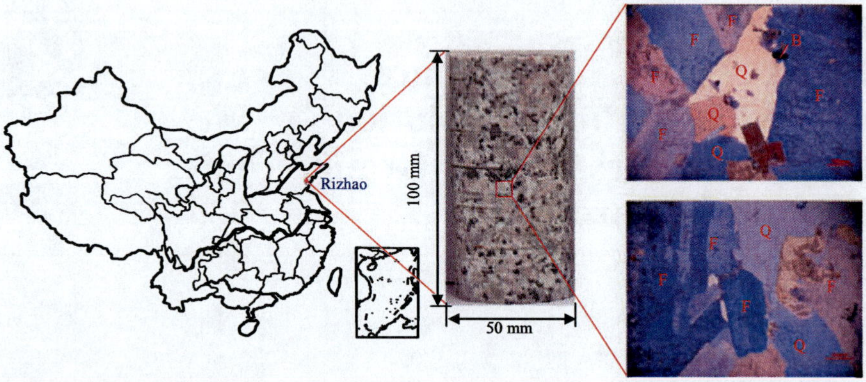

Fig. 15.1 Granite material with coarse-grained structure at room temperature in this study. Q-Quartz, F-Feldspar, B-Biotite

are quartz (11.12%), feldspar (59.85%), biotite (21.56%), amphibole (6%), chlorite (1.01%), and dolomite (0.46%). Figure 15.1 also shows the results of polarised light and optical microscopy of the granite. The granite has a crystalline structure with a connected porosity of 0.828% (Yang et al. 2017). Following the grain size terminology of Williams et al. (1954), the grain size of the granite ranges between 6 and 18 mm and is thus classified as a coarse-grained granite. At room temperature, the average density of the tested granite is 2594 kg/m^3, and the P-wave velocity is approximately 4200 m/s. In this experiment, the cores of all specimens were obtained from the same block of material. The cores have a diameter of 50 mm and length of ~ 100 mm, resulting in a length: diameter ratio of 2:1 for the mechanical tests, which is in accordance with the International Society for Rock Mechanics (ISRM; Fairhurst and Hudson 1999) suggested method.

15.1.2 Permeability Testing System and Procedure

Before beginning the permeability tests, the granite specimens were first placed in a furnace and heated at a rate of 5 °C/min (Shao et al. 2015; Zhu et al. 2016). The predetermined temperature was kept constant for 2 h to ensure temperature homogeneity across the specimen. For the permeability tests, the target treatment temperatures were set to 25, 100, 200, 300, 400, 500, 600, 700, and 800 °C. After the thermal treatment, the heated specimens were left in the furnace to cool down to room temperature before measuring the permeability.

It is difficult to measure the coefficient of permeability of granite using steady-state liquid flow tests because the measurement time for one specimen is too long, ranging from several weeks to several months (Davy et al. 2007; Consenza and Ghoreychi

1999). Therefore, in this study, we used steady-state gas flow tests to measure the granite permeability.

The gas permeability test system (see Fig. 15.2) includes a gas pressure loading system, pressure control panel, gas pressure sensors, pressure loading system, and triaxial pressure chamber. The gas pressure loading system consists of a gas tank, flow pipes and valves in addition to a gas pumping system. Gas pressures may be elevated in the ranges 0–3 MPa to 0–40 MPa. The gas pressure sensors have three different scales to improve measurement accuracy. Furthermore, a high-precision pressure-controlled metering pump provides constant prescribed axial, lateral and triaxial stresses. Note that the maximum confining pressure of the test system is 60 MPa, which is smaller than the compressive strength of the granite tested in this experiment. The system can measure permeability to > 10^{-25} m^2 with this limitation resulting only from the enhance errors entrained in the excessive duration of the experiment.

The measurement of permeability under different triaxial stresses is described as follows. First, the cooled-down specimen is imaged by high-magnification electron microscopy and its weight and dimensions are measured. For the permeability experiment, the cooled-down specimen is placed into the triaxial pressure chamber and a small (0.1–0.2 MPa) axial stress is applied to ensure mating and sealing of the specimen ends with the platens. Subsequently, a confining pressure with a specified value is applied at a constant rate of 0.5 MPa/s. The confining pressure valve is then closed and the axial stress valve is opened to apply the designed axial stress on the specimen. Subsequently, the confining and axial stresses are kept constant, while a gas pressure is applied on the specimen using the pressure loading system. After each stress increment, the gas mass flow at the outlet end rapidly increases and then stabilises. The gas mass flow at the outlet side is recorded every minute. When

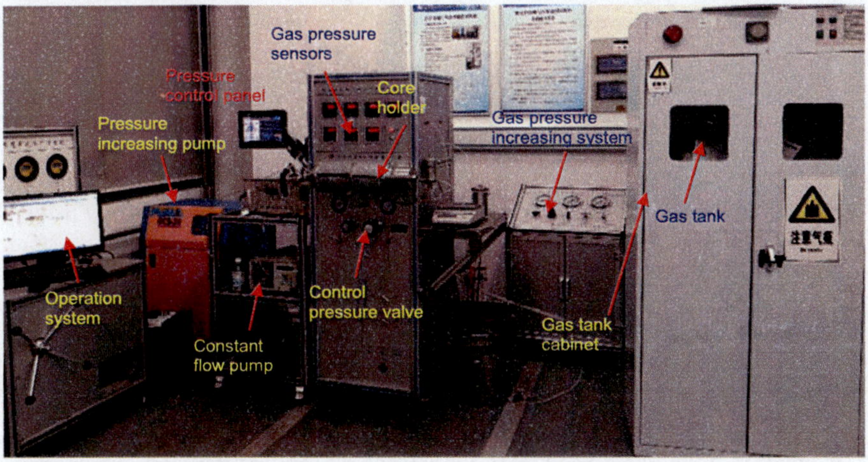

Fig. 15.2 Gas permeability test system developed by State Key Laboratory for Geomechanics and Deep Underground Engineering, China University of Mining and Technology

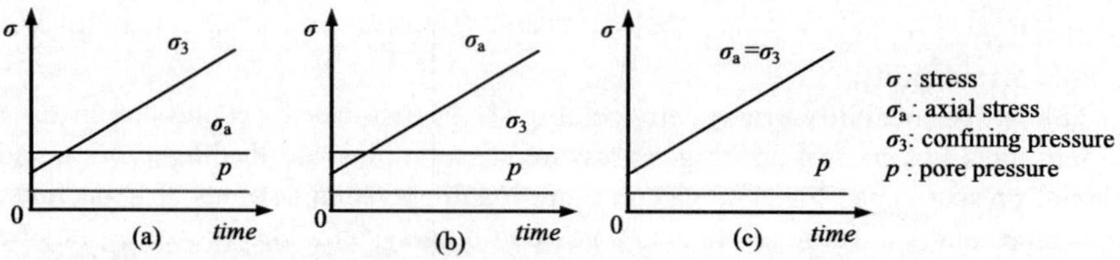

Fig. 15.3 Different compression stress paths in the permeability tests of granite specimens. **a** Constant axial stress and pore pressure; **b** constant confining pressure and pore pressure; **c** constant pore pressure but different hydrostatic pressure

the difference between the current and former values is lower than 2%, the value is regarded as the steady state magnitude and the permeability under a certain triaxial stress can be calculated as:

$$Q = \frac{KAP_1^2}{2\mu L P_a}, \tag{15.1}$$

where A and L are the cross-sectional area and length of the specimen, respectively; μ is the dynamic viscosity of the fluid ($\mu \approx 1.8 \times 10^{-5}$ Pa · s for pure nitrogen at room temperature under the applied pore pressure); P_1 is the pressure on the upstream side; P_a is the standard atmospheric pressure at room temperature; K is the permeability of the specimen; and Q is the mass flow of the gas at the downstream outlet.

To ensure the stability and safety of deep nuclear waste disposal together with the influence of groundwater, it is important to identify the relationship between the permeability and effective stress of the rock. Therefore, a series of gas seepage tests was carried out on granite specimens with the same pore pressure under constant axial stress, confining pressure, and hydrostatic pressure. The compressive stress paths of the permeability tests of the granite specimens are shown in Fig. 15.3.

15.1.3 Mechanical Testing System and Procedure

Conventional triaxial compression was applied to the granite specimens by using a GCTS RTX–4000 high-temperature and high-pressure rock triaxial testing system, as shown in Fig. 15.4a. The maximum axial loading capacity, maximum confining pressure, and maximum pore pressure were 4000 kN, 140 MPa, and 140 MPa, respectively. The axial and circumferential deformations were measured with an axial Linear Variable Differential Transducer (LVDT) with a maximum displacement range of 5 mm, as shown in Fig. 15.4b. During loading, the AE was monitored using a micro-II measuring system manufactured by the Physical Acoustic Corporation (PAC) and the amplitude threshold of the sensors was set to 60 dB.

The target treatment temperatures for the triaxial mechanical tests were set to 25 °C, 150 °C, 300 °C, 450 °C, 600 °C, and 750 °C. The heated granite specimens

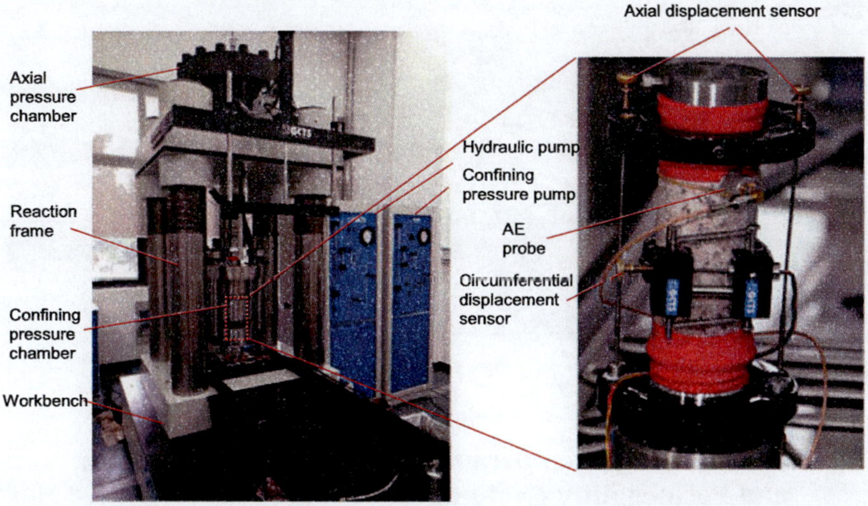

Fig. 15.4 RTX–4000 high-temperature and high-pressure triaxial rock testing system. **a** Overview of the testing system; and **b** detailed axial and lateral linear variable differential transducers

were left in the furnace to cool down to room temperature. After the thermal treatment, the granite was tested using a GCTS RTX-4000 high-temperature and high-pressure rock triaxial testing system with the confining pressure set to 0, 10, 20, 30, and 40 MPa, respectively.

The conventional triaxial experiments consisted of the following two steps: First, the confining pressure was increased to the desired value at a constant rate of 4.0 MPa/min to ensure that the specimen was under uniform hydrostatic stress. Deviatoric stress (σ_1–σ_3) was then applied to the specimen at a constant axial displacement rate of 0.04 mm/min (~6.67 × 10^{-6}/s) until the post-peak failure stage was reached (in accordance with the method suggested by ISRM).

The Transient Plane Source (TPS) method introduced by Gutafsson (1991) and Log and Gustafsson (1995) was utilised to measure the thermal conductivity of the specimens. The hot disk sensor (a hot source and dynamic temperature sensor) was sandwiched between two specimens, as shown in Fig. 15.5. The hot disk sensor was first heated and then used to monitor the variation in the temperature (Zhao et al. 2016). This method has been widely applied to measure the thermal conductivity of rocks (Urquhart and Bauer 2015).

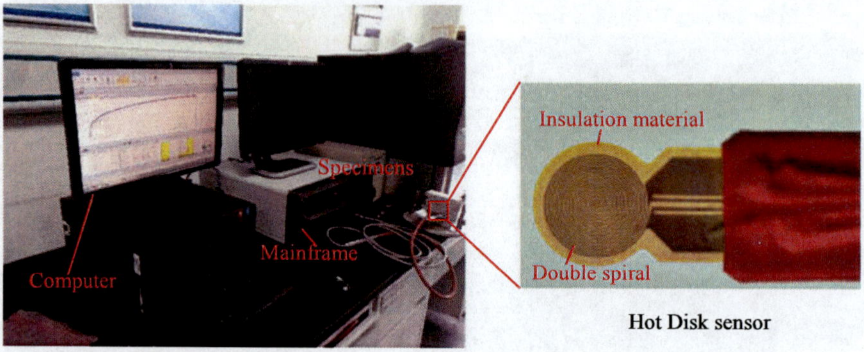

Fig. 15.5 The hot disk TPS-500 system

15.2 Effect of High Temperature on the Physical Behavior and Permeability Evolution of Granite

15.2.1 Effect of High Temperature on the Physical Behavior of Granite

We analyse the effect of high temperature on the physical behaviour of granite by using the volume and mass variation rates of granite in response to changes in temperature, as shown in Fig. 15.6. The volume and mass variation rates are defined as the ratio between the changed and initial volumes and masses, respectively.

Figure 15.6a shows that the volume variation rate first increases at 150 °C and then remains relatively constant at 300 °C. This result is because the volume of the mineral grains in the granite increases with increasing temperature at $T \leq 300$ °C;

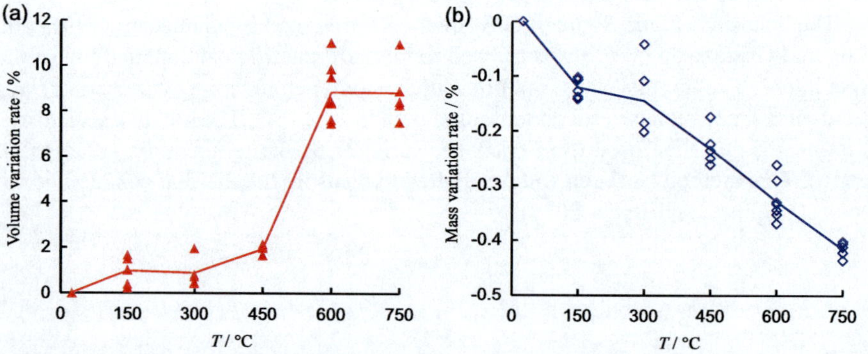

Fig. 15.6 Effect of high temperature on the volume and mass behavior of granite in this research. **a** Volume variation rate; **b** mass variation rate

but the residual strain of the mineral grains is smaller after cooling, resulting in an insignificant changes in the volume variation rate. At 450 °C, the volume variation rate starts to increase as a result of phase changes and the oxidation of biotite at 400 °C. Natural biotite exists in quasi-stable crystalline forms of 3 T and 1 M, whereas the stable crystal form of biotite is in a 2M1 state after heat treatment at temperatures above 400 °C. In addition, the Mg^{2+} in biotite oxidises during the heating (Sun 1987; Chen et al. 2017a, b):

$$K_2Mg_6^{2+}[Si_6Al_2O_{20}](OH)_4 + O_2 = K_2Mg_2^{2+}Mg_4^{3+}[Si_6Al_2O_{20}]O_4 + 2H_2O \uparrow$$

$$K_2Mg_6^{2+}[Si_6Al_2O_{20}](OH)_4 = K_2Mg_2^{2+}Mg_4^{3+}[Si_6Al_2O_{20}]O_4 + 2H_2 \uparrow$$

At 600 °C, the volume change rate of the granite specimen drastically increases. The quartz transforms from α phase to β phase. When the temperature exceeds 573 °C, the quartz grains will expand. However, the quartz retransforms from β phase to α phase as the temperature decreases to < 573 °C during cooling. The expansion of the quartz grains leads to the dislocation of mineral grains, initiation of cracks, and residual deformation above a temperature of 573 °C. At 750 °C, the volume variation rate remains relatively constant due to the formation of macrocracks at 600 °C and thermal stress release.

Heating can also lead to the dehydration of granite: at 25 °C $\leq T \leq$ 100 °C as the absorbed water evaporates. At 100 °C $\leq T \leq$ 300 °C, the bound and crystal water escape; and at 300 °C $\leq T \leq$ 400 °C, the crystal and structural water escape (Sun et al. 2015). Therefore, the mass of the granite specimens decrease with an increase in temperature from 25 °C to 150 °C, as shown in Fig. 15.6b. However, at 300 °C, the mass decrease slows down because the amount of bounded and crystal water in the granite specimen is relatively low. As a result of the oxidation of biotite and the escape of crystal and structural water, the mass of the granite quickly decreases again at 450 °C. At 600 °C, the mass of granite also decreases quickly because of the phase transformation of quartz and fragments separating from the specimen. At 750 °C, the mass continuously decreases. This may be due to continuous separation of fragments—suggested by the observation that the values of the volume variation rate are generally larger than the values of the mass variation rate.

The variation in the thermal conductivity caused by thermal damage is related to heat dissipation and stability of the nuclear waste repository. Knowledge of the thermal conductivity as a function of the temperature is critical for the stable operation of the repository (Sundberg and Hellström 2009; Zhao et al. 2016). The change in thermal conductivity with temperature was measured using a TPS-500 system (Fig. 15.5), as shown in Fig. 15.7a. At $T \leq$ 300 °C, the thermal conductivity decreases almost linearly with increasing temperature; at 450 °C, the thermal conductivity decreases with increasing temperature again approximately linear; and at 600 °C, the thermal conductivity notably decreases, indicating that the large amount of microcracks caused by the quartz phase transition leads to a notable decrease in the thermal conductivity of granite. However, as the temperature increases from 600

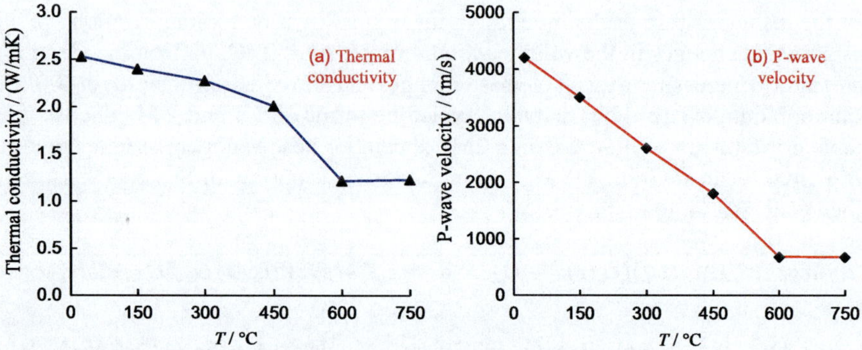

Fig. 15.7 Effect of high temperature on thermal conductivity and P-wave velocity of granite in this research. **a** Thermal conductivity; (b) P-wave velocity

to 750 °C, fewer new thermal cracks form and the thermal conductivity remains relatively constant. The thermal conductivity at $T = 750$ °C is 1.22 W/mK, which is 50% of the room temperature value of 2.53 W/mK, indicating that high temperature has a significant effect on the thermal conductivity of granite.

Figure 15.7b shows the effect of high temperature on the P-wave velocity of the granite, providing a potential method to evaluate the variation in thermal damage. Figure 15.7b shows that the P-wave velocity linearly decreases with increasing temperature at $T \leq 450$ °C; at $T = 600$ °C, the P-wave velocity rapidly decreases, indicating significant damage to the granite specimen under the influence of high temperature. At 750 °C, the P-wave velocity remains essentially unchanged over the range 600 °C to 750 °C with the increase in temperature resulting in little new damage. These trends in behaviour with temperature are similar to those previously observed for Beishan granite (Zhao et al. 2018).

Heating can induce thermal microcracks in granite, leading to changes in the physical and mechanical behaviour of granite specimens. Figure 15.8 shows the microstructure of granite after thermal treatment obtained from optical microscopy. At $T \leq 300$ °C, very few initial microcracks are observed and the mineral grains are tightly cemented. At 450 °C, intergranular cracks can be observed, which means that the microstructure is weakened by thermal stress. Figures 15.6 and 15.7 show that the temperature of 450 °C represents an inflection point with respect to changes in volume and thermal conductivity. At 600 °C, an intergranular network of cracks forms around each grain, which is due to the phase transformation of quartz that causes the polygonal grains to become isolated. In addition, several intragranular cracks form. At 750 °C, the distribution of the microcracks is similar to that at 600 °C, but the resulting microcracks are wider.

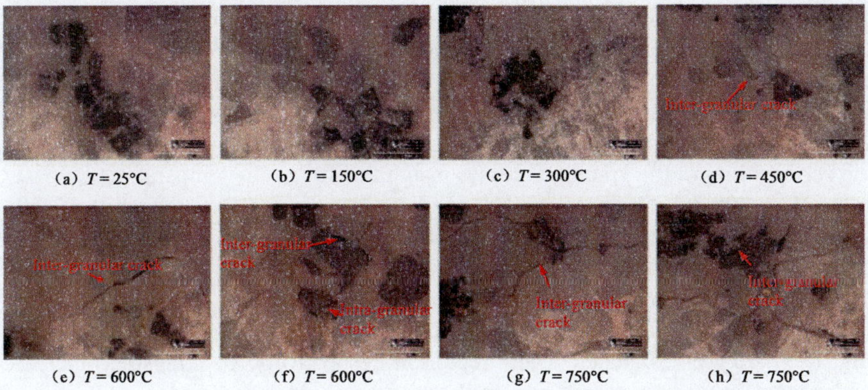

Fig. 15.8 Micro-structure of granite after thermal treatment obtained by optical microscopy (×50)

15.2.2 *Effect of High Temperature on Permeability Evolution of Granite Under Triaxial Compression*

A series of permeability measurements with various gas pressures was completed with the permeability calculated using Eq. (15.1). The mass flow rate of the gas is proportional to the gas pressure enabling permeability to be accurately obtained.

Figure 15.9a shows the relationship between the gas pressure and mass flow rate (Q–P_1 curve) after the specimens were treated at different temperatures. From Fig. 15.9a, it can be seen that the curve steepens with increasing temperature, which means that the permeability increases notably. Figure 15.9b shows that the permeability varies with increasing temperature because the hydrostatic stress significantly influences the permeability of rocks. The volumetric stress (σ_m) is plotted

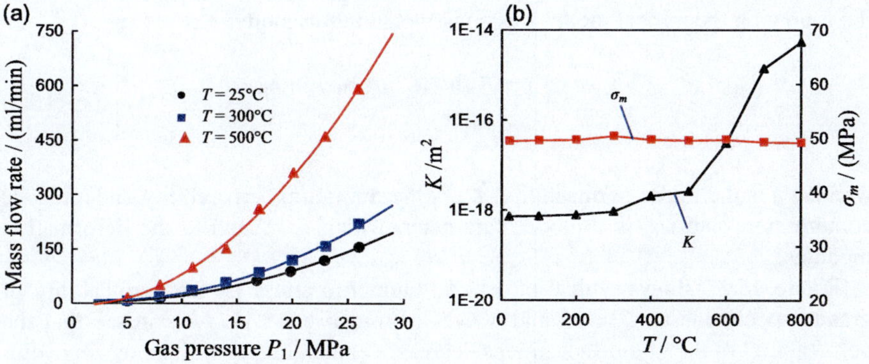

Fig. 15.9 Effect of high temperature on mass flow rate and permeability of granite specimen. **a** Relationship between mass flow rate and gas pressure; **b** relationship between permeability and temperature

in Fig. 15.9b. The figure shows that the temperature has a significant effect on the permeability at a volumetric stress of 50 MPa. Figure 15.9b shows that the permeability increases from 10^{-18} to 10^{-14} (0.001 to 10 mD) when the temperature increases from 25 to 800 °C. In other words, the permeability is 10,000 times as large as the initial value when the temperature increases from 25 to 800 °C. Furthermore, note that the permeability of the granite increases insignificantly as the temperature increases from 25 to 300 °C. The permeability coefficient of granite then increases slightly when the temperature increases from 300 to 500 °C, but increases rapidly from 500 to 800 °C.

Figure 15.10 shows the permeability evolution of granite after different high-temperature treatments. In Fig. 15.10, three different loading paths, as shown in Fig. 15.3, are pre-designated. These represent, respectively, an axial stress increase under constant confining pressure, a confining pressure increases under constant axial stress, and an increase in axial stress and confining stress at the same rate (hydrostatic pressure). The maximum axial stress and confining pressure is limited to 60 MPa by the experimental apparatus. Importantly, this is significantly smaller than the UCS according to the results of uniaxial compression experiments on granite (Yang et al. 2017). Accordingly, the granite specimen experiences elastic compression only.

Figure 15.10 shows that the permeability of granite decreases with increasing axial and confining pressure, while the confining pressure (σ_3) has a larger influence than the axial stress (σ_a) in the range of 10–50 MPa. Taking the experimental result for 25 °C as an example, when the axial stress increases from 10.13 to 49.04 MPa ($\sigma_3 =$ 10 MPa), the permeability coefficient of granite decreases from 0.0031 to 0.0026 mD, while the permeability of granite sharply decreases from 0.00426 mD (4.26×10^{-18} m^2) to 0.000668 mD (6.68×10^{-19} m^2) with the increase in the confining pressure from 9.86 to 49.3 MPa ($\sigma_a = 10$ MPa). This may be related to the difference between the directions of the permeability measurement and axial stress. The permeability of granite under different hydrostatic pressures is also shown in Fig. 15.10. The influence of hydrostatic pressure on the permeability varies with temperature.

To analyse the effect of the volumetric stress (σ_m) on the permeability of granite, the following theoretical model was constructed in this study:

$$K = (K_0 - K_u)\exp(-\alpha_m \sigma_m) + K_u, \tag{15.2}$$

where K_0 is the initial permeability; K_u is the minimum permeability under elastic compression; and α_m is a model parameter, which is related to the deformation modulus.

Figure 15.11 shows typical effect of volumetric stress on the permeability of granite specimens at T = 25 and 800 °C. From Fig. 15.11, we can see that the permeability of the granite specimens decreases exponentially with increasing volumetric stress, while the deviatoric stress is zero. The theoretical curves calculated using Eq. (15.2) were compared with the experimental results. The theoretical curves agree very well with the experimental results. With increasing volumetric stress, the

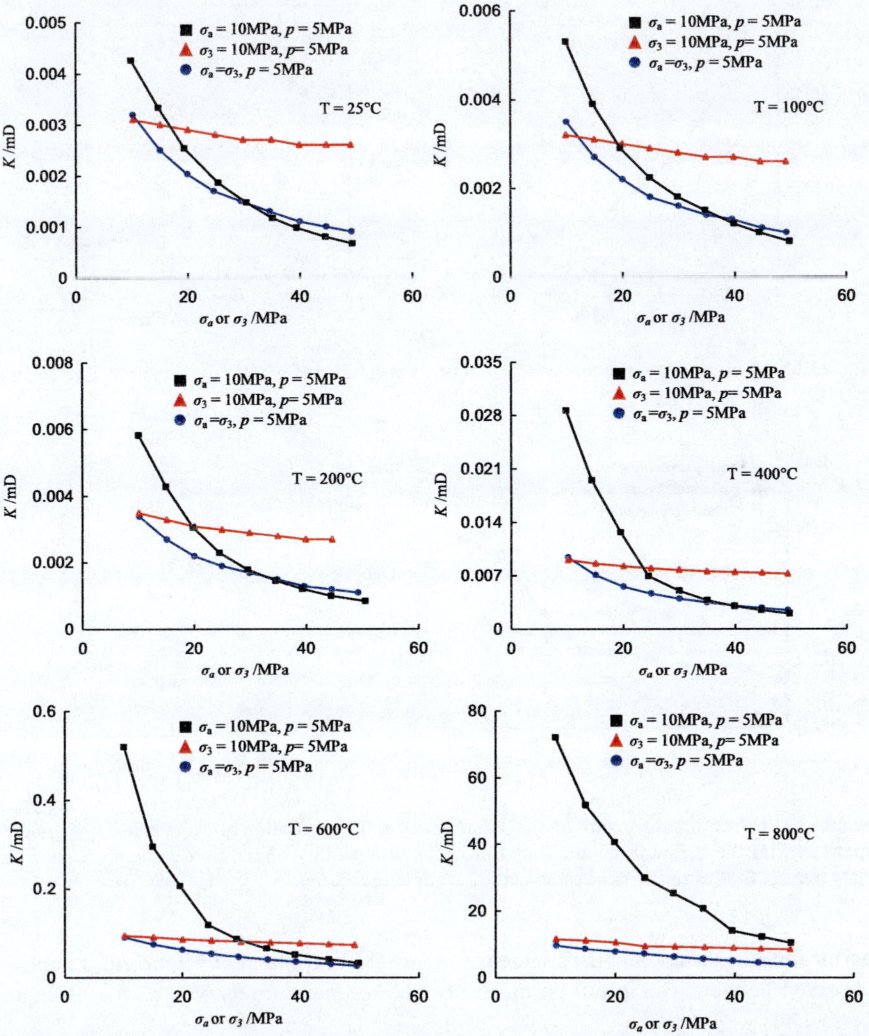

Fig. 15.10 Effect of effective stress on permeability of granite specimens after different high temperatures

permeability of the granite specimens first rapidly decreases and then gradually stabilises.

The initial permeability (K_0) is that at zero compressive stress. With increasing volumetric stress, the permeability gradually decreases to a constant value under elastic compression, which is regarded as the ultimate permeability of rock. Figure 15.12a shows that the initial and ultimate permeabilities of granite have similar correlations with temperature. With increasing temperature, the initial and ultimate

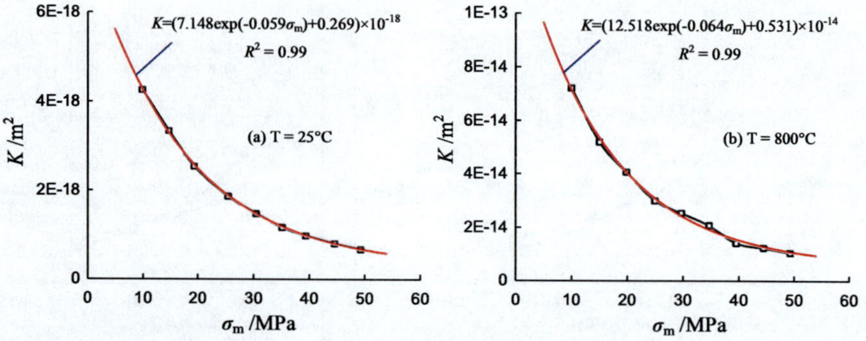

Fig. 15.11 Typical effect of volumetric stress on permeability of granite specimens at T = 25 and 800 °C

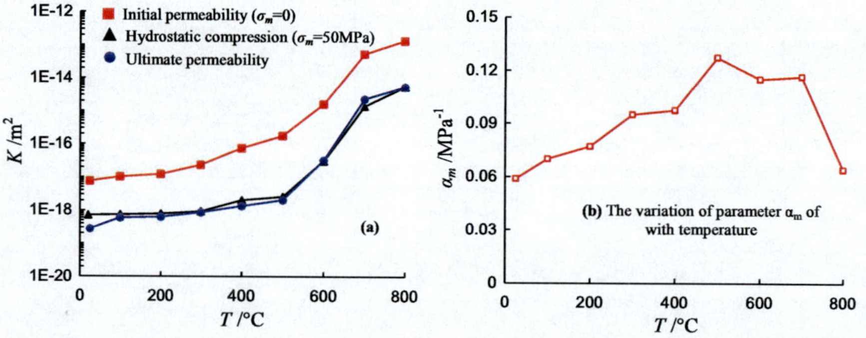

Fig. 15.12 Effect of temperature on the permeability and the variation of parameter α_m of granite specimen. (a) the variation of initial and ultimate permeability under elastic compression with temperature; (b) the variation of parameter α_m with temperature

permeabilities simultaneously increase. A notable exponential relationship can be observed between the initial permeability and temperature as well as the ultimate permeability.

Based on the theoretical curves shown in Fig. 15.11, we can obtain the variation in the parameter α_m with temperature, as shown in Fig. 15.12b. Figure 15.12b shows that the α_m value increases as the temperature increases from 25 to 500 °C, but it decreases sharply at temperatures above 500 °C. The physical meaning of parameter α_m is that the volumetric strain under compression increases notably with the increase in parameter α_m (Xu and Yang 2019), which also means that the bulk modulus is inversely proportional to α_m.

15.3 Effect of High Temperature on Triaxial Deformation and AE Behavior of Granite

15.3.1 Effect of High Temperature on Triaxial Stress–Strain Curves of Granite

Figure 15.13 shows triaxial deviatoric stress–axial strain and circumferential strain curves of granite after different high temperature treatments. The triaxial stress–strain curve of granite can be divided into five stages: (I) initial pore and fissure closure stage, (II) elastic deformation stage, (III) plastic deformation stage, (IV) post-peak failure stage, and (V) residual strength stage.

(I) Initial pore and fissure closure stage: The axial strain of granite increases quickly, while the axial deviatoric stress increases slowly. The deviatoric stress–axial strain curve is concave. The initial pore and fissure closure stage becomes less notable with increasing confining pressure due to the closure of the micro-fissures and micro-pores. Note that, at the same confining pressure, more micro-fissures and micro-pores in the specimens initiated by thermal damage can be observed with increasing temperature and thus initial pore and fissure closure stage becomes more notable.

(II) Elastic deformation stage: The deviatoric stress increases linearly with increasing axial and circumferential strain. The slope between the deviatoric stress and axial strain increases with increasing confining pressure and becomes more sensitive to the confining pressure after undergoing higher temperature treatment. After high temperature treatment, the structure of granite specimens becomes looser due to thermal damage. The confining pressure restrains the displacement and enhances the friction among grains and the slope becomes more sensitive to the confining pressure under the condition of higher temperature.

(III) Plastic deformation stage: The slope between the deviatoric stress and axial strain decreases with increasing axial strain, and the circumferential strain increases nonlinearly. Under uniaxial compression, the stress concentration leads to local failure in granite specimens because they lose their ability to resist the deformation. The slope continues to decrease before the peak strength is reached, which corresponds to the rapid increase in the circumferential strain. With increasing temperature, more thermal micro-cracks are initiated and the bearing structure of granite specimen is damaged. Therefore, the ability to store strain energy in granite specimens decreases and the mineral grains have more spaces to adjust. When local stress failure occurs, less strain energy is released and fewer cracks propagate. Thus, the decrease of slope is not notable. Under the influence of the confining pressure, more adjustment abilities are observed among mineral grains and it is difficult to produce the stress concentration nearby the grains. Therefore, a stress drop

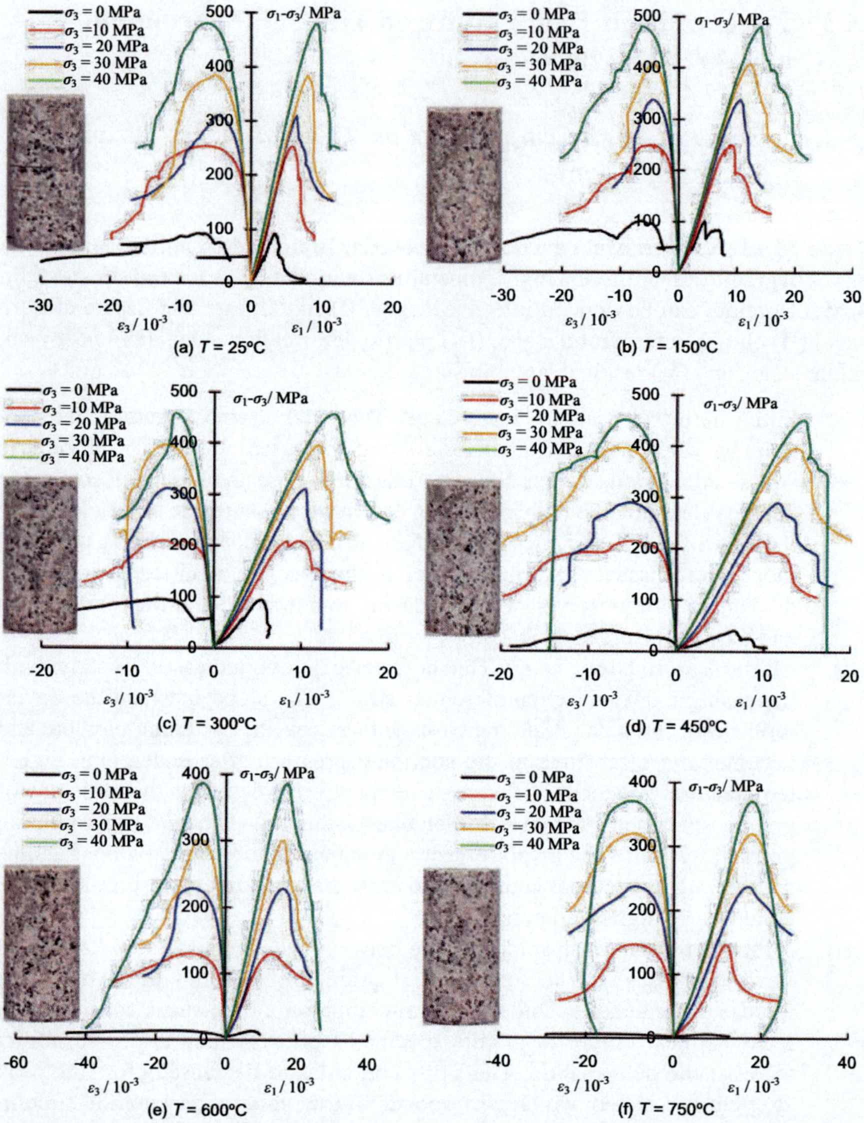

Fig. 15.13 Deviatoric stress-axial strain and circumferential strain curves of granite specimens after high temperature treatment under different confining pressures

cannot be observed in the stress–strain curve of the granite under the confining pressure.

(IV) Post-peak failure stage: The increase of microcrack initiation, propagation, and coalescence are observed during the loading process. Therefore, macro-cracks form and the stress decreases quickly with increasing axial strain. At $T \leq 450$ °C, the granite specimens experience more local damage under uniaxial compression; the deviatoric stress–strain curves thus show more stress drops and the granite specimens fail slowly. At $T \geq 600$ °C, less strain energy is stored in the granite specimen and the deviatoric stress decreases gradually after reaching the peak strength. Under the influence of confining pressure, the granite specimens fail generally along the shear fracture plane. The tested granite specimens have coarse-grained structures and the intra-granular failure involves more stress than intergranular failure. Therefore, the cracks can propagate more easily along intergranular fracture planes than intragranular fracture planes. At $T \leq 450$ °C, the deviatoric stress decreases gradually with increasing axial strain due to intergranular shear plane propagation. At $T \geq 600$ °C, the peak strength of the granite specimens decreases due to heat-induced intragranular cracks. At a higher confining pressure ($\sigma_3 = 40$ MPa), intragranular cracks form more easily with the decrease of axial stress, the shear plane is smoother, and the stress decreases dynamically, as shown in Fig. 15.13e, f.

(V) Residual strength stage: under the influence of confining pressure, inter-locking and friction forces exist between the mineral grains of the granite specimens. Therefore, the deviatoric stress remains relatively constant with continuous loading after the post-peak failure stage. The interlocking and friction forces increase with increasing confining pressure, which also leads to the increase in the residual strength with increasing confining pressure. At $T \geq 600$ °C and a higher confining pressure ($\sigma_3 = 40$ MPa), the shear plane is smoother, which leads to a lower residual strength of the granite specimen.

The evolution of the volumetric strain (ε_v) after high-temperature treatment and for different confining pressures is shown in Fig. 15.14. The parameter ε_v refers to the volumetric strain, which is calculated from the sum of the axial strain (ε_1) and twice the radial strain (ε_3; i.e. $\varepsilon_v = \varepsilon_1 + 2\varepsilon_3$). The volumetric strain is characterised by an initial phase of compaction-dominated behaviour followed by a phase of dilatancy-dominated behaviour, as shown in Fig. 15.14c. After point *A*, the compaction is in direct competition with dilatancy (although the specimen is still in a state of net compaction). Eventually, the rates of compaction and dilatancy become equal at point *B*, resulting in an inflection point in the volumetric strain curve. The crack damage threshold (σ_{cd}) of the specimen (Wong et al. 1997; Fairhurst and Hudson 1999; Heap et al. 2009; Yang et al. 2015) is defined as the corresponding axial deviatoric stress at which the volumetric deformation of the specimen switches from compaction-dominated to dilatancy-dominated. As the stress further increases, more microcracking occurs and the granite specimen continues to dilate. The volumetric

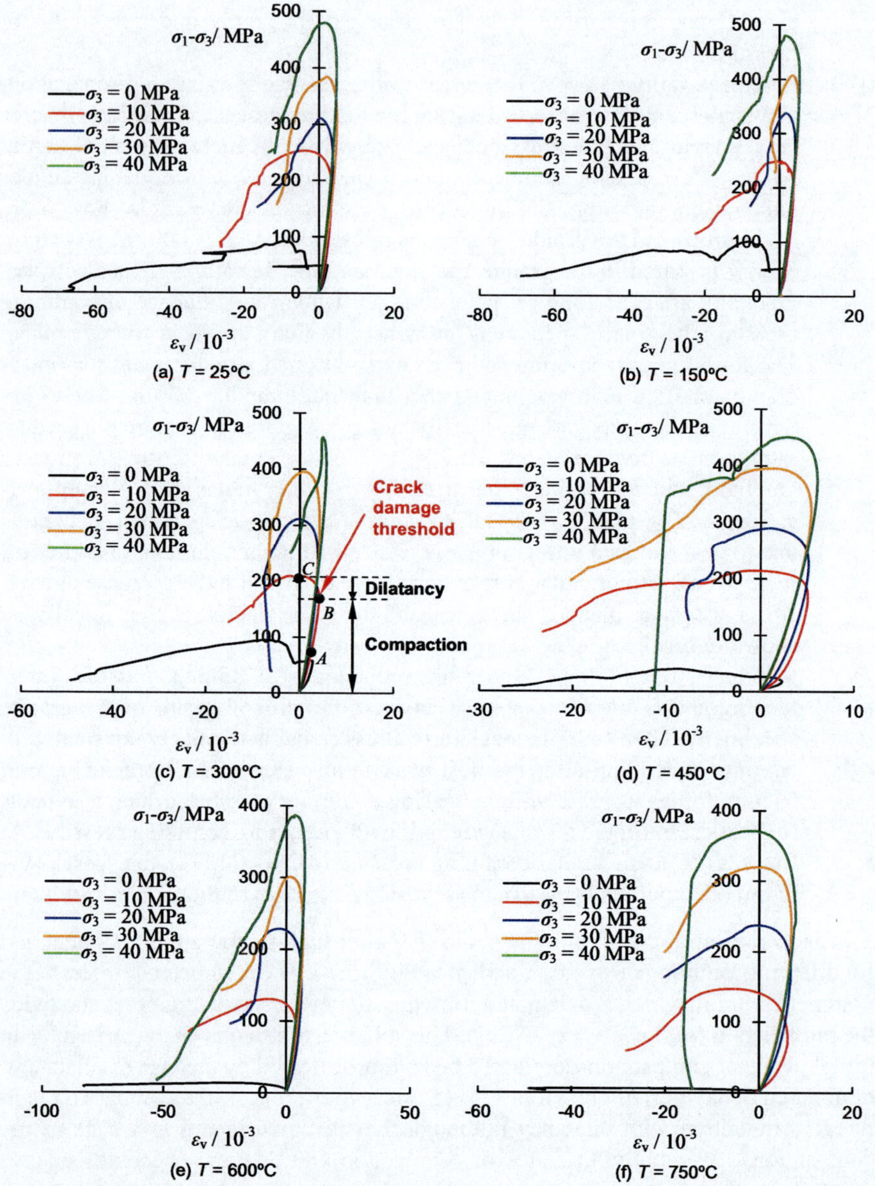

Fig. 15.14 Deviatoric stress-volume strain curves of granite specimens after high temperature treatment under different confining pressures

strain of the granite specimen eventually returns to zero strain (Point *C*) and then continues to increase until the specimen fails.

At $T \leq 450$ °C, the potential for volume shrinkage during subsequent loading is small due to the closure of partial cracks under the confining pressure. Therefore, under the same axial stress, the amount of shrinkage of the specimen decreases with increasing confining pressure before the crack damage threshold is reached. The corresponding volume strain is also smaller. Under uniaxial compression, the local expansion of the cracks causes a sudden volume expansion, which is due to the axial splitting of the specimen under uniaxial compression. Due to the restrained confining pressure, the specimen expands rapidly following damage. Under the confining pressure, the volume expansion of the specimen is restricted. The volume expansion rate of the specimen decreases as the confining pressure increases. At $T \geq 600$ °C, the crystal grains are liable to slip during the compression because the bearing structure of the specimen is broken such that the specimen expands in the beginning of uniaxial compression. The thermal cracks in the specimen close under the influence of the confining pressure, the friction between the crystal grains increases, and the bearing capacity of the specimen rapidly increases. High confining pressure also causes a decrease in the volumetric shrinkage potential of the specimen. Therefore, the volumetric compression rate of the specimen decreases with increasing confining pressure before the damage threshold is reached. Similarly, the confining pressure limits the volume expansion of the specimen. After the crack damage threshold is reached, the volume expansion rate of the specimen decreases with increasing confining pressure.

15.3.2 Effect of High Temperature on Triaxial Deformation Parameters of Granite

Figure 15.15a, b, respectively, present the variation of the elastic modulus (E_S) of granite depending on the temperature and confining pressure. The elastic modulus is defined by the average slope of the approximately straight portion (i.e. from the elastic initial value to the elastic limit) of the axial deviatoric stress–axial strain curve before the peak strength is reached.

From Fig. 15.15a, it can be seen that the elastic modulus of granite under uniaxial compression slightly increases when the temperature increases from 25 to 150 °C and slowly decreases when the temperature increases from 150 to 300 °C. This elastic modulus then quickly decreases as the temperature increases from 300 to 600 °C and then remains relatively constant when the temperature increases continuously to 750 °C. Note that the elastic modulus of granite under uniaxial compression increases from 25 to 150 °C for the following reasons. The absorbed and bound water in the specimen evaporates and escapes, which increases the friction forces between the crystal grains. Additively, thermal residual deformation, the bond between the crystal grains becomes higher. These two effects are additive and result in an increase in the elastic modulus of the granite under uniaxial compression as the temperature

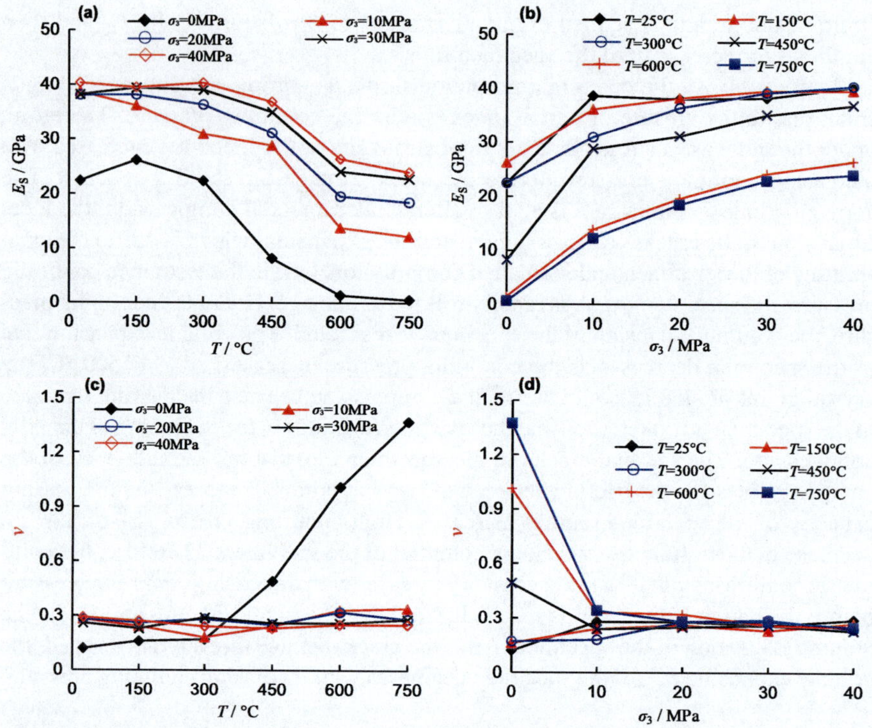

Fig. 15.15 Deformation parameters of granite specimens after high temperature treatment with temperature and confining pressure. **a** Variation of elastic modulus with temperature; **b** Variation of elastic modulus with confining pressure; **c** Variation of Poisson ratio with temperature; **d** Variation of Poisson ratio with confining pressure

increases from 25 to 150 °C. Thermal cracks continue to appear as the temperature increases from 150 to 600 °C, resulting in a decrease in the elastic modulus of the specimen under uniaxial compression. However, the increase in the confining pressure leads to the closure of some of the thermal cracks such that the elastic modulus decreases with decreasing temperature at the same confining pressure. When the temperature increases from 600 to 750 °C, the thermal stress is released due to the occurrence of through-going macroscopic cracks and the formation of later cracks becomes difficult. Therefore, the elastic modulus changes insignificantly.

From Fig. 15.15b, it can be seen that the elastic modulus of the granite quickly increases when σ_3 increases from 0 to 10 MPa at 25 °C and 150 °C. Subsequently, the E_S of the granite remains relatively constant with continuously increasing confining pressure. At 300 °C and 450 °C, the elastic modulus of granite also quickly increases as σ_3 increases from 0 to 10 MPa, but only slowly increases with the confining pressure when σ_3 is greater than 10 MPa. At 600 °C and 750 °C, the elastic modulus continuously and nonlinearly increases with increasing confining pressure. Based on the above-mentioned analysis, the elastic modulus of the granite rapidly increases as the confining pressure increases from 0 to 10 MPa, which is independent of the temperature. With the continuous increase in the confining pressure from 10 to 40 MPa, the rate of increase in the elastic modulus also increases with increasing temperature. At $T = 25$ °C and 150 °C, almost no microcracks can be observed in the granite specimens, except for the initial pores. Therefore, the increase in the confining pressure from 10 to 40 MPa cannot result in the closure of microcracks and pores. However, at $T = 600$ °C and 750 °C, more thermal cracks are induced and the potential for microcrack and pore closure is large. With increasing confining pressure, the microcracks and pores continue to close and the frictional force greatly increases. This leads to the continuous increase in the elastic modulus.

Figure 15.15c, d show the variation in the Poisson ratio (ν) of the granite specimens after high-temperature treatment at different confining pressures. The Poisson ratio is the average of the absolute ratios of the circumferential strain and axial strain in the range of 30% to 70% of the triaxial compressive strength. Figure 15.15c shows that the Poisson ratio under uniaxial compression first increases gradually when the temperature increases from 25 to 300 °C and then rapidly increases when the temperature increases from 300 to 750 °C. Under the influence of the confining pressure, the temperature has almost no effect on the Poisson ratio of granite. Following high-temperature treatment, the granite specimens contain more microcracks and the mineral grains in the granite specimens rotate more easily. Therefore, the granite specimens more easily expand in the circumferential direction under uniaxial compression. However, the confining pressure restrains the rotation of the mineral grains and thus the temperature only has a small effect on Poisson ratio under triaxial compression. Figure 15.15d depicts the variation in Poisson ratio of the granite specimens depending on the confining pressure. From Fig. 15.15d, we can see that the confining pressure also has only a small effect on Poisson ratio at $T \leq 300$ °C. At $T \geq 450$ °C, the Poisson ratio significantly decreases when the confining pressure is applied and then remains relatively constant with continuously increasing confining pressure. This means that thermal cracks are initiated in the specimen at 450 °C and the mineral grains rotate more easily. The confining pressure can restrain the rotation of the grains and thus the Poisson ratio quickly decreases under the increased confining pressure. However, while the confining pressure can restrain the grain rotation, closure of thermally induced cracks can also occur, leading to a decrease in the adjustment space of mineral grains. Under axial loading, the expansion in the circumferential direction is easier than in axial direction. Therefore, the Poisson ratio of granite remains relatively constant when the confining pressure increases.

15.3.3 Effect of High Temperature on the AE and Failure Behavior of Granite

Figures 15.16 and 15.17 show the typical AE characteristics of granite specimens after high-temperature treatment under uniaxial and triaxial ($\sigma_3 = 30$ MPa) compression. The temperature has a significant effect on the AE characteristics. The AE and AAE represent the acoustic emission and accumulated acoustic emission, respectively. Under uniaxial compression at $T \leq 300$ °C, a stress decrease can be observed before the peak strength is reached, which corresponds to a significant AE event. This means that local damage occurs before the peak strength is reached under uniaxial compression. However, at $T \geq 450$ °C, AE can be observed when axial compression is applied because more thermal cracks are induced and the grains immediately begin to adjust. Under triaxial compression, the temperature has little effect on the AE characteristics. However, at $T \geq 600$ °C, intragranular cracks more easily form and the stress quickly decreases after the peak strength is reached. The AE does not increase after the peak strength is reached.

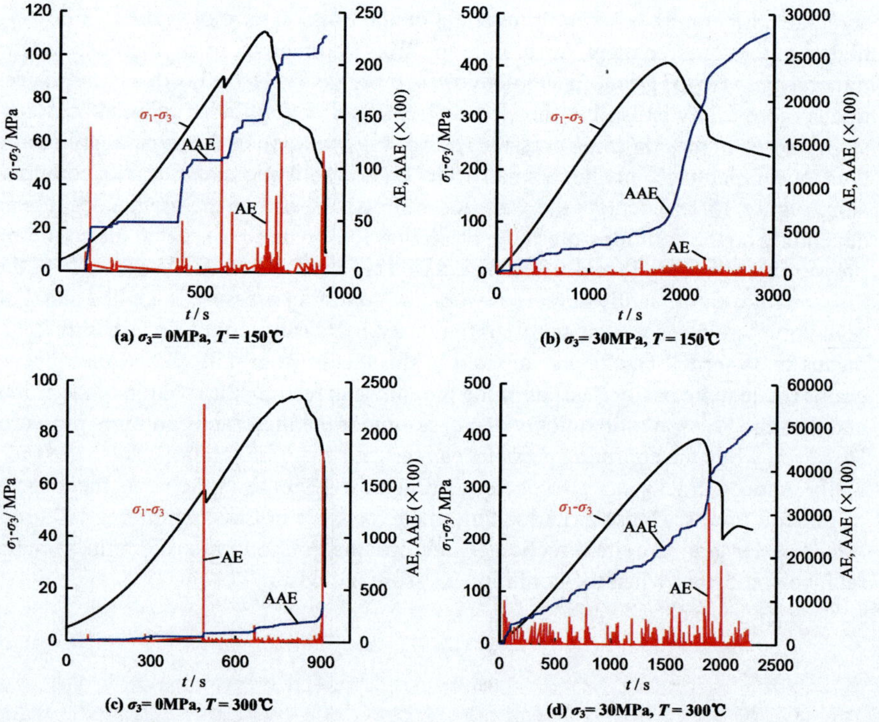

Fig. 15.16 Typical AE characteristic of granite specimens after high temperature treatment (T = 150 °C and 300 °C) under uniaxial and triaxial ($\sigma_3 = 30$ MPa) compression

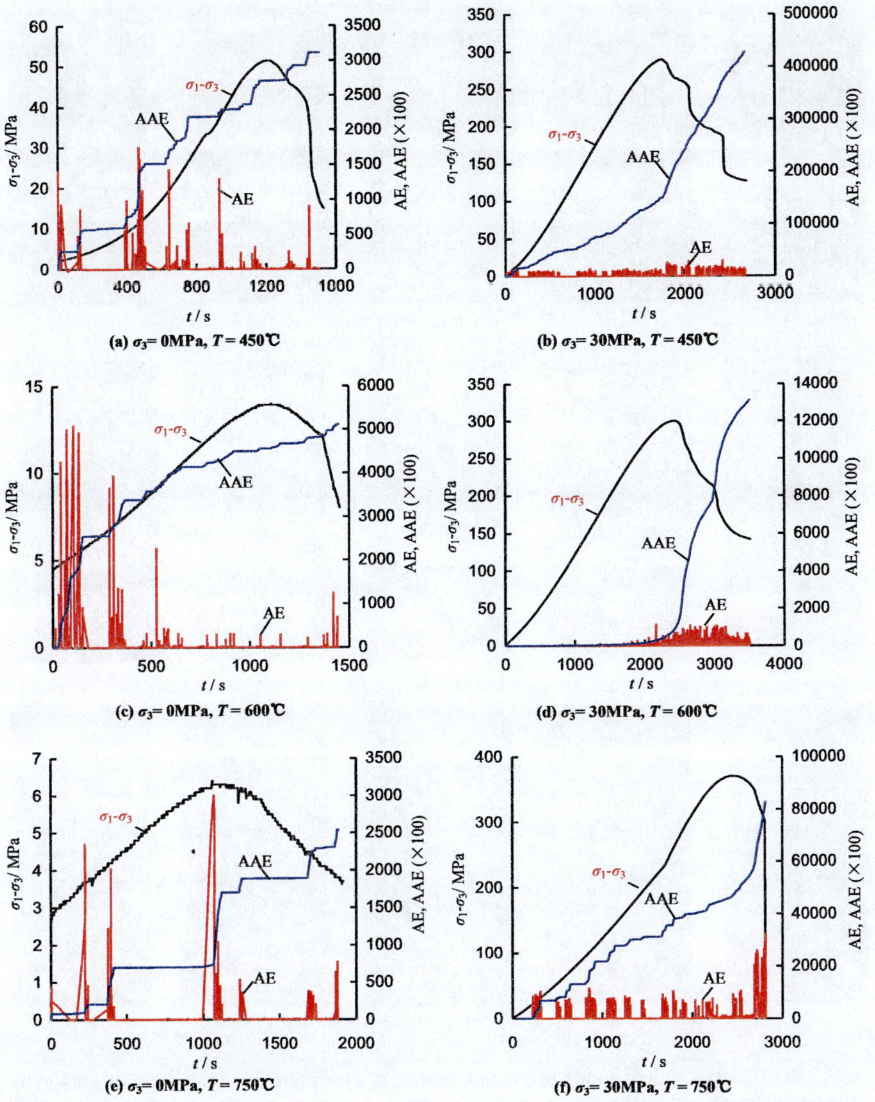

Fig. 15.17 Typical AE characteristic of granite specimens after high temperature treatment (T = 450°C, 600°C and 750°C) under uniaxial and triaxial ($\sigma_3 = 30$ MPa) compression

Figure 15.18 presents the ultimate failure modes of granite after thermal treatment under triaxial compression. The granite specimens fail in an axial splitting tensile mode under uniaxial compression, while they fail in a shear mode under triaxial compression. Under uniaxial compression at $T \leq 450$ °C, the splitting tensile cracks are numerous and almost parallel to the loading direction although few bifurcate.

Fig. 15.18 Ultimate failure mode of granite specimen after high temperature treatments under different confining pressures

The multiple splitting events correspond to multiple stress drops during loading. At $T \geq 600$ °C, more thermal macrocracks are induced in the granite specimens and the ultimate failure mode is controlled by these thermally induced macrocracks. Tensile cracks are densely distributed in the specimen and not completely parallel to the loading direction. During loading, the thermally induced macrocracks coalesce. The cracks slowly propagate and the stress–strain curves show increased ductility.

Under triaxial compression, the failure mode of the granite specimens at room temperature transforms from a single shear plane to rhombhedral shear plane with increasing confining pressure, and the shear fracture plane widens with increasing confining pressure. At $T = 150$ °C and 300 °C, the widening of the cracks is more notable. At $T = 450$ °C, the shear plane widens with increasing confining pressure, while two parallel shear planes exist at $\sigma_3 = 40$ MPa. Because intergranular fractures form more easily than intragranular cracks and also interlocking and frictional forces increase with increasing confining pressure, grains close to the shear plane exert a greater influence on the grains neighbouring when the shear plane forms—resulting in a widening of the shear plane. At $T \geq 600$ °C, the grains weaken and thermally induced and intragranular cracks more easily form under high shear stress. Therefore, the shear plane is smoother under a high confining pressure ($\sigma_3 = 40$ MPa), which corresponds to a sudden stress decrease after the peak strength is reached.

15.4 Effect of High Temperature on Strength Behavior of Granite Under Triaxial Compression

15.4.1 Effect of High Temperature on Peak Strength Behavior of Granite

Figure 15.19a shows the variation in the peak strength of granite under triaxial compression as a function of temperature. At $T \leq 300$ °C, absorbed, bound, and crystal water escape, but smaller residual strain and thermally induced microcracks occur. At 300 °C $\leq T \leq$ 500 °C, the crystal and structural water escape and Mg^{2+} in biotite oxidises, resulting in several heating-induced microcracks. At $T = 573$ °C,

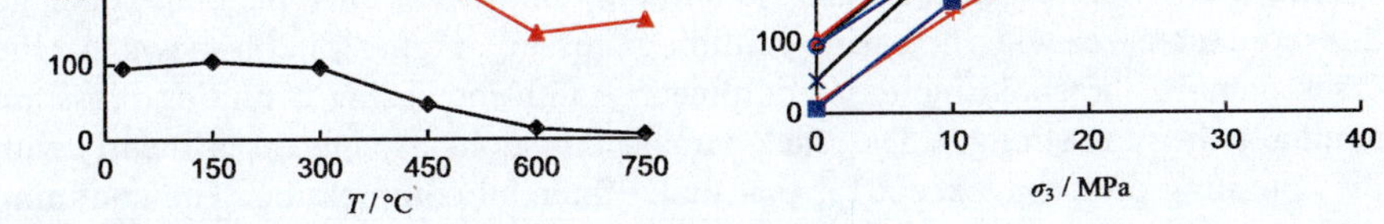

Fig. 15.19 Variation of the peak strength of granite specimens after high temperature treatment with temperature and confining pressure. **a** Variation with temperature, **b** variation with confining pressure

more thermal microcracks are initiated and macrocracks form due to the phase transformation of quartz. Because of the formation of macrocracks, thermal stress is released and the continuous increase in the temperature does not lead to the formation of more microcracks. Therefore, the peak strength of the granite remains relatively constant under uniaxial compression at $T \leq 300$ °C. The peak strength of granite quickly decreases at 300 °C $\leq T \leq$ 600 °C. Subsequently, the peak strength stabilises at 600 °C $\leq T \leq$ 750 °C. With increasing confining pressure, the interlocking and frictional forces between the grains increase, which can lead to an increase in the peak strength. At the same time, the confining pressure can lead to the closure of heat-induced cracks. The peak strength therefore remains relatively constant at $T \leq$ 450 °C. At $T = 600$ °C, more cracks are induced by heating and the peak strength of granite notably decreases. The effect of the temperature on the peak strength of granite decreases with increasing confining pressure.

Figure 15.19b depicts the variation in the peak strength of granite after different high-temperature treatments with increasing confining pressure. The peak strength of the granite increases with increasing confining pressure, regardless of the temperature. Note that the peak strength nonlinearly increases with increasing confining pressure. Furthermore, the nonlinearity becomes more pronounced with increasing temperature. The amount of thermally induced cracks increases with increasing temperature and the confining pressure can induce the closure of these cracks. Therefore, if the amount of thermally induced cracks is larger, the increase in the peak strength becomes more notable after the confining pressure is applied. If the uniaxial compressive strength is not used for the regression, the peak strength of the granite increases near linearly with increasing confining pressure.

15.4.2 Effect of High Temperature on the Crack Damage Threshold of Granite

Figure 15.20a presents the crack damage threshold under triaxial compression with increasing temperature. Similar to the peak strength, the crack damage threshold under uniaxial compression remains relatively constant at $T \leq 300$ °C and quickly decreases when the temperature increases from 300 °C to 600 °C. At $T \geq 600$ °C, the bearing structure of the granite is damaged, the granite specimens immediately expand under uniaxial loading, and the crack damage threshold is close to zero.

Confining pressure can restrain the initiation of cracks; thus, the crack damage threshold increases with increasing confining pressure. Figure 15.20b shows that the crack damage threshold increases nonlinearly with increasing confining pressure, similar to the peak strength. The crack damage threshold also increases linearly with the confining pressure, except for that under uniaxial compression. The confining pressure can result in the closure of thermally induced cracks and reduction of the stress concentration at the tips of cracks. Therefore, the propagation of thermally induced cracks is restrained. Furthermore, the confining pressure can change the

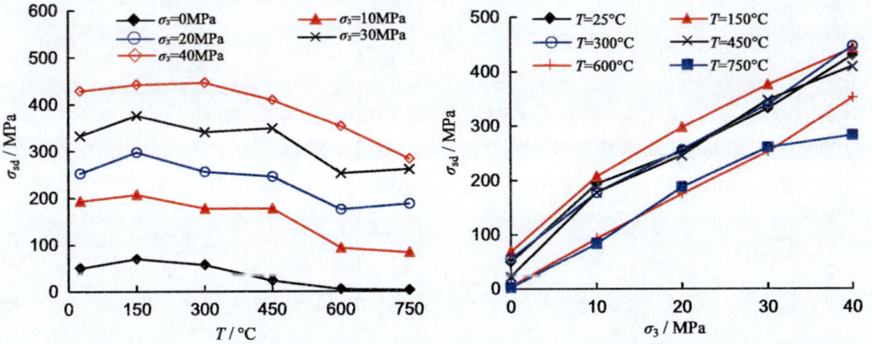

Fig. 15.20 Variation of the crack damage threshold of granite specimens after high temperature treatment with temperature and confining pressure. **a** Variation with temperature, **b** variation with confining pressure

directions of crack initiation and propagation. The failure mode of granite specimens can transform from tensile splitting to shear fracture. The crack damage threshold of granite quickly increases with increasing confining pressure. The confining pressure reduces the stress concentration but also restrains the crack propagation, whereas it does not notably change the crack initiation and propagation directions. Therefore, the crack damage threshold linearly increases with increasing confining pressure.

Note that the peak strength and crack damage threshold of granite slightly increase when the temperature increases from 25 to 150 °C. In theory, cracks are induced by heating and the strength decreases. At $T = 150$ °C, grain expansion may not induce thermal cracks and absorbed, bound, and crystal water escape, which leads to an increase in the frictional resistance between the grains. Based on direct shear tests, Mitchell et al. (2013) investigated the variation in the frictional coefficient depending on the temperature, as shown in Fig. 15.21a. The frictional coefficient increases linearly with temperature. Therefore, the peak strength and crack damage threshold increase at 150 °C.

Figures 15.19a and 15.20a show that the confining pressure can reduce the effect of temperature excursions on the peak strength and crack damage threshold. To quantify the effect of temperature on the peak strength and damage threshold, the ratio between the mean squared error and average value of the peak strength and crack damage threshold after the same temperature treatment was introduced. Figure 15.21b shows that the ratio is maximal. This means that the temperature has the greatest effects on the peak strength and crack damage threshold. At $\sigma_3 = 10$ MPa, the ratio quickly decreases, suggesting that the application of the confining pressure notably reduces the effect of the temperature on the peak strength and crack damage threshold. However, the ratio slowly decreases with increasing confining pressure, suggesting that an increase in the confining pressure only slightly decreases the effect of the temperature on the peak strength and crack damage threshold.

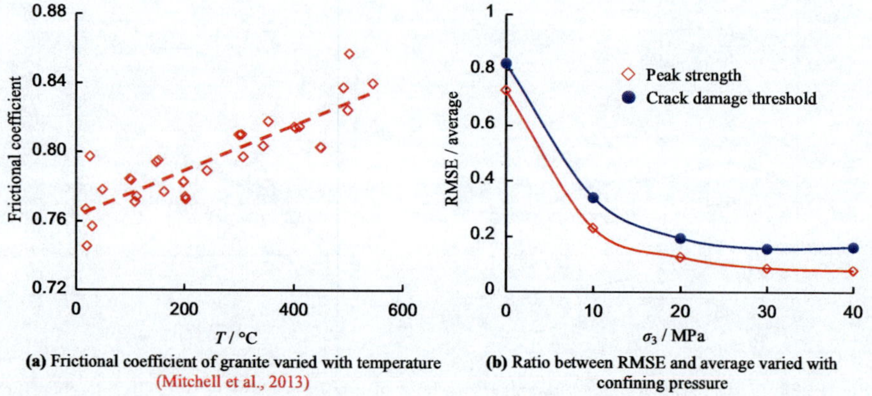

Fig. 15.21 Frictional coefficient of granite with the temperature and ratio between RMSE and average with the confining pressure. **a** Frictional coefficient of granite varied with temperature (Mitchell et al. 2013); **b** Ratio between RMSE and average varied with confining pressure

15.4.3 *Effect of High Temperature on Hoek–Brown Criterion Strength Parameters of Granite*

To illustrate the nonlinear response of the peak strength and crack damage threshold of granite to the confining pressure, the Hoek–Brown criterion (Hoek and Brown 1980; Hoek 1990) can be used. It describes the relationship between the peak strength, crack damage threshold, and confining pressure. The Hoek–Brown criterion can be expressed as:

$$\sigma_S = \sigma_3 + \sigma_c (m\frac{\sigma_3}{\sigma_c} + s)^{0.5}, \tag{15.4}$$

where σ_c is the uniaxial compressive strength and m and s are basic parameters representing the rock, which vary depending on the rock type. The parameter m reflects the weakness of the rock. The bigger the –m coefficient the stronger the rock. The parameter s reflects the extent of pre-existing (or evolving) fractures and ranges from 0 to 1 with unity representing intact rock.

To describe the strength and fracture of granite after thermal treatment, σ_c is set to the uniaxial compressive strength of granite at room temperature. When s is larger than 1, s is set to 1, and when s is smaller than 0, s is set to 0. Based on the regression of the peak strength and crack damage threshold under different confining pressures, s and m values of the granite specimens after different temperature treatments are obtained. Figure 15.22a shows that the parameter s is obtained from the peak strength and is relatively constant with an increase in temperature from 25 to 150 °C. However, s quickly decreases to 0 with the increase in the temperature from 150 to 600 °C. Subsequently, it remains zero when the temperature increases to 750 °C. The parameter s is obtained from the crack damage threshold and increases slightly with

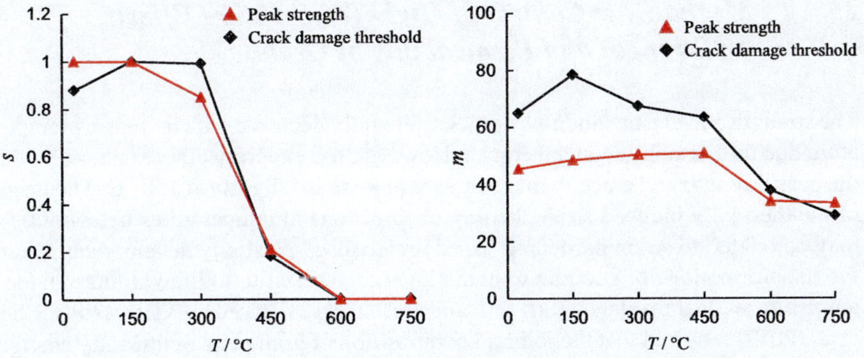

Fig. 15.22 Variation of the Hoek–Brown parameters of the peak strength and crack damage threshold of granite specimens after high temperature treatment with temperature. **a** *s*, **b** *m*

an increase in the temperature from 25 to 150 °C. The *s* parameter obtained from the crack damage threshold quickly decreases to zero with an increase in the temperature from 150 to 600 °C. Finally, this remains zero when the temperature increases to 750 °C. This analysis shows that almost no thermal cracks are induced in the specimens at $T \leq 300$ °C, but thermally induced cracks propagate quickly at 300 °C $\leq T \leq$ 600 °C. When the temperature increases from 600 to 750 °C, it is difficult for thermally induced cracks to propagate due to the formation of macrocracks.

Figure 15.22b shows the variation in the parameter *m* obtained from the peak strength and crack damage threshold depending on the temperature. The parameter *m* obtained from the peak strength first increases with the increase in the temperature from 25 to 450 °C and then quickly decreases with the increase in the temperature from 450 to 600 °C. Finally, it remains relatively constant with an increase in the temperature to 750 °C. The *m* parameter obtained from the crack damage threshold first increases with the increase in the temperature from 25 to 150 °C and then decreases with increasing temperature. Based on this analysis, the specimen becomes stronger at 150 °C. However, the *m* parameter obtained from the peak strength increases when the temperature increases from 150 to 450 °C, while its crack damage threshold decreases. This indicates that cracks are more easily induced after thermal treatment and the ductility and frictional coefficient of the grains increase. This leads to a decrease in the stronger and weakness extent of the crack damage threshold, while the stronger and weakness extent of the peak strength increases. When the temperature increases from 600 to 750 °C, the stronger and weakness extent of the peak strength and crack damage threshold remain relatively constant.

15.4.4 Mechanistic Control of High-Temperature Effects on Strength and Permeability of Granite

The strength and elastic modulus of rocks generally decrease with increasing temperature due to thermally induced cracks. However, the experimental results show that the peak strength and crack damage threshold increase slightly at 150 °C. The origin of the thermally induced strengthening of sandstone at temperatures below 500 °C may be related to strain hardening based on localised plasticity development and to the transformations of goethite to hematite and/or smectite to illite at more modest temperatures on the order of 300 °C (Ranjith et al. 2012). Based on XRD results, Chen et al. (2017) reported that the mineral composition of granite insignificantly changes after thermal treatment; however, the crystalline forms of the minerals change. Based on SEM observations of granite after thermal treatment, the microcrack closure due to thermal expansion improves the granite structure, which leads to an increase in its strength (Huang et al. 2017). However, SEM only scans the granite specimens locally and thus is not representative for the overall characteristics of the granite samples. The variation in the porosity (Φ) and P-wave velocity (V_p) of granite after thermal treatment is known to depend on the temperature, as shown in Fig. 15.23a. The results show that the porosity slowly increases with increasing temperature at $T \leq 400$ °C and then quickly increases at $T \geq 400$ °C. The rate of decrease of V_p is the ratio between the V_p decrease by thermal treatment to the V_p of granite specimens at room temperature. The rate of decrease of V_p is reflects the damage level of granite based on thermal treatment. Figure 15.23b shows that the V_p decreasing rate almost linearly increases with increasing temperature. Figure 15.23 shows that the amount of thermally induced cracks increases with increasing temperature, which contrasts with the hypothesis that microcrack closure due to thermal expansion improves the structure of the granite (Huang et al. 2017).

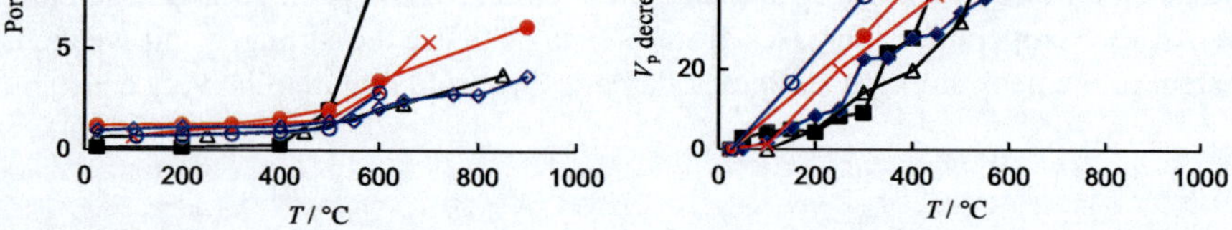

Fig. 15.23 Porosity and V_p decreasing rate with temperature. **a** Porosity, **b** V_p decreasing rate

To verify that the amount of thermally induced cracking increases with increasing temperature, the permeability (K) of the granite after different high-temperature treatments was measured under different effective stresses. Figure 15.11 shows that the permeability of the granite decreases nonlinearly with increasing effective stress, regardless of the temperature. At the same effective stress, the permeability of the granite increases with increasing temperature. The relationship between the permeability and porosity is shown in Eq. (15.2) (McKee et al. 1988). The porosity of granite increases with increasing temperature according to the ratio.

$$K \propto \frac{\phi^3}{(1-\phi)^2} \tag{15.5}$$

As explained in the previous section, the increase in the strength at 150 °C may be due to the increase in the frictional coefficient with increasing temperature (Mitchell et al. 2013). Therefore, a Mohr–Coulomb criterion was used to investigate the effect of the confining pressure on the peak strength and crack damage threshold of granite and thus the internal friction angle (φ) can be obtained. Figure 15.24 shows that the internal frictional angle of the granite obtained from the peak strength and crack damage threshold first increases with increasing temperature at $T \leq 450$ °C. The figure also shows that the strength increase at $T = 150$ °C may be due to the increase in the frictional coefficient with increasing temperature.

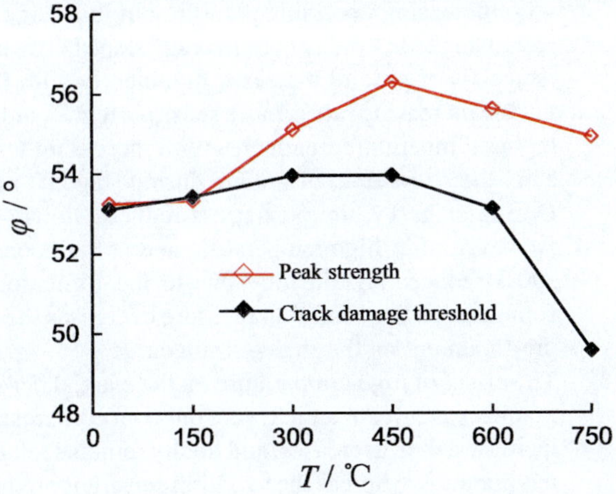

Fig. 15.24 Effect of temperature on internal friction angle of granite obtained by peak strength and crack damage threshold

15.5 Conclusions

Granite is an excellent potential host for HLW repositories because of its low permeability and significant integrity. Based on experimental results for permeability evolution and failure response of granite subjected to thermal cycling, the following conclusions can be drawn.

1. The effect of high temperature on the physical behaviour of granite was analysed by using the volume variation rate, mass variation rate, P-wave velocity, and thermal conductivity of granite as a function of the temperature. The experimental results show that the mass, P-wave velocity, and thermal conductivity of the granite decrease but the volume increases with increasing temperature. High temperature also has a notable effect on the permeability evolution behaviour. The permeability decreases with increasing axial and confining pressure, while the confining pressure has a larger influence than the axial stress in the range of 10–50 MPa. A theoretical model was constructed to analyse the effect of the volumetric stress on the permeability coefficient of granite, which agrees very well with the experimental results. The initial and ultimate permeabilities increase exponentially with increasing temperature.
2. The triaxial stress–strain curve of granite after high-temperature treatment can be divided into five stages: (I) initial pore and fissure closure stage; (II) elastic deformation stage; (III) plastic deformation stage; (IV) post-peak failure stage, and (V) residual strength stage. The range of stage I gradually decreases with increasing confining pressure but this stage becomes more apparent with increasing temperature. The average slope between the stress and the axial strain (stage II) of granite increases nonlinearly with increasing confining pressure, but the increasing rate is more sensitive to the confining pressure under suffering higher temperature treatment. With increasing temperature and confining pressure, the stress drop of granite during stage III becomes less and less obvious. During stage IV, the post-stress reduces slowly with increasing deformation. However, after high temperature and at high confining pressure, the post-peak stress reduces significantly due to the formation of transgranular cracks. To some extent, the confining pressure decreases the effect of the high temperature pre-treatment on the stress–strain curve.
3. The effect of high temperature on the triaxial deformation and AE behaviour of granite was investigated. Under uniaxial compression, the Young's modulus first increases, then decreases, and finally remains relatively constant with increasing temperature; whereas the variation sensitivity reduces with increasing confining pressure. At $T \leq 300$ °C, stress drops can be observed before the peak strength is reached, which correspond to a significant AE event and are caused by the formation of multiple splitting tensile cracks in the loading direction. At $T \geq 450$ °C, AE can be observed once a minor stress is applied because the failure mode is controlled by thermally-induced cracks. However, under triaxial compression, the temperature has little effect on the AE characteristics. The granite fails along the shear fracture plane, which widens with increasing confining pressure.

However, at $T \geq 600$ °C, intragranular cracks are more easily generated and the stress quickly decreases after the peak strength is reached. The shear plane is smoother at higher confining pressure.

4. With increasing temperature, the peak strength and crack damage threshold first remain relatively constant, then decrease, and finally are constant between 600 °C and 750 °C. This can be explained as follows: Mica is oxidised at 400 °C, the α–β phase transformation of quartz occurs at 573 °C, and macroscopic cracks are initiated at 600 °C. Because the thermal stress is released, the propagation of macroscopic cracks becomes more difficult. The peak strength and crack damage threshold of granite nonlinearly increases with the confining pressure. The mechanism based on which the high temperature affects the triaxial strength parameters of granite is discussed. At $T \leq 300$ °C, thermally induced cracks are not notable and the temperature has a limited effect on the behaviour of granite. At 450 °C $\leq T \leq$ 600 °C, thermally induced cracks can be observed and the temperature has a significant effect on the strength behaviour of granite. Because thermal stress is released by the formation of thermal macrocracks, the continuous increase in the temperature has a limited effect on the strength behaviour of granite, even at the maximum temperature of 750 °C.

References

Chaki S, Takarli M, Agbodjan WP (2008) Influence of thermal damage on physical properties of a granite rock: porosity, permeability and ultrasonic wave evolutions. Constr Build Mater 22(7):1456–1461

Chen S, Yang C, Wang G (2017) Evolution of thermal damage and permeability of Beishan granite. Appl Therm Eng 110:1533–1542

Chen YL, Wang SR, Ni J, Azzam R, Fernandez-Steeger TM (2017) An experimental study of the mechanical properties of granite after high temperature exposure based on mineral characteristics. Engineering geology 220:234–242

Consenza P, Ghoreychi M (1999) Effects of very low permeability on the long-term evolution of a storage cavern in rock salt. Int J Rock Mech Min Sci 36:527–533

Davy CA, Skoczylas F, Barnichon JD, Lebon P (2007) Permeability of macro-cracked argillite under confinement: gas and water testing. Phys Chem Earth 32:667–668

Fairhurst CE, Hudson JA (1999) Draft ISRM suggested method for the complete stress-strain curve for intact rock in uniaxial compression. Int J Rock Mech Min Sci 36(3):279–289

Fan LF, Gao JW, Wu ZJ, Yang SQ, Ma GW (2018) An investigation of thermal effects on micro-properties of granite by X-ray CT technique. Appl Therm Eng 140:505–519

Gautam PK, Verma AK, Sharma P, Singh TN (2018) Evolution of thermal damage threshold of Jalore granite. Rock Mech Rock Eng 51(9):2949–2956

Griffiths L, Heap MJ, Baud P, Schmittbuhl J (2017) Quantification of microcrack characteristics and implications for stiffness and strength of granite. Int J Rock Mech Min Sci 100:138–150

Gustafsson SE (1991) Transient plane source techniques for thermal conductivity and thermal difusivity measurements of solid materials. Rev Sci Instrum 62:797–804

Heap MJ, Baud P, Meredith PG, Bell AF, Main IG (2009) Time-dependent brittle creep in Darley Dale sandstone. J Geophys Res 114

Huang YH, Yang SQ, Tian WL, Zhao J, Ma D, Zhang CS (2017) Physical and mechanical behavior of granite containing pre-existing holes after high temperature treatment. Arch Civil Mech Eng 17(4):912–925

Hoek E, Brown ET (1980) Underground excavation in rock. Institution of Mining and Metallurgy, London

Hoek E (1990) Estimating Mohr–Coulomb friction and cohesion values from the Hoek–Brown failure criterion. Int J Rock Mech Min Sci Geomech Abstr 12:227–229

Jiang G, Zuo J, Li L, Ma T, Wei X (2018) The evolution of cracks in Maluanshan granite subjected to different temperature processing. Rock Mech Rock Eng 51(6):1683–1695

Log T, Gustafsson SE (1995) Transient plane source (TPS) technique for measuring thermal transport properties of building materials. Fire Mater 19:43–49

Mckee CR, Bumb AC, Koenig RA (1988) Stress-dependent permeability and porosity of coal and other geologic formations. SPE Form Eval 3(01):81–91

Mitchell EK, Fialko Y, Brown KM (2013) Temperature dependence of frictional healing of Westerly granite: Experimental observations and numerical simulations. Geochem Geophys Geosyst 14(3):567–582

Nasseri MHB, Schubnel A, Young RP (2007) Coupled evolutions of fracture toughness and elastic wave velocities at high crack density in thermally treated Westerly granite. Int J Rock Mech Min Sci 44(4):601–616

Ranjith PG, Viete DR, Chen BJ, Perera MS (2012) Transformation plasticity and the effect of temperature on the mechanical behaviour of Hawkesbury sandstone at atmospheric pressure. Eng Geol 151:120–127

Shao S, Ranjith PG, Wasantha PLP, Chen BK (2015) Experimental and numerical studies on the mechanical behaviour of Australian Strathbogie granite at high temperatures: An application to geothermal energy. Geothermics 54:96–108

Sun Q, Zhang W, Xue L, Zhang Z, Su T (2015) Thermal damage pattern and thresholds of granite. Environ Earth Sci 74(3):2341–2349

Sun SH (1987) The choice of calculating scheme of mica formula. Acta Petrol Sin 4:72–82 (in Chinese)

Sundberg J, Hellström G (2009) Inverse modelling of thermal conductivity from temperature measurements at the Prototype Repository, Äspö HRL. Int J Rock Mech Min Sci 46(6):1029–1041

Urquhart A, Bauer S (2015) Experimental determination of single-crystal halite thermal conductivity, diffusivity and specific heat from 75 C to 300 C. Int J Rock Mech Min Sci 78:350–352

Williams H, Turner FJ, Gilbert CM (1954) Petrography. freeman, San Francisco

Wong TF, David C, Zhu W (1997) The transition from brittle faulting to cataclastic flow in porous sandstones. Mech Deformat J Geophys Res 102(B2):3009–3025

Xu P, Yang SQ (2019) Influence of stress and high-temperature treatment on the permeability evolution behavior of sandstone. Acta Mech Sin 35(2):419–432

Yang SQ, Ranjith PG, Huang YH, Yin PF, Jing HW, Gui YL, Yu QL (2015) Experimental investigation on mechanical damage characteristics of sandstone under triaxial cyclic loading. Geophys J Int 201:662–682

Yang SQ, Ranjith PG, Jing HW, Tian WL, Ju Y (2017) An experimental investigation on thermal damage and failure mechanical behavior of granite after exposure to different high temperature treatments. Geothermics 65:180–197

Yang SQ, Tian WL, Huang YH (2018) Failure mechanical behavior of pre-holed granite specimens after elevated temperature treatment by particle flow code. Geothermics 72:124–137

Yang SQ, Tian WL, Derek Elsworth, Wang JG, Fan LF (2020) An experimental study of effect of high temperature on the permeability evolution and failure response of granite under triaxial compression. Rock Mech Rock Eng 53:4403–4427

Yin T, Li X, Cao W, Xia K (2015) Effects of thermal treatment on tensile strength of Laurentian granite using Brazilian test. Rock Mech Rock Eng 48(6):2213–2223

Zhang F, Zhao J, Hu D, Skoczylas F, Shao J (2018) Laboratory investigation on physical and mechanical properties of granite after heating and water-cooling treatment. Rock Mech Rock Eng 51(3):677–694

Zhao XG, Wang J, Chen F, Li PF, Ma LK, Xie JL, Liu YM (2016) Experimental investigations on the thermal conductivity characteristics of Beishan granitic rocks for China's HLW disposal. Tectonophysics 683:124–137

Zhao XG, Cai M, Wang J, Li PF (2015) Strength comparison between cylindrical and prism specimens of Beishan granite under uniaxial compression. Int J Rock Mech Min Sci 76:10–17

Zhao XG, Zhao Z, Guo Z et al (2018) Influence of Thermal Treatment on the Thermal Conductivity of Beishan Granite. Rock Mech Rock Eng 51(7):1–20

Zhu TT, Jing HW, Su HJ, Yin Q, Du MR, Han GS (2016) Physical and mechanical properties of sandstone containing a single fissure after exposure to high temperatures. Int J Mining Sci Technol 26(2):319–325

Chapter 16
Experiment and Grain-Based Modelling on Damage Failure Behaviour of Granite Under Different Confining Pressures

As it is well-known to us, nuclear power plant can provide efficient and clean energy; however, during its long-term operation, it also produces nuclear waste. Even though the high-level radioactive nuclear waste is lesser than low- and medium-level radioactive nuclear waste, it requires more restrictive disposition environment, and deep geological disposal is widely considered as the most suitable option, as shown in Fig. 16.1. Granite material is an excellent medium for high-level nuclear waste disposition rock projects because of its lower permeability and better integrity (Read 2004; Yang et al. 2019; Bi et al. 2016; Zhao et al. 2015, 2016). Therefore, many researchers have focused on the failure mechanical behaviour of granite under all kinds of stress conditions and external environments.

The mechanical behaviour of disposition rock significantly influences the safety and stability of high-level nuclear waste reservation, and many experiments have been conducted to explore the crack growth and coalescence behaviour of all kinds of rock material (Zhou et al. 2008, 2014 and 2018), especially for granite material. Tapponnier and Brace (1976) revealed new details of the development of dilatant micro-crack in stressed Westerly granite at a confining pressure of 50 MPa (500 bar) using ion-thinned samples under scanning election microscope. Haimson and Chang (2000) used the true triaxial cell to conduct an extensive series of tests in Westerly granite and obtained a new true triaxial strength criterion for the rock, which took into account the effect of the intermediate principal stress. Zhao et al. (2015) compared the strength between the cylindrical and prism specimens of Beishan granite under uniaxial compression, and showed that the average characteristic stress threshold associated with the deformation process of the two shaped specimens was close to each other. Liu et al. (2014) identified five phases in the stress–strain curve of granite under triaxial compression, and put forward a coupled elasto-plastic damage model for simulating the mechanical behaviour of granite under the compression.

Particle flow code (PFC) can be used to model the movement and interaction of particles by distinct element method (2004), and has been widely used to explore the failure mechanical behaviour of granite. Cho et al. (2007) used the clumped-particle geometry to calibrate the stress–strain behaviour and the complete nonlinear failure

S.-Q. Yang, *Mechanical Behavior and Damage Fracture Mechanism of Deep Rocks*,
https://doi.org/10.1007/978-981-16-7739-7_16

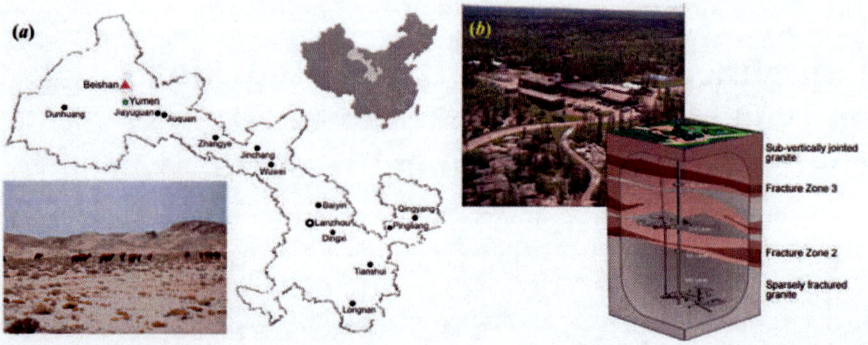

Fig. 16.1 **a** Geographical location of the Beishan granite area and a photo showing its typical topography (Zhao et al. 2015); **b** Underground research laboratory (URL) surface facilities and generalized granite mass geology in the vicinity of the underground excavations (Read 2004)

envelop of Lac du Bonnet granite under compression and tension tests. Jiang et al. (2015) simulated crack propagation and rock fracture in single- and double-flawed specimens with different flaw inclinations using the normal force dependent bond contact model in PFC. Considering the micro-structure of crystalline granite, the grain-based modelling (GBM) was developed to explore the micro-cracking behavior of Bukit Timah granite (Peng et al. 2018), and then investigated the influence of material heterogeneity on the strength and deformation behaviour and the associated microcracking process (Peng et al. 2017). Yang et al. (2018) used the cluster model to explore the meso-mechanics of granite specimens containing pre-existing holes after different temperature treatments. Liang et al. (2018) established a microgeometric model of granite particles to simulate both uniaxial and triaxial compression tests. Shu et al. (2018) used a flat-joint model (FJM) in PFC^{3D} to examine the effect of heterogeneity on the macro-mechanical properties of granite. Li et al. (2018) also used FJM to simulate the uniaxial and Brazilian test of Lac du Bonnet granite.

X-ray computed tomography (CT) technique is shown to be an efficient measuring tool to reflect the evolution and propagation of defects in materials and is widely applied to investigate the damage and fracture behaviour within the rock material in many past studies (Ge et al. 2001; Hirono et al. 2003; Feng et al. 2004; Zhou et al. 2008; Sufian and Russell 2013; Zhao et al. 2014; Meier et al. 2015; Yang et al. 2015; Yang and Huang 2017; Fan et al. 2018; Yang 2018). Using a conventional medical X-ray CT scanning system, Kawakata et al. (1997, 1999) analysed 3-D observations of faulting in Westerly granite under uniaxial and triaxial conditions. Ikeda et al. (2000) investigated 3-D interconnections of crystals by CT, and showed that at least 89.9% of the quartz rods were connected in three dimensions. Yang and Huang (2017) analysed the internal damage mechanism of granite specimens with respect to various fissure angles under triaxial compression by X-ray micro-CT observation, and dedicated obvious effects of confining pressure and fissure angle on the crack system of granite. Cartwright-Taylor et al. (2017) presented 4-D X-ray micro-topographic data collected during a triaxial deformation experiment and documented damage localization and fracture propagation in a micro-granite. Nakashima et al. (2017) suggested a new method for detecting fracture surfaces from CT images using an edge detection algorithm and successfully measured the fracture surface geometry, the aperture distribution and the contact ratio.

3-D AE technique provides an ideal non-destructive method for studying crack growth at its actual state (2014). Zhao et al. (2015) used 3-D AE technique in combination with the stress–strain relationships to characterise the complete evolution of crack damage in the rock during rock deformation and observed the typical failure modes such as splitting in uniaxial compression and shear failure in triaxial compression. Liu et al. (2014) used 3-D AE technique to capture the microcracking process under three widely used tensile tests. It is revealed that in direct tensile testing, the AE events accumulated mainly along the failure surface, whereas in indirect tests the location and variation of AE events are much more complex. Chen et al. (2014, 2015) experimentally studied the cracking process of granite under compressive stress condition with the aid 3-D AE monitoring system and proposed a coupled elasto-plastic damage model to simulate the mechanical behaviour. Because of severely limited by prevailing visualization technique, Hohl et al. (2018) introduced a novel approach to visualize 3-D date through time (4-D) using unfiltered AE event data.

However, previous triaxial compression experiments and numerical simulations have provided no satisfactory answers on the damage failure mechanism of granite. In addition, X-ray micro-CT and spatial AE observations have rarely considered real-time damage failure mechanism of granites under various confining pressures. Therefore, in this chapter, we report the results of a series of triaxial compression tests on granite subjected to various confining pressures ranging from 2 to 50 MPa. Based on the experimental results, we first investigate the influence of confining pressure on the crack damage, peak strength and deformation parameters of granite material. Then, based on clustering algorithm, a new GBM method is put forward to simulate the damage mechanical behaviour of granite under different confining pressures. The numerical simulation results (including the axial deviatoric stress–strain curve, peak strength, crack damage threshold and ultimate failure mode) of granite under triaxial compression are compared quantitatively with the experimental results. Finally, using an X-ray micro-CT scanning system and spatial acoustic emission (AE) monitoring, the internal crack and failure process of deformed granite specimens with different confining pressures are analysed in detail.

16.1 Granite Material and Testing Procedure

16.1.1 Tested Granite Material

The material used throughout this research was granite collected from Rizhao city in Shandong province, China, which was used to investigate the deformation and cracking behaviour of rock material subjected to various confining pressures. In this experiment, all specimens were cored from the same block of material to an actual diameter of 50 mm and approximately 100 mm in length, which resulted in a length: diameter ratio of 2: 1 for mechanical tests in accordance with the ISRM (International Society for Rock Mechanics) suggested method (Fairhurst and Hudson 1999). According to the results of X-ray diffraction, the minerals in the granite material are quartz (Q), feldspar (F), biotite (B), amphibole, chlorite and dolomite. The detailed composition of this granite material is as follows: 11.12% quartz, 59.85% feldspar, 21.56% biotite, 6% amphibole, 1.01% chlorite and 0.46% dolomite (Yang et al. 2017). The tested granite has a crystalline and blocky structure; which is macroscopically very homogeneous with a connected porosity of 0.828% and an average density of approximately 2638 kg/m^3.

16.1.2 Mechanical Testing, AE and X-ray CT Scanning Procedure

All experiments were performed in an MTS815 Flex Test GT triaxial rock mechanics experimental system (Yang et al. 2012). The maximum loading capacity of the servo-controlled system is 4,700 kN, and the maximum confining pressure is 140 MPa. During experimentation, we measured the axial force (using a load cell with a loading capacity of 2,600 kN) and the axial deformation using an axial linear variable differential transducer (LVDT) with a range of 5 mm. The circumferential deformation was measured using an LVDT attached to a chain wrapped tightly around the specimen, for which the maximum displacement capacity was 8 mm. To eliminate the end friction effects (Mogi 1966; Hawkes and Mellor 1970), the axial and circumferential LVDTs were located in the central part of the specimen.

As it is well-known to us, in the past, most triaxial experiments on rock have been performed using the axial displacement rate as the controlling feedback signal (Yang et al. 2015, 2012; Yang and Huang 2017). However, since specimen failure in the brittle regime is often very rapid under this mode of control, it is often difficult to obtain reliable post-peak deformation data. Therefore, to reiterate, we used the circumferential displacement as the controlling feedback signal (Yang et al. 2012; Amann et al. 2011; Youn and Tonon 2010) rather than the axial displacement feedback commonly used in rock deformation experiments (Paterson and Wong 2005). In this research, we performed a series of triaxial compression experiments using a circumferential displacement rate of 0.06 mm/min. The conventional triaxial compression experiments were conducted under different confining pressures of 2, 10, 20, 30, 40 and 50 MPa, and consisted of the following two steps. First, the confining pressure was increased to the desired value at a constant rate of 0.1 MPa/s to ensure that the specimen was under uniform hydrostatic stresses. Second, the deviatoric stress was applied to the surface of the specimen at a constant circumferential displacement rate of 0.06 mm/min until failure.

During experimentation, to study the fracture evolution characteristics of the deforming specimens, the output of AEs was continuously monitored. AEs are transient elastic waves generated by the rapid release of energy within a material, such as the strain energy released during microcrack propagation. Monitoring AE during deformation has become an increasingly important diagnostic tool in material science which has provided a wealth of information regarding the failure process in brittle materials (Fortin et al. 2006; Lockner et al. 1991; Brantut et al. 2011) and can be used in failure forecasting modelling (Bell et al. 2011a, 2011b). In this study, the output of AE was recorded using a PCI-2 AE 3D measuring system made by Physical Acoustic Corp. (PAC) with eight AE sensors at a frequency of 6 kHz. The amplitude threshold of the sensors was set to 40 dB.

In this research, X-ray CT scanning of the granite specimens is conducted using a Nanotom 160 high-resolution micro-CT at a spatial resolution of 30 μm (Yang et al. 2017). The X-ray micro CT scanning procedure is the same as that in Chap. 3. The X-ray micro CT scanning parameter in this chapter is as follows: the electron current was 90 μA, the accelerating voltage was 130 kV, and the scanning time was 4 s.

16.2 Experimental Results on Damage Failure Behaviour of Granite

The deviatoric stress–strain curves for granite under different confining pressures are presented in Fig. 16.2, where, σ_1 and σ_3 represent the major principal stress and the confining pressure, respectively; ε_1, ε_3 and ε_v are the axial, circumferential and volumetric strain, respectively; σ_{cd} and σ_p are the crack damage threshold and peak strength, respectively, which are confirmed in accordance with the deviatoric stress–strain curves of granite specimens, as shown in Fig. 16.2.

In accordance with Fig. 16.2, it can be seen that generally, with the increase in confining pressure, the post-peak behaviour of granite specimen changes from the brittle characteristics to plastic deformation behaviour. Based on the experimental results, Table 16.1 lists the strength and deformation parameters of granite under different confining pressures. In Table 16.1, E_S refers to the elastic modulus, which is calculated by the average slope of the approximately straight line portion of the deviatoric stress–axial strain curve before the peak strength. μ refers to the Poisson's ratio, which is calculated by the average value of the radial-to-axial strain ratio at the stage of elastic deformation.

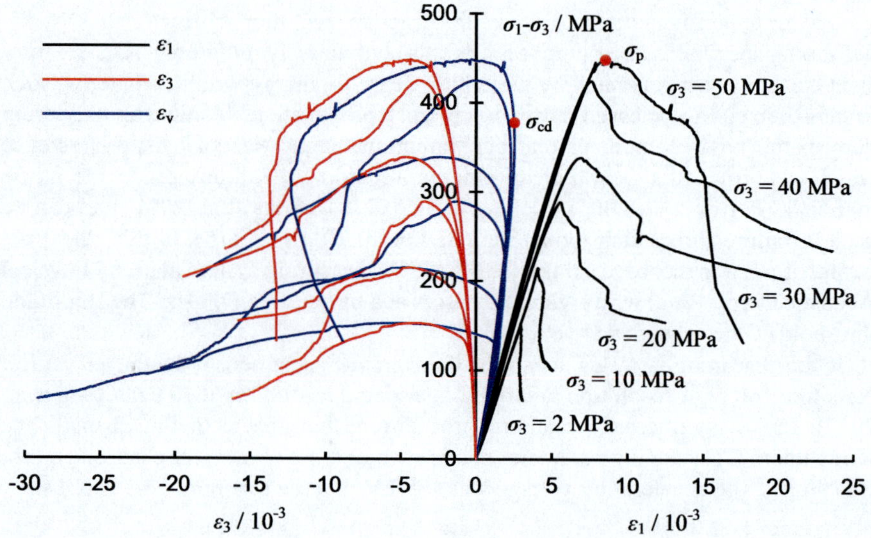

Fig. 16.2 Axial deviatoric stress–strain curves for granite under different confining pressures. The volumetric strain (ε_v) is calculated by the sum of the axial strain and twice the circumferential strain (i.e., $\varepsilon_v = \varepsilon_1 + 2\,\varepsilon_3$)

Table 16.1 Strength and deformation parameters of granite under different confining pressures

σ_3 / MPa	E_S / GPa	ε_{1p} / 10^{-3}	μ	σ_{cd} / MPa	σ_{sd} / MPa	σ_P / MPa	σ_S / MPa
0	37.35	2.970	0.127	68.30	68.30	80.06	80.06
2	55.02	2.921	0.127	112.51	114.51	152.19	154.19
10 (GA-49#)	54.57	4.347	–	–	–	222.74	232.74
10 (GA-25#)	54.05	4.086	0.173	144.23	154.23	215.03	225.03
20	59.14	5.546	0.177	210.71	230.71	288.97	308.97
30	56.39	6.895	0.207	244.92	274.92	338.65	368.65
40	61.96	7.583	0.228	335.99	375.99	424.72	464.72
50	59.38	8.414	0.225	355.69	405.69	448.36	498.36

Note $\sigma_{sd} = \sigma_{cd} + \sigma_3$; $\sigma_S = \sigma_P + \sigma_3$. The experimental data of granite at $\sigma_3 = 0$ MPa are from the reference (Yang et al. 2017)

16.2.1 Effect of Confining Pressure on Peak Strength and Crack Damage Threshold

In this section, the Mohr–Coulomb criterion is used to investigate the effect of the confining pressure on the peak strength and crack damage threshold of granite. The linear Mohr–Coulomb criterion can be expressed with the following equation (Yang et al. 2011; Mohr 1914):

$$\sigma_S = \sigma_0 + q\sigma_3 = \frac{2C\cos\phi + \sigma_3(1+\sin\phi)}{1-\sin\phi} \tag{16.1}$$

where, σ_S is the maximum axial supporting capacity of rock, σ_0 is usually regarded as the uniaxial compressive strength (UCS) of rock material, and q is an influence coefficient of confining pressure on the σ_S of rock. The σ_0 and q are related to the cohesion C and the internal friction angle ϕ.

The Hoek–Brown criterion describing the peak triaxial compressive strength of a wide range of isotropic rock materials can be written as follows (Hoek and Brown 1980, 1997):

$$\sigma_S = \sigma_3 + (m\sigma_c\sigma_3 + s\sigma_c^2)^{0.5} \tag{16.2}$$

where σ_c is the UCS of the intact rock material and m and s are material constants for a specific rock. When parameter m is larger, the rock is stronger. The parameter s reflects the fracture extent, ranging from zero to one. When the parameter s is closer to one, the rock is more intact.

Based on the strength data listed in Table 16.1, the influence of σ_3 on the peak strength and crack damage threshold of granite is presented in Fig. 16.3. From Fig. 16.3a, it can be seen that the peak strength of granite has a distinct nonlinear behaviour with an increase in σ_3. Furthermore, the non-linear Hoek–Brown criterion reflects the peak strength behaviour of granite better than the linear Mohr–Coulomb criterion. Moreover, the sensitivity of the peak strength on the confining pressure is clearly higher than that for the crack damage threshold. By using the linear Mohr–Coulomb criterion, the strength parameters of granite are calculated and listed in Table 16.2. The peak cohesion of granite is 23.32 MPa, which is considerably higher than 6.4 MPa obtained by using the crack damage threshold. However, the peak internal friction angle of granite is 50.9°, which is a little higher than that obtained by using the crack damage threshold.

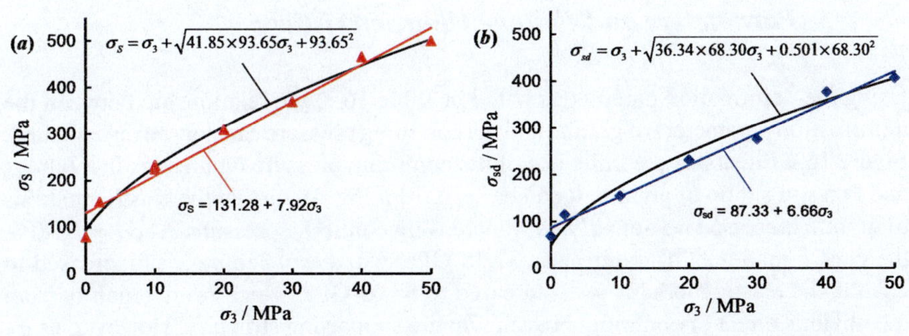

Fig. 16.3 Effect of confining pressure on the peak strength and crack damage threshold of granite specimen by the nonlinear Hoek–Brown criterion and the linear Mohr–Coulomb criterion. **a** Peak strength; **b** crack damage threshold

Table 16.2 Strength parameters of granite obtained by the linear Mohr–Coulomb criterion

Different regression analysis	σ_0 / MPa	q	C / MPa	ϕ/ (°)	R
Using peak strength (σ_S)	131.28	7.92	23.32	50.9	0.984
Using crack damage threshold (σ_{sd})	87.33	6.66	16.92	47.6	0.992

Table 16.3 Strength parameters of granite obtained by the nonlinear Hoek–Brown criterion

Different regression analysis	σ_c / MPa	m	s	a	R
Using peak strength (σ_S)	80.06	48.95	1.368	0.5	0.994
Using peak strength (σ_S)*	93.65	41.85	1.000	0.5	0.994
Using crack damage threshold (σ_{sd})	68.30	36.34	0.501	0.5	0.981

Note The nonlinear Hoek–Brown strength parameters m and σ_c with *symbol are back-calculated by giving $s = 1.000$

Table 16.3 lists the strength parameters of granite obtained by the nonlinear Hoek–Brown criterion. It should be noted that when carrying out the regression analysis for the peak strength of granite by the non-linear Hoek–Brown criterion, parameters m and s are first back-calculated by setting $\sigma_c = 80.06$ MPa based on the experimental value of UCS of granite. However, $s = 1.368$, which is greater than 1.0, but s in Hoek–Brown criterion must be in the range of 0~1.0. Therefore, σ_c and m are back-calculated by assuming $s = 1.000$ in this study.

From Fig. 16.3b, it can be seen that the linear Mohr–Coulomb criterion reflects the crack damage threshold behaviour of granite better than the nonlinear Hoek–Brown criterion even though the peak strength has obvious nonlinear behaviour with the confining pressure, which is also similar to the previous conclusion on the red sandstone (Yang et al. 2012).

16.2.2 Effect of Confining Pressure on Deformation Parameters and Failure Characteristics

Using the deformation parameters listed in Table 16.1, the relationship between the deformation parameters of granite and the confining pressure can now be investigated. Figure 16.4 illustrates the influence of the confining pressure on the elastic modulus and Poisson's ratio of granite. It can be seen from Fig. 16.4a that the elastic modulus of granite increased nonlinearly with increasing confining pressure. At $\sigma_3 = 0$ MPa, the elastic modulus of granite was 37.35 GPa, whereas when σ_3 was increased to 2 MPa, the elastic modulus was increased to 55.02 GPa, which results mainly from the obvious effect of confining pressure on the supporting structure. However, as σ_3 increased from 2 to 50 MPa, the elastic modulus increased from 55.02 to 59.38 GPa, which shows a slight increase. From Fig. 16.4b, Poisson's ratio of granite increased obviously from 0.127 to 0.225 as the confining pressure increased from 0 to 40 MPa;

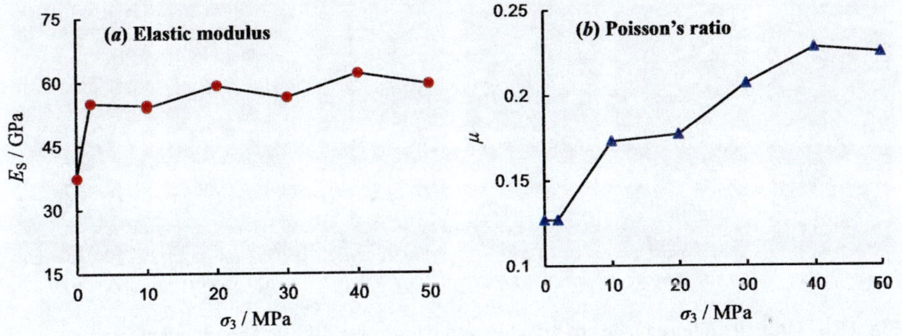

Fig. 16.4 Effect of confining pressure on the elastic modulus and Poisson's ratio of granite specimen. **a** Elastic modulus; **b** Poisson's ratio

however, at the confining pressure above 40 MPa, Poisson's ratio of granite remained constant.

The influence of confining pressure on the peak axial and circumferential strain (listed in Table 16.1) is presented in Fig. 16.5. It should be noted that circumferential strain is plotted by the absolute value. At $\sigma_3 = 0$ MPa, the peak axial strain of granite equals to 2.970×10^{-3}, whereas, as σ_3 was increased to 50 MPa, the peak axial strain of granite increased to 8.414×10^{-3}. A theoretical relationship between the peak axial strain and confining pressure is put forward, i.e. $\varepsilon_{1p} = 3.027 + 0.115\,\sigma_3$. From the comparison between the experiment results and theoretical line, we can conclude that the peak axial strain of granite increased linearly, which agrees very

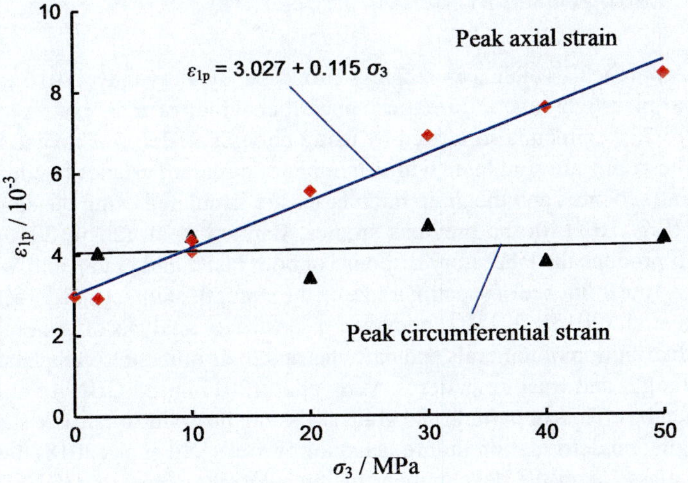

Fig. 16.5 Effect of confining pressure on the peak axial and circumferential strain of granite specimen

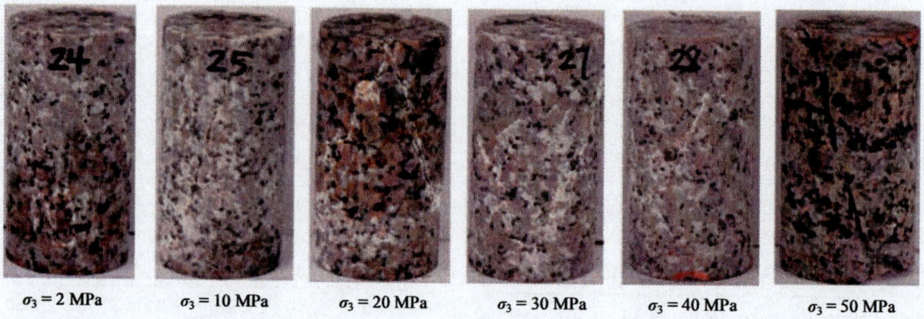

Fig. 16.6 Ultimate failure modes of granite specimens under different confining pressures

well with the previous conclusion on red sandstone (Yang et al. 2012). However, the peak circumferential strain of granite has almost no any relation with the confining pressure.

Figure 16.6 shows the influence of confining pressure on the ultimate failure mode of granite specimens. From Fig. 16.6, it can be seen that with the increase in confining pressure, the ultimate failure mode of granite specimens transmits from axial splitting tensile failure to shear failure.

16.3 Grain-Based Modeling on Damage Failure Behaviour of Granite

16.3.1 Grain-Based Modelling Procedure

A GBM approach, which was recently proposed by Potyondy (2010), has been adopted to investigate the failure mechanical behaviour of rock extensively. In the GBM, a rock specimen is simulated by using polygonal, deformable and breakable grains. The grains are simulated with a number of circular particles bonded together using parallel bonds, and the grain boundaries are simulated using the smooth-joint contacts (Fig. 16.7). In the previous studies, Bahrani et al. (2014, 2016) adopted GBM to reproduce the laboratory response of both intact and granulated Wombeyan marble, and the influence of specimen size on the strength of intact and defected rocks. Hofmann et al. (2015a, 2015b) conducted a sensitivity analysis of mineral size and random distribution of minerals and particles and their influence on the elastic properties, strength and fracture patterns. Peng et al. (2017) used GBM to numerically investigate the effects of particle and grain sizes and the grain-to-particle size ratio on the strength and deformation failure behavior of rocks. Xu et al. (2018) investigated the rock masses and the development of micro-cracks under thermal–mechanical coupling effect. It can be seen from Fig. 13.5 in Chap. 13 that the interface between the grains in the granite specimen is not a line, which has many zigzags. However, the

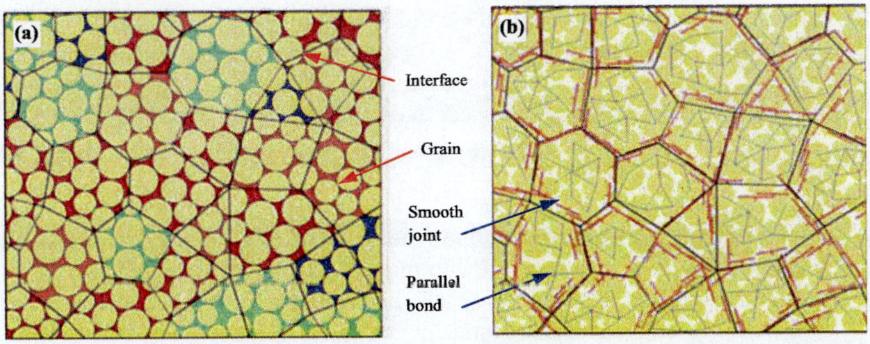

Fig. 16.7 GBM composed with parallel bond and smooth joint. **a** Bonded particle model overlaid on the grain structure, **b** GBM consisting of grains or minerals (parallel-bond disks, grey) and interfaces between the grains (smooth-joint contacts, red) (Potyondy 2010; Hofmann et al. 2015a)

previous method to generate GBM does not actually reflect the zigzag characteristics of grain boundary (Hofmann et al. 2015a). Therefore in this research, a new GBM method of simulating the damage failure behaviour of granite is presented to form the grained rock vividly.

The new method, which is different from the GBM put forward by Potyondy (2010) is put forward to construct GBM based on clustering algorithm.

① The densely packed bonded material was generated, as shown in Fig. 16.8a. The densely packed bonded material-genesis procedure employs the following five-step process: (a) Compact initial assembly, (b) install specified isotropic stress, (c) reduce the number of "floating" particles, (d) install parallel bonds, (e) remove from material vessel. Furthermore, it should be noted that the radius of the ball particle distributes in the range of 0.3–0.48 mm, and each mineral grain contains ten ball particles.

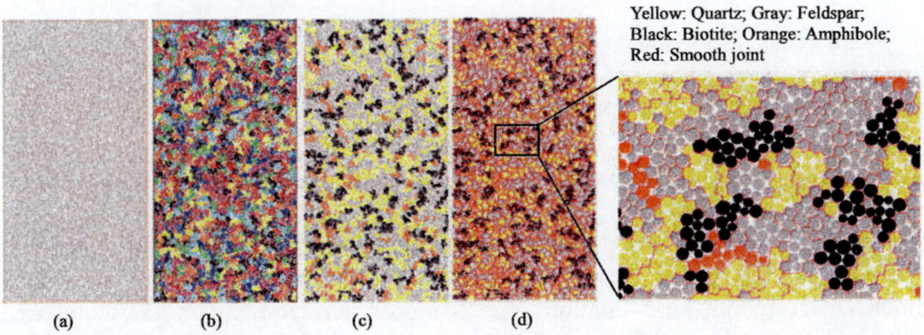

Fig. 16.8 A new method to construct GBM. **a** The densely packed bonded specimen, **b** the cluster model specimen, **c** numerical specimen with different minerals, and **d** the inter-granular bonded replaced by Smooth joint

② The maximum number of particles in a cluster was defined as 10, and the cluster model was constructed, as shown in Fig. 16.8b. The size of numerical mineral grains is approximately same to the experimental values. A cluster is defined as a set of particles that are adjacent to one another. By adjacent it means that a connected path can be constructed between any two particles in a cluster by traversing bonded particle–particle contacts.

③ Each cluster is randomly specified to a kind of mineral and the content corresponding to the mineral proportion in the granite, and modifying particles and contact properties to correspond with the minerals, as shown in Fig. 16.8c. In this numerical simulation, we only take into account four main minerals in the granite material, i.e. feldspar, biotite, quartz and amphibole. The particle radius is uniformly distributed and approximately 10,197 particles were generated in a numerical specimen (i.e. 100 mm in height and 50 mm in width), and the number of cluster in the specimen is 1033. Thus, 618 clusters are defined as feldspar, 222 clusters are defined as biotite, 115 clusters are defined as quartz and 62 clusters are defined as amphibole.

④ The inter-granular parallel bonds were replaced by Smooth joint, as shown in Fig. 16.8d. First, the location and dip angle inter-granular parallel were recorded and then a smooth joint at this location with same dip angle was generated. The stiffness properties of smooth joint are inherited from the contact and the two contact balls, according to the following equations:

$$\overline{k_n} = (k_n/A) + \overline{k^n} \tag{16.3}$$

$$\overline{k_s} = (k_s/A) + \overline{k^s} \tag{16.4}$$

$$A = 2\overline{R}t, t = 1.0 \tag{16.5}$$

$$\overline{R} = \overline{\lambda}\min(R^A, R^B) \tag{16.6}$$

where $\overline{k_n}$ and $\overline{k_s}$ are the smooth joint contact normal stiffness and shear stiffness, respectively. k_n and k_s are the contact normal and shear stiffness, respectively. $\overline{k^n}$ and $\overline{k^s}$ are the parallel bond normal and shear stiffness, respectively. A is the cross sectional area of the smooth joint contact, t the ball thickness, and $\overline{R}$ is the smooth joint contact radius, which is a multiple $\overline{\lambda}$ of the minimum radii of two particles (Bahrani and Kaiser 2016).

By using this new method to construct a GBM model, the boundary zigzag and mineral grain inter-locking can be reflected more obviously. However, the precious method used smooth-joint 'dilation angle' to represent fracture roughness. Compared with the optical microscopic observation of granite specimen (Fig. 13.5a in Chap. 13), it can be seen that the grain micro-structure of new constructed GBM is more vivid to mimic the granitic basement rock.

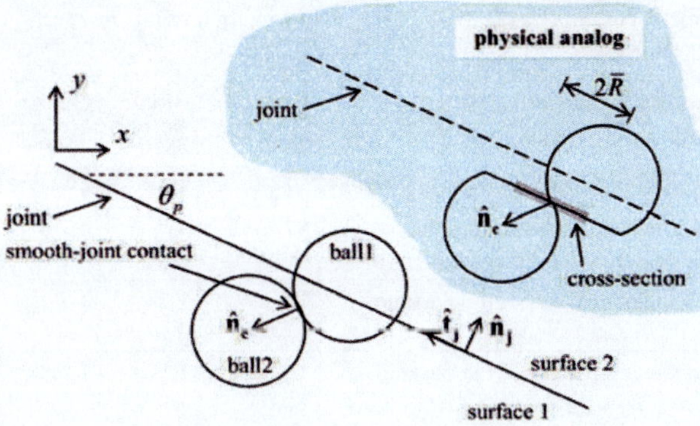

Fig. 16.9 Schematic diagram of smooth-joint contact. (Ivars et al. 2008)

The smooth-joint model can simulate the behaviour of an interface regardless of the local particle contact orientations along the interface. Particle pairs joined by a smooth-joint contact may overlap and 'slid' past each other, instead of being forced to move around one another, as shown in Fig. 16.9. The effective joint geometry consists of two planar surfaces. During each time step, the relative translational displacement increment between the two particle surfaces is decomposed into components normal and tangential to the joint surfaces. These components are multiplied by the smooth-joint normal and shear stiffness to produce increments of joint force. The force–displacement law operates in the joint coordinate system and provides Coulomb sliding with either dilation or bonded behaviour (Ivars et al. 2008).

16.3.2 Calibrating Micro-parameters by Experimental Results

Using the previous calibration procedure (Yang et al. 2014, 2019), a group of smooth joint micro-parameters are obtained, as listed in Table 16.4. SJ contact normal and shear stiffness factor are 0.55 and 0.65, respectively. SJ bond friction coefficient is 0.75. SJ bond tensile strength and cohesion are 20 and 50 MPa, respectively. To validate the feasibility of the micro-parameters, the stress–strain curves, the peak strength, deformational parameters and failure modes of granite specimens under different confining pressure were adopted.

Figure 16.10 shows the comparison of axial deviatoric stress–strain curves of the granite specimens under different confining pressures obtained by experiment and particle flow simulation. From Fig. 16.10, it can be seen that, in general, the numerically simulated axial deviatoric stress–strain curves of granite specimens under different confining pressures are in a good agreement with the experimental curves, especially the pre-peak deformation region, which also means that the elastic modulus

Table 16.4 The micro-parameters of granite using GBM method in this research

Micro-parameter	Values			
Micro-properties of the minerals	Feldspar	Biotite	Quartz	Amphibole
Density, ρ (kg/m^3)	2600	2850	2650	2400
Effective Young's modulus of the particle, E_c (GPa)	57	37	49	69
Effective Young's modulus of the parallel bond, $\overline{Ec}$ (GPa)	57	37	49	69
Ratio of normal to shear stiffness of the particle, k_n/k_s	1.5	1.5	1.5	1.5
Ratio of normal to shear stiffness of the parallel bond,$\overline{k}n/\overline{k}s\overline{k}_n/\overline{k}_s$	1.5	1.5	1.5	1.5
Particle friction coefficient, μ	1.2	1.2	1.2	1.2
Mean and standard deviation, parallel-bond normal strength σ_n(MPa)	470	360	410	490
Mean and standard deviation, parallel-bond shear strength τ_n(MPa)	470	360	410	490
Micro-properties of the inter-granular contacts				
SJ contact normal stiffness factor, sj_nsf	0.55			
SJ contact shear stiffness factor, sj_ssf	0.65			
SJ bond friction coefficient, sj_fric	0.75			
SJ bond tensile strength, sj_ten (MPa)	15			
SJ bond cohesion, sj_coh (MPa)	90			

and Poisson's ratio of granite obtained by numerical simulation agree very well with those obtained by experiment.

However, at the post-peak region, it should be noted that the larger difference of circumferential deformations between experiment and simulation may mainly result from the difference of macroscopic crack evolution speed along the circumferential direction. Under the lower confining pressures, the macroscopic crack evolution speed along the circumferential direction obtained by experiment is faster than that obtained by numerical simulation due to larger energy dissipation during the numerical simulation. However, under the higher confining pressures ($\sigma_3 = 30$ and 50 MPa), the axial deformation after the peak strength obtained by numerical simulation is more brittle than that obtained by experiment; furthermore, the confining pressure restrains the macroscopic crack evolution speed of specimen obtained by numerical simulation resulting in an approximate circumferential deformation compared with the experimental results.

Figure 16.11 compares the peak strength and crack damage threshold of granite specimens under different confining pressures obtained by experiment and simulation. From Fig. 16.11a, we can conclude that the numerical peak strength of granite under different confining pressures is quantitatively equal to that obtained by experiment. The maximum difference in peak strength between experiment and simulation

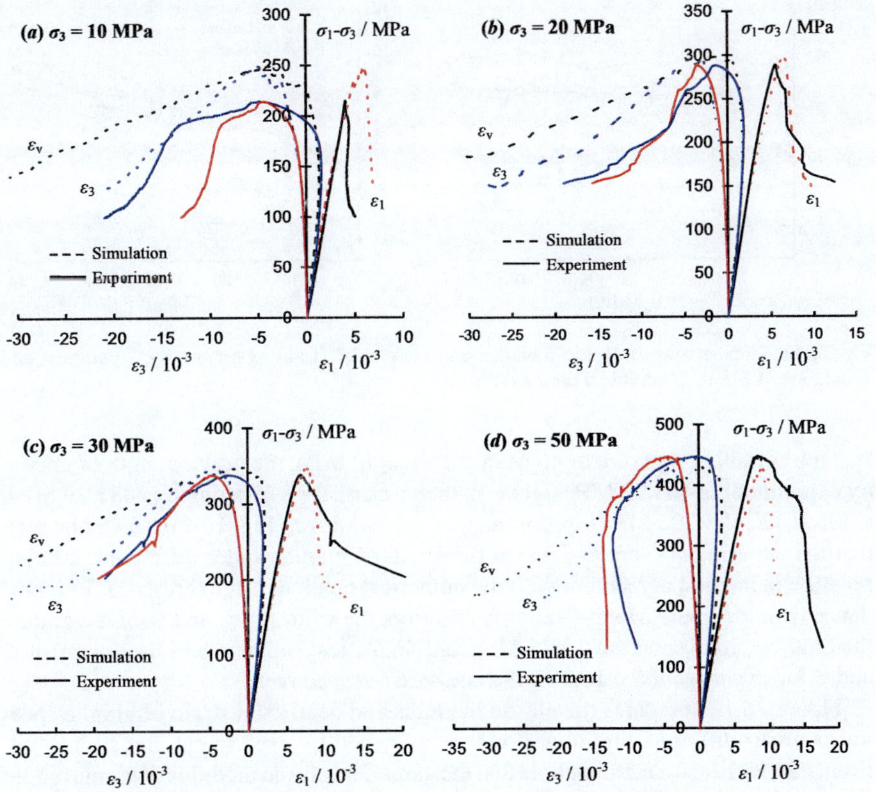

Fig. 16.10 Comparison of axial deviatoric stress–strain curves for granite under different confining pressures by experiment and simulation

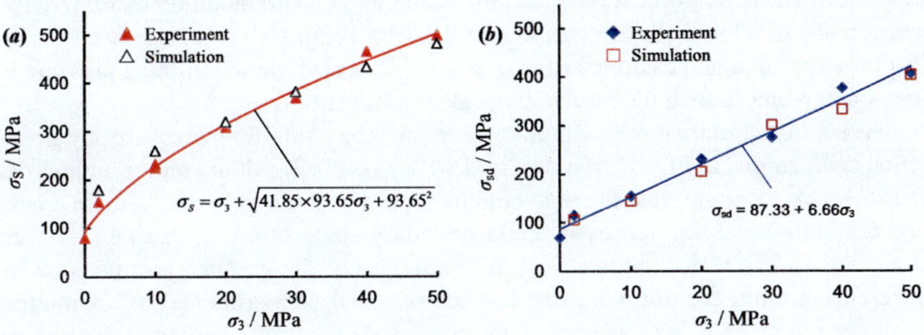

Fig. 16.11 Comparison of peak strength and crack damage threshold of granite by experiment and simulation. **a** Peak strength; **b** crack damage threshold

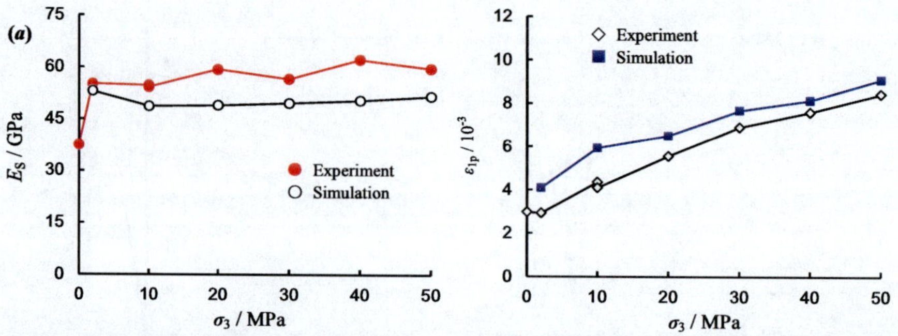

Fig. 16.12 Comparison of elastic modulus and peak axial strain of granite by experiment and simulation. **a** Elastic modulus; **b** peak axial strain

is at the confining pressure of 40 MPa. At $\sigma_3 = 40$ MPa, the peak strength of granite by experiment is 464.72 MPa, whereas the numerical peak strength is 431.79 MPa, a difference of 32.93 MPa (approximately 7.09%). From Fig. 16.11b, it can be seen that the experimental crack damage threshold of granite under different confining pressure is in good agreement with the numerical crack damage threshold. In accordance with the linear Mohr–Coulomb criterion, the values of C and Φ of the granite material by simulation are 17.75 MPa and 46.6°, respectively, and the values of C and Φ by experiment are 16.92 MPa and 47.6°, respectively.

Figure 16.12 compares the elastic modulus and peak axial strain of granite specimens under different confining pressures obtained by experiment and simulation. From Fig. 16.12a, we can conclude that the numerical elastic modulus of granite under different confining pressures is quantitatively equal to that obtained by experiment. The maximum difference in elastic modulus between experiment and simulation is at the confining pressure of 40 MPa. At $\sigma_3 = 40$ MPa, the elastic modulus of granite by experiment is 61.96 GPa, whereas the numerical elastic modulus is 50.27 GPa, a difference of 32.93 MPa (approximately 18.7%). From Fig. 16.12b, it can be seen that the experimental peak axial train of granite under different confining pressure is in good agreement with the numerical peak axial strain.

The ultimate failure mode of granite under different confining pressures by simulation is illustrated in Fig. 16.13a. In Fig. 16.13a, the black colours represent internal tensile crack within grains, the blue colours represent grain boundary tensile crack, and the yellow colours represent grain boundary shear crack. It can be observed from Fig. 16.13a that, the numerical failure modes of the granite specimens under different confining pressures are very approximate to the experimental failure modes, as shown in Fig. 16.6. We can see that the confining pressure also affects the ultimate failure modes of the granite specimens. Under lower confining pressures, the granite specimens are mainly demonstrated with axial splitting fracture. However, under the action of confining pressure, the granite specimens are mainly demonstrated with single shear fracture.

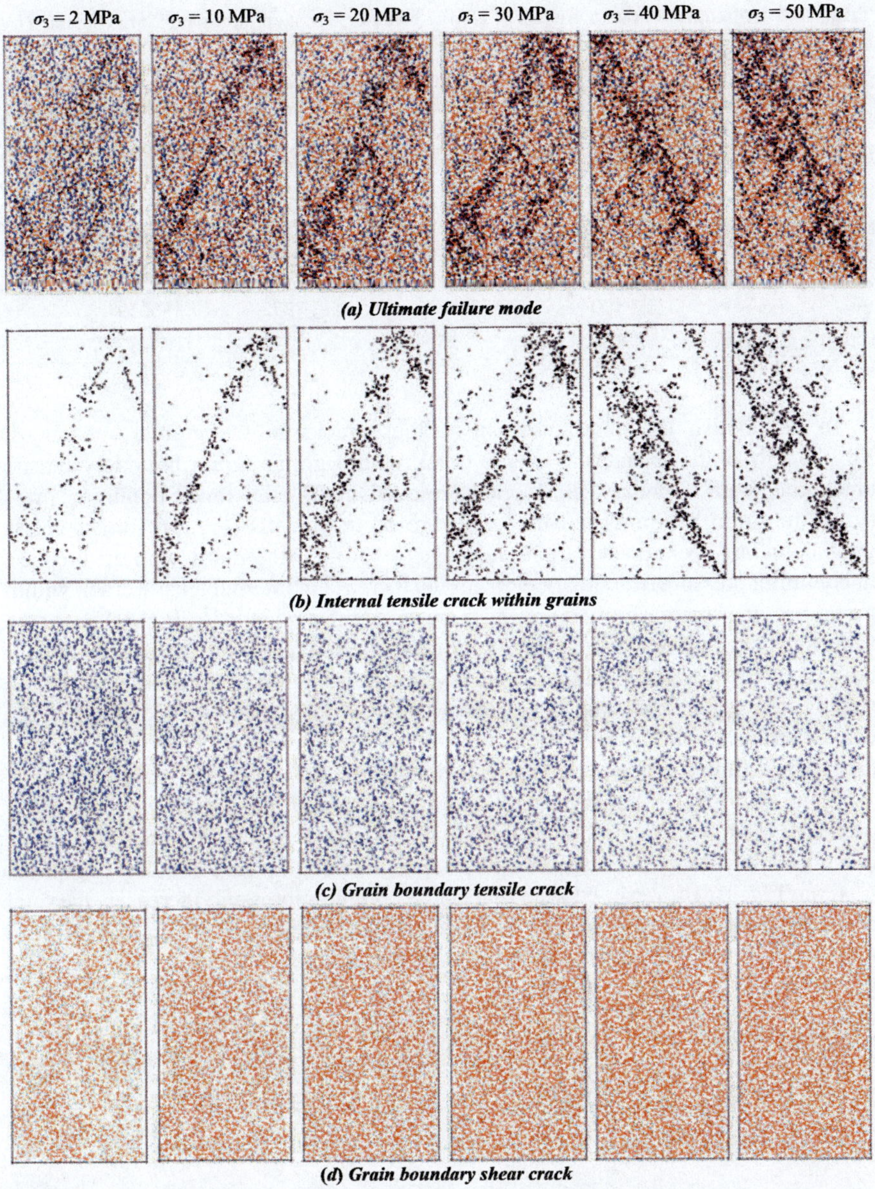

Fig. 16.13 Ultimate failure mode of granite under different confining pressures by simulation. **a** Ultimate failure mode; **b** internal tensile crack within grains; **c** grain boundary tensile crack; **d** grain boundary shear crack

Table 16.5 All kinds of crack numbers of granite under different confining pressures obtained by simulation

σ_3 / MPa	Total crack numbers	Internal tensile crack within grains	Internal shear crack within grains	Grain boundary tensile crack	Grain boundary shear crack
2	5176	290	0	2713	2173
10	6146	506	0	2269	3371
20	6736	714	0	1911	4111
30	7513	1033	14	1653	4813
40	7451	1004	16	1372	5059
50	7957	1260	25	1235	5437

To distinguish the distribution of various cracks in the granite specimens, Fig. 16.13b–d shows internal tensile crack within grains, grain boundary tensile crack and grain boundary shear crack, respectively. Under lower confining pressures, the grain boundary tensile cracks are more obvious (Fig. 16.13c), but less grain boundary shear cracks are found (Fig. 16.13d). Furthermore, with the increase in confining pressure, the shear belt from the increase in internal tensile crack within grains becomes increasingly more and more apparent, as shown in Fig. 16.13b. At the same time, because of the restraining action of confining pressure, the grain boundary tensile cracks decrease gradually whereas the grain boundary shear cracks increase step by step as the confining pressure increases. Besides, it should be noted that the crack distribution in Fig. 16.13 is more discrete, which is mainly because we did not take into account the heterogeneity of bonding strength among the grains.

Table 16.5 lists all kinds of crack numbers of granite under different confining pressures obtained by simulation. Based on the data listed in Table 16.5, Fig. 16.14 illustrates the influence of confining pressure on all kinds of crack numbers of granite.

From Fig. 16.14, it can be seen that as the confining pressure increases, the total crack numbers, the grain boundary shear crack and the internal tensile crack within grains all increases gradually, whereas the grain boundary tensile crack decreases gradually.

16.4 Discussion on Damage Fracture Mechanism of Granite

16.4.1 Spatial AE Behaviour of Granite Under Different Confining Pressures

Figures 16.15 and 16.16 shows the relationship between axial deviatoric stress, AE counts and the circumferential strain of granite at a confining pressure of 10 and

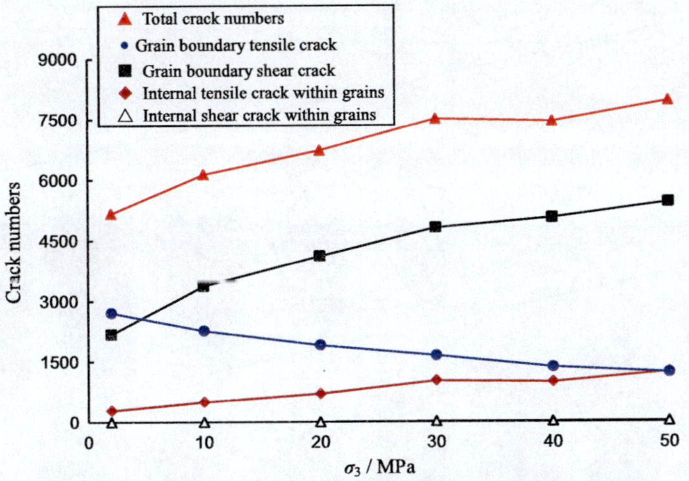

Fig. 16.14 Influence of confining pressure on all kinds of crack numbers of granite by simulation

20 MPa, respectively. At the same time, in Figs. 16.15 and 16.16, the spatial AE locations behaviour for the same experiment is also shown. The denoted letters shown in the axial deviatoric stress-circumferential strain of granite correspond to those on spatial AE evolution pictures.

Using Fig. 16.15, we can analyse the coalescence process of internal cracks in granite under the confining pressure of 10 MPa. When the axial deviatoric stress was loaded to points a-c, the AE counts in the specimen were very rare, and they resulted from small micro-failure occurrences during elastic deformation. With increasing circumferential deformation, the axial deviatoric stress–strain curve began to depart from the domain of elastic deformation. When the axial deviatoric stress reached point d (σ_1-σ_3 = 172.18 MPa = 81.5% σ_p, $\varepsilon_3 = 0.983 \times 10^{-3}$), the AE counts in the specimen reached 224, which meant that the specimen contained a small amount of internal damage. When the specimen was loaded to point e (σ_1-σ_3 = 193.42 MPa = 90.0%σ_p, $\varepsilon_3 = 1.764 \times 10^{-3}$), the AE counts within the specimen reached 652, indicating that more microcracking occurred within the specimen. The AE counts were located in the central region of the specimen.

When the axial deviatoric stress was loaded to point g ($\sigma_1 - \sigma_3$ = 215.03 MPa = 100%σ_p, $\varepsilon_3 = 4.449 \times 10^{-3}$) from point e, the AE counts increased to 1186 from 652 in the specimen. One obvious shear fracture plane can be observed, as shown in elliptical region in Fig. 16.15g. After the peak strength, from point g to k, the axial deviatoric stress on the specimen slowly dropped from 215.03 to 191.80 MPa as the circumferential strain increased from 4.449×10^{-3} to 8.724×10^{-3}, but the axial strain of the specimen only increased to 4.449×10^{-3} from 4.086×10^{-3}. During the reduction of axial deviatoric stress from point g to k, even though it involved a small stress drop, the circumferential deformation of the specimen was increasing faster than the axial deformation (indicating that there was a large amount of dilatancy). The

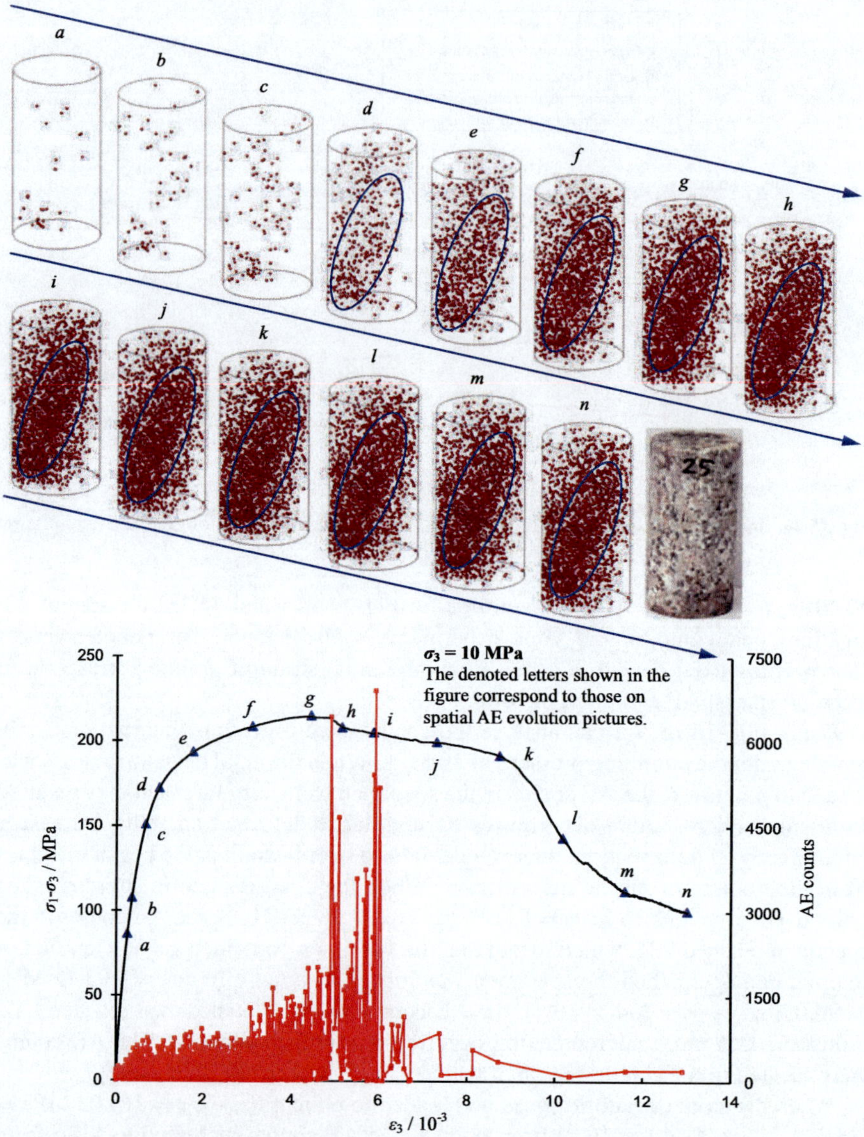

Fig. 16.15 Spatial AE locations behavior for granite at a confining pressure of 10 MPa

AE counts increased to 1245 at point k and more AE events were observed throughout the specimen. It should be noted that the maximum AE counts were at point i after the peak strength, which indicated that the granite specimen caused the macroscopic shear fracture nearby point i. After point k, even though the axial deviatoric stress on the specimen dropped from 191.08 to 100.50 MPa as the circumferential strain

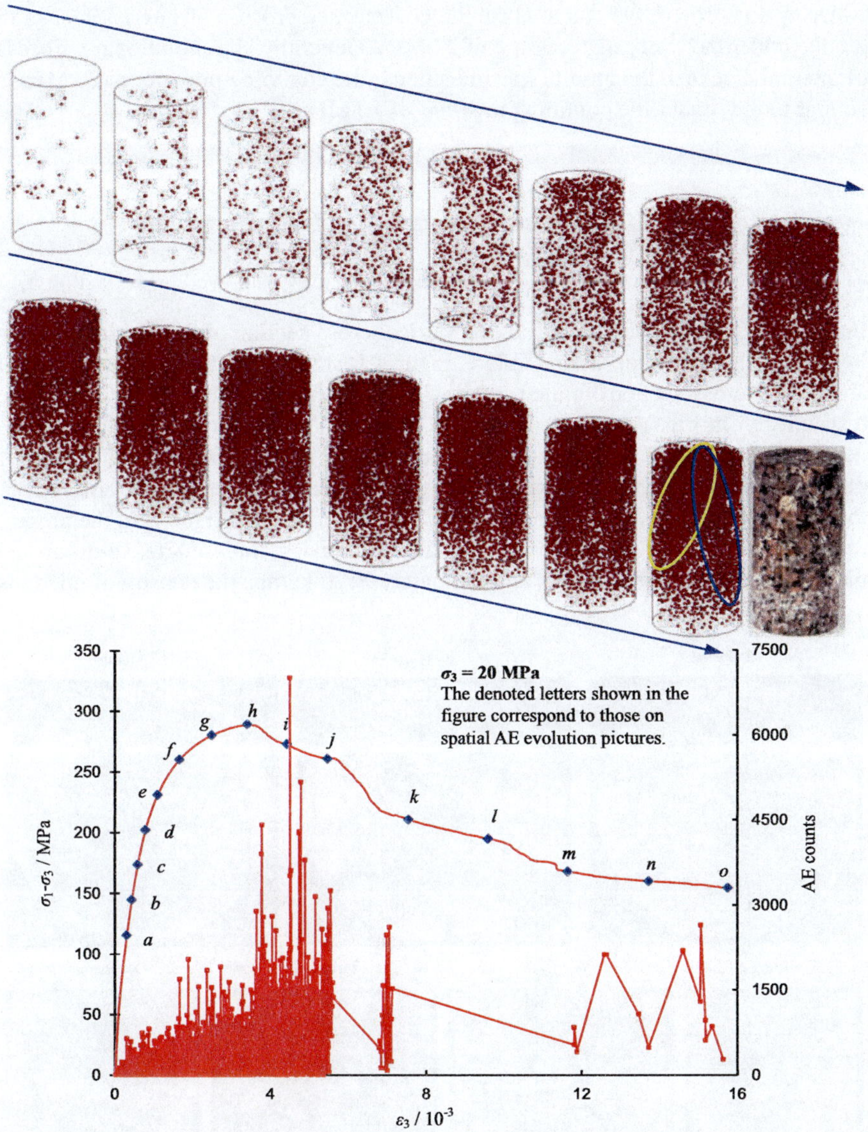

Fig. 16.16 Spatial AE locations behavior for granite at a confining pressure of 20 MPa

increased from 8.724×10^{-3} to 1.3×10^{-2}, almost no AE accounts can be further initiated, which indicated that the granite specimen supported the axial stress by the friction slippage. The failure mode of the specimen formed by spatial AE location analysis showed a good agreement with the experimental failure mode.

Using Fig. 16.16, we can analyse the coalescence process of internal cracks in granite under the confining pressure of 20 MPa. Generally, the evolutionary process of internal cracks in the granite specimen under the confining pressure of 20 MPa is similar to that under the confining pressure of 10 MPa.

16.4.2 Internal Crack Characteristics of Granite Under Triaxial Compression

Figure 16.17 shows the horizontal and vertical cross-section of failed granite specimen after uniaxial compression failure. Figures 16.18 and 16.19 show the horizontal and vertical cross-section of failed granite specimen after triaxial compression failure, respectively. In Figs 16.17, 16.18 and 16.19 , the CT number is a function of the density and chemical composition of the material in a voxel (Hirono et al. 2003). If the CT number of air is defined as 0, the CT number of water is 1000. Moreover, the density of grain is larger; and the colour of CT image is brighter. In the present research, the granite contained mainly three mineral grains: quartz, feldspar and biotite. The density of biotite is approximately 3120 kg/m^3, the density of quartz is

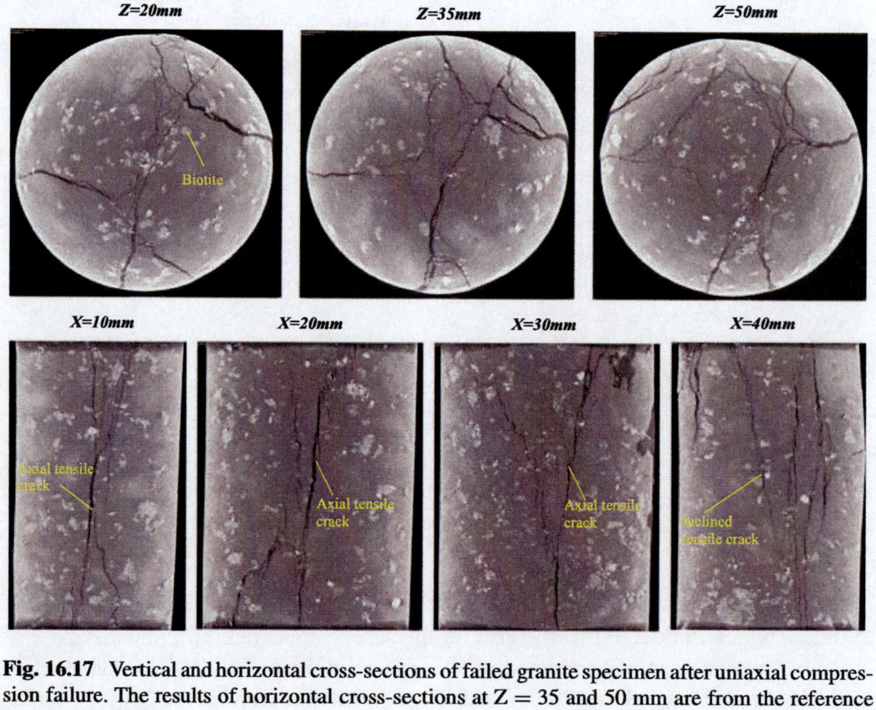

Fig. 16.17 Vertical and horizontal cross-sections of failed granite specimen after uniaxial compression failure. The results of horizontal cross-sections at Z = 35 and 50 mm are from the reference (Yang et al. 2017a)

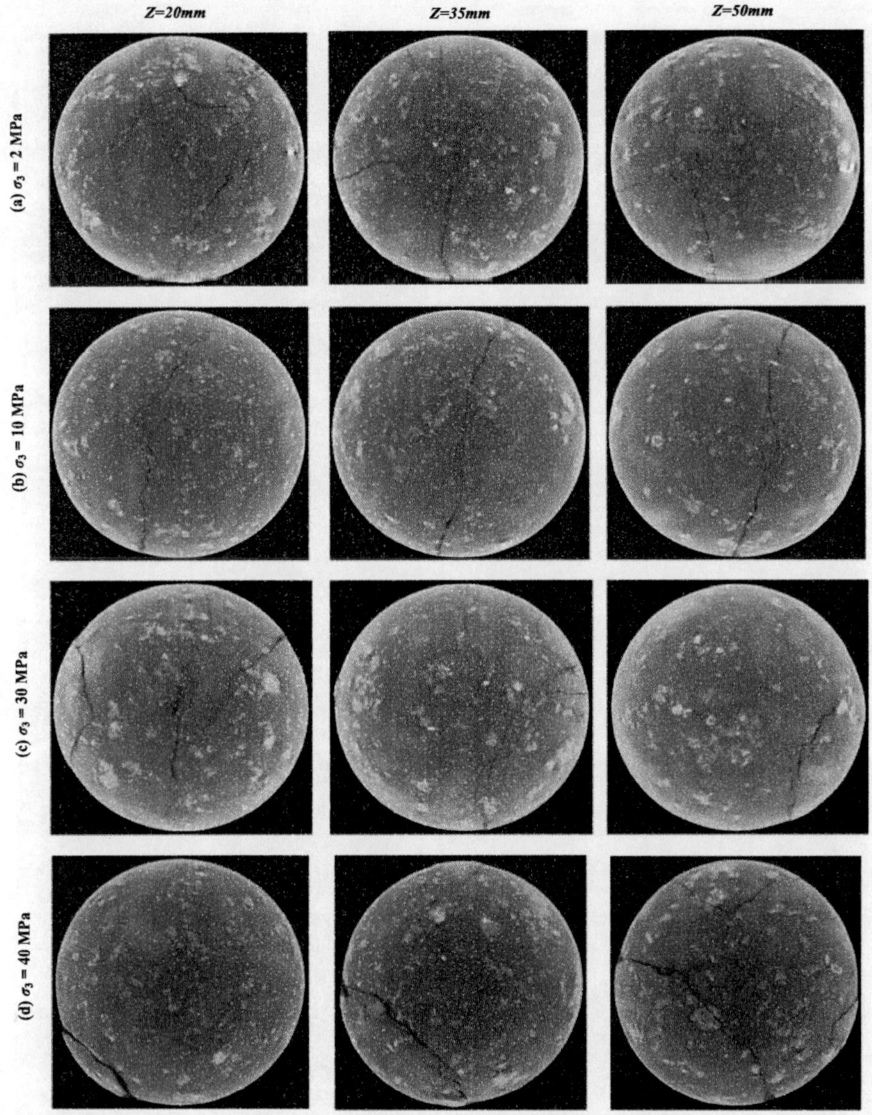

Fig. 16.18 Horizontal cross-sections with different height of failed granite specimen after different confining pressures

approximately 2650 kg/m^3, and the density of feldspar is approximately 2570 kg/m^3. Therefore, the biotite in the CT image is brighter than quartz and feldspar, as denoted in Fig. 16.17. However, it is very difficult to distinguish the feldspar and quartz because of their approximate density in the CT image. In this research, it should be

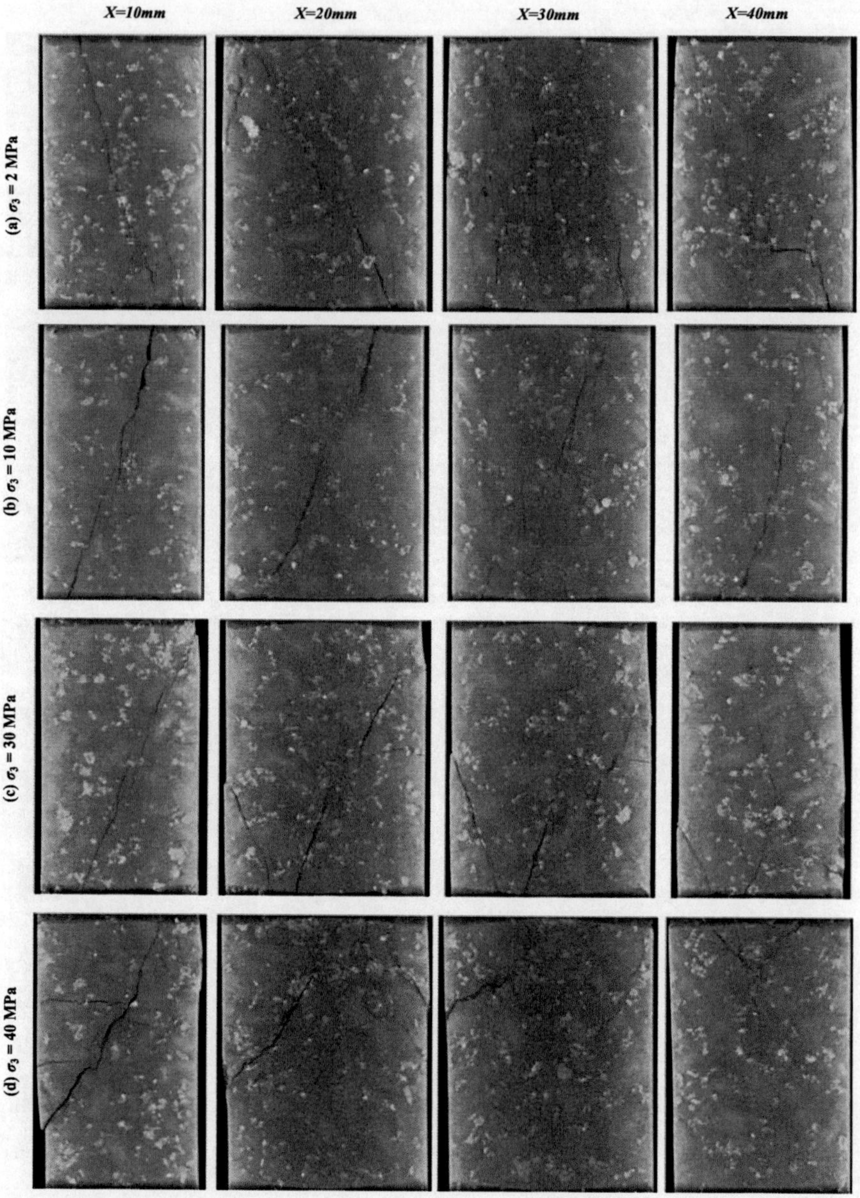

Fig. 16.19 Vertical cross-sections with different depths of failed granite specimen after different confining pressures

noted that black regions imply cracks or surface spalling, and other regions indicate no surface failure.

From Fig. 16.17, it can be seen that the granite under uniaxial compression is a kind of typically brittle rock material, showing axial splitting tensile failure mode and many axial tensile cracks are observed, both at the surface and internally, which also results in a typical brittle deformation after peak stress (Yang et al. 2017a). It is clear that the cross-section with various depths and heights has different crack system modes. However, because of the effect of three mineral grains on the crack propagation path, the splitting tensile cracks are not linear. Furthermore, one tensile crack has different apertures at different regions. At $x = 10$ mm, one obvious axial tensile crack is observed, and the cracks pass through the entire specimen height. At the same time, nearby the main tensile crack, there are many tensile cracks at the top and bottom regions of this specimen. Furthermore, the coalescence path of these cracks is not linear. At $x = 20$ and 30 mm, one big axial tensile crack is observed, but on the left regions, near this large axial crack, there are clusters of many tensile cracks. But, the direction of these tensile cracks is not always along the axial stress. At $x = 40$ mm, the crack system mode of granite is very complicated, consisting of one inclined tensile crack and multiple axial tensile cracks passing through the entire specimen. Furthermore, the crack aperture of each tensile crack is different at various regions. Furthermore, some tensile cracks (e.g. nearby the left top region of the specimen) only propagate a short distance, which can also be verified by horizontal cross-sections with different heights.

However compared with the results of granite after uniaxial compression failure (Fig. 16.17), the internal crack characteristics of granite will change obviously with the increase in confining pressure, as shown in Figs. 16.18 and 16.19. Based on the triaxial compressive curve of granite shown in Fig. 16.2, it can be seen that the granite specimen reaches the stage of residual strength. Then, with the increase in deformation, the system of cracks internal to the specimen will become stable. To explore the extent of internal damage in specimens after triaxial compression failure, Figs. 16.18 and 16.19, respectively, show vertical and horizontal cross-sections with different depths of granite specimens after triaxial compression failure ($\sigma_3 = 2$, 10, 30 and 40 MPa).

Generally, in accordance with Figs. 16.17, 16.18 and 16.19, it can be seen that under triaxial conditions, the system of cracks internal granite specimens is less complicated than that under uniaxial conditions, which is also in a good agreement with the findings obtained by a macro CT (the resolution was approximately 0.3–1.0 mm) observations of faulting in Westerly granite under uniaxial and triaxial conditions (Kawakata et al. 1999). Furthermore, the fracture planes under triaxial compression are inclined to the major principal stress with lower angles than under uniaxial compression, as demonstrated in Fig. 16.6. With the increase in confining pressure, the fracture angle of granite specimen (i.e. the angle between the shear fracture plane and confining pressure) will gradually become smaller, as shown in Fig. 16.19. Under the action of confining pressure, the mail failure and observed shear cracks all occur at local regions. It should be noted that even though under higher confining pressure, the granite is mainly the shear fracture mode, but on the

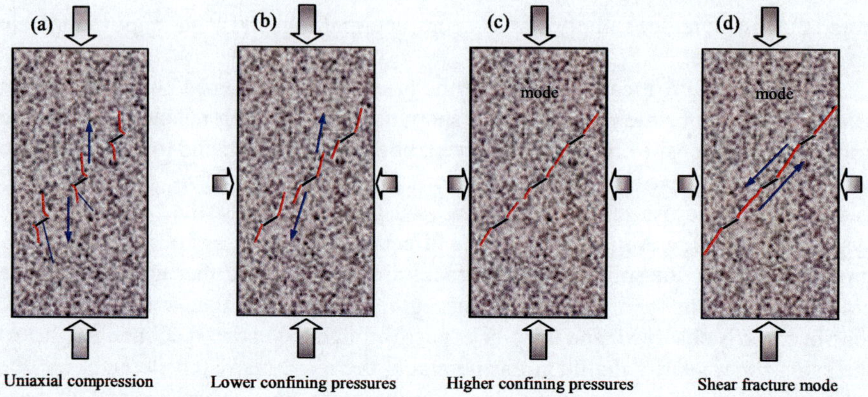

Fig. 16.20 A conceptual model of granite material under different confining pressures: axial splitting tensile fracture is due to the coalescence of wing cracks starting from pre-existing fissures. With the increase in confining pressure, the direction of crack initiation can occur the deviation towards the direction of confining pressure

location, there are also some axial and lateral tensile cracks, as shown in Figs. 16.18 and 16.19.

Under the action of uniaxial compression and lower confining pressures, the main failure and observed tensile cracks all occur at local regions. The propagation and coalescence of tensile cracks cause the ultimate failure mode, but in the other regions of the specimens, no macrocracks are observed, even though the granite specimen has undergone unstable failure. This reason can be explained as follows. Rock is a kind of natural heterogeneous material. In a rock specimen, the strength of local material is unequal for various regions. Once the applied stress concentration exceeds the strength of the local material, cracking will begin to initiate (Fig. 16.20a, b). After the occurrence of local cracking, the stress concentration transfers to the tip of the crack (Yang et al. 2014) and promotes the failure to propagate continuously in the region. It should be noted that the tensile cracks all propagate along the direction of axial stress. Under lower confining pressure, the direction of crack initiation can cause the deviation towards the direction of confining pressure, but they are still in tensile fracture mode.

Under the action of higher confining pressures, with the increase in axial deformation, the cracks are initiated from the tips of fissures. Most of the mesoscopic material elements with low strength and stiffness first reach their maximum supporting capacity, yield softening and produce the plastic deformation. Then, the axial stress must be increased continuously to allow the specimen to fail macroscopically. When the specimen is loaded to peak strength, the cracks begin to propagate. The direction of crack propagation is different under lower confining pressures, and the cracks will propagate towards the left and right boundaries of the specimen (Fig. 16.20c, d). At this time, the mesoscopic material elements with higher strength and stiffness in the specimen will also reach their maximum supporting limit, yielding failure and initiating plastic deformation. Therefore the total deformation in the specimen that tends to uniform and the plastic deformation of the specimen will increases, which results in a distinct yielding platform in the macroscopic stress–strain curves close to the peak stress. Furthermore, the yielding platform of the specimen with higher confining pressures is more obvious. When the stress is loaded to post-peak drop stress, the specimen will form the shear fracture due to the crack coalescence. When the specimen reaches the residual strength, the shear fiction slippage will control the axial capacity.

16.4.3 Internal Crack Evolution Mechanism of Granite During Triaxial Compression

Figures 16.21 and 16.22 show the simulated stress–strain behavior and micro-cracking processes of GBM under the confining pressure of 20 MPa and 40 MPa, respectively. The letters in deviatoric stress–strain curve shown correspond to different crack evolution process.

Using Figs. 16.21 and 16.22, we can analyse the evolution process of cracks of specimen under the confining pressure of 20 MPa. When the axial deviatoric stress is loaded to point a, the microcracks in the specimen are all very rare, and they results from small micro-failure occurrences during the elastic deformation. Moreover, it can be concluded that the microcracks are randomly distributed throughout the specimen. After point a, with the increase in axial deformation, the microcracks continue to increase, propagate and coalesce at the local positions, and when the axial deviatoric stress reaches point d (peak strength, i.e. $\sigma_1 - \sigma_3 = 100\% \, \sigma_p$), a number of microcracks are observed and some microcracks begin to concentrate and interact. Note that, before the peak strength, the evolution behaviour of micro-cracks in the specimen seems to be independent of the confining pressure, as the crack evolution mode shown at points a-c in Figs. 16.21 and 16.22. But after the peak strength, it can be seen that the micro-cracks increase sharply, propagate and coalesce, which forms the main fracture surface and results in the final collapse of specimen. It should be noted that the crack evolution process in the specimen after the peak strength depends on the confining pressure. The numerical specimen under different confining pressures has different shear failure characteristics, as shown at points g-h in Figs. 16.21 and 16.22.

From Fig. 16.21, we can also observe that before the initial loading stage (about lower than 2.0×10^{-3}), no obvious micro-cracks can be observed. But, after that, with the increase in axial deformation, the grain boundary tensile crack and grain boundary shear crack all gradually increase. However, the evolution rate of grain boundary shear crack is higher than that of grain boundary tensile crack, which indicates that the grain boundary shear crack dominate the final fracture of the specimen. After the axial strain is increased to 7.28×10^{-3}, the grain boundary shear crack and grain boundary tensile crack all basically keep stable with increasing axial strain.

Moreover, the ratio of grain boundary shear crack to grain boundary tensile crack is approximately equal, i.e. about 2.15 times. Compared with grain boundary shear crack and grain boundary tensile crack, internal tensile crack within grains is very negligible especially before the axial strain is smaller than 6.0×10^{-3}. Afterwards, the internal tensile crack within grains has a slight increase with the increasing axial deformation. It should be noted that during the entire loading, almost no internal shear crack within grains is observed. When the crack density is sufficiently high, the interaction and coalescence of grain boundary crack and internal tensile crack within grains result in the formation of non-persistent macroscopic shear bands, as shown in Fig. 16.21e–h.

Compared with the simulated results under the confining pressure of 20 MPa, the grain boundary shear crack in the specimen under the confining pressure of 40 MPa is much more favoured (Fig. 16.22). Furthermore, very few internal shear cracks within grains in the specimen under the confining pressure of 40 MPa develop during the entire loading stage, which is similar to that under the confining pressure of 20 MPa. However, with the increase in confining pressure, the prominent type of grain boundary cracks can take place as the changes from tensile to shear due to the high applied confining pressure (Peng et al. 2018), which also indicates that the main mechanism for crack initiation changes from tension to shear, as shown in Fig. 16.20. We should also note that when the axial deviatoric stress approaches the peak strength, the number of internal tensile cracks within grains increases rapidly, especially after the peak strength.

16.5 Conclusions

To explore the damage failure characteristics of granite material under different confining pressures, a series of triaxial compression experiment and discrete element simulation were conducted. By comparing the damage failure behaviour of granite under different confining pressures obtained by experiment and numerical simulation and analyzing the results of experiment and numerical simulation, the following conclusion can be obtained.

(1) Based on triaxial compression experimental results of granite with respect to different confining pressures, the effect of confining pressure on the strength and deformation parameters of granite is investigated. The experimental results showed that the Young's modulus and Poisson's ratio of granite increased nonlinearly with increasing confining pressure. The nonlinear Hoek–Brown criterion can better reflect the peak strength properties of granite than the linear Mohr–Coulomb criterion. The peak axial strain of granite increased linearly with the confining pressure.

(2) A new GBM method is put forward to simulate the damage mechanical behaviour of granite under different confining pressures. The micro-parameters of the numerical model are calibrated to match the macro-properties of

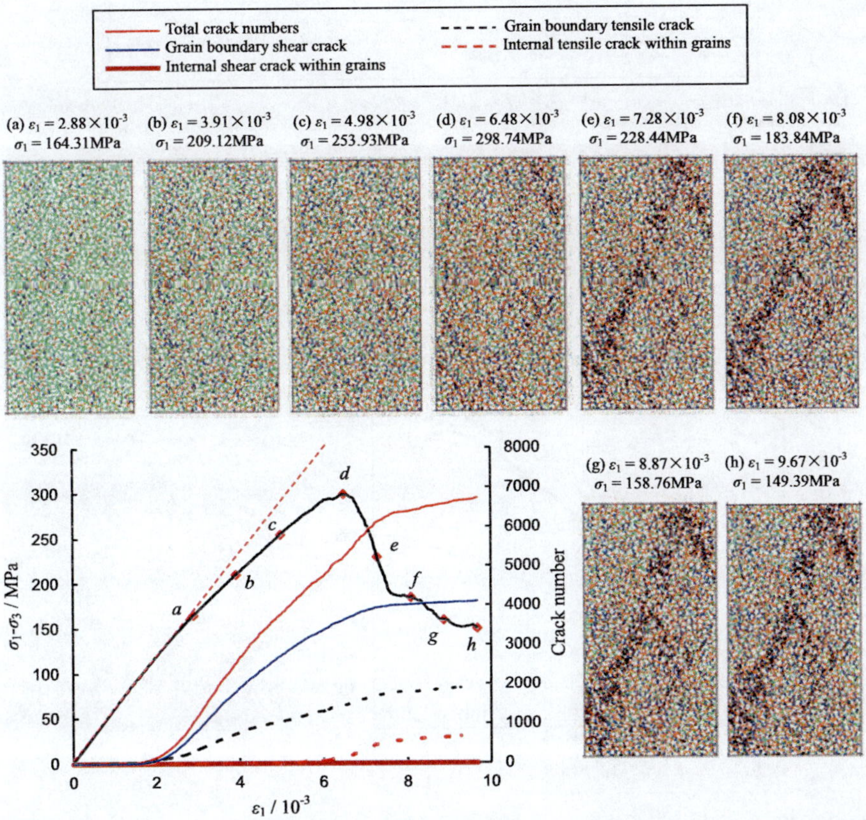

Fig. 16.21 Stress–Strain behaviour and micro-cracking processes of grain-based modelling under the confining pressure of 20 MPa. The letters in deviatoric stress–strain curve correspond to different crack evolution process

the granite tested in the laboratory. The numerical deviatoric stress–strain curve, peak strength, crack damage threshold and ultimate failure mode of granite under triaxial compression are all compared with the experimental results, which show that the numerical results are in good agreement with the experimental results.

(3) On the basis of numerical simulated results, under lower confining pressures, the grain boundary tensile cracks are more obvious, but less grain boundary shear cracks are found. Furthermore, with the increase in confining pressure, the shear belt from the increase in internal tensile crack within grains becomes more and more apparent. At the same time, because of the restraining action of confining pressure, the grain boundary tensile cracks decrease gradually whereas the grain boundary shear cracks increase step by step as the confining pressure increases. The total crack numbers, the grain boundary shear crack

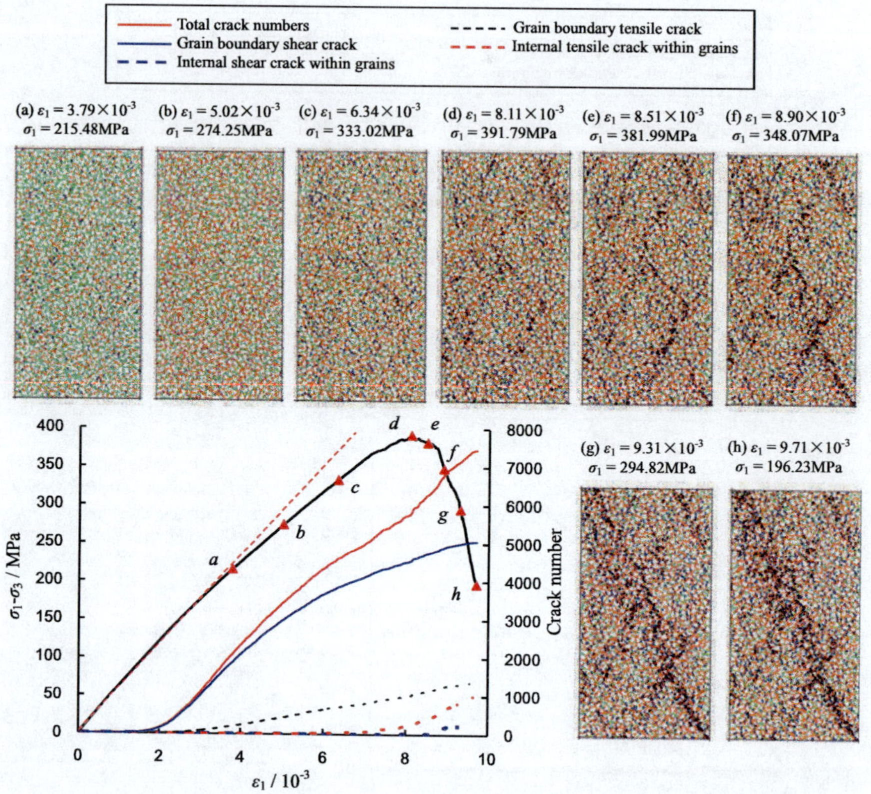

Fig. 16.22 Stress–strain behaviour and micro-cracking processes of grain-based modelling under the confining pressure of 40 MPa. The letters in deviatoric stress–strain curve correspond to different crack evolution process

and the internal tensile crack within grains all increases gradually and the grain boundary tensile crack decreases gradually with the increase in confining pressure.

(4) By analysing the cross-sections from X-ray micro CT, spatial AE evolution processes and the relationship between stress–strain curves and micro-crack numbers by GBM method, the damage mechanism of granite under triaxial compression is deeply revealed. A conceptual model of granite material under different confining pressures is put forward. With the increase in confining pressure, the direction of crack initiation can cause the deviation towards the direction of confining pressure. The relationship between stress–strain behaviour and micro-cracking processes of GBM under various confining pressure is constructed, which can distinguish the evolution of all kinds of cracks.

References

Amann F, Button EA, Evans KF, Gischig VS, Blumel M (2011) Experimental study of the brittle behavior of clay shale in rapid unconfined compression. Rock Mech Rock Eng 44(4):415–430

Bahrani N, Kaiser PK (2016) Numerical investigation of the influence of specimen size on the unconfined strength of defected rocks. Comput Geotech 77:56–67

Bahrani N, Kaiser PK, Valley B (2014) Distinct element method simulation of an analogue for a highly interlocked, non-persistently jointed rockmass. Int J Rock Mech Min Sci 71:117–130

Bell AF, Greenough J, Heap MJ, Main IG (2011a) Challenges for forecasting based on accelerating rates of earthquakes at volcanoes and laboratory analogues. Geophys J Int 185(2):718–723

Bell AF, Naylor M, Heap MJ, Main IG (2011b) Forecasting volcanic eruptions and other material failure phenomena: an evaluation of the failure forecast method. Geophys Res Lett. https://doi.org/10.1029/2011GL048155

Bi J, Zhou XP, Qian QH (2016) The 3D numerical simulation for the propagation process of multiple pre-existing flaws in rock-like materials subjected to biaxial compressive loads. Rock Mech Rock Eng 49(5):1611–1627

Brantut N, Schubnel A, Gueguen Y (2011) Damage and rupture dynamics at the brittle-ductile transition: the case of gypsum. J Geophys Res 116:B01404. https://doi.org/10.1029/2010JB007675

Cartwright-Taylor A, Fusseis F, Butler I et al (2017) Damage localisation and fracture propagation in granite: 4D synchrotron x-ray microtomographic observations from an in-situ triaxial deformation experiment at SOLEIL. EGU Gen Assembly Conf Abstr 19:17515

Chen L, Liu JF, Wang CP et al (2014) Characterization of damage evolution in granite under compressive stress condition and its effect on permeability. Int J Rock Mech Min Sci 71:340–349

Chen L, Wang CP, Liu JF et al (2015) Damage and plastic deformation modeling of beishan granite under compressive stress conditions. Rock Mech Rock Eng 48(4):1623–1633

Cho N, Martin CD, Sego DC (2007) A clumped particle model for rock. Int J Rock Mech Min Sci 44(7):997–1010

Fairhurst CE, Hudson JA (1999) Draft ISRM suggested method for the complete stress-strain curve for the intact rock in uniaxial compression. Int J Rock Mech Min Sci 36(3):279–289

Fan LF, Gao JW, Wu ZJ, Yang SQ, Ma GW (2018) An investigation of thermal effects on micro-properties of granite by X-ray CT technique. Appl Therm Eng 140:505–519

Feng XT, Chen SL, Zhou H (2004) Real-time computerized tomography (CT) experiments on sandstone damage evolution during triaxial compression with chemical corrosion. Int J Rock Mech Min Sci 41(2):181–192

Fortin J, Stanchits S, Dresen G, Gueguen Y (2006) Acoustic emission and velocities associated with the formation of compaction bands in sandstone. J Geophys Res. https://doi.org/10.1029/2005JB003854

Ge XR, Ren JX, Pu YB, Ma W, Zhu YL (2001) Real-in time CT test of the rock meso-damage propagation law. Science in China (series e) 44(3):328–336

Haimson B, Chang C (2000) A new true triaxial cell for testing mechanical properties of rock, and its use to determine rock strength and deformability of Westerly granite. Int J Rock Mech Min Sci 37(1–2):285–296

Hawkes I, Mellor M (1970) Uniaxial testing in rock mechanics laboratories. Eng Geol 4:177–285

Hirono T, Takahashi M, Nakashima S (2003) In situ visualization of fluid flow image within deformed rock by X-ray CT. Eng Geol 70:37–46

Hoek E, Brown ET (1980) Underground excavation in rock. Institution of Mining and Metallurgy, London

Hoek E, Brown ET (1997) Practical estimates of rock mass strength. Int J Rock Mech Min Sci 34(8):1165–1186

Hofmann H, Babadagli T, Yoon JS, Zang A, Zimmermann G (2015) A grain based modeling study of mineralogical factors affecting strength, elastic behavior and micro fracture development during compression tests in granites. Eng Fract Mech 147:261–275

Hofmann H, Babadagli T, Zimmermann G (2015) A grain based modeling study of fracture branching during compression tests in granites. Int J Rock Mech Min Sci 77:152–162
Hohl A, D Griffith A, Eppes MC et al (2018) Computationally enabled 4D visualizations facilitate the detection of rock fracture patterns from acoustic emissions. Rock Mech Rock Eng 1–14
Ikeda S, Nakano T, Nakashima Y (2000) Three-dimensional study on the interconnection and shape of crystals in a graphic granite by X-ray CT and image analysis. Mineral Mag 64(5):945–959
Ivars DM, Potyondy DO, Pierce M et al (2008) The smooth-joint contact model. In: World Congress on computational mechanics and, European congress on computational methods in applied sciences and engineering
Jiang M, Chen H, Crosta GB (2015) Numerical modeling of rock mechanical behavior and fracture propagation by a new bond contact model. Int J Rock Mech Min Sci 78:175–189
Kawakata H, Cho A, Yanagidani T et al (1997) The observations of faulting in Westerly granite under triaxial compression by X-ray CT scan. Int J Rock Mech Min Sci 34(3–4):151
Kawakata H, Cho A, Kiyama T, Yanagidani T, Kusunose K, Shimada M (1999) Three-dimensional observations of faulting process in Westerly granite under uniaxial and triaxial conditions by X-ray CT scan. Tectonophysics 313:293–305
Li K, Cheng Y, Fan X (2018) Roles of model size and particle size distribution on macro-mechanical properties of Lac du Bonnet granite using flat-joint model. Comput Geotech 103:43–60
Liang M, Kong C, Wang H, et al. Microcrack growth behavior and CWFS criterion parameters optimization of granite with PFC. Journal of Testing and Evaluation, 2018, 46(6)
Liu J, Chen L, Wang C et al (2014) Characterizing the mechanical tensile behavior of Beishan granite with different experimental methods. Int J Rock Mech Min Sci 69:50–58
Lockner D, Byerlee J, Kuksenko V, Ponomarev A, Sidorin A (1991) Quasi-static fault growth and shear fracture energy in granite. Nature 350:39–42
Meier T, Rybacki E, Backers T, Dresen G (2015) Influence of bedding angle on borehole stability: a laboratory investigation of transverse isotropic oil shale. Rock Mech Rock Eng 48(4):1535–1546
Mogi K (1966) Some precise measurements of fracture strength of rocks under uniform compressive stress. Rock Mech Eng Geol 4:41–55
Mohr O (1914) Abhandlungen aus dem Gebiete der Technischen Mechanik, 2nd edn. Ernst, Berlin
Nakashima S, Sakamoto T, Yasuhara H et al (2017) Observation and quantification of fracture aperture in granite core using X-Ray tomography and edge detection technique. In: 51st US Rock mechanics/geomechanics symposium. American rock mechanics association
Paterson MS, Wong TF (2005) Experimental rock deformation—the brittle field. Springer, New York
Peng J, Wong LNY, Teh CI (2017) Influence of grain size heterogeneity on strength and microcracking behavior of crystalline rocks. Journal of Geophysical Research: Solid Earth 122(2):1054–1073
Peng J, Wong LNY, Teh CI et al (2018) Modeling micro-cracking behavior of Bukit Timah granite using grain-based model. Rock Mech Rock Eng 51(1):135–154
Potyondy DO (2010) A grain-based model for rock: approaching the true microstructure. In: Proceedings of the rock mechanics in the Nordic Countries. Kongsberg, Norway
Read RS (2004) 20 years of excavation response studies at AECL' s underground research laboratory. Int J Rock Mech Min Sci 41(8):1251–1275
Shu B, Liang M, Zhang S, et al. Numerical modeling of the relationship between mechanical properties of granite and microparameters of the flat-joint model considering particle size distribution. Mathematical Geosciences, 2018: 1–18
Sufian A, Russell AR (2013) Microstructural pore changes and energy dissipation in Gosford sandstone during pre-failure loading using X-ray CT. Int J Rock Mech Min Sci 57:119–131
Tapponnier P, Brace W F. Development of stress-induced microcracks in Westerly granite. Int. J. Rock Mech. Min. Sci. Geomech. Abstr, 1976, 13(4): 103–112
Xu Z, Li T, Chen G, Ma C, Qiu S, Li Z (2018) The grain-based model numerical simulation of unconfined compressive strength experiment under thermal-mechanical coupling effect. KSCE J Civ Eng 2:1–12

Yang SQ (2018) Fracturing mechanism of compressed hollow-cylinder sandstone evaluated by X-ray micro-CT scanning. Rock Mech Rock Eng 51:2033–2053

Yang SQ, Huang YH (2017) An experimental study on deformation and failure mechanical behavior of granite containing a single fissure under different confining pressures. Environ Earth Sci 76(10):364

Yang SQ, Jing HW, Li YS, Han LJ (2011) Experimental investigation on mechanical behavior of coarse marble under six different loading paths. Exp Mech 51(3):315–334

Yang SQ, Jing HW, Wang SY (2012) Experimental investigation on the strength, deformability, failure behavior and acoustic emission locations of red sandstone under triaxial compression. Rock Mech Rock Eng 45:583–606

Yang SQ, Huang YH, Jing HW, Liu XR (2014) Discrete element modeling on fracture coalescence behavior of red sandstone containing two unparallel fissures under uniaxial compression. Eng Geol 178:28–48

Yang SQ, Ranjith PG, Huang YH, Yin PF, Jing HW, Gui YL, Yu QL (2015) Experimental investigation on mechanical damage characteristics of sandstone under triaxial cyclic loading. Geophys J Int 201:662–682

Yang SQ, Ranjith PG, Jing HW, Tian WL, Ju Y (2017) An experimental investigation on thermal damage and failure mechanical behavior of granite after exposure to different high temperature treatments. Geothermics 65:180–197

Yang SQ, Tian WL, Huang YH (2018) Failure mechanical behavior of pre-holed granite specimens after elevated temperature treatment by particle flow code. Geothermics 72:124–137

Yang SQ, Huang YH, Tian WL, Yin PF, Jing HW (2019) Effect of high temperature on deformation failure behavior of granite specimen containing a single fissure under uniaxial compression. Rock Mechanics and Rock Engineering, 52:2087–2107

Youn H, Tonon F (2010) Multi-stage triaxial test on brittle rock. Int J Rock Mech Min Sci 47(4):678–684

Zhao GF, Russell AR, Zhao XB, Khalili N (2014) Strain rate dependency of uniaxial tensile strength in gosford sandstone by the distinct lattice spring model with X-ray micro CT. Int J Solids Struct 51:1587–1600

Zhao XG, Cai M, Wang J et al (2015) Strength comparison between cylindrical and prism specimens of Beishan granite under uniaxial compression. Int J Rock Mech Min Sci 76:10–17

Zhao XG, Wang J, Chen F, Li PF, Ma LK, Xie JL, Liu YM (2016) Experimental investigations on the thermal conductivity characteristics of Beishan granitic rocks for China's HLW disposal. Tectonophysics 683:124–137

Zhou XP, Zhang YX, Ha QL (2008) Micromechanical modelling of the complete stress-strain relationship for crack weakened rock subjected to compressive loading. Rock Mech Rock Eng 41(5):747–769

Zhou XP, Chen H, Feng YF (2014) An experimental study of crack coalescence behaviour in rock-like materials containing multiple flaws under uniaxial compression. Rock Mech Rock Eng 47(6):1961–1986

Zhou XP, Zhang JZ, Wong LNY (2018) Experimental study on the growth, coalescence and wrapping behaviors of 3D cross-embedded flaws under uniaxial compression. Rock Mech Rock Eng 51(5):1379–1400

Chapter 17
Strength and Failure Behavior of Coal Specimens with Different Diameters Under Conventional Triaxial Compression

Mechanical behaviors of rock material are very important for geotechnical engineering design. However, for reason of size effect of rock (Hudson et al. 1972; Hoek and Brown 1980; Liu et al. 1998; Tang et al. 2000; Pan et al. 2002a, b; Yang et al. 2005a; Yoshinaka et al. 2008), the strength, deformation and failure properties of rock with constant size may not be applied directly as part of a constitutive relation of rock for geotechnical engineering design (Yang et al. 2005b). The standard core specimens recommended by ISRM (Fairhurst and Hudson 1999) can not always be extracted, especially from weak, stratified and fractured rock mass, therefore carrying out the experiments using the specimens with normal sizes becomes more difficult. Where rock cores divided into many discs due to the presence of the closely spaced weakness planes are available, the size of rock specimen may be too small and can not reach the standard size recommended by ISRM for laboratory experiments to get the strength and deformation parameters of rock. Therefore in order to construct the relations between mechanical parameters and size of rocks, many experiments, theoretical and numerical investigations (Hudson et al. 1972; Hoek and Brown 1980; Liu et al. 1998; Tang et al. 2000; Pan et al. 2002a, b; Yang et al. 2005a, b; Yoshinaka et al. 2008) have been conducted on scale effect for various rock types in recent decades.

Because of the importance of size effect of rock material, many researchers have tried to construct relations between mechanical parameters and size of rocks, and especially the relations between the strength of in-situ rock mass, the strength of specimens tested in laboratory and the size of rocks. More typical studies are described in Hudson et al. (1972) for the influence of size on the strength and deformation of rock. Based on many laboratory experimental results of rock specimens with different sizes for various rock types, Hoek and Brown (1980) obtained the effect of diameter of rock specimen on the uniaxial compressive strength (UCS), which is shown in Fig. 17.1. The equation is best fitted on the experimental data and the parameter ξ equals to 0.18. Notice, although the equation in Fig. 17.1 is constucted on basis of a lot of experimental results of rock specimens with different diameters, this equation is not applicable to all various rock types. Where, the parameter ξ has

S.-Q. Yang, *Mechanical Behavior and Damage Fracture Mechanism of Deep Rocks*,
https://doi.org/10.1007/978-981-16-7739-7_17

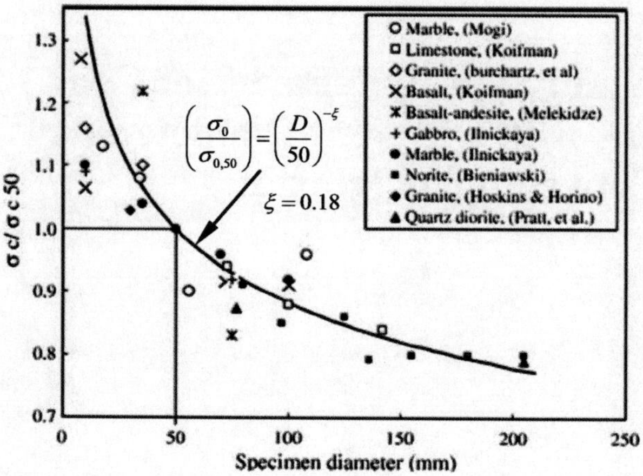

Fig. 17.1 The influence of the diameter on uniaxial compressive strength (UCS) of rock specimen (Hoek and Brown 1980)

a close relation with rock type. The effect of rock type on the parameter ξ is made a detailed discussion by Yoshinaka et al (2008). The results show that the obtained values for ξ range from 0.1 to 0.8, but in smaller specimens the range is 0.15–0.3. Besides, Guo (1996) developed a relation between the strength and size by carrying on an experimental study with uniaxial compression.

Some theoretical investigations has also been carried out on the effect of size on the mechanical behaviors of rock material, which are described as follows. Liu et al. (1998) constructed an exponential equation describing the decrease of rock mass strength with size by analyzing the uniaxial compressive strengths of seven rocks. Zhou et al. (2001) proposed a new strength criterion of considering the heterogeneity of rock material and localized deformation, which was used to account for the size effect of rock. Based on the experimental results of sandstone specimens of the same diameter but of different lengths, Pan et al. (2002a, b) studied the size effect on rock strain softening by using the gradient plastic theory. Yang et al. (2005a, b) investigated the relation between mechanical parameters of marble material and size based on the experimental results of marble specimens with different length-to-diameter ratio. And then based on the strain strength theory and the random statistical distribution hypothesis of rock strength, a statistical constitutive model for rock damage under uniaxial compression, which considers the nonlinear relation of unit volume failure or elastic modulus and the size, was established by adopting the damage constitutive theory of continuous medium.

In recently several years, with the rapid development of information technology, a lot of numerical simulation methods are widely applied to rock mechanics. The most typical study is that CRISR of Northeastern University develops a Rock Failure Process Analysis code (RFPA^{2D}) by considering the heterogeneity of rock material, which can reproduce some conventional phenomena of rock mechanics in the laboratory (Tang et al. 1998, 2000). Tang et al. (2000) carried out numerical studies for the constraint, slenderness and size effect of rock specimens in uniaxial compression by RFPA^{2D}. Besides, an elasto-plastic cellular automaton (EPCA) with the associated code was developed to simulate the non-linear fracturing process of rocks under uniaxial compression (Feng et al. 2006). Using the EPCA code, the effect of specimen sizes and height/width ratios on mechanical behaviors were investigated (Feng et al. 2006), which showed that the numerical simulation reproduced some of the well-known phenomena by previous researchers in uniaxial compression tests.

However, the above investigations only limit in discussing the effect of size on the strength of rock specimen under uniaxial compression, and do not analyze size effect of strength of rock specimen under triaxial compression. In situ rock mass is usually located in three-dimensional stress state. In order to investigate the following several problems:

- What kinds of effect of the size on mechanical behavior of specimen under triaxial compression?
- Whether there are the same strength criterion for specimen with different sizes or not?
- How to take place the variance on strength parameters (cohesion and internal friction angle, etc.) of specimen with the increase of size?

Therefore, we must carry out triaxial compression experiments for rock specimen with different sizes. But unfortunately up to now, fewer investigations have been performed on the scale effect of rock under triaxial compression in previous studies. Yang and Xu (2004) performed numerical simulation tests on rock materials with different sizes under different confining pressures by use of RFPA^{2D}. The numerical results showed that the dependence of the strength of rock materials on their sizes became less evident as the confining pressure increased, i.e. the confining pressure could reduce the sensitivity of the size of rock materials to their strength. Coal is a kind of special rock, which has distinct scale effect. Mehurst and Brown (1998) investigated the effect of the diameter on the mechanical behavior of coal specimen under different confining pressures, and discussed the strength behavior of coal specimen with different diameters by adopting generalized Hoek–Brown strength criterion. However, the above studies did not investigate the method to confirm the cohesion and internal friction angle of rock specimen with different sizes.

When carrying out the numerical analysis for rock engineering, the strength parameters (i.e. cohesion and internal friction angle) of rock material often needs to be brought forward in many numerical soft wares (such as FLAC^{3D}, etc.). Up to now, the mechanical behavior of coal specimen with constant size has been fully investigated (Bieniawski 1968; George 1997; Xu et al. 2006; Nathan et al. 2008). But the investigation results of coal specimen with constant size are difficult and also impossible to be applied directly to engineering practice. In order to bring forward the reference for the choose of the cohesion and internal friction angle of coal specimen in mining engineering, we will investigate the strength and failure behaviors of coal specimens with different diameters in this chapter. The emphasis of this research is focused on to investigate the influence of the diameter on strength and failure behavior of coal material, to put forward a new evaluation criterion of confirming the cohesion and internal friction angle of coal material, to analyze the effect of the diameter on the cohesion and internal friction angle of coal specimen.

17.1 Experimental Results of Coal Specimens with Different Diameters

Using the servo-controlled testing machine, Mehurst and Brown (1998) carried out many triaxial compression tests on 61, 101, 146 and 300 mm diameter specimens to investigate the effects of diameter on the mechanical behavior of coal. The range of confining pressure was expected 0~10 MPa to cover a large proportion of the confinement conditions typically found in coal pillar loading (Mehurst and Brown 1998). Generally, the experiments were undertaken in accordance with ISRM suggested method for determining the strength of coal specimens with different diameters under conventional triaxial compression. In all tests, the strain rate of between 1 and 2% axial strain per hour were used to load axially for the coal specimens with different diameters. However, unfortunately, only a group of experimental results could be used to analyze the effect of the diameter on the strength and deformation behaviors of coal specimen, as shown in Table 17.1. In the Table, D is denoted as the diameter of coal specimen, mm; σ_3 and σ_S were the confining pressure and triaxial compressive strength (TCS), respectively; the elastic modulus E_{av} is defined as the slope of approximation linear part in the axial stress-axial strain curve; υ is the Poisson's ratio of coal specimen.

Coal is a kind of special rock and there are plenty of original fissures and weak faces distributed randomly in the coal specimen, which result in distinct heterogeneity. From Table 17.1, it can be concluded that the strength and deformation behaviors of coal specimens are found depending not only the confining pressure, but also on the diameter. Even though under the condition with the same diameter and confining pressure, the mechanical behaviors among different coal specimens have larger difference due to distinct heterogeneity of coal specimen. Take the coal specimen with the diameter of 146 mm as a example, under the same confining pressure of 0.2 MPa, the strengths of four coal specimens are 15.17, 19.37, 17.18 and 20.17 MPa, respectively, with a mean 17.97 MPa and a higher dispersion (i.e. the percentage of the D-value between the maximum and minimum value to the mean value) ~27.8%.

In accordance with the strength data listed in Table 17.1, we can see that the strength of coal specimen is the dual function of both the confining pressure and the diameter. Under conventional triaxial compression, the elastic modulus of coal specimen with the same diameter but different confining pressures appears not to be dependent distinctly to the triaxial compressive strength (Fig. 17.2), which can be concluded from listed data of the coal specimens with the diameters of 61 and 101 mm in Fig. 17.2. However, overall, the elastic modulus of coal specimens with different diameters has a positive linear relation with the triaxial compressive strength (i.e.

Table 17.1 Experimental results for Moura mid-brightness coal specimens with different diameters (Data from Medhurst and Brown 1998)

Coal specimen no	D / mm	σ_3 / MPa	σ_S / MPa	E_{av} / GPa	Poisson's ratio (υ)
PS-01	61	0.2	28.92	4.57	0.25
PS-02	61	0.2	33.80	5.16	N/A
PS-03	61	0.2	40.00	3.80	0.22
PS-04	61	2.0	52.03	5.07	0.25
PS-05	61	5.0	78.69	N/A	N/A
PS-06	61	10.0	90.10	4.88	N/A
PS-07	101	0.2	26.06	3.18	0.22
PS-08	101	0.2	27.38	3.09	0.21
PS-09	101	2.0	43.16	2.97	N/A
PS-10	101	5.0	51.56	3.17	N/A
PS-11	146	0.2	15.17	1.98	0.21
PS-12	146	0.2	19.37	2.02	0.19
PS-13	146	0.2	17.18	N/A	N/A
PS-14	146	0.2	20.17	3.06	0.24
PS-15	146	1.0	25.66	2.26	0.17
PS-16	146	2.0	30.90	2.20	0.24
PS-17	146	3.0	40.28	3.02	0.21
PS-18	146	4.0	42.23	2.61	0.25
PS-19	300	0.2	12.60	2.11	0.25
PS-20	300	0.2	8.81	1.73	0.29
PS-21	300	0.4	13.33	2.40	0.26
PS-22	300	0.8	20.52	2.48	0.24

$E_{av} = 0.038\sigma_s + 1.864, R = 0.663$), which is a better agreement with the findings by Arslan et al. (2007) and Basu et al. (2008).

Figure 17.3 shows the relation between the elastic modulus of coal specimen and the diameter, in which the denoted number is the confining stress. The test data are best fitted with the equation of $E_{av}/E_{av,61} = (D/61)^{-0.60}$, where $E_{av,61}$ is the mean of elastic modulus of all the coal specimens with the diameter of 61 mm, ~4.70GPa. From Fig. 17.3, it can be seen that the elastic modulus of coal specimen decreases approximately as a negative power function with the increasing diameter, but after D increases to a certain critical value, the E_{av}-D curve has a tendency to converge. Furthermore, the confining pressure has a smaller effect on the E_{av}-D curve. The above relation can be explained as follows. The random fissures and weak surfaces in the coal specimen with the same diameter have taken place the closure under the action of hydrostatic pressure, which heightens the homogeneity of coal material, therefore the elastic modulus of coal specimen with the same diameter is not related distinctly to the confining pressure. However, there are more fissures and

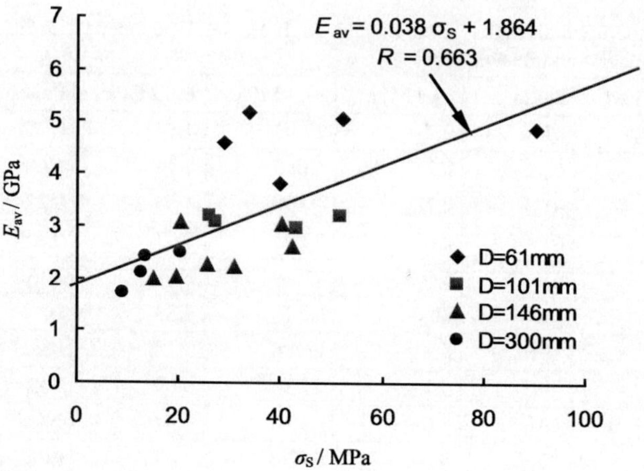

Fig. 17.2 Relation between the elastic modulus and triaxial compressive strength of coal specimen with different diameters

Fig. 17.3 Relation between the elastic modulus and diameter of coal specimen under different confining pressures

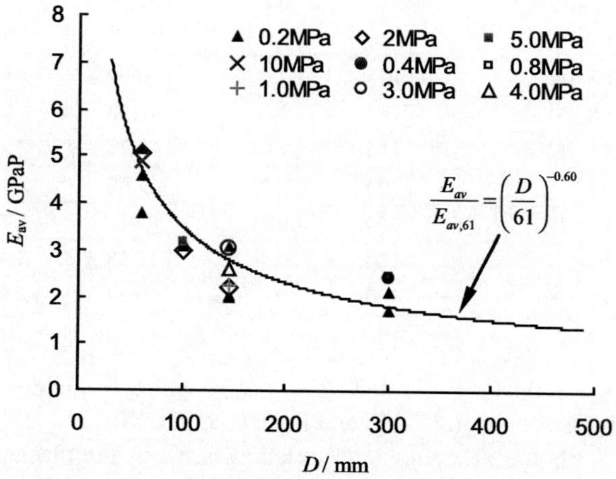

weak surfaces in the coal specimens with the bigger diameters, which results in the distinct heterogeneity. So the elastic modulus of coal specimen decreases with the increase of the diameter.

From the Poisson's ratio of coal specimens with different diameters listed in Table 17.1, it can be seen that Poisson's ratio of coal specimen with the same diameter is not dependent to the confining pressure, and the range of Poisson's ratio of coal specimen is between 0.21 and 0.25, with a mean ~ 0.23. Thus we can believe that there is almost no scale effect on Poisson's ratio of coal specimen, which is a better agreement with the conclusions for intact rock specimen by Yoshinaka et al. 2008.

17.2 Numerical Model and Micro-parameters for Coal Specimen

17.2.1 Numerical Model for Coal Specimen

In this section, the numerical model and micro-parameters for coal specimen will be investigated. A rock failure process analysis code (RFPA^{2D}), developed by Northeastern University, China (Tang 1997), is chosen to simulate the mechanical behaviors of coal specimens with different diameters. Although RFPA^{2D} is only two-dimensional, it can be used to simulate many problems in mining and geotechnical engineering, such as laboratory uniaxial and triaxial compression experiments on brittle rock material (Tang et al. 2000; Chen et al. 2006), coal and gas outbursts in underground collieries (Xu et al. 2006), failure process in brittle rocks (Zhu and Tang 2004), crack coalescence and propagation in the rock specimen containing pre-existing flaws (Tang et al. 2001; Wong et al. 2002), etc. In many published papers (Tang et al. 1998; Zhu and Tang 2004; Wong et al. 2002; Xu et al. 2006), the principle of RFPA^{2D} code can be found. Therefore in this research, we will not describe the principle of the RFPA^{2D} code, and only the numerical model and micro-parameters for coal specimen are described as follows.

Typical numerical model for coal specimen by RFPA^{2D} is shown in Fig. 17.4, which is discredited into a large number of meso-level elements. In order to model the heterogeneity of coal material, the micro-parameters, including the elastic modulus and compressive strength, are assumed to conform separate Weibull's distribution, $\theta_1(\omega, E_0)$ or $\theta_2(\omega, \sigma_0)$, where E_0 and σ_0 are the mean elastic modulus and the mean

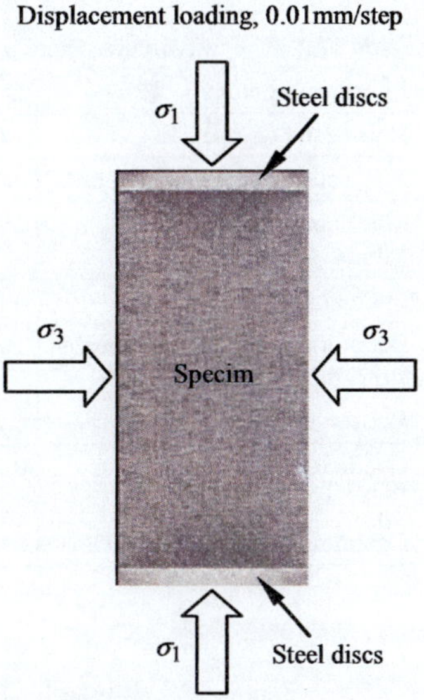

Fig. 17.4 Numerical model for coal specimen (by RFPA^{2D})

compressive strength of all the meso-level elements, respectively; ω is the homogeneity index. According to the definition (Zhu and Tang 2004), a bigger ω implies a more homogeneous material and vice versa. In the modeling, the Mohr–Coulomb criterion with a tensile cut-off (Zhu and Tang 2004; Xu et al. 2006) is chosen as the failure criterion for meso-level elements: The numerical simulation for coal specimens with different diameters are carried out on 2-D plane strain models. The upper surface of numerical coal specimens are imposed the axial stress progressively at the 0.01 mm/step displacement control until the failure took place, and the lower boundary of numerical specimens being kept fixed, as is generally done during laboratory tests. In order to better simulate the experimental results, two steel discs are also taken into account in the numerical simulation, which are completely the same as the laboratory tests.

17.2.2 Micro-parameters for Coal Specimen

In accordance with the experimental results of coal specimens with different diameters listed in the above section, micro-parameters for coal specimens are confirmed in Table 17.2. In Table 17.2, the E_0, υ and Φ values are drawn directly from the experimental results of coal specimen with the diameter of 61 mm. Furthermore, υ and Φ is assumed as a constant during numerical simulation. Other micro-parameters of coal specimen are obtained after some simulations. It needs to be noticed that the homogeneity index ω tends to decrease with the increase of the diameter, which results from more fissures and weak surfaces in the coal specimens with the bigger diameters and heightens the heterogeneity of coal material.

Table 17.2 The micro-parameters of RFPA^{2D} model for coal specimens with different diameters

Micro-parameters	D = 61 mm	D = 101 mm	D = 146 mm	D = 300 mm
Homogeneity index, ω	3.0	2.2	1.8	1.5
Mean elastic modulus, E_0, GPa	4.70	4.70	4.70	4.70
Mean compressive strength, σ_0, MPa	150	150	150	150
Internal friction angle, Φ°	47	47	47	47
Ratio of compressive to tensile strength, C/T	10	10	10	10
Poisson's ratio, υ	0.23	0.23	0.23	0.23
Coefficient of stress influence, β	0.1	0.1	0.1	0.1
Confining pressure, σ_3, MPa	0~10 MPa	0~10 MPa	0~10 MPa	0~10 MPa

17.3 Failure Behaviors of Coal Specimens with Different Diameters

In order to validate the numerical model and micro- parameters of coal specimen proposed in above section, the comparison of axial stress-axial strain curve between experimental result (Medhurst and Brown 1998) and numerical simulation for coal specimen with the diameter of 300 mm is shown in Fig. 17.5. Unfortunately, we can not compare other numerically simulated axial stress-axial strain curves with the experimental results of coal specimens for $D = 61$, 101 and 146 mm because Medhurst and Brown (1998) only gave the experimental stress- strain curves of coal specimen with the diameter of 300 mm under the confining pressure of 0.2~0.8 MPa. From Fig. 17.5, it can be seen that the numerically simulated curve under the confining pressure of 0.8 MPa agrees well with the experimental curve, except for the initial phase of the curve. In Fig. 17.5, the experimental stress–strain curve of coal specimen at low stress levels shows the downward concave and the initial non-linear deformation, which mainly results from the closure of some primary cracks, pores and voids in the coal specimen. Overall, the numerical simulation method can replicate better the complete process of deformation and failure of coal specimen, especially the localization of deformation and failure (Xu et al. 2006).

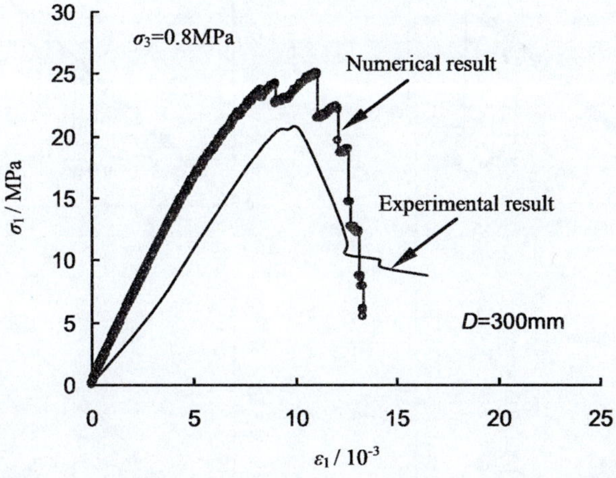

Fig. 17.5 Comparison of stress–strain curve between experimental result and numerical result for coal specimen ($D = 300$ mm)

17.3.1 Numerical Stress–Strain Curves of Coal Specimen with Different Diameters

Numerically simulated axial stress-axial strain curves for coal specimens with all different diameters (D = 61, 101, 146 and 300 mm) are shown in Fig. 17.6, in which the denoted number is the confining stress, MPa. In accordance with Fig. 17.6, we can conclude that the yielding stress (the stress value that begins to depart the linear elastic phase in the axial stress-axial strain curve) and TCS (triaxial compressive strength) of coal specimens under the condition of the same diameter increase gradually with the confining pressure. Moreover for the coal specimens with the same diameter, the elastic modulus increases with the confining pressure. The above simulated phenomena agree well with many laboratory experiment results on rock specimen (Yang et al. 2005c; Yang et al. 2008). Besides, we can believe that coal specimen has a scale effect even though under higher confining pressure of 10 MPa. Under the same confining pressure, the bigger the diameter of coal specimen is, the more plastic and ductile deformation the coal specimen takes on nearby the peak strength.

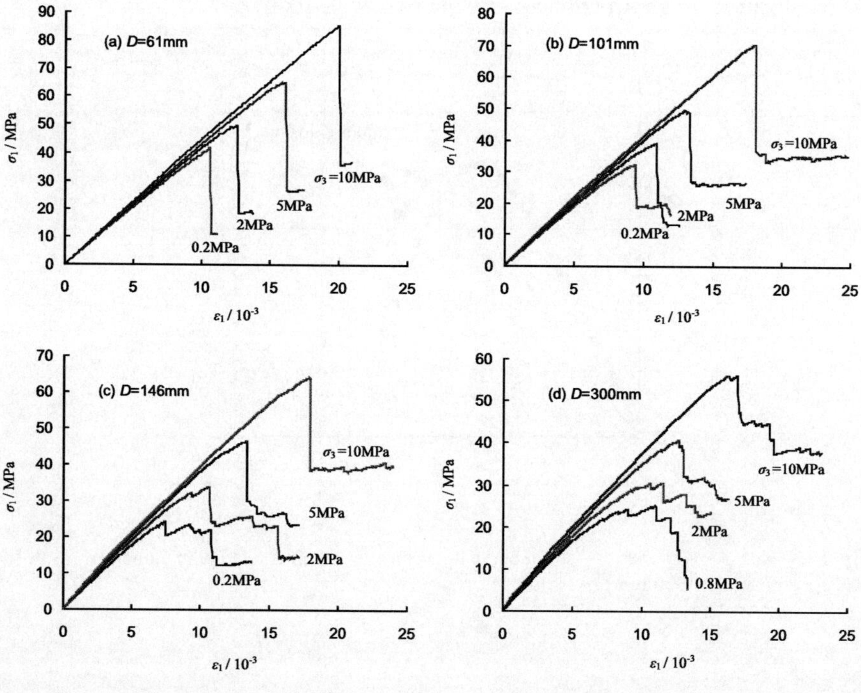

Fig. 17.6 Numerical stress–strain curves of coal specimens with different diameters under different confining pressures (by RFPA^{2D})

As it is well-known to us, coal is a kind of typical brittle material, which can also be seen out from the post failure region, as shown in Fig. 17.6. The simulated axial stress-axial strain behavior can be divided into four typical stages, i.e. elastic deformation, crack initiation and stable crack growth, crack damage and unstable crack growth, failure and post-peak behavior (Martin and Chandler 1994; Eberhard et al. 1999; Alkan et al. 2007), which will be made a detailed analysis in accordance with microscopic failure process shown in Fig. 17.7.

Figure 17.7 shows simulated microscopic failure process of coal specimen with the diameter of 146 mm during triaxial compression loading. From the simulated axial stress-axial strain curve shown in Fig. 17.7, we can see that a substantial nonlinearity exists in the pre-peak range of simulated curve, and there is also an obvious post-peak stage. Notice that all the symbols (a-i) denoted in the simulated axial stress-axial strain curve are corresponding with the macroscopic failure process. The locations of shear stress and acoustic emission (AE) events are also shown in Fig. 17.7a–i, in which the gray scale represents the relative shear stress and the diameter of the circle represents the magnitude of the released energy. The stress abrupt coincides with the propagation of cracks in the coal specimen.

From Fig. 17.7, it is also very clear that four typical stages, i.e. elastic deformation (before *a* point), crack initiation and stable crack growth (*ac*), crack damage and unstable crack growth (*ce*), failure and post-peak behavior (after *e* point), can be observed. Before *a* point, the elastic deformation dominates the stress–strain curve of coal specimen. The linear stress–strain behavior of the specimen does not happen to change although there are some failed elements in this stage (Fig. 17.7a). With the increase of axial deformation, the specimen begins to enter the stage of crack initiation and stable crack growth. At this stage, the stress–strain curve departs from the elastic behavior and comes out nonlinear deformation. Failed elements in the specimen develop towards stress concentration from random distribution, and the number of failed elements increases gradually, which can be seen from Fig. 17.7b, c. The above characteristics results in the increase of mesoscopic heterogeneity and therefore reduces the macroscopic elastic modulus of specimen. Besides, with the increase of axial stress, on one hand, the interaction among failed elements in the specimen begins to intensify; on the other hand, due to the difference of transmission rate of each failed element on the force and the deformation of each failed element, the stress field in the specimen takes place non-uniform distribution, which causes the stress concentration region on the location of specimen.

After *c* point, the specimen begins to come into being the stage of crack damage and unstable crack growth with the increase of axial deformation. From *c* point to *d* point, the stress occurs to a minor abrupt (32.49 MPa → 31.22 MPa), which results in rapid and sudden increase of damaged element number in the specimen. The nonlinear deformation of specimen will increase continually. Up to *e* point, the specimen comes out the peak strength. For the moment, the specimen has come into being a large stress concentration region, as shown in the elliptical part in Fig. 17.7. After peak strength *e* point, the macroscopic crack comes out rapidly (Fig. 17.7f). With the increase of axial deformation, the post-peak supporting capacity reduces and softens gradually, and the macroscopic crack widens and lengths in the specimen.

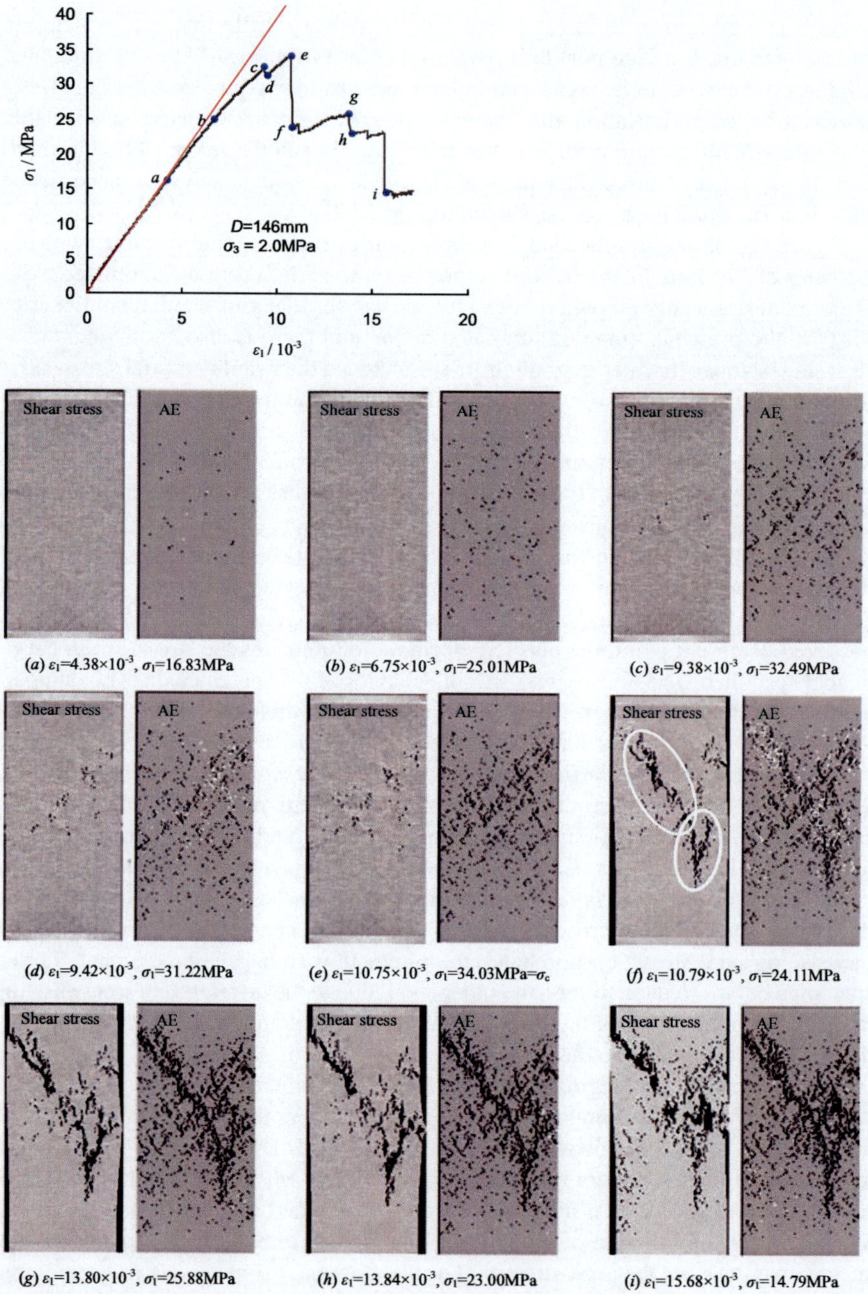

Fig. 17.7 Microscopic failure process of coal specimen with the diameter of 146 mm under confining pressure of 2.0 MPa (Numerical results simulated by RFPA^{2D})

17.3.2 *Failure Behavior of Coal Specimens Under Different Confining Pressures*

Figure 17.8 shows a typical comparison between experimental and simulated failure modes of coal specimens under different confining pressures, in which the experimental failure modes were obtained by Medhurst and Brown (1998). From Fig. 17.8, we can see that the simulated failure modes of coal specimens are in good agreement with the experimental failure modes. Under lower confining pressures (less than 1 MPa), the coal specimens with different diameters takes on a axial splitting failure mode; however under higher confining pressures (greater than 1 MPa), the coal specimens with different diameters all lead to a shearing failure mode with a single fracture surface, which results from the friction of macroscopic crack surfaces.

The effect of confining pressure on the failure mode of coal specimen with the diameter of 146 mm is shown in Fig. 17.9. From Fig. 17.9, it is very clear that the coal specimen shows a axial splitting failure mechanism at a confining pressure of less than 1.0 MPa ($\sigma_3 = 0.2$ MPa) and a shearing failure mechanism at a confining pressure of

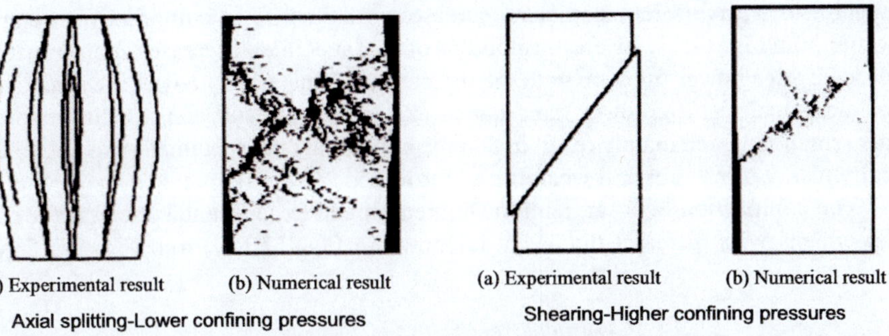

Fig. 17.8 Comparison between experimental and simulated failure modes of coal specimens (experimental failure modes of coal specimens were cited from Medhurst and Brown 1998)

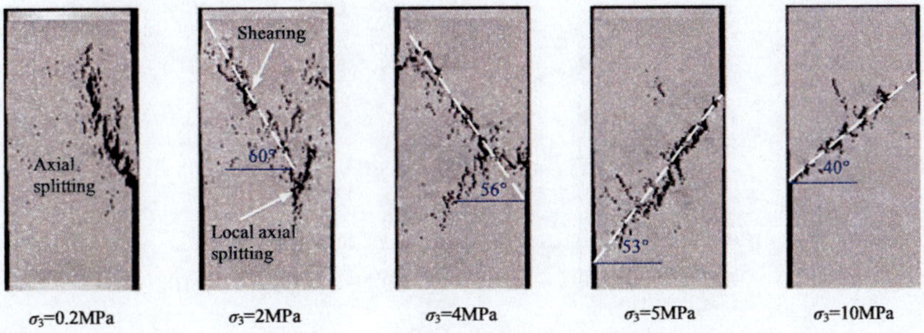

Fig. 17.9 Typical failure modes of coal specimens ($D = 146$ mm) under different confining pressure (numerical results)

greater than 1.0 MPa, but the specimen at $\sigma_3 = 2.0$ MPa also comes out a local axial splitting failure. Moreover the angle between the macroscopic failure plane and the minimum principal stress direction (i.e. σ_3 direction) gradually decreases with the increase confining pressure, which is in a good agreement with theoretical prediction. Under different confining pressures, the failure modes of coal specimens with other diameters ($D = 61$, 101 and 300 mm) are similar to those shown in Fig. 17.9.

17.3.3 Comparison Between Numerical and Experiment Results of Coal Specimens

Figure 17.10 gives the comparison of numerical and experimental results on the relation between the elastic modulus and triaxial compressive strength, and Fig. 17.11 gives the comparison of numerical and experimental results on the relation between the elastic modulus and the diameter. In accordance with Figs. 17.10 and 17.11, we can conclude that numerical simulation results agree qualitatively well with experimental results. From Fig. 17.10, it can be seen that the elastic modulus of coal specimens with different diameters increases with the triaxial compressive strength, while from Fig. 17.11, the elastic modulus of coal specimen decreases approximately as a negative power function with the increasing diameter. Moreover, the sensitivity of numerical result is lower than that of experimental result and the dispersion is also smaller, which mainly result in that the numerical coal specimens can not reflect fully plenty of fissures and weak faces distributed randomly in the coal specimen.

The comparison between numerical strength and experimental strength for coal specimens with different diameters is shown in Fig. 17.12. From Fig. 17.12, we

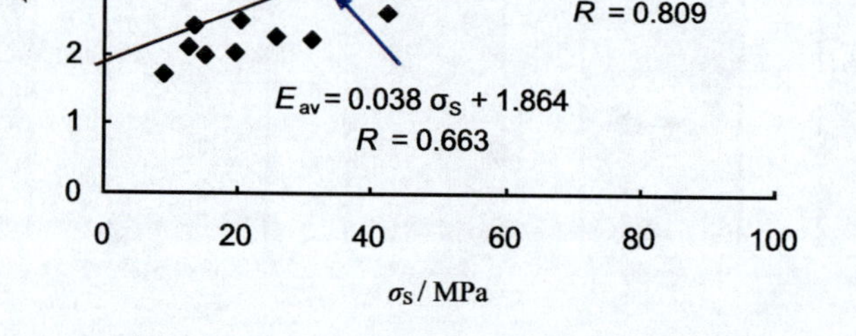

Fig. 17.10 Numerical and experimental results on the relation between the elastic modulus and triaxial compressive strength

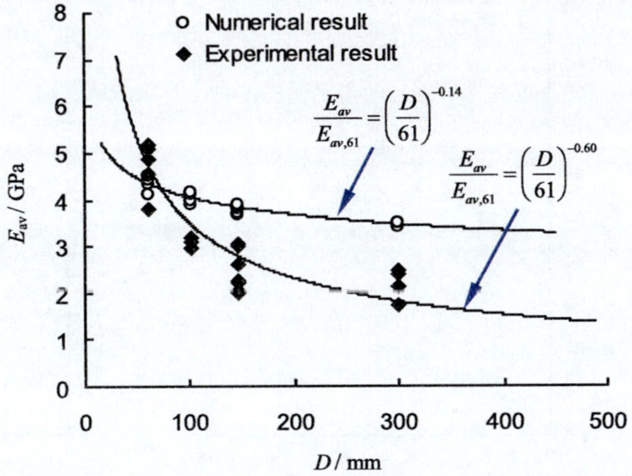

Fig. 17.11 Numerical and experimental results on the relation between the elastic modulus and the diameter

can see that there is a good agreement on the strength of coal specimens between numerical and experimental result, which shows that the micro-parameters used to simulate the mechanical behaviors of coal specimens by RFPA^{2D} are more reasonable and feasible. In next section, we will investigate the strength behaviors of coal specimens with different diameters in detail.

17.4 Strength Behaviors of Coal Specimens with Different Diameters

The generalized Hoek–Brown criterion was derived from a lot of test data. Up to now, the generalized Hoek–Brown criterion has been applied widely to rock mechanics and engineering practice. The basic equation describing the peak triaxial compressive strength of a wide range of isotropic rock materials can be written as Hoek and Brown (1980, 1997).

$$\sigma_S = \sigma_3 + \sigma_c \left(m \frac{\sigma_3}{\sigma_c} + s \right)^a \tag{17.1}$$

where σ_c is UCS of intact rock material, m, s and a are all material parameters, which vary with rock type. The parameter m reflects the soft and hard extent. While the parameter m is larger, the rock is harder. However, the parameter s reflects the fractured extent ranging from 0 to 1. When the parameter s is closer to zero, the rock is more broken.

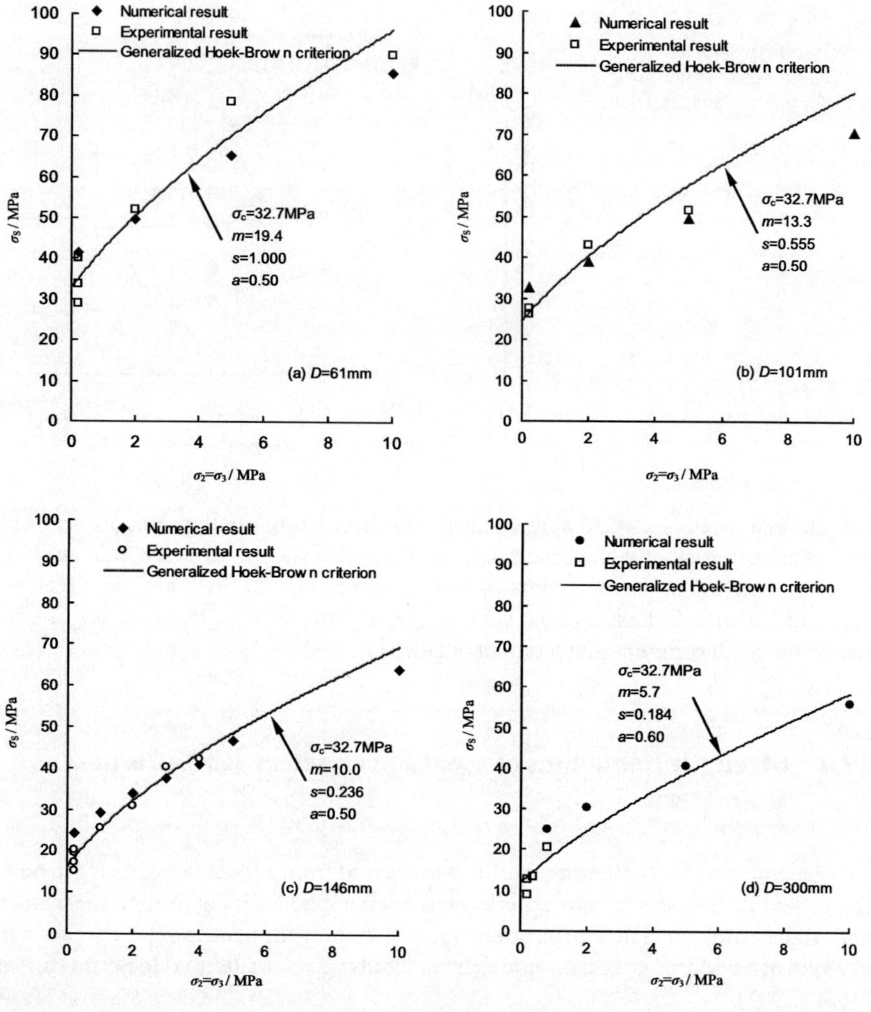

Fig. 17.12 Experimental results, numerical results and generalized Hoek–Brown criterion for coal specimens with different diameters (Generalized Hoek–Brown criterion parameters were cited from Medhurst and Brown 1998)

17.4.1 Strength Parameters of Coal Specimens with Different Diameters

Medhurst and Brown (1998) had analyzed the relation between the compressive strength of coal specimens with different diameters and the confining pressure in accordance with generalized Hoek–Brown criterion, which obtained the generalized Hoek–Brown criterion parameters, as shown in Fig. 17.12. Notice that σ_c is UCS of

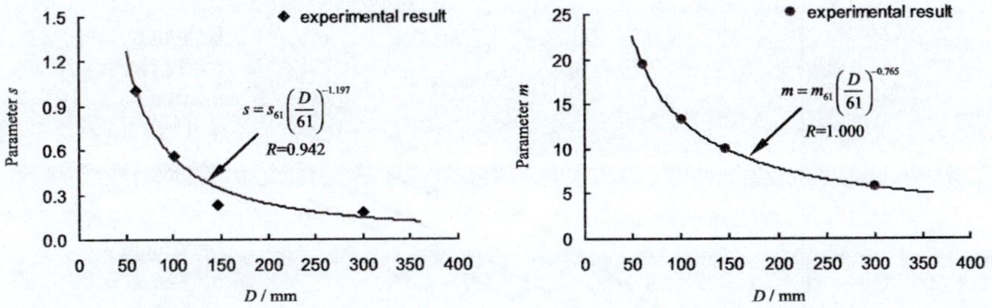

Fig. 17.13 Relation between of generalized Hoek–Brown criterion parameters of coal specimen and the diameter

coal specimen with the diameter of 61 mm, i.e. $\sigma_c = 32.7$ MPa. From Fig. 17.12, it is very clear that the strength of coal specimens with different diameters is in good agreement with nonlinear generalized Hoek–Brown criterion.

Based on generalized Hoek–Brown criterion parameters listed in Fig. 17.12, we can see that there exists a certain function relation between *m*, *s* and the diameter of coal specimen. By taking into account this kind of relation, the strength of coal specimens with different diameters and different confining pressures can be predicted approximately in the range of $0 \leq \sigma_3 \leq 10$ MPa and 61 mm $\leq D \leq$ 300 mm. Figure 17.13 shows the relation between generalized Hoek–Brown criterion parameters of coal specimen and the diameter. In Fig. 17.13, the experimental results on the parameters *m* and *s* were fitted best by Eqs. (17.2) and (17.3), respectively.

$$m = m_{61}\left(\frac{D}{61}\right)^{-n} \tag{17.2}$$

$$s = s_{61}\left(\frac{D}{61}\right)^{-b} \tag{17.3}$$

where, m_{61} and s_{61} are generalized Hoek–Brown criterion parameters of coal specimen with the diameter of 61 mm, and $m_{61} = 19.4$, $s_{61} = 1.000$. In Fig. 17.13, the parameters *n* and *b* are 0.765 and 1.197, respectively. And thus by Eqs. (17.2) and (17.3), we obtain the generalized Hoek–Brown criterion parameters of coal specimens with other diameters in the range of $0 \leq \sigma_3 \leq 10$ MPa and 61 mm $\leq D \leq$ 300 mm. The predicted relations between the strength of coal specimen with other diameters ($D = 80$, 125 and 225 mm) and confining pressure are shown in Fig. 17.14.

In most of the geotechnical software, the strength parameters (C and ϕ) of rock often need to be provided when carrying out the numerical analysis. However, for a kind of rock that has been evaluated as a generalized Hoek–Brown material, the determination of the values of C and ϕ is a difficult problem (Hoek and Brown 1997). There is no direct correlation between linear Mohr–Coulomb criterion and nonlinear generalized Hoek–Brown criterion, but the tangent method theory, which was put

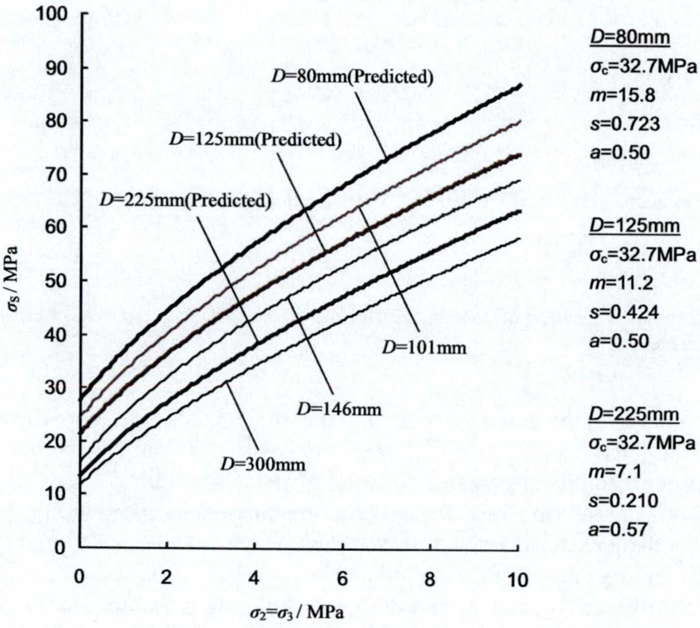

Fig. 17.14 Predicted relations between the strength of coal specimen with different diameters and confining pressure

forward by Hoek and Brown (1990, 1997), can be used to confirm approximately the equivalent C and ϕ of rock.

For a given confining pressure σ_3, the tangent line of Eq. (17.1) is actually linear Mohr–Coulomb criterion $\sigma_S = M + N\sigma_3$, where M and N can be expressed in the following forms, respectively.

$$M = \frac{2C\cos\phi}{1-\sin\phi} \tag{17.4}$$

$$N = \frac{1+\sin\phi}{1-\sin\phi} \tag{17.5}$$

If Eq. (17.1) is differentiated with respect to σ_3, the following equation can be obtained.

$$\frac{\partial\sigma_S}{\partial\sigma_3} = 1 + am\left(m\frac{\sigma_3}{\sigma_c} + s\right)^{a-1} \tag{17.6}$$

Thus in accordance with Eq. (17.5), Eq. (17.7) can be obtained.

$$N = \tan^2(45 + \frac{\phi}{2}) = \frac{\partial \sigma_S}{\partial \sigma_3} = 1 + am\left(m\frac{\sigma_3}{\sigma_c} + s\right)^{a-1} \tag{17.7}$$

According to Eq. (17.7), the internal friction angle can be expressed in generalized Hoek–Brown criterion parameters *m*, *s* and *a*, i.e.

$$\phi = 2\arctan\sqrt{\left[1 + am\left(m\frac{\sigma_3}{\sigma_c} + s\right)^{a-1}\right]} - 90^\circ \tag{17.8}$$

If Eq. (17.1) is substituted into linear Mohr–Coulomb criterion $\sigma_S = M + N\sigma_3$, Eq. (17.9) can be obtained.

$$M = \frac{2C\cos\phi}{1 - \sin\phi} = (1 - N)\sigma_3 + \sigma_c\left(m\frac{\sigma_3}{\sigma_c} + s\right)^a \tag{17.9}$$

Therefore in accordance with Eqs. (17.7) and (17.9), the cohesion can be expressed in the following form.

$$C = \frac{[s\sigma_c + (1 - a)m\sigma_3]\left(m\frac{\sigma_3}{\sigma_c} + s\right)^{a-1}}{2\sqrt{1 + am\left(m\frac{\sigma_3}{\sigma_c} + s\right)^{a-1}}} \tag{17.10}$$

According to Eqs. (17.8) and (17.10), *C* and ϕ for a kind of rock that has been evaluated as a generalized Hoek–Brown material can be confirmed. It can be seen that the values of *C* and ϕ are very sensitive to the range of values of σ_3 used to generate the simulated full-scale triaxial test results. On basis of trial and error method, Hoek and Brown (1997) found that the most consistent results can be approximately obtained in the range $0 < \sigma_3 < 0.25\sigma_c$. Therefore, Eqs. (17.8) and (17.10) can be transferred into the following forms by denoting $\sigma_3 = k\sigma_c$ ($k = 0 \sim 0.25$).

$$\phi = 2\arctan\sqrt{[1 + am(mk + s)^{a-1}]} - 90^\circ \tag{17.11}$$

$$C = \frac{[s + (1 - a)mk]\sigma_c(mk + s)^{a-1}}{2\sqrt{1 + am(mk + s)^{a-1}}} \tag{17.12}$$

Equations (17.11) and (17.12) shows the effect law of *m*, *s* and *a* values on the relation between *C*, ϕ and confining pressure $\sigma_3 = k\sigma_c$. From Eqs. (17.11) and (17.12), we can conclude that the *C* and ϕ of rock are actually the function of σ_3, in which, the *C* values increase with increasing σ_3, but the ϕ values decrease with increasing σ_3. Tables 17.3 and 17.4 lists the cohesion and internal friction angle ($k = 0.06 \sim 0.22$) of coal specimens with different diameters, respectively, which are calculated based on the tangent method theory by Eqs. (17.11) and (17.12). In

Table 17.3 The cohesion (MPa) of coal specimens with different diameters in accordance with generalized Hoek–Brown criterion parameters by adopting the tangent method theory

$k = \sigma_3/\sigma_c$	D = 61 mm	D = 80 mm	D = 101 mm	D = 125 mm	D = 146 mm	D = 225 mm	D = 300 mm
0.06	6.38	5.68	5.17	4.71	3.77	3.21	2.91
0.08	6.84	6.12	5.60	5.12	4.20	3.53	3.17
0.10	7.28	6.54	6.01	5.52	4.62	3.83	3.43
0.12	7.72	6.96	6.41	5.90	5.02	4.13	3.68
0.14	8.14	7.37	6.80	6.28	5.40	4.42	3.93
0.16	8.56	7.77	7.18	6.64	5.77	4.71	4.17
0.18	8.97	8.16	7.56	7.00	6.13	4.98	4.41
0.20	9.37	8.54	7.92	7.35	6.48	5.26	4.64
0.22	9.77	8.91	8.28	7.69	6.83	5.52	4.87

Table 17.4 The internal friction angle (°) of coal specimens with different diameters in accordance with generalized Hoek–Brown criterion parameters by adopting the tangent method theory

$k = \sigma_3/\sigma_c$	D = 61 mm	D = 80 mm	D = 101 mm	D = 125 mm	D = 146 mm	D = 225 mm	D = 300 mm
0.06	50.11	48.89	47.80	46.71	47.07	45.30	43.52
0.08	48.78	47.48	46.32	45.16	45.29	43.84	42.18
0.10	47.63	46.26	45.05	43.84	43.80	42.62	41.06
0.12	46.61	45.18	43.94	42.69	42.53	41.57	40.10
0.14	45.69	44.23	42.95	41.67	41.42	40.64	39.25
0.16	44.86	43.36	42.06	40.76	40.43	39.82	38.49
0.18	44.10	42.57	41.26	39.93	39.54	39.08	37.81
0.20	43.40	41.85	40.52	39.18	38.74	38.41	37.19
0.22	42.75	41.18	39.83	38.48	38.00	37.79	36.62

the process of obtaining the cohesion and internal friction angle of coal specimens with different diameters, the generalized Hoek–Brown criterion parameters *m*, *s* and *a* values as shown in Figs. 17.12 and 17.14 were adopted. In accordance with Tables 17.3 and 17.4, the cohesion and internal friction angle of coal specimen are found depending not only on the confining pressure, but also on the diameter. The effect of the diameter on the cohesion and internal friction angle of coal specimen will be discussed in the following sections.

17.4.2 A New Evaluation Criterion Based on Optimal Approximation Polynomial Theory

Based on the above tangent method theory, many cohesions and internal friction angles for a kind of rock that has been evaluated as a generalized Hoek–Brown material can be obtained by different σ_3 values. Therefore, when adopting the tangent method theory to confirm C and ϕ values, the most critical step is the selection of the range of σ_3 values, but up to now, there are no theoretically correct methods for choosing the range (Hoek and Brown 1997). Moreover, the tangent method theory (See Fig. 17.15) is actually to approximate generalized Hoek–Brown criterion by many polygonal lines. Even though the approximate order is very high, it is impossible to reduce the whole approximation error by tangent method theory. Therefore, here, a new evaluation criterion for a kind of rock that has been evaluated as a generalized Hoek–Brown material is put forward by adopting optimal approximation polynomial theory in functional analysis, which can be used to confirm more precisely the strength parameters (cohesion and internal friction angle) of coal specimens, and to avoid effectively the shortcoming of good approximation in a certain testing point but bad approximation on the whole.

Optimal approximation polynomial theory is described as follows (Rivlin 1974, 1987). Let $C_{[a,\,b]}$ be the space of real-valued continuous functions on the real interval $[a, b]$. We denote by H_n ($H_n \subset C_{[a,\,b]}$) the space of algebraic polynomials of degree at most n, i.e., $H_n = \text{span}\,\{1, x, \ldots, x^n\}$. 1, x, ..., x^n are a group of function set without linear correction. Element $P_n(x)$ on H_n space can be denoted as follows.

$P_n(x) = a_0 + a_1 x + \ldots + a_n x_n$, $a_0, a_1, \ldots, a_n$ are arbitrary real number.

We set

$$\|f\|_\infty = \max_{x\in[a,b]} |f(x)|, \quad \forall f \in C_{[a,b]} \tag{17.13}$$

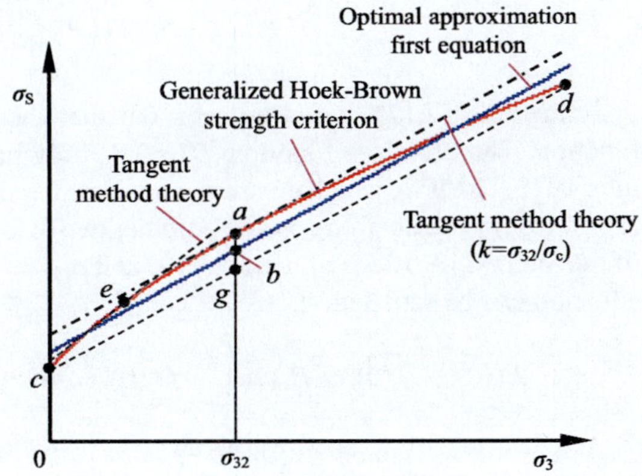

Fig. 17.15 The tangent method theory and optimal approximation first equation based on generalized Hoek–Brown criterion

$P_n^*(x) \in H_n$ is called a best approximation polynomial to f on the real interval [a, b], which makes $P_n^*(x)$ satisfies the following equation.

$$\Delta(f, P_n) = \max_{a \le x \le b} \left| f(x) - P_n^*(x) \right| = \min_{p_n \in H_n} \| f(x) - P_n(x) \| \tag{17.14}$$

In Eq. (17.14), such an $P_n^*(x)$ always exists, and is, indeed, unique. When $n =$ 1, $P_1(x) = a_0 + a_1 x$ is called as optimal approximation first equation. We can now state the famous Chebyshev Theorem (Rivlin 1974).

$P_n^*(x) \in H_n$ is a best approximation polynomial to $f \in C_{[a,b]}$ if, and only if, there exist $n + 2$ distinct points,$a \le x_1 < x_2 < \cdots < x_{n+2} \le b$, for which

$$P(x_k) - f(x_k) = (-1)^k \lambda \| P(x) - f(x) \|_\infty, \lambda = \pm 1 \quad k = 1, 2, 3, \ldots, n+2 \tag{17.15}$$

The main themes of the Chebyshevian approach are existence, uniqueness and characterization of best approximations.

Based on the optimal approximation polynomial theory, a new evaluation criterion for a kind of rock that has been evaluated as a generalized Hoek–Brown material is obtained to confirm the cohesion and internal friction angle. That is to say, on the basis of generalized Hoek–Brown criterion (Eq. (17.1)), we need to calculate the optimal approximation first equation P1 $(\sigma_3) = M + N\sigma_3$ on the real interval [0, 10 MPa]. Here, M and N are given in Eqs. (17.4) and (17.5), respectively.

In accordance with Chebyshev Theorem (Eq. (17.15)), there exists 3 distinct points at least, $0 \le \sigma_{31} < \sigma_{32} < \sigma_{33} \le 30MPa$ when $n = 1$, for which

$$P_1(\sigma_{3k}) - f(\sigma_{3k}) = (-1)^k \lambda \| P(\sigma_3) - f(\sigma_3) \|_\infty, \quad \lambda = \pm 1, k = 1, 2, 3 \tag{17.16}$$

According to Eq. (17.1), i.e. $\sigma_S = \sigma_3 + \sigma_c (m\frac{\sigma_3}{\sigma_c} + s)^a$, we can get

$$f'(\sigma_3) = 1 + am\left(m\frac{\sigma_3}{\sigma_c} + s \right)^{a-1} \tag{17.17}$$

From Eq. (17.17), it can be seen out that the function $f'(\sigma_3)$ is a monotonic function. Therefore the function $f'(\sigma_3)$-N only has single zero point on the open interval (0, 10 MPa), which this zero point is denoted as σ_{32}. Thus $P_1(\sigma_{32}) - f'(\sigma_{32}) = N - f'(\sigma_{32}) = 0$, i.e. $f'(\sigma_{32}) = N$. Another two distinct points must be at the endpoint of the interval [0, 10 MPa], i.e. $\sigma_{31} = 0$ and $\sigma_{33} = 10$ MPa. Moreover the following equation can be satisfied.

$$P_1(0) - f(0) = P_1(10) - f(10) = -[P_1(\sigma_{32}) - f(\sigma_{32})] \tag{17.18}$$

According to Eq. (17.18), we can obtain.

$$\begin{cases} M + N \times 0 - f(0) = M + N \times 10 - f(10) \\ M + N \times 0 - f(0) = f(\sigma_{32}) - (M + N\sigma_{32}) \end{cases} \tag{17.19}$$

Solving Eq. (17.19), Eq. (17.20) can be obtained.

$$\begin{cases} N = \frac{f(10) - f(0)}{10} = f'(\sigma_{32}) \\ M = \frac{f(0) + f(\sigma_{32})}{2} - \frac{f(10) - f(0)}{10} \frac{\sigma_{32}}{2} \end{cases} \tag{17.20}$$

Thus we get the optimal approximation first equation P1(σ_3) in accordance with Eq. (17.20), and then by Eqs. (17.4) and (17.5), the C and ϕ values of rock can be confirmed. The special geometrical significance is shown in Fig. 17.15, in which the new evaluation criterion line is parallel to the straight-line *cd*. The maximum deviation equals to the length of line segment *ab*.

According to the generalized Hoek–Brown strength criterion parameters listed in Figs. 17.12 and 17.14, the cohesion and internal friction angle of coal specimens with different diameters are confirmed in Table 17.5 based on above optimal approximation polynomial theory. From Table 17.5, we can see that the cohesion and internal friction angle of coal specimen with the diameter of 61 mm given by optimal approximation first equation are 7.09 MPa and 46.68°, respectively, which have a slight difference with that by best fit Mohr–Coulomb envelope for triaxial test data (i.e. $C = 7.33$ MPa and $\phi = 45.82°$). However for the coal specimens with bigger diameters such as 300 mm, due to little triaxial test data, the cohesion and internal friction angle confirmed directly by linear Mohr–Coulomb strength criterion can not reflect reasonable the actual strength behavior of coal specimen. Therefore the cohesion and internal friction angle confirmed by optimal approximation first equation are more precise, which are listed in Table 17.5.

In accordance with Table 17.5, it can be seen that k is approximate ~0.12. Therefore, the cohesion and internal friction angle obtained by tangent method theory ($k = 0.12$) in Tables 17.3 and 17.4 are used to analyze the influence of the diameter on the strength parameter, which is shown in Fig. 17.16. At the same time, the cohesion and internal friction angle obtained by optimal approximation first equation in Table

Table 17.5 The cohesion and internal friction angle of coal specimens with different diameters in accordance with generalized Hoek–Brown criterion parameters by adopting optimal approximation polynomial theory

D / mm	Optimal approximation first equation	σ_{32} / MPa	$k = \sigma_{32} / \sigma_c$	C/MPa	ϕ/ (°)
61	$35.70 + 6.34\sigma_3$	3.88	0.119	7.09	46.68
80	$30.70 + 5.93\sigma_3$	3.83	0.117	6.30	45.35
101	$27.15 + 5.59\sigma_3$	3.79	0.116	5.74	44.15
125	$23.98 + 5.29\sigma_3$	3.75	0.115	5.21	43.00
146	$19.02 + 5.35\sigma_3$	3.56	0.109	4.11	43.24
225	$15.76 + 5.02\sigma_3$	3.71	0.114	3.52	41.90
300	$13.75 + 4.66\sigma_3$	3.78	0.115	3.18	40.29

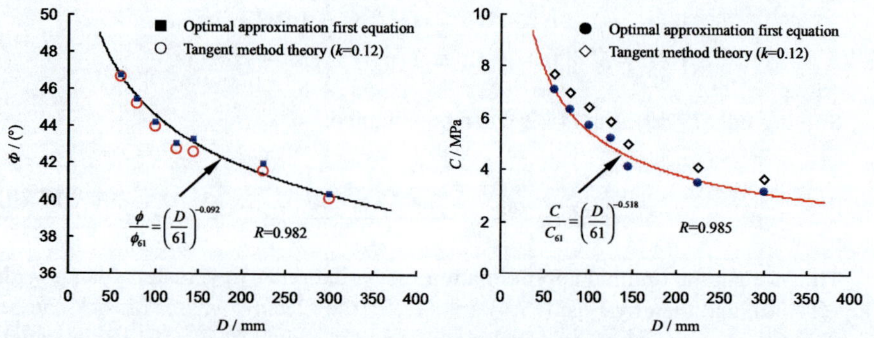

Fig. 17.16 The influence of the diameter on the cohesion and internal friction angle of coal specimen

17.5 are also plotted in Fig. 17.16. From Fig. 17.16, we can see that the diameter has a key effect on the cohesion and internal friction angle of coal specimen. With the increasing diameter, the cohesion and internal friction angle of coal specimen decrease step by step and the variance tends to slow down.

The cohesion and internal friction angle of coal specimen obtained by optimal approximation first equation can be fitted by the following equation:

$$\frac{\lambda}{\lambda_0} = \left(\frac{D}{D_0}\right)^{-\alpha} \tag{17.21}$$

where λ represents the cohesion or internal friction angle of coal specimens with different diameters; λ_0 represents the cohesion or internal friction angle of coal specimen with normal diameter D_0 ($D_0 = 61$ mm in this research). α is a parameter, which is related to rock character, experimental condition (such as loading rate, etc.). From Fig. 17.16, α equals to 0.092 for the internal friction angle of coal specimen; while α is 0.518 for the cohesion, which shows that the sensitivity of the diameter on the cohesion of coal specimen is higher than that on the internal friction angle. The internal friction angle confirmed by optimal approximation polynomial theory (first equation) has no great difference with that confirmed by tangent method theory ($k = 0.12$), however the cohesion obtained by optimal approximation polynomial theory (first equation) is lower than that obtained by tangent method theory ($k = 0.12$). Therefore the cohesion obtained by tangent method theory ($k = 0.12$) is a little greater than actual cohesion of coal specimen, which should be careful of applying the cohesion to the engineering practice.

Medhurst and Brown (1998) estimated the generalized Hoek–Brown criterion parameters of in situ coal seam, i.e. $m = 2.6$, $s = 0.052$ and $a = 0.65$. Based on the above parameters, the cohesion and internal friction angle of in situ coal seam can be obtained by adopting tangent method theory ($k = 0.12$) and optimal approximation polynomial theory (first equation), respectively, which are shown in Fig. 17.17. At

Fig. 17.17 The cohesion and internal friction angle of in situ coal seam obtained by several kinds of methods

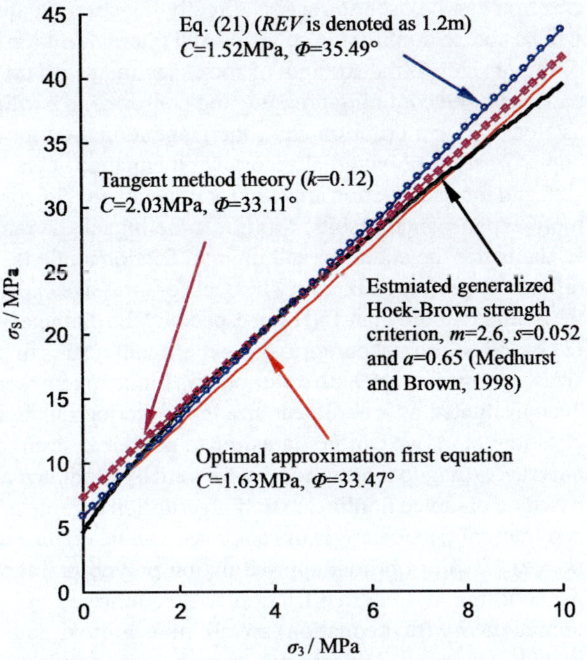

the same time, the cohesion and internal friction angle confirmed by Eq. (17.21) are also given in Fig. 17.17, and the represented element volume (*REV*) of in situ coal seam is assumed to D = 1.2 m. In accordance with obtained strength parameters of in situ coal seam by several kinds of methods in Fig. 17.17, it can be seen that the $C = 2.03$ MPa confirmed by tangent method theory ($k = 0.12$) is higher ~ 19.7% than $C = 1.63$ MPa confirmed by optimal approximation first equation, and ~ 25.1% than $C = 1.52$ MPa confirmed by Eq. (17.21). However for internal friction angle, there are no distinct difference between tangent method theory ($k = 0.12$) and optimal approximation polynomial theory (first equation), which are approximately 33°. The internal friction angle obtained by Eq. (17.21) is 35.49°, which is only higher ~5.7% than the value obtained by optimal approximation first equation.

The above analysis brings forward two kinds of methods to confirm the strength parameters (i.e. the cohesion and internal friction angle) of in situ rock mass in accordance with experimental results of laboratory rock block. (1) Triaxial compression experiment is carried out for the rock specimen with different size ~ S (S may be the diameter of cylindrical specimen or the length of side of cubic specimen). Based on the experimental results, the triaxial strength of rock specimen with different sizes can be obtained. Thus, we can construct the relation between the cohesion, internal friction angle and size of rock specimen, which can be used to predict approximately the cohesion and internal friction angle of in situ rock mass (*REV*). But the following contents have to be noticed. If the strength of rock specimen with the same size increases linearly with the confining pressure, linear Mohr- Coulomb criterion can be used to regress directly the strength and confining pressure, which will be able to confirm the cohesion and internal friction angle of rock specimen with different sizes. If the strength of rock specimen with the same size increases nonlinearly with the confining pressure and conforms to nonlinear strength criterion (such as Hoek–Brown criterion etc.), the tangent method theory ($k = \sigma_{32}/\sigma_c$) or optimal approximation polynomial theory (first equation) can be used to confirm the cohesion and internal friction angle of rock specimen. However this method needs to exist more experimental results. Moreover the bigger the range of size of rock specimen is, the better the cohesion and internal friction angle is. While it has still been more difficult to perform laboratory triaxial compressive experiment for the specimen with the diameter more than 150 mm, especially the diameter ~ 300 mm (Zhao et al. 2008). (2) Based on triaxial compressive experiment results of rock specimen with different sizes, appropriate strength criterion was found to represent. For a kind of rock that has been evaluated as a nonlinear strength criterion (such as generalized Hoek–Brown criterion, etc.), we can firstly estimate nonlinear strength parameters of in situ rock mass according to the method put forward by Medhurst and Brown (1998). And then, based on obtained nonlinear strength criterion parameters of rock mass, the cohesion and internal friction angle of rock mass can be confirmed by tangent method theory ($k = \sigma_{32}/\sigma_c$) or optimal approximation polynomial theory (first equation). Notice, the cohesion and internal friction angle confirmed by optimal approximation polynomial theory (first equation) are all more approximate to the strength parameter of in situ rock mass. The internal friction angle confirmed by tangent method theory ($k = \sigma_{32}/\sigma_c$) is more approximate to the engineering practice of in situ rock mass, while the cohesion confirmed by tangent method theory ($k = \sigma_{32}/\sigma_c$) will not be able to apply directly to the engineering practice until reduced cohesion with a certain extent.

17.5 Conclusions

In mining engineering, the mechanical behavior of coal is very important and necessary for understanding the stability and safety of pillar and in situ coal seam. In this paper, on basis of the experimental result, theoretical analysis and numerical simulation of coal specimens with different diameters, the influence of the diameter on strength and failure behaviors of coal specimens under conventional triaxial compression were investigated. The following conclusions in this chapter can be drawn.

(1) The mechanical behavior of coal with a given size is only a kind of macroscopical manifestation, which can not be applied directly to the pillar design. Coal is a kind of distinctly heterogeneous material, there are plenty of fissures and weak faces distributed randomly, which results in great difference of strength and failure behaviors of coal specimens with different diameters. In accordance with the experimental data, the micro-parameters of material were confirmed

to simulate the strength and failure behaviors of coal specimens with different diameters by RFPA^{2D}. Numerical results reappear better the experimental phenomena of coal specimens with different diameters.

(2) Based on the numerical and experimental results, the strength of coal specimen is found depending not only on the confining pressure, but also on the diameter. With the increase of the diameter, the elastic modulus of coal specimen decreases approximately as a negative power function, but after the diameter increases to a certain critical value, the elastic modulus has a tendency to converge. However for the coal specimen with the same diameter, the elastic modulus has no a clear relation with the confining pressure, which results from distinct heterogeneity of coal material. The elastic modulus of coal specimen with different diameter has a positive linear relation with the triaxial compressive strength, i.e. $E_{av} = A\sigma_s + B$.

(3) The strength of coal specimens with different diameters shows distinct nonlinear behavior, which agrees well with generalized Hoek–Brown criterion. In order to confirm more precisely the cohesion and internal friction angle, a new evaluation criterion for a kind of rock that has been evaluated as a generalized Hoek–Brown material is put forward by adopting optimal approximation polynomial theory, which can avoid the shortage of tangent method theory that has good approximation in a certain testing point but bad approximation on the whole.

(4) The diameter has a great effect on the cohesion and internal friction angle of coal specimen, which is made a detailed discussion. With the increase of the diameter, the cohesion and internal friction angle all decrease gradually, but after the diameter increases to a certain critical value; the cohesion and internal friction angle have a tendency to converge. The sensitivity of the cohesion of coal specimen on the diameter is higher than that of the internal friction angle on the diameter. The internal friction angle obtained by tangent method theory ($k = \sigma_{32} / \sigma_c$) is basically equal to that obtained by optimal approximation polynomial theory (optimal approximation first equation), but the cohesion obtained by tangent method theory ($k = \sigma_{32} / \sigma_c$) is higher than that obtained by optimal approximation first equation. In the end, two kinds of methods are put forward to confirm the cohesion and internal friction angle of in situ rock mass in accordance with the experimental result of laboratory rock block.

References

Alkan H, Cinar Y, Pusch G (2007) Rock salt dilatancy boundary from combined acoustic emission and triaxial compression tests. Int J Rock Mech Min Sci 44:108–119

Arslan AT, Koca MY, Aydogmus T, Klapperich H, Yilmaz HR (2007) Correction of unconfined compressive strength with Young's modulus and Poisson's ratio in gypsum from Sivas (Turkey). Rock Mech Rock Eng https://doi.org/10.1007/s00603-007-0145-8

Basu A, Celestino TB, Bortolucci AA (2008) Evaluation of rock mechanical behaviors under uniaxial compression with reference to assessed weathering grades. Rock Mech Rock Eng https://doi.org/10.1007/s00603-008-0170-2

Bieniawski ZT (1968) In situ strength and deformation characteristics of coal. Eng Geol 2(5):325–340

Chen ZH, Tham LG, Yeung MR, Xie H (2006) Confinement effects for damage and failure of brittle rocks. Int J Rock Mech Min Sci 43:1262–1269

Eberhard E, Stead D, Stimpson B (1999) Quantifying progressive pre-peak brittle fracture damage in rock during uniaxial compression. Int J Rock Mech Min Sci 36:361–380

Fairhurst CE, Hudson JA (1999) Draft ISRM suggested method for the complete stress-strain curve for the intact rock in uniaxial compression. Int J Rock Mech Min Sci 36:279–289

Feng XT, Pan PZ, Zhou H (2006) Simulation of the rock microfracturing process under uniaxial compression using an elasto-plastic cellular automaton. Int J Rock Mech Min Sci 43:1091–1108

George JDSt (1997) Structural effects on the strength of new zealand coal. Int J Rock Mech Min Sci. 34(3/4), paper No.299

Guo Z (1996) Application mechanics of rock mass. Earthquake Press, Beijing, pp 21–25

Hoek E, Brown ET (1980) Underground excavation in rock. Institution of Mining and Metallurgy, London

Hoek E (1990) Estimating Mohr-Coulomb friction and cohesion values from the Hoek-Brown failure criterion. Int J Rock Mech Min Sci Geomech Abstr 12:227–229

Hoek E, Brown ET (1997) Practical estimates of rock mass strength. Int J Rock Mech Min Sci 34:1165–1186

Hudson JA, Crouch SL, Fairhurst C (1972) Soft, stiff and servo-controlled testing machines. Eng Geol 6(3):155–189

Liu BC, Zhang JS, Du QZ, Tu JF (1998) A study of size effect for compression strength of rock. Chin J Rock Mech Eng 17(6):611–614

Martin D, Chandler NA (1994) The progressive fracture of Lac du Bonnet Granite. Int J Rock Mech Min Sci Geomech Abstr 31:643–659

Medhurst TP, Brown ET (1998) A study of the mechanical behaviour of coal for pillar design. Int J Rock Mech Min Sci 35:1087–1105

Nathan D, Gentzis T, Chalaturnyk RJ (2008) Unconventional geomechanical testing on coal for coalbed reservoir well design: the alberta foothills and plains. Int J Coal Geology 75:15–26

Pan YS, Wang XB, Li ZH (2002) Analysis of the strain softening size effect for rock specimens based on shear strain gradient plasticity theory. Int J Rock Mech Min Sci 39:801–805

Pan YS, Wei JM (2002) Experimental and theoretical study on size effect on strain softening of rock materials. Chin J Rock Mech Eng 21(2):215–218

Rivlin TJ (1974) The Chebyshev Polynomials. John Wiley, New York

Rivlin TJ (1987) A view of approximation theory. IBM J Res Dev 31:162–168

Tang CA (1997) Numerical simulation of progressive rock failure and associated Seismicity. Int J Rock Mech Min Sci 34:249–261

Tang CA, Kou SQ (1998) Crack propagation and coalescence in brittle materials under compression. Eng Fract Mech 61:311–324

Tang CA, Liu H, Lee PKK, Tsui Y, Tham LG (2000) Numerical studies of the influence of microstructure on rock failure in uniaxial compression part II: constraint, slenderness and size effect. Int J Rock Mech Min Sci 37(4):571–583

Tang CA, Lin P, Wong RHC, Chau KT (2001) Analysis of crack coalescence in rock-like materials containing three flaws-part II: numerical approach. Int J Rock Mech Min Sci 38:925–939

Xu T, Tang CA, Yang TH, Zhu WC, Liu J (2006) Numerical investigation of coal and gas outbursts in underground collieries. Int J Rock Mech Min Sci 43:905–919

Wong RHC, Tang CA, Chau KT, Lin P (2002) Splitting failure in brittle rocks containing pre-existing flaws under uniaxial compression. Eng Fract Mech 69:1853–1871

Yang SQ, Xu WY (2004) Numerical simulation of strength-size effect of rock materials under different confining pressures. J Hohai Univ (Nat Sci Ed) 32(5):578–582 (in Chinese)

Yang SQ, Su CD, Xu WY (2005) Experimental and theoretical study of size effect of rock material. Eng Mech 22(4):112–118 (in Chinese)

Yang SQ, Xu WY, Su CD (2005) Study on statistical damage constitutive model of rock considering scale effect. Chin J Rock Mech Eng 24(24):4484–4490 (in Chinese)

Yang SQ, Su CD, Xu WY (2005) Experimental investigation on strength and deformation properties of marble under conventional triaxial compression. Rock Soil Mech 26(3):475–478 (in Chinese)

Yang SQ, Jiang YZ, Xu WY, Chen XQ (2008) Experimental investigation on strength and failure behavior of pre-cracked marble under conventional triaxial compression. Int J Solids Struct 45:4796–4819

Yoshinaka R, Osada M, Park H, Sasaki T, Sasaki K (2008) Practical determination of mechanical design parameters of intact rock considering scale effect. Eng Geol 96(3/4):1087–1105

Zhao YS, Wan ZJ, Zhang Y, Qu Fang, Xie GY, Wei XJ, Ma W (2008) Research and development of 20MN servo-controlled rock triaxial testing system with high temperature and high pressure. Chinese J Rock Mech Eng 27(1):1–8

Zhou GL, Than LG, Lee PKK et al (2001) New strength criterion for rock with shear failure mode. Chin J Rock Mech Eng 20(6):753–762

Zhu WC, Tang CA (2004) Micromechanical model for simulating the fracture process of rock. Rock Mech Rock Eng 37(1):25–56